AIRCRAFT ENGINE DESIGN

Originally published in 1942

ISBN: 1940001420

The Aviation Collection
by
Sportsman's Vintage Press
2015

AIRCRAFT ENGINE DESIGN

BY

JOSEPH LISTON, M.E.
Associate Professor of Aeronautical Engineering
Purdue University; Member S.A.E.

PREFACE

This text has been assembled to aid technical students in bridging the gap between the point (*a*) where they have a fairly complete knowledge of the fundamentals of mathematics, mechanics, and machine design and (*b*) the point where they are sufficiently familiar with the application of these fundamentals to the design of aircraft engines to enable them to be of value to the aircraft-engine building industry.

Usually students entering this field of study are totally lacking in the experience so essential to deciding a logical order of procedure of engine design. They also lack the accumulated information upon which experienced designers can call for making the innumerable assumptions that must precede or parallel the analyses of various parts. Hence, an outline of procedure and a considerable accumulation of more or less rational data have been included. However, it is pointed out that although the Suggested Design Procedure is one way of carrying through the analysis, it is not the only way, or even the best possible way in a particular instance. Students are usually encouraged to select a "conventional" type of engine for a first design because there are more "signposts" to guide them, but this should not be misinterpreted as implying a negative attitude toward new ideas and possible improvements over present practice. Rather, it is based on the belief, founded largely on teaching experience, that a student cannot very well design an improved or unconventional engine until he is familiar with the shortcomings and weaknesses of conventional engines.

The author is greatly indebted to the various stated sources for illustrative data, and in each case he has endeavored to give proper credit. The author is also indebted to G. D. Angle, P. M. Heldt, various staff technicians at Wright Field, the NACA, the engine industry, and his associates at Purdue, particularly Dean A. A. Potter, Prof. G. A. Young, and especially Prof. K. D. Wood for valuable suggestions, criticisms, and assistance.

LAFAYETTE, INDIANA,
April, 1942.

JOSEPH LISTON.

CONTENTS

APPENDICES

AIRCRAFT ENGINE DESIGN

CHAPTER 1

REQUIREMENTS, POSSIBILITIES, AND LIMITATIONS

1-1. Sources of Power.—The source of power for all present-day aircraft is the internal-combustion engine. So far, this type of prime mover is the only one that has proved capable of meeting all the exacting requirements of powered flight successfully. Other prime movers such as steam plants have been considered and even tried experimentally in a few instances, but none has, to the writer's knowledge, survived to the production stage.

1-2. Basic Requirements and Limitations.—Some of the more important basic requirements of the airplane engine are

1. Adequate power.
2. Very low weight-power ratio.
3. High specific power output.
4. High thermal efficiency.
5. Compactness.
6. Reliability and long life.
7. Relative ease of maintenance.
8. Reasonable initial cost.
9. Ability to operate under adverse conditions.

Considering these items briefly:

1. Historians now generally agree that the first successes in powered flight were delayed several years because of lack of a suitable engine. Present large aircraft designers are continually clamoring for more and more powerful engines. Reductions in parasite drag have contributed markedly to the improvements in performance attained during the last 10 years, but further improvement from this source appears to be following the law of diminishing returns.

The power necessary for any given proposed airplane is usually determined from an estimate of parasite drag, combined with wing drag and propeller characteristics. This enables the designer to estimate the brake horsepower necessary for the maximum speed at which he desires to fly.

From the fundamental relations of drag, velocity, horsepower, and propeller efficiency, the maximum brake horsepower required for any given airplane may be expressed by

$$P = \frac{D \times V_{\max}}{375\eta} = \frac{C_D \frac{\rho}{2} S \left(\frac{88}{60}\right)^2 V_{\max}^3}{375\eta}$$

where P = brake horsepower needed at $V_{\max}$.
D = drag, lb.
C_D = coefficient of drag (for the entire airplane).
ρ = mass density.
S = a representative area (usually the wing area) corresponding to C_D.
$V_{\max}$ = maximum speed of the airplane, m.p.h.
η = propeller efficiency.

Letting $S = f$, the total equivalent flat-plate area of the airplane $(= f_{\text{parasite}} + f_{\text{wing profile}})$ of C_D = unity, assuming standard density, and collecting constants,

$$P = \frac{f}{\eta}\left(\frac{V_{\max}}{52.8}\right)^3 \tag{1-1}$$

At maximum speed, the wings will be at or very near an angle of attack at which the wing drag is a minimum. If the corresponding wing-drag coefficient is increased to unity, the drag equation will still hold if S is decreased by the inverse ratio. The new value of area is called flat-plate area of minimum wing-profile drag and may be designated f_{wp}. In symbols,

$$D_{\min} = C_{D\,\min} \frac{\rho}{2} S V^2 = C_D(=1) \frac{\rho}{2} f_{wp} V^2$$

or

$$\frac{C_{D\,\min}}{C_D(=1)} = \frac{f_{wp}}{S}$$

from which

$$f_{wp} = C_{D\,\min} \times S$$

Since the minimum drag coefficient of most airfoils is very near 0.01, $f_{wp} = 0.01S$, approximately. Then for estimating power requirements,*

$$P = \frac{f_p + 0.01S}{\eta}\left(\frac{V_{max}}{52.8}\right)^3 \tag{1-2}$$

where P = brake horsepower needed at V_{max}.
f_p = square feet of parasite flat-plate area of $C_D = 1.0$.
S = wing area, sq. ft.
η = propeller efficiency.
V_{max} = maximum speed of the airplane, m.p.h.

Example.—A company plans to develop an engine for military student-training planes of the following general characteristics: Well streamlined biplanes with retractable landing gear and cowled engines, 2,500 to 3,500 lb. gross weight, wing loadings around 12 to 16 lb. per sq. ft., and top speeds of 130 to 140 m.p.h. Approximately what rated engine horsepower should the company design for?

Solution.—Assuming mean values, $S = 3{,}000/14 = 214$ sq. ft., f_p = about 6 sq. ft. (Fig. 1-1), and η will probably be about 80 per cent. Therefore

$$P = \frac{6 + 0.01 \times 214}{0.8}\left(\frac{140}{52.8}\right)^3 = 189 \text{ b.hp.}$$

Correlation of brake horsepower, wing area, and maximum speed for existing planes can be used as a basis of estimating brake horsepower necessary. This has been done for 68 American airplanes (Fig. 1-2). These planes included all types from light sport planes to large flying boats, and as is indicated in the figure, fair correlation exists. The slope of the mean line (Fig. 1-2) is 2.56, and from its equation

$$P = S\left(\frac{V_{max}}{155}\right)^{2.56} \tag{1-3}$$

where the symbols are the same as in Eq. (1-2). Equation (1-3) is useful in approximating the maximum brake horsepower, but Eq. (1-2) is more accurate and should be used when f_p and η are known or can be determined.

2. The weight-power ratio is an important criterion to the value of an engine for airplane use. Figure 1-3 shows the weight-

* Equation (1-2) gives reasonably close values of brake horsepower, but some additional minor factors should be considered for precise calculations. See reference 1, Chap. 1, reference 2, p. 122, and *NACA Tech. Rept.* 408.

power ratios for 36 representative American engines. The position of any given engine with respect to the mean line may be taken as a measure of the degree of excellence of the design. However, excessively low weight-power ratios are usually

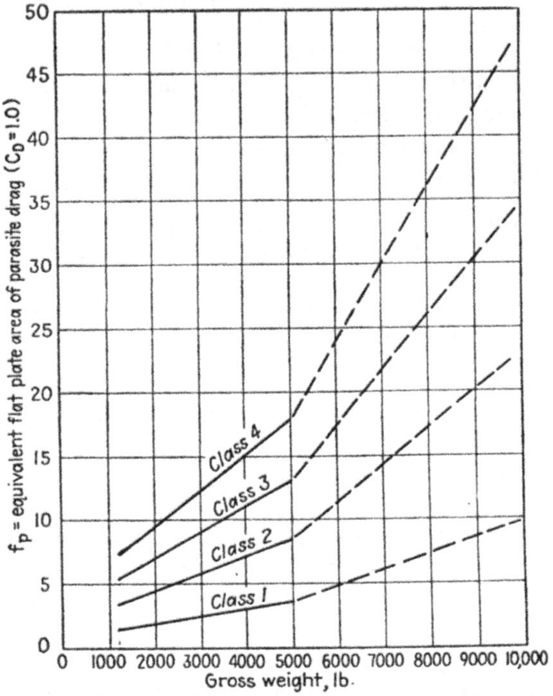

FIG. 1-1.—Flat-plate equivalent of parasite drag of airplanes.

Class 1. Cantilever monoplane with retractable chassis, streamline fuselage, well-cowled engine, and no external bracing.

Class 2. (*a*) Cantilever or wire-braced monoplane with cantilever or wire-braced chassis, wheel pants, streamlined fuselage, engine cowl or ring. (*b*) Biplane or externally braced monoplane with retractable chassis, streamline fuselage, engine cowl or ring.

Class 3. Biplane or externally braced monoplane with streamline fuselage, engine cowl or ring.

Class 4. Airplanes having excessive parasite drag. (*From Civil Aeronautics Manual* 04.)

attained either with the aid of very high octane fuels or at a sacrifice in operating life. For example, racing engines have very low weight-power ratios, but they require special fuels and usually have a very short life between overhauls.

Nevertheless, it is essential that the weight be low, as additional plane weight simply means less useful load. Figure 1-4 shows the weight of engines in terms of gross weight of the plane for 22 American aircraft in the gross weight range between 1,000

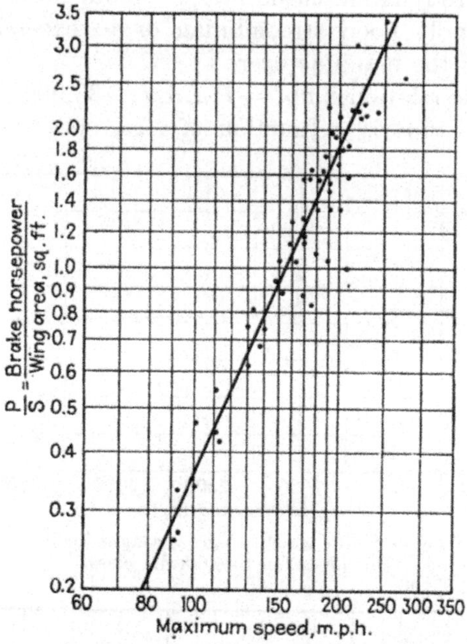

FIG. 1-2.—Variation of maximum speed with brake horsepower per square foot of wing area for 68 American airplanes.

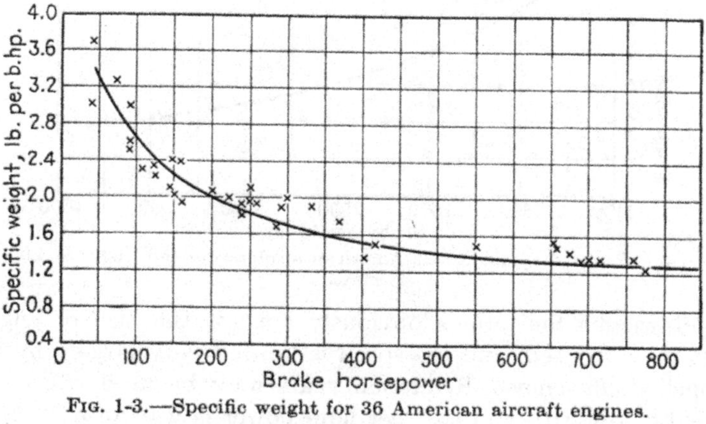

FIG. 1-3.—Specific weight for 36 American aircraft engines.

and 4,000 lb. The power loadings for typical American airplanes in the 1,000- to 4,000-lb. class are shown in Fig. 1-5.

The effect of power loading on maximum speed is shown in Fig. 1-6. Points well below the trend line are evidence of ineffective aerodynamic cleanliness. Points far above may be due to unusually good streamlining or to excessive optimism on the part of the manufacturer.

3. From the relation b.hp. = $P_B L A N_e n/33{,}000$, it is apparent that power output is a function of size, speed, and pressure.

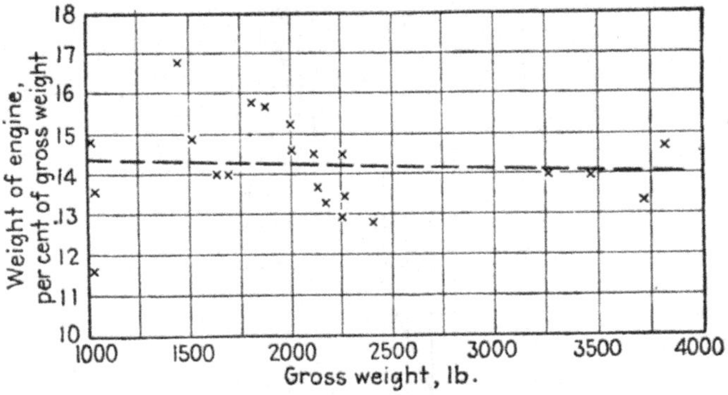

FIG. 1-4.—Proportion of engine weight to gross weight for 22 American airplanes in the 1,000- to 4,000-lb. class.

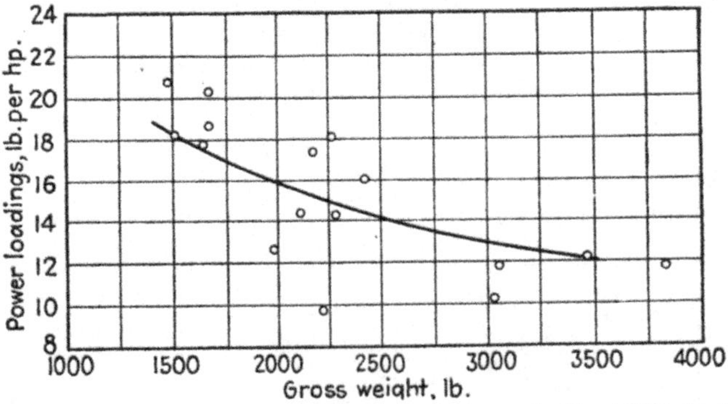

FIG. 1-5.—Power loadings for 22 American airplanes in the 1,000- to 4,000-lb. class.

Rigid weight limitations obviously control the size of engine that may be used, and the speed is limited very largely by the propeller efficiency. Reduction gears may be used where the added complexity and cost per horsepower is warranted. With reduction gearing, speed limitations are imposed by the valve gear and by crankpin loadings. Increase in the effective working pressure is one of the most valuable methods of increasing the specific power output. Some increase in b.m.e.p. (= P_B) can be obtained by increasing the compression ratio (Fig. 1-7*A*). Greater increases are possible by supercharging (Fig. 1-7*B*). The

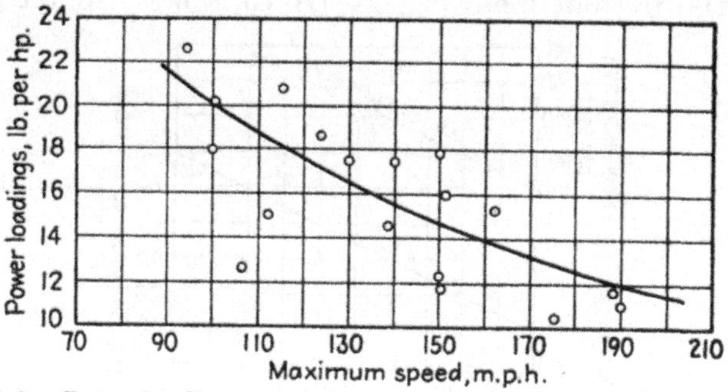

FIG. 1-6.—Power loading vs. maximum speed for 22 American airplanes.

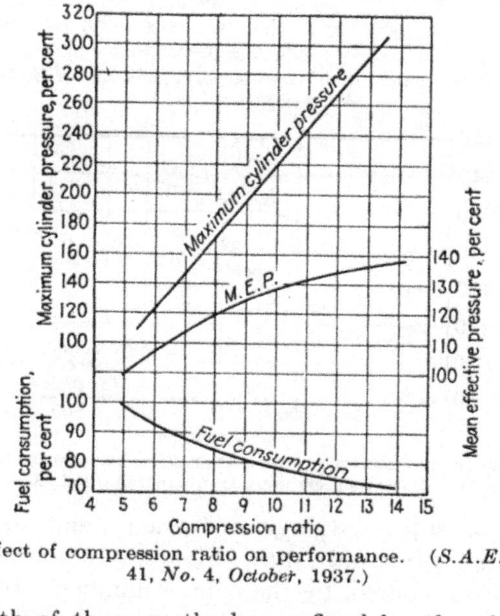

FIG. 1-7A.—Effect of compression ratio on performance. (*S.A.E. Journal, Vol.* 41, *No.* 4, *October*, 1937.)

limits on both of these methods are fixed by the ability of the fuel to withstand detonation, *i.e.*, by the octane number of the fuel to be used, and by the maximum allowable cylinder pressures. Higher maximum pressures mean heavier cylinder construction, hence increased specific weight. This incidentally, is an important obstacle to successful use of the Diesel-type aircraft engine. Figure 1-8 shows the b.m.e.p. vs. brake horsepower for 42 American engines. The b.m.e.p.$_{cruising}$ was taken from manufacturer's data; the b.m.e.p.$_{max}$ was determined (see Table A1-13) from b.m.e.p.$_{max}$ = $1/0.75 \times 0.9 \times$ b.m.e.p.$_{cruising}$.

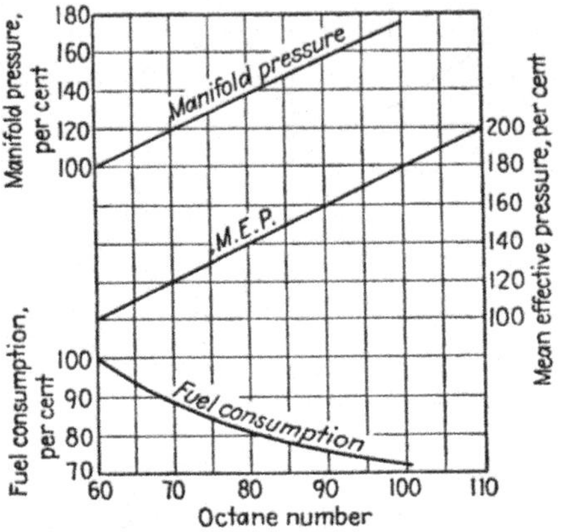

FIG. 1-7B.—Effect of octane number on performance. (*S.A.E. Journal, Vol.* 41, *No.* 4, *October*, 1937.)

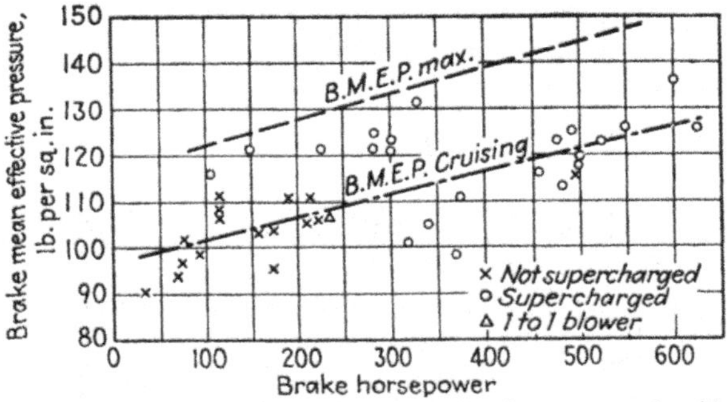

FIG. 1-8.—Brake mean effective pressure at cruising horsepower for 42 American aircraft engines. *X*, not supercharged; 0, supercharged, Δ – 1:1 blower.

The effect of supercharging on b.m.e.p. and horsepower is shown in Figs. 1-9*A* and 1-9*B*. Figure 1-10 shows the limitations placed on b.m.e.p. by the octane number. In this figure, points above the mean line *EF* indicate good combustion-chamber and cooling design. Points far below this line indicate either poor design or use of unnecessarily expensive fuels. Specified octane number vs. compression ratio for 23 American aircraft engines is shown in Fig. 1-11.

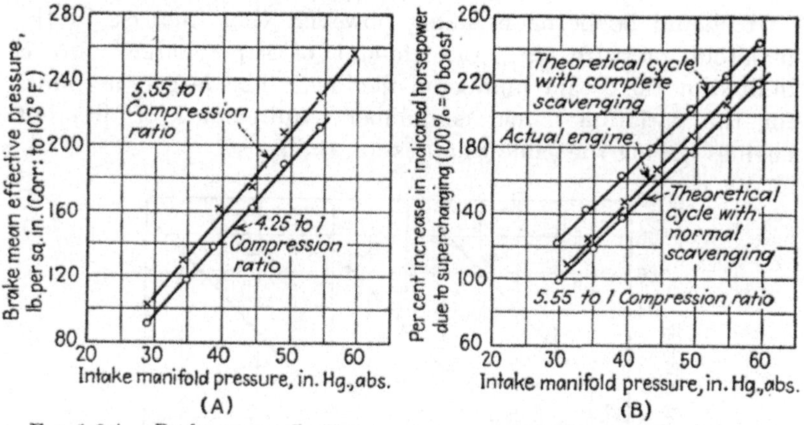

FIG. 1-9*A*.—Brake mean effective pressure vs. manifold boost for a 233 cu. in. supercharged engine at 1,000 r.p.m. (Air supplied by a separately driven blower.) (*From Sneed, An Investigation of Some of the Fundamentals of Supercharging, S.A.E. Annual Meeting Paper*, 1938.)

FIG. 1-9*B*.—Effect of manifold boost on indicated horsepower. Actual engine data from a 233 cu. in. supercharged engine at 1,000 r.p.m. (Air supplied by a separately driven blower.) (*From Sneed, An Investigation of Some of the Fundamentals of Supercharging, S.A.E. Annual Meeting Paper*, 1938.)

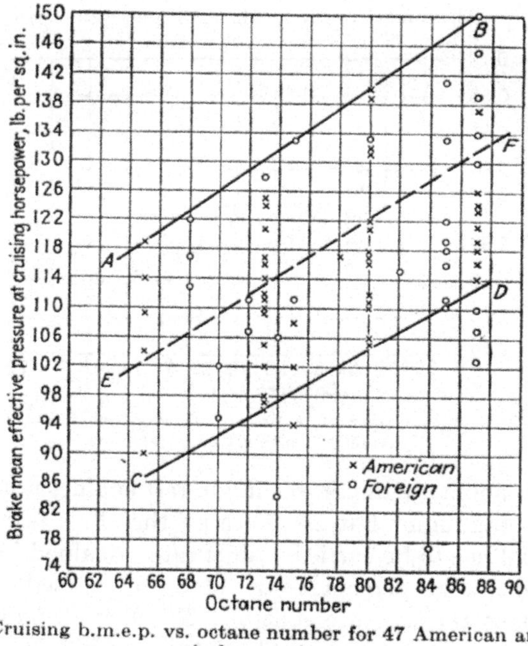

FIG. 1-10.—Cruising b.m.e.p. vs. octane number for 47 American and 33 foreign airplane engines.

It should be borne in mind, however, that increase in the manifold pressure, *i.e.*, the amount of supercharge, also is limited by the octane number of the fuel. This largely accounts for the variation in octane-number requirements of different engines having the same compression ratio.

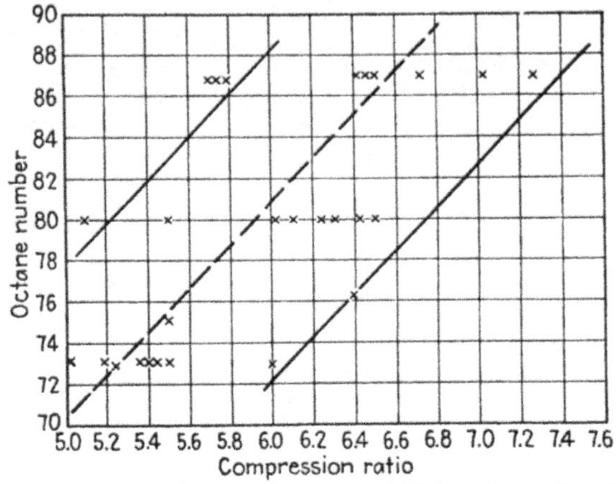

FIG. 1-11.—Octane-number requirements of 23 American aircraft engines.

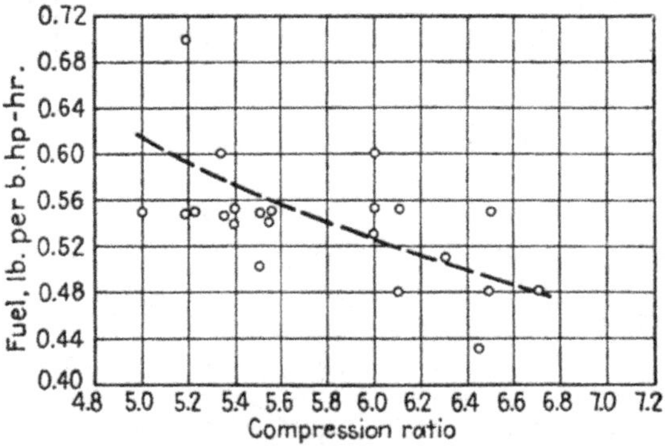

FIG. 1-12.—Effect of compression ratio on specific fuel consumption for 22 American aircraft engines.

4. Fuel economy is important in aircraft engines not only from a fuel-cost standpoint but also because the effect is cumulative, since the fuel has to be carried. Fuel rates attained in operation are affected by the percentage of rated power used and by the adjustment of the mixture control. Recently, much attention has been focused on economical mixtures owing in part to the

use of exhaust-gas analyzers as a means of indicating good mixture adjustment. However, limitations are imposed by permissible cylinder temperatures. Higher compression ratios (Fig. 1-12) are a means of improving economy, but this necessitates more expensive fuels or resort to "fuel cooling" by richer mixtures. Currently attained economy of operation is indicated in Fig. 1-13. These values represent manufacturers' guarantees rather than best possible performance, however.

5. Engine compactness is essential to low parasite drag, but for direct cooling (air cooling), limitations are imposed by the necessity of getting rid of a very large amount of heat through

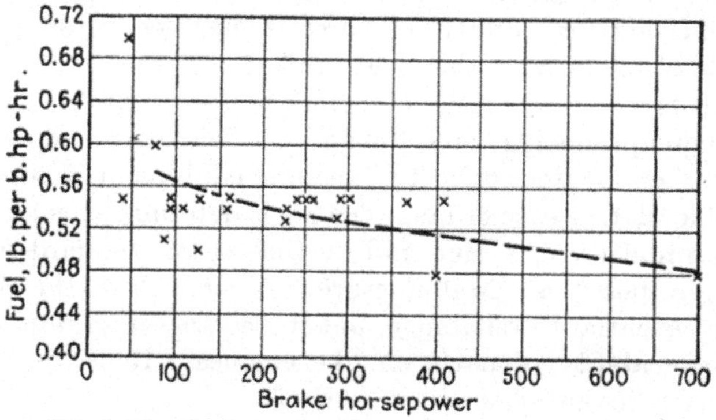

FIG. 1-13.—Fuel rate for 25 American aircraft engines.

the fins per unit of time. In liquid-cooled engines, the problem is merely shifted from the engine to the radiator. All the heat liberated in the cylinder except that converted into work must ultimately be removed and given up to the surrounding air. Thermal efficiencies vary from 25 to 35 per cent; hence 65 to 75 per cent of the heat energy in the fuel must be disposed of at a rate sufficient to prevent temperature rises above allowable limits. It is apparent that improving the thermal efficiency simplifies the cooling problem. In the larger sizes of engines, cooling becomes a limiting factor to cylinder dimensions.

Weight limitations are also closely related to compactness. Obviously, the more elongated and spread out an engine is, the more difficult it becomes to keep the specific weight within allowable limits. In this respect, radial engines are generally conceded to have some advantage over other types.

6. In commercial operation, reliability is of great importance. Forced landings and crashes with resulting newspaper headlines are about the best negative publicity an airline can have. More directly, they are tremendously expensive in both equipment and personnel.

In military work, reliability is equally important. Failure of equipment at a crucial moment may mean the difference between victory and defeat. The old saying "Because of a nail the horseshoe was lost, etc., down to the loss of the Army" is just as applicable to the working parts of an airplane engine. In the air, no item, however minor, is entirely unimportant.

Operating life between overhaul periods directly bears on the revenue-producing ability of aircraft engines. For a given schedule of operations, the longer the life between major servicing periods, the fewer reserve engines necessary and the less the nonrevenue-producing investment. Skilled-labor requirements are also a major item of cost in keeping equipment in service.

7. The relative ease of maintenance is determined very largely by the simplicity of design and by the use of standard parts in so far as possible. Manufacture to very close dimensional limits is essential to reliability, but it also aids in permitting the interchangeability of parts on like engines. Reduction in the repair-parts inventory is most desirable.

8. In the ultimate, the usefulness of all commercial products is determined by their cost. Regardless of the perfection of the design and manufacture, the product cannot compete if it is too costly. In the case of the airplane engine, the raw materials iron, nickel, aluminum, etc., in their initial states are relatively inexpensive. Fabrication is the major item, and that is largely determined by complexity of the design. The simplest design that will meet the requirements is generally the most satisfactory when all items are considered.

9. Ability to operate under adverse conditions is probably more important in the aircraft engine than in any other type of prime mover. In addition to the requirement of unfailing reliability, the engine is called upon to operate in widely varying positions and extreme altitudes. These requirements dictate dual ignition, a dry-sump crankcase, special carburetor design, and numerous other special features that are relatively much less important in engines for land or marine craft.

1-3. Government Requirements.—Primarily for the purpose of maintaining reasonable safety standards, the Federal Civil Aeronautics Authority has been given extensive control over American aviation. In the engine field, the control consists in forbidding the use of engines not approved by the CAA.

To receive approval, an engine must successfully pass a number of exacting tests and meet certain other requirements. Hence, it is important for the designer to know the nature of these requirements. Current requirements for approval of engines are known as Civil Air Regulations, Part 13, "Aircraft Engine Airworthiness." In addition, the CAA has prepared *Civil Aeronautics Manual* 13, "Aircraft Engine Airworthiness" which is "intended to interpret, explain and suggest methods of compliance with the regulations. . . . " Both of these publications should be available to the designer.

References

1. Wood: "Airplane Design."
2. Wood: "Technical Aerodynamics."

Problems

1. An airplane of aerodynamic cleanness halfway between class 1 and class 2 (Fig. 1-1), has a gross weight of 2,250 lb., a landing speed of 47 m.p.h. with flaps, and a maximum lift coefficient of 2.19. What brake horsepower is developed by the engine when the plane is flying at its maximum speed of 162 m.p.h.? (Assume the propeller efficiency at maximum speed is 83 per cent.)

2. For a wing area of 180 sq. ft., a wing loading of 15 lb. per sq. ft., a propeller efficiency of 81 per cent, and a maximum speed of 150 m.p.h., what equivalent flat-plate area of parasite resistance is implied in Eq. (1-3)? To what class of airplane does this correspond (Fig. 1-1)? If the wing loading were 10 lb. per sq. ft., what class of airplane would be implied?

3. *a.* What values of f/S are implied by Eq. (1-3) at values of P/S of 0.3, 1.0, and 3.0 if the propeller efficiency is assumed to be 80 per cent?

b. What kinds (classes) of 5,000-lb. airplanes do the above (part *a*) P/S values represent if the wing loadings are 12 lb. per sq. ft.?

4. *a.* For values of f/S of 0.015, 0.025, 0.035, and 0.045, find the value of K in the equation $P = S\left(\frac{V_{max}}{K}\right)^3$. Assume $\eta = 0.8$.

b. Plot curves of $P = S\left(\frac{V_{max}}{K}\right)^3$ for the values of K found in part *a* on logarithmic paper using P/S as the ordinate and V_{max} as the abscissa. Superimpose the mean line in Fig. 1-2 for comparative purposes.

5. Plot Fig. 1-3 on logarithmic paper, and determine an expression for variation of specific weight with size (*i.e.*, b.hp.).

CHAPTER 2

OUTLINE OF THE PROJECT

2-1. Selection of the Potential Market.—The logical first step in designing an airplane engine is to determine the size, *i.e.*, the performance desired. This will, of course, depend upon the use to which the engine will be put and the type, size, and performance of the plane that the engine will power. For practically all engines, with the possible exception of designs for special racing purposes, the ultimate object is to build a unit that can be sold profitably and that will produce results in competition with other engines sufficient to attract further orders. The decision of first cost vs. operating cost, reliability, and operating life must be made in order to establish a "policy" in selecting materials, dimensions, compression ratios, accessories, etc.

For instance, an engine designed for the low-cost light plane would incorporate good materials sufficient for reasonable reliability, but the designer could not specify the very best known high-strength alloys because that would push the cost above competitive levels. Instead, he would be obliged to use heavier sections of less expensive materials to provide the necessary strength, and that would increase the weight per horsepower. To keep the cost down, he would be obliged to omit special features such as superchargers and automatic mixture controls and, in extreme cases, possibly reduction gearing or the dry-sump crankcase type of lubrication. To enable the purchaser of the engine to use low-cost fuels, the designer will have to keep the compression ratio down, and this will mean a sacrifice in performance.

On the other hand, if an engine is to be designed for a high-performance military plane, first cost (within reason) becomes secondary and every means at the disposal of the designer should be used to attain maximum power with reasonable economy. For use in a large transport plane, first cost is important, but not to the extent of a light-plane engine. Low weight per horsepower is essential, and fuel economy is vital. This will mean the use of stronger alloys, higher compression ratios, and greater supercharging. The resulting higher first cost will be justified by reduced operating and maintenance cost and by the ability to increase the specific pay load.

Obviously the potential market for the engine must be well defined before the design is begun.

2-2. Selection of Rated Power.—The market having been selected, the rated power may be decided upon. Here again cost vs. performance enters. For instance, any given airplane will operate with quite a range of engine sizes. The engine builder must decide whether he can make more money by building an engine that will appeal to the user who is willing to sacrifice high speed for lower operating cost, or vice versa. This decision is in no essential different from that which must be made by the manufacturers of any other product. Essentially, the decision becomes "will the company cater to a large market with small profit per unit or to a more limited market with greater profits per unit?" At best, by the very nature of things, some factors must be left to chance.

2-3. Preliminary Specifications.—Preliminary specifications may now be drawn as indicated in Table 2-1. In general, for each item the decision of what value to use is largely a compromise between the best possible value and the one that will keep costs within allowable limits. Here the accumulated experience of the designer becomes very important.

The fundamental principles of the basic sciences are very useful in the design of complex machines, but derived formulas are either not available or are too complicated to be practical in many of the more intricate parts. Hence, empirical rules based on accumulated experience must be used. Many of these rules are merely the judgment of the designers based on methods that have been found to be satisfactory. To a considerable extent, they are the results of the trial-and-error method.

Unfortunately, progress is slow in the school of experience, and for the beginner, the way forward is anything but clear. Specifically, if the inexperienced designer of airplane engines is to short-cut the long winding road, he must rely heavily upon landmarks established by others. He must base his decisions regarding the major portion of the items upon the findings and experiences of his predecessors. In proceeding with the design, the student should assume the attitude that his "company" is going to spend a lot of money on his recommendations. If he departs too widely from convention, he is shifting the project from an investment to a gamble.

This recommended adherence to convention should not be construed as a reactionary attitude toward new methods and progress. Rather, it is based on the belief that improvement in an existing art cannot very logically be made until current methods, limitations, and shortcomings are thoroughly understood.

2-4. Justification of Values in Example 1.—1, 2. Selection of values for Table 2-1 probably can be best discussed by means

TABLE 2-1.—GENERAL SPECIFICATIONS

Specification Item	Example 1
1. Brake horsepower:	
a. Maximum for take-off	125 at sea level
b. Cruising	94
c. Maximum except take-off	125
2. Revolutions per minute:	
a. For take-off	2,000
b. For cruising	1,800
c. For maximum except take-off hp	2,000
3. Octane number of fuel used:	
a. For take-off	73
b. Cruising and maximum except take-off	73
4. B.m.e.p., lb. per sq. in.:	
a. Maximum for take-off	120
b. Cruising	100
c. Maximum except take-off	120
5. Compression ratio	5.3
6. Type of cycle (two or four stroke)	4
7. Type of cooling	Air
8. Arrangement of cylinders	Radial
9. Number of cylinders and tenative size	Five 4½ by 5⅜ in.
10. Connecting rod—crank (L/R) ratio	4/1
11. Valve arrangement	Overhead, inclined, valve in head, rocker arms, and push rods
12. Method of supercharging	None
13. Ignition system	Dual magnetos
14. Reduction-gear ratio	1:1

of the example. In this particular case, the "engine company" plans to build a unit for the medium-light plane class. Usual power loadings in this class (1,500 to 2,500 lb. gross weight) range from 10 to 20 lb. per hp. (Fig. 1-5) with a fair average at about 16 for a 2,000-lb. plane. Hence the engine will have to develop 2,000/16 = 125 b.hp. at full throttle and rated speed. This rating is based on sea-level performance since, for the purpose intended, most of the operation will be at relatively low altitudes.

Some engines are given a special horsepower rating for take-off above that which they can safely develop for extended periods. Under these conditions, take-off horsepower is for limited periods of a few minutes only, as the engine would overheat if operated continuously at take-off horsepower. A special take-off high-octane fuel may be used when take-off power is developed. Under these special conditions, item 1*c* may be considered as the maximum safe horsepower for extended operation. For the example, however, the added cost of providing for extra take-off power, *i.e.*, high supercharge and special fuel, is not considered justifiable, and item 1*a* is specified as equal to the maximum full-throttle power for continuous operation.

The specified speed of 2,000 r.p.m. is selected because (*a*) propeller efficiencies drop rapidly at speeds above this figure and (*b*) reduction gearing would add too much to the cost of the engine. From Fig. 1-6, the corresponding top speed of the plane may be assumed to approximate 135 m.p.h.

There is an increasing tendency in the aviation industry toward designing the engine to fit a specific plane. When potential sales warrant such a procedure, the power of the engine is determined by the particular requirements of the plane. In such cases, sufficient data will be available to permit the use of Eq. (1-2).

The cruising horsepower (Table A1-13) should be about 75 per cent of the take-off horsepower, and the corresponding cruising speed will be about 90 per cent of take-off speed, hence the selection of 94 cruising hp at 1,800 r.p.m.

3. To avoid the extra cost of premium fuels, an octane number of 73 is specified.

4. Brake mean effective pressures in the 75- to 150-hp. (cruising) class range around 100 lb. per sq. in. (Fig. 1-8). For a

73 octane number fuel, a higher value might be used (Fig. 1-10), but it might be difficult to attain without supercharging (see item 12, Table 2-1). Since b.m.e.p. may be expressed by the relation

$$P_B = \frac{\text{b.hp.} \times 33{,}000}{LAN_e n}$$

the b.m.e.p. for take-off conditions is

$$P_B\ (\text{take-off}) = \frac{\text{b.hp. (take-off)}}{\text{b.hp. (cruising)}} \times \frac{\text{r.p.m. (cruising)}}{\text{r.p.m. (take-off)}} \times P_B\ (\text{cruising})$$

or

$$P_B\ (\text{take-off}) = \frac{125}{94} \times \frac{1{,}800}{2{,}000} \times 100 = 120 \text{ lb. per sq. in.}$$

Referring to Fig. 1-10, it is seen that this value is still within the range of 73 octane number fuel. Hence, a special fuel for take-off will not be necessary.

5. Compression ratio is limited by the octane number of the fuel, and for a knock rating of 73, a value of 5.3 for the *CR* should be satisfactory (Fig. 1-11).

6. Two-stroke-cycle engines for aircraft are still largely in the experimental stage; hence much greater assurance of success will be had by adhering to the more conventional four-stroke-cycle principle.

7. Direct air cooling eliminates the cost of radiators and troublesome piping. The absence of any water-cooled engines in the power class in which this engine falls (Table A1-1) is good evidence that previous attempts at water or liquid cooling have not been able to meet the competition of the air-cooled engines. Profiting by the experiences of others is a good way to avoid red ink on the ledgers.

8. As regards arrangement of cylinders, Table A1-1 indicates that radials predominate in the power class under consideration. However, the inverted in-line engine is also in considerable evidence, and the flat-opposed or 180-deg. V-engine has much to recommend it for certain types of installations such as in the wings of bimotored ships. In the final decision, the company designer will probably be influenced by the president's ideas on cylinder arrangement, but for the present purpose, it is of interest to list some of the important items as follows:

Item	Air-cooled radial	Air-cooled in-line	Air-cooled opposed
Crankcase	Compact and rigid	Must be heavier for necessary rigidity	Intermediate for same powered engine
Cylinders	All equally exposed to cooling air	Careful cowling necessary to cool rear cylinders adequately	Some cowling to deflect air on rear cylinders except in the smallest sizes
Crankshaft	Short and rigid, heavily loaded crankpin. Counterweights necessary	Heavier for necessary stiffness. Usually no counterweights	Intermediate for same powered engine
Valve gear	Push rod and rocker arm limit speed of geared engine	Overhead camshaft may be used, less noisy, less maintenance, but tends to limit valve size	Push rods and rocker arms or two overhead camshafts
Parasite drag	Considerable even with cowling, especially in wing engines. Nose engines increase fuselage drag because of slip stream	Less frontal area, but necessary air scoop for cooling adds to total drag	More adaptable for cowling in wing, but cooling-air scoop adds to total drag
Visibility in single-engine tractor-type plane	Relatively obstructed	Excellent for inverted type	Better than radial

Many other items, varying in importance between the types, will come to mind, but the foregoing comparison is sufficient to indicate that none of the three arrangements of cylinders is outstandingly superior. For example 1, a radial has been selected.

9. Firing order in a single-bank radial very nearly dictates an odd number of cylinders. The greater the number of cylinders, the greater the overlap of power impulses, hence the smoother the torque curve; but fewer cylinders means a smaller number of parts and usually lower cost. Increasing the number of cylinders beyond seven in a single-bank radial increases the over-all

diameter and parasite drag. Fewer and larger cylinders are more difficult to cool, since for a given cooling-fin design, the volume increases as the cube of the dimensions, whereas the surface area of the cylinder increases only as the square. However, this will probably not be a limiting factor so soon as torque-curve variation in the size of engine under consideration. For the example, 5 cylinders have been selected as the compromise of the logical possibilities, 3, 5, or 7.

From the relation,

$$\text{b.hp.} = \frac{P_B L A N_e n}{33{,}000}$$

the displacement per cylinder is

$$D = 12LA = \frac{125 \times 33{,}000 \times 12}{120 \times \dfrac{2{,}000}{2} \times 5}$$

$$= 82.5 \text{ cu. in. for the example}$$

Ratios of stroke to bore vary rather widely (Table A1-1), and they are quite often dictated by the desire to increase the number of interchangeable parts in models of similar design but different power output. A low stroke-bore ratio reduces the over-all diameter, hence the parasite resistance, but it increases the distance the heat has to flow to escape from the center of the piston. This generally means a heavier piston and a greater weight of reciprocating parts. A stroke-bore ratio of 1.2 is tentatively selected as representing good practice. This will permit the later development of a larger engine in which many of the parts in the present model can be used. The larger unit will doubtless have larger diameter cylinders, but by using a fairly high stroke-bore ratio in the present model, a reasonable stroke-bore ratio can still be had in the larger model with the same crank-arm radius. The cylinder dimensions are found from

$$D = S \times d^2 \times \frac{\pi}{4} = \frac{1.2\pi}{4} d^3$$

$$d = \left(\frac{82.5 \times 4}{1.2 \times \pi}\right)^{1/3} = 4.45 \text{ in., bore}$$

$$S = 1.2 \times 4.45 = 5.34 \text{ in., stroke}$$

As cylinder dimensions are usually specified to the nearest eighth of an inch, the bore and stroke may be taken as 4½ by 5⅜ in.

Referring to Table A1-1, it is seen that these values compare favorably with bore and stroke values for similar sizes of engines.

10. Ratios of center-to-center length of connecting rods to crank-arm radii (L/R ratios) vary from about 3.3 or less to as high as 4.5. The longer the connecting rod in proportion to the crank radius, the less the angularity between the connecting-rod axis and the cylinder center line, hence the less the side pressure, *i.e.*, friction against the cylinder wall. But large values of L/R mean greater over-all transverse dimensions of the engine, hence greater parasite drag. A compromise value of $L/R = 4$ has been selected for Example 1.

11. For a radial engine, rocker arms and push rods are about the only feasible means of transmitting the cam-follower motion to the valves. Inclined valves in the head are specified to permit maximum valve port openings and as direct a flow as possible for the gases. To attain the b.m.e.p. specified, a high volumetric efficiency will be necessary, but two intake valves per cylinder would be objectionable because of complexity and cost.

12. Gear-driven centrifugal blowers are conventional for supercharging radial engines, but supercharging is omitted in this example to keep down the cost. To attain equal mixture distribution to the various cylinders in a radial engine, a centrifugal-type blower directly connected to the rear end of the crankshaft is desirable. For later more powerful models, this blower may be converted to a supercharger by gearing it for a speed higher than the crankshaft r.p.m.

13. Two magnetos are specified for safety and reliability and to conform with government requirements. Dual ignition will also increase the power output slightly.

14. A reduction gear between the propeller and crankshaft will not be necessary (1:1 indicates direct drive) as no appreciable loss in propeller efficiency will be had at the crankshaft speed specified.

2-5. Preparation of Design Data and Drawings.—Design data to be of value must not only be accurate but also be in logical form and neatly prepared. A jumbled array of illegible calculations and incomplete penciled drawings is of little value no matter how accurate. Your employer will judge the quality of your work by its neat appearance in the same general way that you are influenced toward a new car by the appeal of streamlined contours and glistening finish. In either case, the quality of the product may or may not be high, but to sell it, it must *look* right. The inner quality will determine the repeat orders.

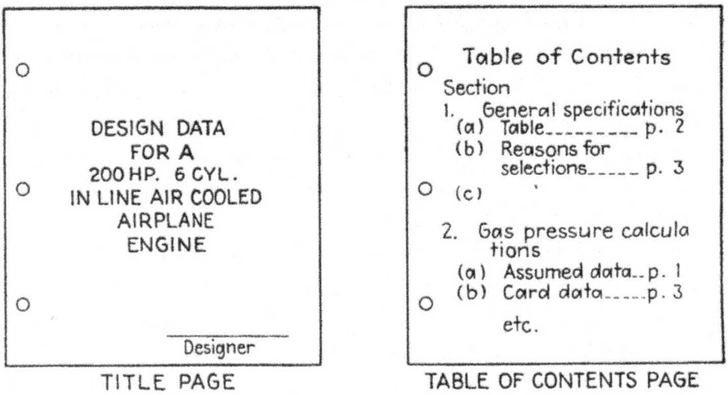

FIG. 2-1.—Suggested arrangement of title and table-of-contents pages.

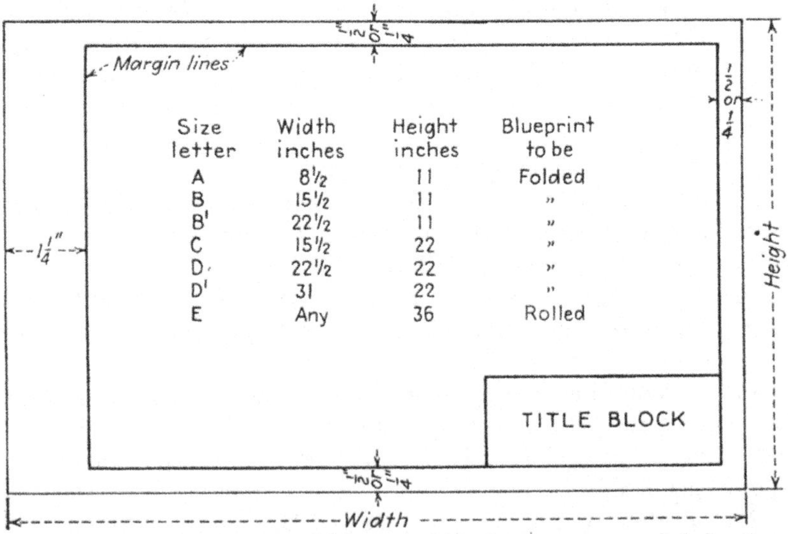

FIG. 2-2.—American Society of Mechanical Engineers recommended drawing-paper size and arrangement.

It is recommended that design notes and completed drawings be kept in a standard 8½- by 11-in. three-ring notebook. The first page of the notebook should be a title sheet, the second a table of contents (Fig. 2-1), and each section of the work should be identifiable by a small tab attached to the first sheet of the section. This tab should have a title or section number conforming to the title and section number in the table of contents. Subtitles should follow each section title, and their location in the section should be by section page number. Each drawing should be identifiable by a suitable designating letter or number

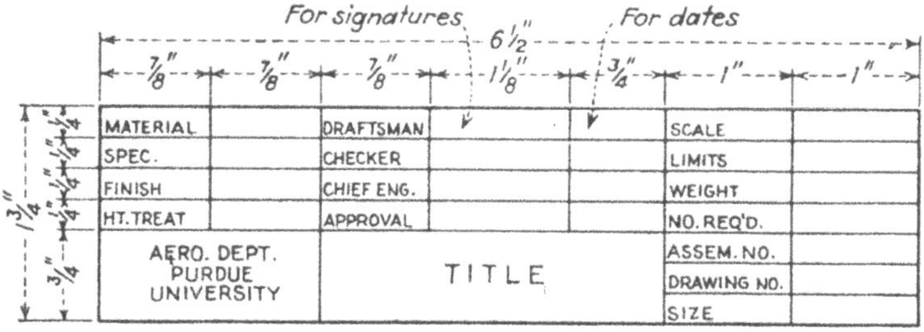

FIG. 2-3.—Suggested title block for working drawings.

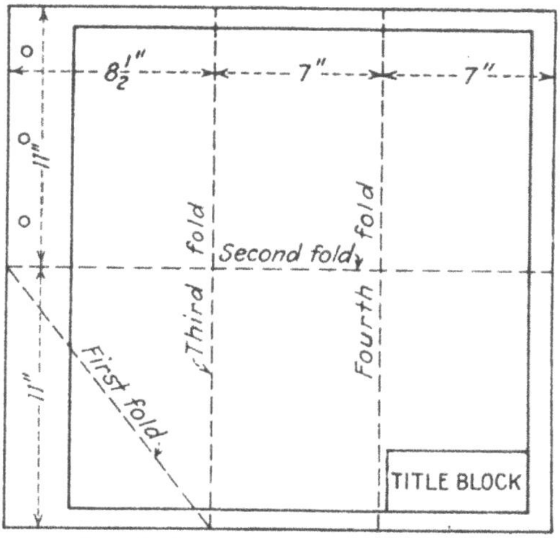

FIG. 2-4.—Suggested method of folding drawings for insertion in design notebook. An alternate method of folding so that the title block shows when folded is standard practice with many companies.

and title both on the drawing and in the table of contents. In all cases, tabulated data, calculations, and graphical constructions should be titled, accompanied by adequate explanations, and, as applies, by sample calculations.

Whenever possible, drawings should be on standard sizes of paper and in all cases properly titled (Fig. 2-3). For insertion in the notebook, drawings should be blueprinted and folded as indicated in Fig. 2-4. Drawings should be *complete*. A drawing returned by the production department for lack of adequate dimensions or other data is a direct reflection on the engineering department and the designer. Repeated offenses usually result in the individual's failure to receive "promotion."

Suggested Design Procedure

1. Select the class of plane for which an engine is to be designed. Tabulate the approximate weight, performance, parasite drag (if available), and other pertinent data.

It is recommended that, for undergraduate students, the selection be confined to the medium- or light-plane class as large planes require engines of more cylinders, greater complexity, etc., and this greatly increases the detail work necessary in designing the engine without adding proportionately to the fundamental knowledge gained. Also in most undergraduate courses, the time available for the design work is insufficient for those added details.

2. Prepare a table of general specifications similar to Table 2-1.

Adhere closely to current practice in deciding upon the type of engine to be designed. Confine the selection to (*a*) a single-bank radial of preferably not more than seven cylinders, (*b*) an in-line type of not more than six cylinders, or (*c*) a V type of not more than eight cylinders.

In limiting the selections to the preceding types, it is not the purpose to suppress originality or potential inventive genius, but rather to require the selection of a problem that the beginner can have a fair chance of completing. As an instance of the pitfalls of allowing unlimited selections, a case is recalled in which a student was permitted, in his first undergraduate course in engine design, to select any type of engine he desired, and, without knowing of the difficulties ahead, he selected a three-lobe-cam engine. Almost immediately, innumerable questions arose concerning logical values for this detail and proper sizes for that part. Without precedent to guide him, he soon became hopelessly lost and little was accomplished. The inexperienced designer will have problems enough with a conventional type of engine. Later, when he has acquired experience, he can depart from the conventional if he chooses.

3. Justify each specification item by reference to current practice wherever possible and by reference to the allowable cost decided upon.

Hasty selection of the general specifications is false economy of time, as specifying impossible values may mean the repeating of a great deal of tedious calculation farther on.

4. Make a line layout of (*a*) a transverse section and (*b*) a longitudinal section through the engine at the cylinder center lines. Show the location and desired sizes of the principal parts. Indicate desired dimensions of important parts, positions of center lines, etc., but do not try to make a complete *detailed* drawing* at this stage of the design. Check closest posi-

* In preparing this preliminary layout, the designer may be likened to a topographer preparing a map of a little known region. The first step is to

tion of connecting-rod center line to lower end of cylinder. (Cylinder must extend down approximately as far as the lowest position of the bottom of the piston skirt.) Check all features of desired arrangement to be reasonably certain of adequate mechanical clearance of parts. Inspect available blueprints, drawings, engine parts, etc., of similar designs for assistance in selecting logical sizes for the various parts.

This preliminary layout drawing should be developed with the idea in mind that it is the general arrangement desired. At best, some detail changes will be necessary before the final design is completed, but if too radical an arrangement is attempted, major changes may have to be made. This will greatly increase the work necessary later. Hence, it is very desirable to give careful study to the proposed arrangement. The layout need not be blueprinted at this stage, but it should be to a large enough scale to permit close study and on standard-size paper properly titled (size *D* or *E*, Fig. 2-2, is recommended).

5. When items 1 to 4 have been completed and put in proper form (Par. 2-5), submit for checking and approval. Keep a record of the man-hours required on each item.

---

bound the region as accurately as possible and to insert the position of important features such as rivers, lakes, and mountain chains (*i.e.*, major dimensions, center lines, etc.). Obviously, the details will have to be added gradually as the information becomes available, but it should be borne in mind that unnecessary carelessness or poor judgment in the preliminary layout will result either in doubtful accuracy of the finished product or a great deal of time-consuming revision that might have been partly avoided.

The successful designer also has something in common with the successful artist who, in preparing the layout for a painting, is able to visualize in his mind's eye the appearance of the finished product. It takes years of practice to perfect this ability, but the greatest attainments always have been made by men (engineers as well as artists) who put everything they had into *every* job at the beginning as well as at the height of their careers.

CHAPTER 3

GAS-PRESSURE FORCES

3-1. Forces in the Cylinder.—Forces on the piston represent a combination of gas pressure and inertia forces. These forces are usually determined separately at increment angular positions of the crankshaft, plotted as unit or total force against crank angle, and then, by adding ordinates, a curve of the net force parallel to the cylinder axis is obtained. As the dimensions of the various parts of the engine are largely determined by the stresses resulting from the maximum forces, it is obviously necessary to investigate the case causing these extreme conditions, *i.e.*, full throttle and highest speed.

3-2. Construction of the Indicator Card.—For the gas-pressure forces, it is necessary to construct an indicator card representing full-throttle conditions. Very elaborate procedures have been developed[1,2] for analyzing the phenomena in engine cylinders with the idea in mind of more closely approaching actual conditions. However, even with the most complex of these methods, some discrepancies exist and must be accounted for by a "card factor." It is believed that simpler methods of determining values for the indicator card, although admittedly less rational, are more practical and may be applied just as effectively by using a somewhat larger card factor. In short, instead of endeavoring to account for variable specific heats, dissociation, chemical equilibrium, heat flow back and forth between the gases and the cylinder, and then applying a small card factor, use values for the exponents of compression and expansion consistent with actual measured results from engines that have been indicated, calculate the pressures and volumes by the older and much simpler thermodynamic relations of the modified "air-standard" cycle, and then apply a slightly larger card factor. The values for plotting may be found as follows:

Determine the i.m.e.p. from

$$\text{i.m.e.p.} = \frac{\text{b.m.e.p.}}{e_m} \tag{3-1}$$

where i.m.e.p. = indicated mean effective pressure, lb. per sq. in.
b.m.e.p. = brake mean effective pressure at maximum horsepower (= take-off horsepower), lb. per sq. in.
e_m = mechanical efficiency at take-off horsepower.

The absolute pressure at the end of expansion may be found from[3,4]

$$P_D = \frac{(n - 1)(R - 1)}{R^n - R} \times \frac{\text{i.m.e.p.}}{F_D} + P_A \qquad (3\text{-}2)$$

where P_D = pressure at end of expansion, lb. per sq. in. abs.
n = exponent of the compression and expansion curves.
R = compression ratio.
F_D = card factor representing the ratio of actual to theoretical card areas.
P_A = pressure at the beginning of compression, lb. per sq. in. abs.

$$P_C = P_D \times R^n \qquad (3\text{-}3)$$

$$P_B = P_A \times R^n \qquad (3\text{-}4)$$

where P_C = pressure at the beginning of expansion, lb. per sq. in. abs.
P_B = pressure at the end of compression, lb. per sq. in. abs.

Cylinder volumes corresponding to the foregoing pressures may be found by combining the relations

$$V_A - V_B = D \qquad (3\text{-}5)$$

$$\frac{V_A}{V_B} = R \qquad (3\text{-}6)$$

where V_A = cylinder volume at the beginning of compression, cu. in.
V_B = clearance volume, cu. in.
D = piston displacement for one cylinder, cu. in.

Also

$$V_A = V_D \quad \text{and} \quad V_B = V_C$$

where V_D = cylinder volume at the end of expansion, cu. in.
V_C = cylinder volume at the beginning of expansion, cu. in.

Intermediate pressures along the compression line may be found from

$$P_{1,2,3,\text{ etc.}} = \frac{P_B V}{V^n_{1,2,3,\text{ etc.}}} \qquad (3\text{-}7)$$

$V_{1,2,3,\text{ etc.}}$ being found by measuring along the abscissa or volume line to scale. Similarly, intermediate pressures along the expansion line may be found.

An alternate method of finding intermediate pressures is to plot points P_AV_A, P_BV_B, P_CV_C, and P_DV_D on logarithmic cross-sectional paper and connect successive points by straight lines. Pressures corresponding to any intermediate volumes may be read directly from the ordinate scale.

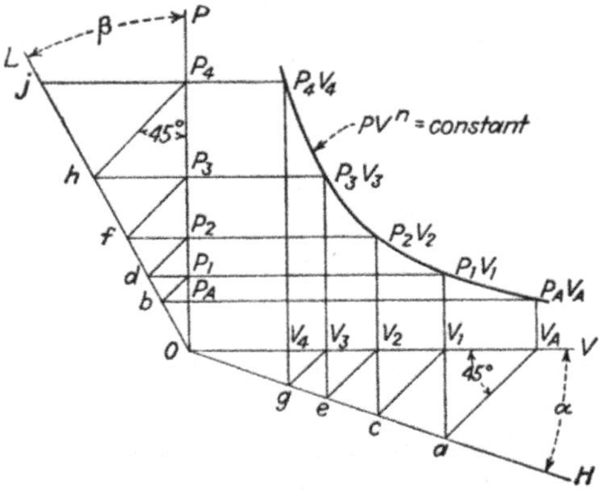

FIG. 3-1.—Graphical construction of a PV^n = constant line.

A graphical method of plotting compression and expansion lines is illustrated in Fig. 3-1. For instance, to construct a compression line, lay off coordinates OV and OP and locate point P_AV_A. Select α and β such that $(1 + \tan \alpha)^n = 1 + \tan \beta$. Draw OH, making the angle α with OV, and OL, making the angle β with OP. Erect a line perpendicular to OV at V_A. Construct angle OV_Aa equal to 45 deg. Erect a line through a perpendicular to OV. Construct P_Ab perpendicular to OP. Construct angle bP_1O equal to 45 deg. Draw a line through P_1 parallel to OV. The intersection of the perpendicular line through a and the horizontal line through P_1 locates point P_1V_1. Construct angle OV_1c equal to 45 deg. Erect a line through c perpendicular to OV. Extend the horizontal line through P_1 to d. Construct angle dP_2O equal to 45 deg. Draw a line through P_2 parallel to OV. The intersection of the perpendicular line through c and the horizontal line through P_2 locates point P_2V_2. Proceed in the same way to locate points P_3V_3, P_4V_4, etc., and connect the points thus located with a smooth curve. The equation of the curve is PV^n = a constant.

3-3. Example.—Construct an indicator card for the engine selected in Example 1, Table 2-1.

Procedure.—The b.m.e.p. for take-off horsepower represents maximum conditions. Mechanical efficiency will be that for full load and speed. For

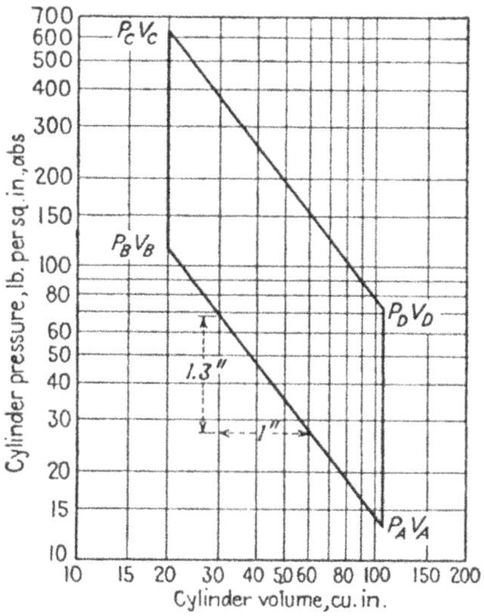

FIG. 3-2.—Logarithmic plot for determining indicator-card data.

these conditions, the mechanical efficiency may be taken as equal to 85 per cent (Fig. A1-2). Then

$$\text{i.m.e.p.} = \frac{120}{0.85} = 141 \text{ lb. per sq. in.}$$

The average exponent of compression and expansion may be taken as $n = 1.3$, the card factor $F_D = 0.9$, and the pressure at the beginning of compression, $P_A = 90$ per cent of atmospheric pressure (for a nonsupercharged engine)[3,4]. Then, for the example,

$$P_D = \frac{(1.3 - 1)(5.3 - 1)}{5.3^{1.3} - 5.3} \times \frac{141}{0.9} + 0.9 \times 14.7 = 72.7 \text{ lb. per sq. in. abs.}$$

Let $P_D = 73$ lb. per sq. in. abs.

$P_C = 73 \times 5.3^{1.3} = 635$ lb. per sq. in. abs.

$P_B = 13.2 \times 5.3^{1.3} = 115$ lb. per sq. in. abs.

The displacement per cylinder is

$$D = 4.5^2 \times 0.785 \times 5.375 = 85.5 \text{ cu. in.}$$

Hence

$$V_B = \frac{85.5}{5.3 - 1} = 19.9 \text{ cu. in.} = V_C$$
$$V_A = 85.5 + 19.9 = 105.4 \text{ cu. in.} = V_D$$

Plotting the four points thus determined on logarithmic paper and connecting gives Fig. 3-2. Table 3-1 is obtained by selecting intermediate volumes and reading the corresponding pressures. The completed indicator card

TABLE 3-1.—INDICATOR-CARD DATA FROM FIG. 3-2

Volumes, cu. in.	Compression-line pressures, lb. per sq. in. abs.	Expansion-line pressures, lb. per sq. in. abs.
105.4	13.2	73
100	14	76
95	14.95	82
90	16	87.5
85	17.1	94
80	18.4	100
75	20	110
70	22	121
65	24.5	132
60	27	148
55	30	165
50	34.5	187
45	39	215
40	46	250
35	55	295
30	66	365
25	85	460
22	100	550
19.9	115	635

(Fig. 3-3) is constructed from the data in Table 3-1. The area of this card to the scale drawn is 6.23 sq. in., the length is 4.1 in., and the spring scale is 100 lb. per in. of ordinate, hence

$$\text{i.m.e.p. (theoretical)} = \frac{6.23 \times 100}{4.1} = 152 \text{ lb. per sq. in.}$$

The i.m.e.p. for the assumed actual conditions was 141 lb. per sq. in. Therefore, the card factor should be

$$F_D = {}^{141}\!/_{152} = 0.925$$

This value checks the originally assumed value of 0.9 reasonably well.

In superimposing the actual card by rounding the corners of the theoretical card, P_{max} may be taken as about 75 per cent of $P_{C^{3,4}}$. Actually, maximum cylinder pressures vary over a considerable range and depend upon numerous factors, including ignition timing, air-fuel ratio, etc. When a low-octane fuel is used, the maximum pressures are much higher owing to detonation of a part of the charge. In extreme cases of knocking, these pressures can cause serious damage in an engine.

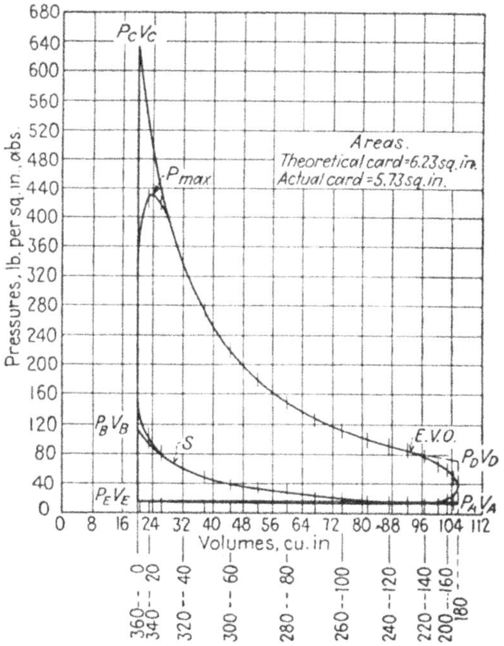

FIG. 3-3.—Indicator card for Example 1 (Table 3-1).

The pressure near the end of the expansion line will drop owing to opening the exhaust valve before bottom dead center (point E.V.O., Fig. 3-3). Usually the pressure will not have expanded to exhaust-stroke pressure until the piston has moved an appreciable distance on the exhaust stroke. This justifies rounding the "toe" of the card.

If the spark occurs too early, the pressure will rise above the compression line before top dead center is reached. This is undesirable, but a greater loss will occur due to P_{max} being farther from top dead center if the spark is retarded too much. Hence, some rounding of the card near the end of the compression stroke should be made.

The area of the superimposed "actual" card is 5.73 sq. in. for the example, and

$$F_D = \frac{5.73}{6.23} = 0.92$$

This again checks the originally assumed value reasonably well.

3-4. Gas-pressure–Crank-angle Diagrams.—To utilize the gas-pressure data from the indicator card, it is necessary to convert to a pressure–crank-angle basis. This may be done most conveniently by a graphical construction as illustrated in Fig. 3-4. With reference to this figure, the constructed indicator card is tacked to a drawing board and the atmospheric line or a line

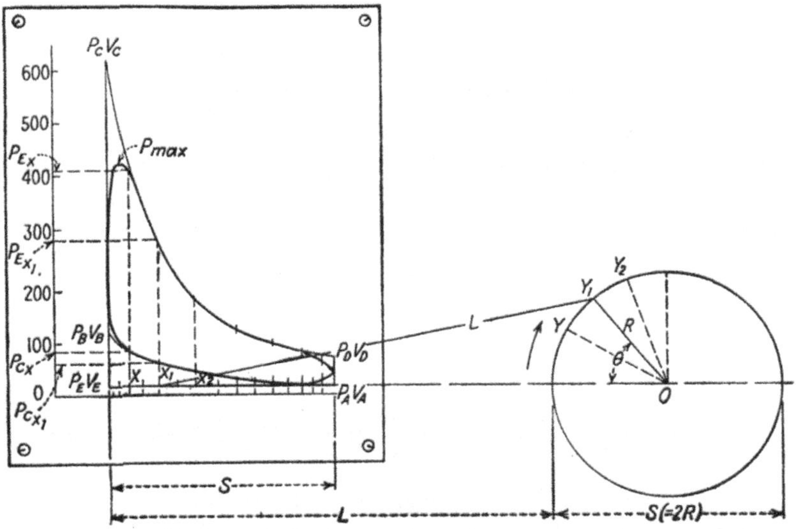

FIG. 3-4.—Graphical method of converting the pressure-volume card to a pressure–crank-angle basis.

parallel to it is extended as shown. The length of the card parallel to the atmospheric line is directly proportional to the stroke of the piston. Hence, by using this length as a base, all other dimensions may be scaled down accordingly. The atmospheric line may be considered as the center line of the cylinder, and the extension will pass through the center of the crankshaft.* The location of this center may be found by scaling down the connecting-rod length to the scale of the card using the relation

* This is true except in offset-cylinder engines, but the construction may be modified readily for such cases.

$$\frac{S}{S'} = \frac{L}{L'} \tag{3-8}$$

where S = length of indicator card, in.
S' = piston stroke, in.
L' = center-to-center length of the engine connecting rod, in.
L = center-to-center length of the connecting rod to the scale of the card, in.

Point O, Fig. 3-4, is distant from point P_EV_E by the amount $L + R$ where $R = S/2$. By using O as a center, construct a

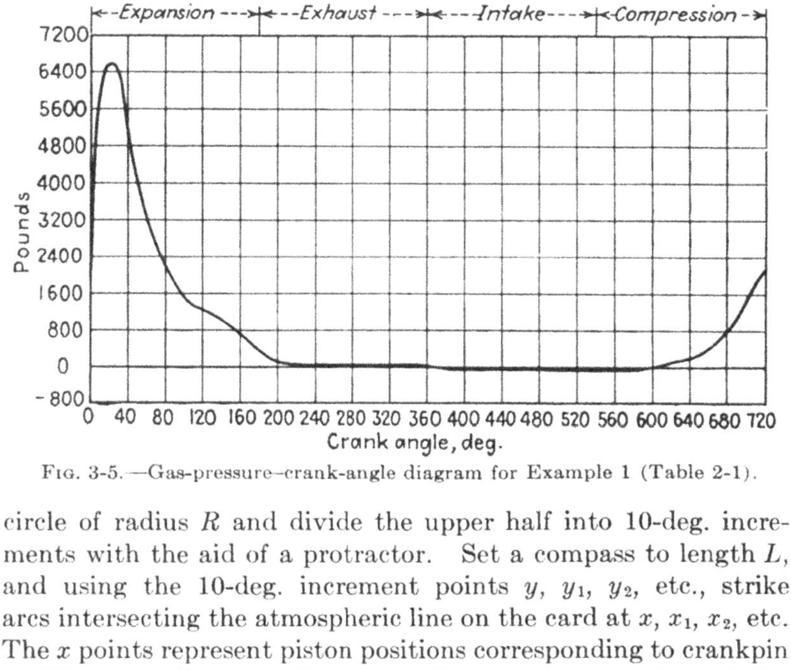

Fig. 3-5.—Gas-pressure–crank-angle diagram for Example 1 (Table 2-1).

circle of radius R and divide the upper half into 10-deg. increments with the aid of a protractor. Set a compass to length L, and using the 10-deg. increment points y, y_1, y_2, etc., strike arcs intersecting the atmospheric line on the card at x, x_1, x_2, etc. The x points represent piston positions corresponding to crankpin positions y. Erect ordinates at the x points, and read the pressures at the intercepts with the card lines directly from the ordinate scale. Convert the pressures to total force by multiplying each by the area of the piston, and plot against crank angle as an uniformly spaced abscissa. As atmospheric pressure is at all times acting on the under side of the piston, the effective gas pressure thrust on the piston is represented by the gage pressure. Hence, it is necessary to convert the ordinate-scale readings to gage readings. This is most easily done by shifting the ordinate scale.

3-5. Example.—Construct a gas-force–crank-angle diagram for the engine selected in Example 1, Table 2-1.

Procedure.—Before locating crank-angle positions on the card (Fig. 3-3), it is necessary to determine the (scale) center-to-center length of the connecting rod from the L/R ratio. By using the selected value of $L/R = 4$ (Table 2-1), scaling down as explained in Par. 3-4, and making the graphical construction, the crank-angle positions were located as shown on Fig. 3-3. Ordinates erected to the compression and expansion lines from these points gave the data from Table 3-2, and these data were used to construct Fig. 3-5.

TABLE 3-2.—GAS-PRESSURE, CRANK-ANGLE FORCES FROM FIG. 3-3
Area of Piston = 15.9 sq. in.

	Crank angle, deg.	Pressure from card, lb. per sq. in. abs.	Pressure, lb. per sq. in. gage	Total net force on piston, lb.		Crank angle, deg.	Pressure from card, lb. per sq. in. abs.	Pressure, lb. per sq. in. gage	Total net force on piston, lb.
Expansion	0	150	135	2,145	Intake	360	16	1.3	20.65
	10	380	365	5,800		370	14.5	−0.2	−3.18
	20	430	415	6,600		380	13	−1.7	−27
	30	420	405	6,440		↓	↓	↓	↓
	40	340	325	5,160		530	13	−1.7	−27
	50	272	257	4,090	Compression	540	13.2	−1.5	−23.8
	60	218	203	3,225		550	13.5	−1.2	−19.1
	70	180	165	2,620		560	13.8	−0.9	−14.3
	80	152	137	2,180		570	14	−0.7	−11.2
	90	130	115	1,830		580	14.5	−0.2	−3.18
	100	112	97	1,540		590	15	0.3	4.76
	110	100	85	1,350		600	17.5	3	47.6
	120	91	76	1,220		610	20	5	79.5
	130	85	70	1,120		620	22	7	112
	140	78	63	1,000		630	24	9	143
	150	68	53	842		640	29	14	222
	160	59	44	700		650	34	19	302
	170	50	35	556		660	39	24	382
Exhaust	180	35	20	318		670	48	33	525
	190	28	13	206.5		680	61	46	781
	200	20	5	79.5		690	80	65	1,032
	210	18	3	47.7		700	105	90	1,430
	220	16	1.3	20.65		710	135	120	1,920
	↓	↓	↓	↓		720	150	135	2,145

References

1. *Univ. Ill., Eng. Expt. Sta., Bull.* 160.
2. Hersey, Eberhardt, and Hottel: Thermodynamic Properties of the Working Fluid in Internal Combustion Engines, *S.A.E. Jour.*, Vol. 39, No. 4, October, 1936.
3. *A.S.I.C.* 421.
4. Angle: "Engine Dynamics and Crankshaft Design."

Suggested Design Procedure

1. For the engine selected for your design, construct a full-throttle–full-speed theoretical indicator card.

If the design is to be supercharged, the effects of the altered inlet pressure must be considered in following the preceding examples.

2. Round the corners of the theoretical card to form the actual card, and determine the card factor.

3. From the actual card thus obtained, determine the indicated horsepower of the engine. Apply the assumed mechanical efficiency, and check the brake horsepower obtained with the originally assumed value of brake horsepower. If a reasonably close agreement is not had, recheck the work for errors.

4. Using data obtained from the actual card, construct a total *net* gas-force–crank-angle (uniform angular spacing) diagram.

5. When items 1 to 4 have been completed and put in proper form, submit for checking and approval. Keep a record of the man-hours required for each item.

Problems

1. Using the basic relations of thermodynamics as applied to the modified air standard Otto cycle, prove that the pressure at the end of the expansion stroke will be as in Eq. (3-2).

2. Referring to Fig. 3-1, prove that when $(1 + \tan \alpha)^n = (1 + \tan \beta)$ the resulting curve has the equation PV^n = a constant.

CHAPTER 4

ANALYSIS OF THE CRANK CHAIN

4-1. Forces Due to the Reciprocating Parts.—In converting the reciprocating motion of the piston into the rotating motion of the crankshaft, the inertia forces of the reciprocating parts play an important part in determining the net turning effort. These parts must be started from rest, accelerated to high velocity, slowed to rest, accelerated again, and stopped a second time during each revolution. At the speed at which airplane-engine crankshafts turn, this process causes quite high inertia forces. Analysis of these forces will be first considered for a single-cylinder engine.

4-2. Piston Velocity and Acceleration.*

In Fig. 4-1, P represents the piston-pin center, C is the crank-pin center, M is the center of crankshaft, L is the center-to-center

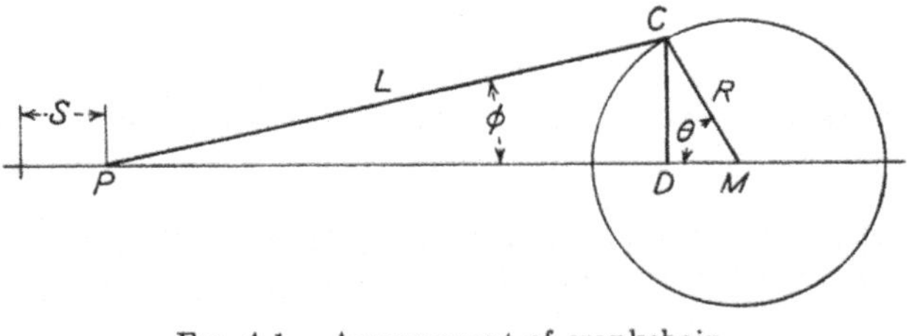

FIG. 4-1.—Arrangement of crankchain.

length of the connecting rod in inches, and R is the crank radius in inches (= 1/2 the stroke). For any given displacement (S) in inches from head-end dead center:

$$S = L + R - L \cos \phi - R \cos \theta$$

$$CD = L \sin \phi = R \sin \theta$$

$$\sin \phi = \frac{R}{L} \sin \theta$$

$$\cos \phi = \sqrt{1 - \sin^2 \phi} = \sqrt{1 - \frac{R^2}{L^2} \sin^2 \theta} = \frac{1}{L} \sqrt{L^2 - R^2 \sin^2 \theta}$$

* A more complete method of analysis is to be found in reference 3.

Therefore

$$S = L + R - (L^2 - R^2 \sin^2 \theta)^{1/2} - R \cos \theta \qquad (4\text{-}1)$$

The velocity corresponding to S is

$$V = \frac{dS}{dt} = \frac{1}{2}(L^2 - R^2 \sin^2 \theta)^{1/2} \times 2R^2 \sin \theta \cos \theta \frac{d\theta}{dt} + R \sin \theta \frac{d\theta}{dt}$$

$$V = R \sin \theta \frac{d\theta}{dt} + \left[\frac{R^2 \sin \theta \cos \theta}{(L^2 - R^2 \sin^2 \theta)^{1/2}} \right] \frac{d\theta}{dt}$$

$$V = R \sin \theta \left[1 + \frac{R \cos \theta}{(L^2 - R^2 \sin^2 \theta)^{1/2}} \right] \frac{d\theta}{dt}$$

In any practical engine, for a given condition of operation, the angular velocity of the crankshaft is very closely uniform. Therefore, $d\theta/dt = 2\pi N$ where N is in revolutions per minute. Substitution of $2\pi N$ in the preceding expression gives V in inches per minute. Dividing by 12×60 gives velocity in feet per second.

$$V_{\text{f.p.s.}} = \frac{2\pi NR \sin \theta}{12 \times 60} \left[1 + \frac{R \cos \theta}{(L^2 - R^2 \sin^2 \theta)^{1/2}} \right]$$

$$V_{\text{f.p.s.}} = 0.00873\ NR \sin \theta \left[1 + \frac{R \cos \theta}{(L^2 - R^2 \sin^2 \theta)^{1/2}} \right] \qquad (4\text{-}2)$$

The acceleration in feet per sec.^{2} corresponding to $V_{\text{f.p.s.}}$ is

$$A = \frac{dV}{dt} = 0.00873NR \Bigg\{ \sin \theta \left[\frac{(L^2 - R^2 \sin^2 \theta)^{1/2} \left(-R \sin \theta \frac{d\theta}{dt} \right) - \frac{1}{2} R \cos \theta (L^2 - R^2 \sin^2 \theta)^{-1/2} \left(-2R^2 \sin \theta \cos \theta \frac{d\theta}{dt} \right)}{L^2 - R^2 \sin^2 \theta} \right]$$

$$+ \left[1 + \frac{R \cos \theta}{(L^2 - R^2 \sin^2 \theta)^{1/2}} \right] \cos \theta \frac{d\theta}{dt} \Bigg\}$$

$$A = 0.00873NR \Bigg\{ \frac{-R \sin^2 \theta \frac{d\theta}{dt}}{(L^2 - R^2 \sin^2 \theta)^{1/2}} + \frac{R^3 \sin^2 \theta \cos^2 \theta \frac{d\theta}{dt}}{(L^2 - R^2 \sin^2 \theta)^{3/2}}$$

$$+ \cos \theta \frac{d\theta}{dt} + \frac{R \cos^2 \theta \frac{d\theta}{dt}}{(L^2 - R^2 \sin^2 \theta)^{1/2}} \Bigg\}$$

but $\cos^2\theta - \sin^2\theta = \cos 2\theta$,

$$\sin^2\theta\cos^2\theta = \frac{2(\sin\theta\cos\theta)2(\sin\theta\cos\theta)}{4} = \frac{\sin^2 2\theta}{4}$$

$$\frac{d\theta}{dt} = 2\pi N \text{ radians per min.} = \frac{2\pi N}{60} \text{ radians per sec.}$$

Therefore,

$$A = 0.000914N^2R\left[\cos\theta + \frac{R\cos 2\theta}{(L^2 - R^2\cos^2\theta)^{1/2}} + \frac{R^3\sin^2 2\theta}{4(L^2 - R^2\sin^2\theta)^{3/2}}\right] \text{ft. per sec.}^2 \qquad (4\text{-}3)$$

The preceding formulas for velocity and acceleration [Eqs. (4-2) and (4-3)] would be cumbersome to use, and the following substitutions can be made without appreciable error:

Let

$$L = (L^2 - R^2\sin^2\theta)^{1/2}$$

$$Z = \frac{R}{L}$$

Then,

$$V_{\text{f.p.s.}} = 0.00873NR\sin\theta(1 + Z\cos\theta)$$

$$= 0.00873NR\left(\sin\theta + Z\,\frac{2\sin\theta\cos\theta}{2}\right)$$

but $2\sin\theta\cos\theta = \sin 2\theta$.

Therefore,

$$V_{\text{f.p.s.}} = 0.00873NR(\sin\theta + \tfrac{1}{2}Z\sin 2\theta) \qquad (4\text{-}4)$$

The term $(\sin\theta + \frac{1}{2}Z\sin 2\theta)$ is called the *piston-velocity factor*. In determining the piston velocity at various crank positions, the calculations may be simplified considerably by taking the value of the piston-velocity factor from Table 4-1.

For the acceleration,

$$A = 0.000914N^2R(\cos\theta + Z\cos 2\theta + \tfrac{1}{4}Z^3\sin^2 2\theta)$$

But the last term $\frac{1}{4}Z^3\sin^2 2\theta$ is small and can be neglected without appreciable error.

Therefore,

$$A = 0.000914N^2R(\cos\theta + Z\cos 2\theta) \text{ ft. per sec.}^2 \qquad (4\text{-}5)$$

The term $(\cos\theta + Z\cos 2\theta)$ is called the *piston-acceleration factor*. Table 4-2 is a convenient aid in determining the piston acceleration at various crank angles.

TABLE 4-1.—TANGENTIAL-FORCE AND PISTON-VELOCITY FACTORS
Values for $(\sin\theta + \frac{1}{2}Z \sin 2\theta)$

θ	$1/Z$					θ
	3.50	3.75	4.00	4.25	4.5	
0	0.000	0.000	0.000	0.000	0.000	360
5	0.112	0.110	0.109	0.108	0.107	355
10	0.223	0.219	0.216	0.214	0.212	350
15	0.330	0.326	0.322	0.318	0.314	345
20	0.434	0.428	0.422	0.417	0.413	340
25	0.532	0.525	0.518	0.513	0.508	335
30	0.624	0.616	0.608	0.602	0.596	330
35	0.708	0.699	0.691	0.684	0.678	325
40	0.784	0.774	0.766	0.759	0.752	320
45	0.850	0.840	0.832	0.825	0.818	315
50	0.907	0.897	0.889	0.882	0.875	310
55	0.954	0.945	0.937	0.930	0.924	305
60	0.990	0.982	0.974	0.968	0.962	300
65	1.016	1.008	1.002	0.997	0.992	295
70	1.032	1.026	1.020	1.015	1.011	290
75	1.037	1.032	1.028	1.025	1.022	285
80	1.034	1.031	1.028	1.025	1.023	280
85	1.021	1.019	1.018	1.017	1.016	275
90	1.000	1.000	1.000	1.000	1.000	270
95	0.971	0.973	0.975	0.976	0.977	265
100	0.936	0.939	0.942	0.945	0.947	260
105	0.894	0.899	0.903	0.907	0.910	255
110	0.848	0.854	0.859	0.864	0.868	250
115	0.797	0.804	0.811	0.816	0.821	245
120	0.742	0.750	0.758	0.764	0.770	240
125	0.685	0.694	0.702	0.709	0.715	235
130	0.625	0.635	0.643	0.650	0.657	230
135	0.564	0.5.4	0.582	0.589	0.596	225
140	0.502	0.512	0.520	0.527	0.533	220
145	0.439	0.448	0.456	0.463	0.469	215
150	0.376	0.384	0.392	0.398	0.404	210
155	0.313	0.321	0.327	0.333	0.338	205
160	0.250	0.256	0.262	0.267	0.271	200
165	0.187	0.192	0.196	0.200	0.203	195
170	0.125	0.128	0.131	0.134	0.136	190
175	0.062	0.064	0.066	0.067	0.068	185
180	0.000	0.000	0.000	0.000	0.000	180

TABLE 4-2.—PISTON-ACCELERATION AND INERTIA FACTORS
Values for ($\cos \theta + Z \cos 2\theta$)

θ	1/Z					θ
	3.5	3.75	4.0	4.25	4.5	
0	1.286	1.267	1.250	1.235	1.222	360
5	1.277	1.259	1.242	1.228	1.215	355
10	1.253	1.235	1.220	1.206	1.194	350
15	1.213	1.197	1.182	1.170	1.159	345
20	1.159	1.144	1.131	1.120	1.110	340
25	1.090	1.078	1.067	1.058	1.049	335
30	1.009	0.999	0.991	0.984	0.977	330
35	0.917	0.911	0.905	0.900	0.895	325
40	0.816	0.813	0.810	0.807	0.805	320
45	0.707	0.707	0.707	0.707	0.707	315
50	0.593	0.596	0.599	0.602	0.604	310
55	0.476	0.482	0.488	0.493	0.498	305
60	0.357	0.367	0.375	0.382	0.389	300
65	0.239	0.251	0.262	0.271	0.280	295
70	0.123	0.138	0.151	0.162	0.172	290
75	0.011	0.028	0.042	0.055	0.066	285
80	−0.095	−0.077	−0.061	−0.048	−0.035	280
85	−0.194	−0.175	−0.159	−0.145	−0.132	275
90	−0.286	−0.267	−0.250	−0.235	−0.222	270
95	−0.368	−0.350	−0.333	−0.319	−0.306	265
100	−0.442	−0.424	−0.409	−0.395	−0.383	260
105	−0.506	−0.490	−0.475	−0.463	−0.451	255
110	−0.561	−0.547	−0.534	−0.522	−0.512	250
115	−0.606	−0.594	−0.583	−0.574	−0.566	245
120	−0.643	−0.633	−0.625	−0.618	−0.611	240
125	−0.671	−0.665	−0.659	−0.654	−0.650	235
130	−0.692	−0.689	−0.686	−0.683	−0.681	230
135	−0.707	−0.707	−0.707	−0.707	−0.707	225
140	−0.716	−0.720	−0.723	−0.725	−0.727	220
145	−0.722	−0.728	−0.734	−0.739	−0.743	215
150	−0.723	−0.733	−0.741	−0.748	−0.755	210
155	−0.723	−0.735	−0.746	−0.755	−0.763	205
160	−0.721	−0.735	−0.748	−0.760	−0.769	200
165	−0.718	−0.735	−0.749	−0.762	−0.773	195
170	−0.717	−0.734	−0.750	−0.764	−0.776	190
175	−0.715	−0.734	−0.750	−0.764	−0.777	185
180	−0.714	−0.733	−0.750	−0.765	−0.778	180

TABLE 4-3.—PISTON-TRAVEL FACTORS
Values for $(1 - \cos\theta + \frac{1}{2}Z \sin^2\theta)$

θ	1/Z 3.5	3.75	4.0	4.25	4.5	θ
0	0.000	0.000	0.000	0.000	0.000	360
5	0.005	0.005	0.005	0.005	0.005	355
10	0.020	0.019	0.019	0.019	0.018	350
15	0.044	0.043	0.043	0.042	0.042	345
20	0.077	0.076	0.075	0.074	0.073	340
25	0.119	0.118	0.116	0.115	0.114	335
30	0.170	0.167	0.165	0.163	0.162	330
35	0.228	0.225	0.222	0.220	0.217	325
40	0.293	0.289	0.286	0.283	0.280	320
45	0.364	0.360	0.355	0.352	0.348	315
50	0.441	0.435	0.430	0.426	0.422	310
55	0.522	0.516	0.510	0.505	0.500	305
60	0.607	0.600	0.594	0.588	0.583	300
65	0.695	0.687	0.680	0.674	0.669	295
70	0.784	0.776	0.768	0.762	0.756	290
75	0.874	0.866	0.858	0.851	0.845	285
80	0.965	0.956	0.948	0.941	0.934	280
85	1.055	1.045	1.037	1.030	1.023	275
90	1.143	1.133	1.125	1.118	1.111	270
95	1.229	1.220	1.211	1.204	1.197	265
100	1.312	1.303	1.295	1.288	1.282	260
105	1.392	1.383	1.375	1.369	1.363	255
110	1.468	1.460	1.452	1.446	1.440	250
115	1.540	1.532	1.525	1.519	1.514	245
120	1.607	1.600	1.594	1.588	1.583	240
125	1.669	1.663	1.657	1.652	1.648	235
130	1.727	1.721	1.716	1.712	1.708	230
135	1.779	1.774	1.770	1.766	1.763	225
140	1.825	1.821	1.818	1.815	1.812	220
145	1.866	1.863	1.860	1.858	1.856	215
150	1.902	1.899	1.897	1.895	1.894	210
155	1.932	1.930	1.928	1.927	1.926	205
160	1.956	1.955	1.954	1.943	1.953	200
165	1.976	1.975	1.974	1.973	1.973	195
170	1.989	1.989	1.989	1.988	1.988	190
175	1.997	1.997	1.997	1.997	1.997	185
180	2.000	2.000	2.000	2.000	2.000	180

With reference to Eq. (4-1), by expanding the radical

$$(L^2 - R^2 \sin^2 \theta)^{1/2}$$

by means of the binomial theorem and neglecting the unimportant terms, the piston travel may be written as

$$S = R(1 - \cos \theta + \tfrac{1}{2}Z \sin^2 \theta) \qquad (4\text{-}6)$$

The term $(1 - \cos \theta + \tfrac{1}{2}Z \sin^2 \theta)$ is called the *piston-travel factor*. Values of this factor for different crank angles may be calculated, but it is more convenient to use Table 4-3.

4-3. Example.—Determine the velocity and acceleration for the master-rod piston in Example 1, Table 2-1.

Procedure.—For this example, $N = 2{,}000$ r.p.m., $R = 5.375/2 = 2.6875$ in. and $L/R = 4$. Hence, from Eq. (4-4),

$$V_{\text{f.p.s.}} = 0.00873 \times 2{,}000 \times 2.6875\ (\sin \theta + \tfrac{1}{2}Z \sin 2\theta)$$

By using values of the piston-velocity factor as found in Table 4-1, the values of V for increment crank angles are found (Table 4-4).

TABLE 4-4

θ	$V_{\text{f.p.s.}}$	θ	$V_{\text{f.p.s.}}$	θ	$V_{\text{f.p.s.}}$	θ	$V_{\text{f.p.s.}}$
0	0	100	44.3	190	6.15	280	48.3
10	10.17	110	40.4	200	12.3	290	48
20	19.85	120	35.6	210	18.4	300	45.75
30	28.6	130	30.2	220	24.4	310	41.7
40	36	140	24.4	230	30.2	320	36
50	41.7	150	18.4	240	35.6	330	28.6
60	45.75	160	12.3	250	40.4	340	19.85
70	48	170	6.15	260	44.3	350	10.17
80	48.3	180	0	270	47	360	0
90	47						

The acceleration from Eq. (4-5) is

$$A = 0.000914 \times \overline{2{,}000}^2 \times 2.6875\ (\cos \theta + Z \cos 2\theta)$$

By using values of the piston-acceleration factor, the values of A for increment crank angles are found (Table 4-5).

It is also of interest to determine the piston travel. By using Eq. (4-6) and proceeding as above, the values vs. crank angle are found (Table 4-6).

TABLE 4-5

θ	A, ft. per sec.^{2}	θ	A, ft. per sec.^{2}	θ	A, ft. per sec.^{2}	θ	A, ft. per sec.^{2}
0	12,280	100	−4,005	190	−7,350	280	− 599
10	12,000	110	−5,240	200	−7,340	290	1,482
20	11,110	120	−6,140	210	−7,280	300	3,680
30	9,730	130	−6,740	220	−7,100	310	5,875
40	7,950	140	−7,100	230	−6,740	320	7,950
50	5,875	150	−7,280	240	−6,140	330	9,730
60	3,680	160	−7,340	250	−5,240	340	11,110
70	1,482	170	−7,350	260	−4,005	350	12,000
80	−599	180	−7,350	270	−2,455	360	12,280
90	−2,455						

TABLE 4-6

θ	S, in.	θ	S, in.	θ	S, in.	θ	S, in.
0	0	100	3.48	190	5.35	280	2.55
10	0.0511	110	3.91	200	5.25	290	2.062
20	0.202	120	4.29	210	5.1	300	1.596
30	0.444	130	4.61	220	4.89	310	1.158
40	0.77	140	4.89	230	4.61	320	0.77
50	1.158	150	5.1	240	4.29	330	0.444
60	1.596	160	5.25	250	3.91	340	0.202
70	2.062	170	5.35	260	3.48	350	0.0511
80	2.55	180	5.375	270	3.025	360	0
90	3.02						

Graphical representations of Tables 4-4, 4-5, and 4-6 are shown in Fig. 4-2.

4-4. Piston Displacement, Velocity, and Acceleration for Articulated Rods.—When articulated rods are used as in the case of radial engines and in many V-engines, the path of the link-pin center is not a true circle, and the preceding formulas for displacement, velocity, and acceleration of the piston are somewhat in error. Formulas for the articulated system can be derived, but they are too complex for practical use, and a graphical analysis is preferable.

Since $V = dS/dt$, the velocity may be found by drawing tangents to the piston-travel curve at increment positions and measuring the slope. For instance, in Fig. 4-2 at 50 deg. the piston travel is at the rate of 3.2 in. in 78 deg. of crank travel. At a speed of 2,000 r.p.m., the time in seconds corresponding to 78 deg. is 60/2,000 × 78/360 = 0.0065 sec., and the velocity is 3.2/(0.0065 × 12) = 41 ft. per sec., which closely checks the value determined by calculation. Similarly, the acceleration at 50 deg. is 38/0.0065 = 5,850 ft. per sec.[2] By taking a sufficient number of points, velocity and acceleration curves may be plotted.

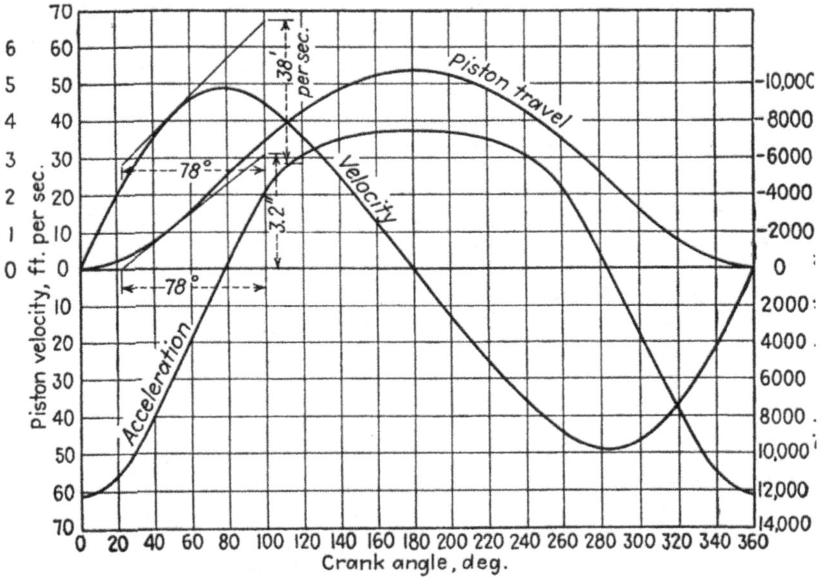

FIG. 4-2.—Piston travel, velocity, and acceleration curves for Example 1.

Figure 4-3 illustrates a method of finding the path of the link-pin center. In this figure, which is based on the dimensions of a Curtiss Conqueror engine, P_M is the master-rod piston-pin center, C_M is the crankpin center, C_L is the link-pin center located at 2.406 in. from C_M, and P_L is the link-rod piston-pin center.

Center-to-center length of master rod = 10 in. Center-to-center length of the link rod is 7.594 in. Obviously, angle $C_LC_MP_M$ is fixed by the design of the master rod.

Plotting the path of the link-pin center consists in locating C_M at increment angles and finding the corresponding position of C_L. The link-rod piston positions are then found by setting the compass to the link-rod length and striking arcs intercepting the link-rod cylinder axis. The corresponding piston-travel

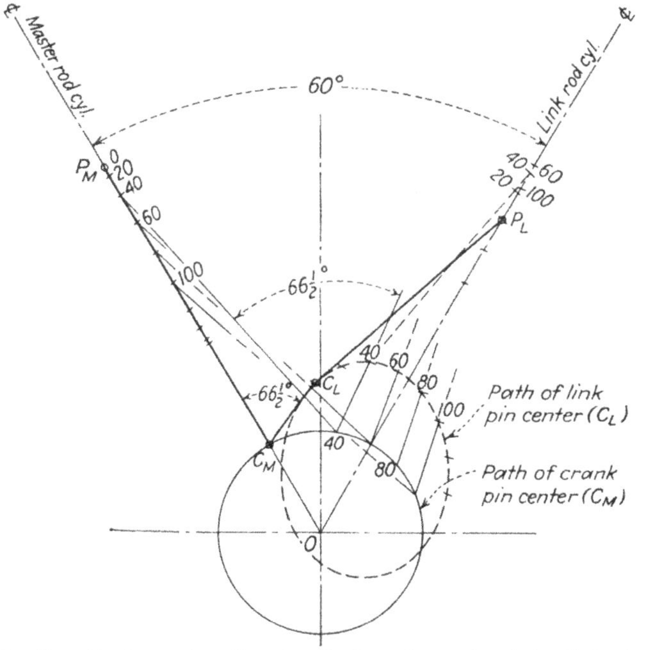

FIG. 4-3.—Graphical construction for finding the path of the link-pin center for a Curtiss V-1570 Conqueror engine.

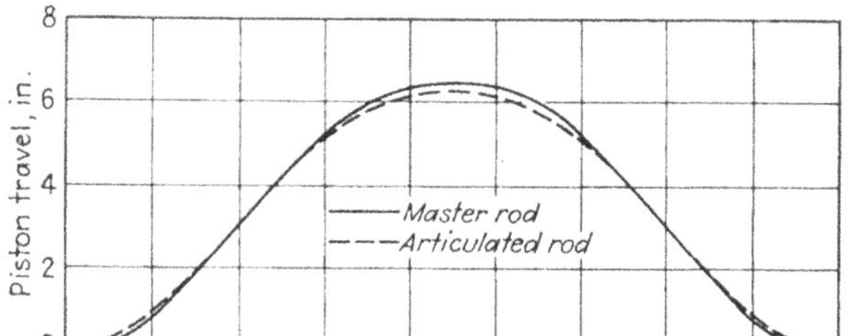

FIG. 4-4.—Piston travel vs. crank angle for a Curtiss V-1570 Conqueror engine.

positions may be found by measuring the distances from the extreme position of the piston pin to these intercepts.

Figure 4-4 shows the piston travel of the master-rod and articulated-rod cylinders for the Curtiss V-1570 engine. As the slope of the articulated-rod curve is nowhere very different from the slope of the master-rod curve, the velocity and acceleration curves will also be closely similar (Fig. 4-5). This fact justifies the usual simplifying procedure of assuming that the acceleration of the articulated-rod pistons may be taken as equal to that of the master-rod pistons. It should be noted, however, that the farther the link-pin center is from the master-rod center, the greater will be the discrepancy. It is also of importance to note that increasing this distance increases the stroke of the

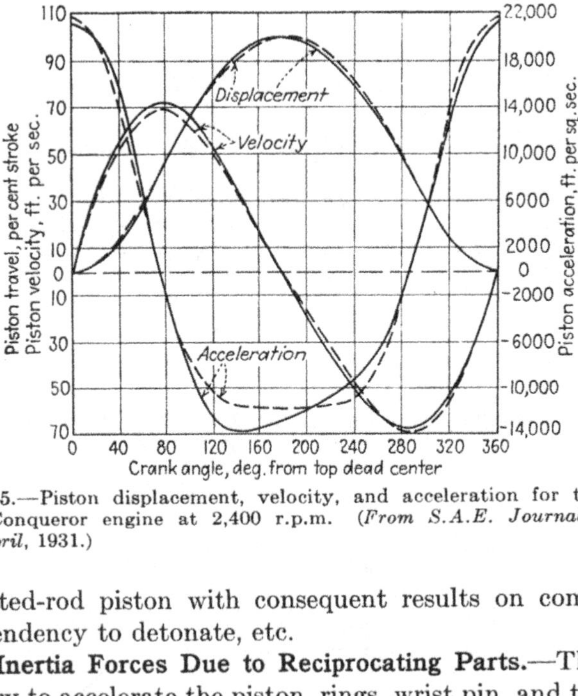

FIG. 4-5.—Piston displacement, velocity, and acceleration for the Curtiss V-1570 Conqueror engine at 2,400 r.p.m. (*From S.A.E. Journal, Vol. 29, No. 4, April, 1931.*)

articulated-rod piston with consequent results on compression ratio, tendency to detonate, etc.

4-5. Inertia Forces Due to Reciprocating Parts.—The forces necessary to accelerate the piston, rings, wrist pin, and the upper end of the connecting rod are directly proportional to the weight of these parts, and in consequence, it is desirable to keep their weights to a minimum consistent with the other functions that they perform. When the accelerations are known, the inertia forces may be calculated from

$$F = MA = \frac{W}{g} A \tag{4-7}$$

where F = inertia force, lb.
M = mass of reciprocating parts.
W = weight of reciprocating parts, lb.
A = acceleration, ft. per sec.^{2}
g = 32.2.

It is convenient, in calculating inertia forces, to combine Eqs. (4-5) and (4-7), thus

$$F_R = 0.0000284N^2WR(\cos\theta + Z\cos 2\theta) \qquad (4\text{-}8)$$

where F_R = inertia force of the reciprocating parts, lb.
N = r.p.m.
W = weight of reciprocating parts, lb.
R = crank radius, in.

The term $(\cos\theta + Z\cos 2\theta)$ is called the *inertia factor*. It is most conveniently found from Table 4-2.

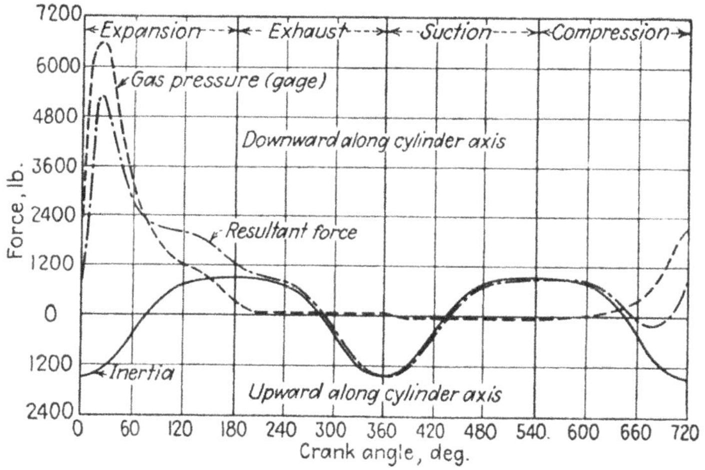

FIG. 4-6.—Gas pressure, inertia, and resultant forces (in respect to direction). (*Method from Angle, "Engine Dynamics and Crankshaft Design."*)

In the analysis of an engine, the weight of the reciprocating parts must be known to determine the inertia forces. For a new unit, this involves practically a complete design of the reciprocating parts. But this is difficult without a knowledge of the stresses involved. Obviously, a preliminary weight estimate is necessary to determine the forces, and an intelligent estimate necessitates a reference to previous attainments. Figures A1-3 and A1-4 are of assistance in this respect.

4-6. Example.—Estimate the inertia force due to reciprocating parts at increment angles through 360 deg. for 1 cylinder of Example 1, Table 2-1.

Procedure.—By using the data of Example 1, and referring to Figs. A1-3 and A1-4, the probable weight of the reciprocating parts will be 0.25 lb. per sq. in. of piston area, or a total of $0.25 \times 4.5^2 \times 0.785 = 4$ lb. per cylinder, approximately. This weight may be substituted in Eq. (4-8), but inasmuch as the accelerations have already been found (Table 4-5), the forces may be found from $F = \frac{4}{32.2} A$ for the various crank angles. Results of these calculations are shown in Table 4-7.

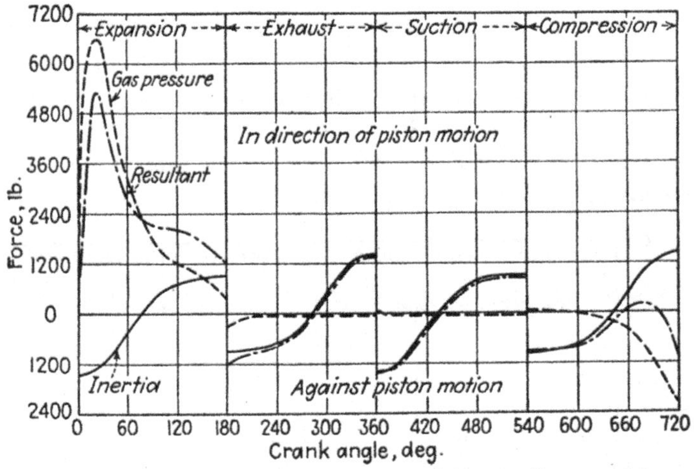

FIG. 4-7.—Resultant forces of gas pressure and inertia (in respect to work). (*Method from Angle, "Engine Dynamics and Crankshaft Design."*)

For combining with the gas-pressure forces, inertia forces may be plotted in either of two ways (Figs. 4-6 and 4-7).

4-7. Torque or Turning Effort per Cylinder.—The part of the force along the cylinder axis which does useful work is the com-

TABLE 4-7

θ	F, lb.	θ	F, lb.	θ	F, lb.	θ	F, lb.
0	1,525	100	−498	190	−912	280	−74.5
10	1,491	110	−650	200	−911	290	184
20	1,381	120	−761	210	−905	300	457
30	1,210	130	−836	220	−881	310	730
40	988	140	−881	230	−836	320	988
50	730	150	−905	240	−761	330	1,210
60	457	160	−911	250	−650	340	1,381
70	184	170	−912	260	−498	350	1,491
80	−74.5	180	−912	270	−325	360	1,525
90	−325						

ponent tending to rotate the crankshaft. This turning force may be expressed in terms of the force parallel to the cylinder axis. Referring to Fig. 4-8, P is the piston-pin center, C is the crankpin center, and M is the axis of the crankshaft.

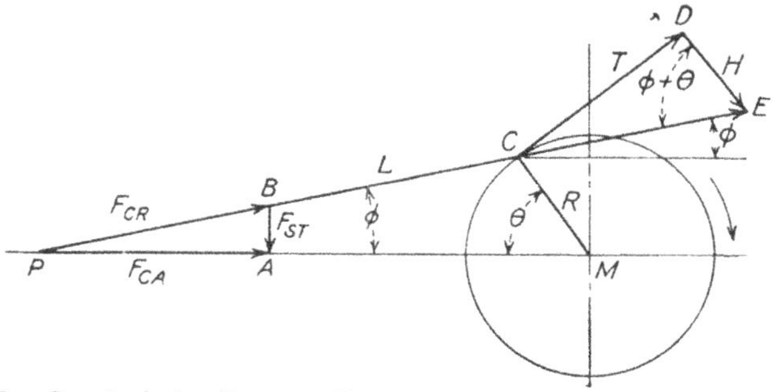

FIG. 4-8.—Crank-chain diagram illustrating the method of determining the torque.

In the figure, the force parallel to the cylinder axis is F_{CA} and the force in the connecting rod is PB ($= F_{CR}$). From the diagram,

(1) $DE = CE \cos(\theta + \phi) = H$, the component tending to bend the crankshaft.

(2) $CD = CE \sin(\theta + \phi) = T$, the component tending to rotate the crankshaft.

(3) $CE = PB = \dfrac{F_{CA}}{\cos \phi}$.

Subsituting (3) in (1),

$$H = \frac{F_{CA}}{\cos \phi} \times \cos(\theta + \phi)$$

and substituting (3) in (2),

$$T = \frac{F_{CA}}{\cos \phi} \times \sin(\theta + \phi)$$

If F_{CA} is in pounds and R is in inches, the torque Q is

$$Q = TR = R \times F_{CA} \frac{\sin(\theta + \phi)}{\cos \phi} \text{ in pound-inches}$$

However, it is best to have the equation for T in terms of θ only as ϕ is difficult to determine; hence

$$T = F_{CA} \frac{\sin(\theta + \phi)}{\cos \phi} = F_{CA} \frac{\sin \theta \cos \phi + \cos \theta \sin \phi}{\cos \phi}$$

but

$$\cos \phi = \sqrt{1 - \frac{R^2}{L^2} \sin^2 \theta} = \sqrt{1 - Z^2 \sin^2 \theta}$$

and

$$\sin \phi = \frac{R}{L} \sin \theta = Z \sin \theta$$

substituting

$$T = F_{CA} \frac{\sin \theta \sqrt{1 - Z^2 \sin^2 \theta} + \cos \theta \, Z \sin \theta}{\sqrt{1 - Z^2 \sin^2 \theta}}$$

$$T = F_{CA} \sin \theta \left(1 + \frac{Z \cos \theta}{\sqrt{1 - Z^2 \sin^2 \theta}}\right)$$

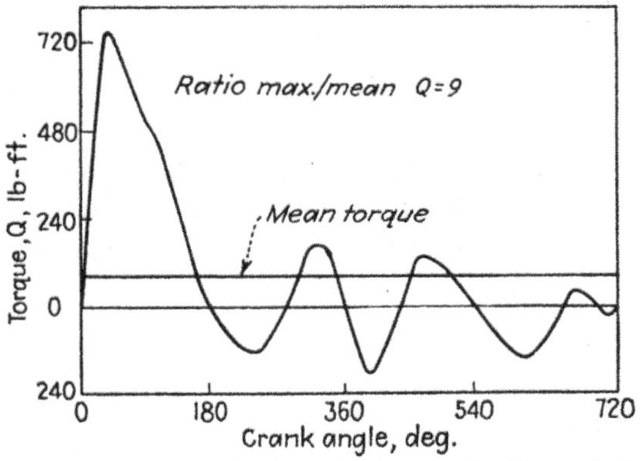

FIG. 4-9.—Single-cylinder torque curve for Example 1.

The expression $(Z^2 \sin^2 \theta)$ is small and may be neglected without appreciable error. Then

$$\begin{aligned} T &= F_{CA} \sin \theta (1 + Z \cos \theta) \\ &= F_{CA} (\sin \theta + Z \sin \theta \cos \theta) \end{aligned}$$

but $2 \sin \theta \cos \theta = \sin 2\theta$

Therefore,

$$T = F_{CA} \left(\sin \theta + \frac{Z}{2} \sin 2\theta\right) \tag{4-9}$$

The term $[(\sin \theta + (Z/2 \sin 2\theta)]$ is called the *tangential-force factor*, and it is most conveniently found from Table 4-1.

The torque,

$$Q \text{ (in lb.-ft.)} = T \text{ (in lb.)} \times R \text{ (in ft.)} \tag{4-10}$$

4-8. Example.—For the engine selected in Example 1, plot a curve of torque per cylinder against crank angle through one complete cycle.

Procedure.—Values of F_{CA} (= the net or resultant force parallel to cylinder axis) are read from Fig. 4-6. The resultant turning effort and torque at increment crank angles obtained from Eqs. (4-9) and (4-10) are given in Table 4-8. The torque per cylinder is shown graphically in Fig. 4-9.

TABLE 4-8

θ	F (lb.)	T (lb.)	Q (lb.-ft.)	θ	F (lb.)	T (lb.)	Q (lb.-ft.)
0	620	0	0	370	−1,488	−321	−71.9
10	4,309	930	208	380	−1,354	−571	−128.0
20	5,220	2,200	493	390	−1,183	−720	−161
30	5,230	3,180	712	400	−961	−736	−165
40	4,172	3,200	716	410	−703	−625	−140
50	3,360	2,980	667	420	−430	−419	−93.8
60	2,768	2,690	603	430	−157	−160	−35.8
70	2,436	2,480	555	440	48	49.4	11
80	2,255	2,320	520	450	352	352	78.9
90	2,155	2,155	482	460	471	444	99.5
100	2,038	1,920	430	470	677	582	130
110	2,000	1,720	385	480	734	556	124.5
120	1,980	1,500	336	490	809	520	116.5
130	1,956	1,258	284	500	834	434	97.1
140	1,880	978	219	510	878	344	77.0
150	1,747	684	153	520	884	232	52.0
160	1,610	422	94.5	530	885	116	26.0
170	1,468	192	43.0	540	888	0	0
180	1,230	0	0	550	−893	−117	−26.2
190	−1,119	−146	−32.7	560	−897	−235	−52.6
200	−990	−259	−58.0	570	−894	−350	−78.4
210	−953	−374	−88.7	580	−884	−460	−103.0
220	−902	−469	−105	590	−841	−541	−121.0
230	−857	−558	−125	600	−808	−613	−137
240	−782	−593	−133	610	−730	−627	−140
250	−677	−582	−130	620	−620	−584	−131
260	−519	−489	−109.5	630	−468	−468	−105
270	−352	−352	−78.8	640	−297	−305	−68.3
280	−95	−97.6	−21.9	650	−118	−120.4	−27.0
290	163	166	37.2	660	75	73	16.3
300	436	424	95.0	670	205	182	40.8
310	709	630	141.0	680	207	159	35.6
320	967	740	166.0	690	178	108.4	24.3
330	1,189	722	162.0	700	−50	−21.1	−4.72
340	1,360	574	128.5	710	−430	−93	−20.8
350	1,470	318	71.2	720	−620	0	0
360	1,504	0	0				

4-9. Torque Reaction.—The reaction to the torque force is the piston side thrust. Referring to Fig. 4-8, the side thrust is represented by the force vector F_{ST} and from the diagram $F_{ST} = F_{CA} \tan \phi$, but

$$\tan \phi = \frac{\sin \phi}{\cos \phi} = \frac{R/L \sin \theta}{\sqrt{1 - (R^2/L^2) \sin^2 \theta}} = \frac{\sin \theta}{\sqrt{(L/R)^2 - \sin^2 \theta}}$$

hence

$$F_{ST} = \frac{F_{CA} \sin \theta}{\sqrt{(L/R)^2 - \sin^2 \theta}} \tag{4-11}$$

It should be noted that the shorter the connecting-rod length L in proportion to the crank radius R, the less the overall dimensions of the engine but the greater the side-thrust component and hence the relative friction and wear in the cylinder.

An example of variation of piston side thrust with crank angle is shown in Fig. 4-10.

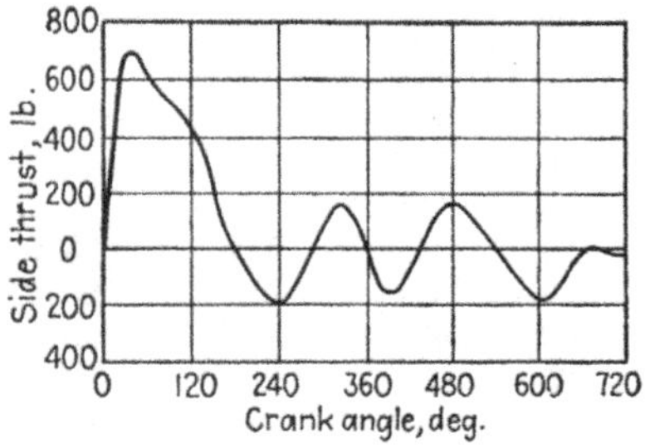

FIG. 4-10.—Variation of piston side thrust for a typical aircraft engine. (*From Angle, "Engine Dynamics and Crankshaft Design."*)

4-10. Total Engine Torque. For the purposes of design, it may be assumed that the torque curves for all cylinders will be the same. Hence, to determine the total turning effort

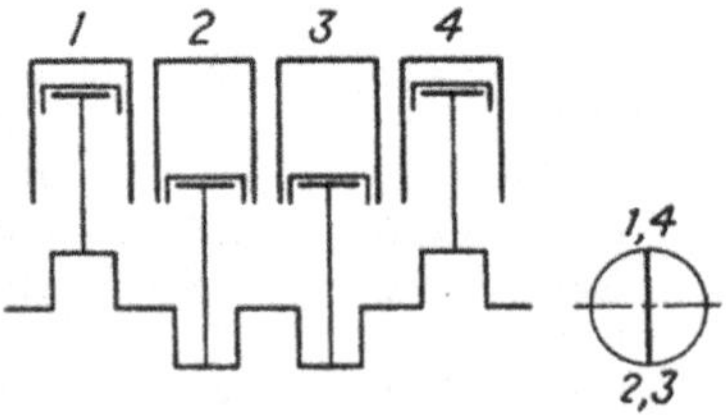

FIG. 4-11.—Usual crank-arm arrangement for four cylinder in-line engines.

Cylinder Numbers

Crank Angle	1	2	3	4
0–180	P	*C*	*E*	*I*
180–360	*E*	P	*I*	*C*
360–540	*I*	*E*	*C*	P
540–720	*C*	*I*	P	*E*

FIG. 4-12.—Diagram for determining firing orders in conventional four-cylinder in-line engines.

of the engine, it is merely necessary to space the individual cylinder curves properly with respect to crank angle and add the ordinates. To determine the angular spacing, it is necessary to know the arrangement of the crankshaft crank arms and the firing order in the cylinders.

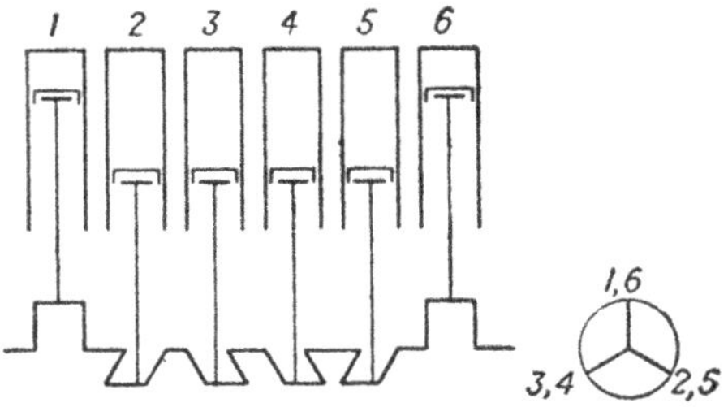

Fig. 4-13.—Usual crank-arm arrangement for six-cylinder in-line engines.

Cylinder Numbers

Crank Angle	1	2	3	4	5	6
0–60	*P*	*E*	*P*	*I*	*C*	*I*
60–120	*P*	*E*	*E*	*C*	*C*	*I*
120–180	*P*	*I*	*E*	*C*	*P*	*I*
180–240	*E*	*I*	*E*	*C*	*P*	*C*
240–300	*E*	*I*	*I*	*P*	*P*	*C*
300–360	*E*	*C*	*I*	*P*	*E*	*C*
360–420	*I*	*C*	*I*	*P*	*E*	*P*
420–480	*I*	*C*	*C*	*E*	*E*	*P*
480–540	*I*	*P*	*C*	*E*	*I*	*P*
540–600	*C*	*P*	*C*	*E*	*I*	*E*
600–660	*C*	*P*	*P*	*I*	*I*	*E*
660–720	*C*	*E*	*P*	*I*	*C*	*E*

Fig. 4-14.—Diagram for determining firing orders in conventional six-cylinder in-line engines.

For four-cylinder in-line engines, the usual method of arranging the crank arms is shown in Fig. 4-11. The firing order may be found from a diagram such as Fig. 4-12. In this figure, the firing order is 1-2-4-3. The other possible firing order for four-cylinder in-line engines having the conventional crank arrangement shown in Fig. 4-11 is 1-3-4-2.

The usual crank arrangement for six-cylinder in-line engines is illustrated in Fig. 4-13.

A method of determining the firing order for six-cylinder engines is illustrated in Fig. 4-14. In this figure, the firing order is 1-5-4-6-2-3. Other firing orders are 1-2-3-6-5-4, 1-2-4-6-5-3, and 1-5-3-6-2-4. The firing order 1-5-3-6-2-4 is usually considered best as no two adjacent cylinders fire in succession.

The firing order for conventional single-bank radial engines is 1-3-5-7-9-2-4-6-8 for nine-cylinder engines, and the same procedure applies for a lesser number of cylinders. The reason for using an odd number of cylinders is obvious.

American airplane engines are usually designed to rotate clockwise when viewed from the end opposite the propeller. Customary methods of numbering the cylinders are:

1. For in-line engines:

2. For V-engines:

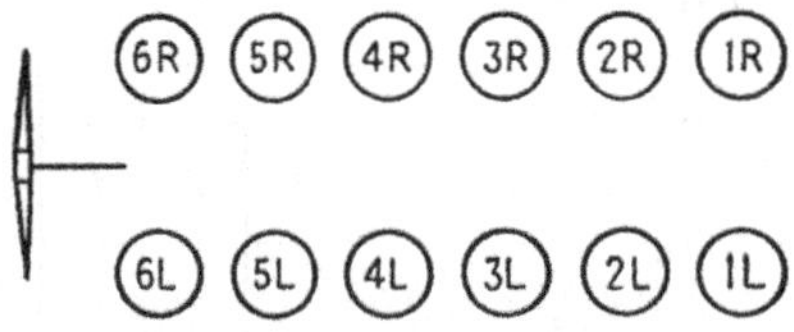

However, numbering of cylinders is largely arbitrary, and many engines differ from the preceding method of numbering.

3. For single-bank radial engines, the cylinders are numbered in the direction of rotation.

4-11. Example.—Determine the firing order, and plot a curve of total engine torque for Example 1, Table 2-1.

Procedure.—In a five-cylinder single-bank radial, the firing order will necessarily be 1-3-5-2-4. The angular spacing of the cylinder center lines is $360/5 = 72$ deg. In spacing the individual torque curves with No. 1 cylinder starting expansion at 0 deg., No. 3 will start at 144 deg., No. 5 will start at 288 deg., No. 2 will start at 432 deg., and No. 4 will start at 576 deg.

Individual torque curves for the five cylinders and the curve of resultant torque for the engine are shown in Fig. 4-15.

The mean torque is found by taking the area under the resultant engine torque curve and dividing by the length. The mean-torque line is located at a height above the zero line equal to the quotient.

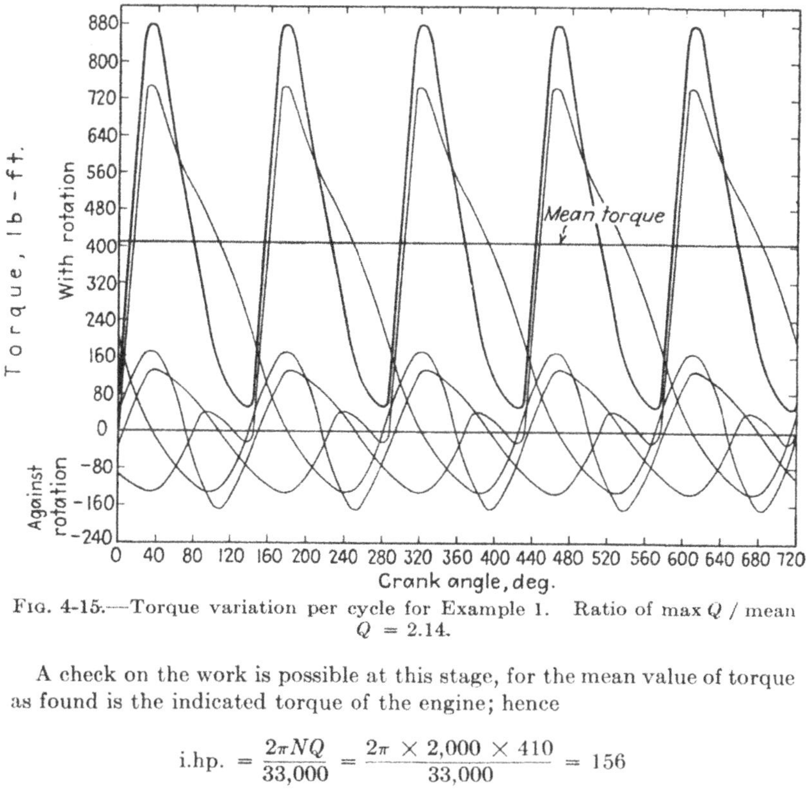

FIG. 4-15.—Torque variation per cycle for Example 1. Ratio of max Q / mean $Q = 2.14$.

A check on the work is possible at this stage, for the mean value of torque as found is the indicated torque of the engine; hence

$$\text{i.hp.} = \frac{2\pi NQ}{33{,}000} = \frac{2\pi \times 2{,}000 \times 410}{33{,}000} = 156$$

For the assumed mechanical efficiency of 85 per cent (Par. 3-3) the brake horsepower is

$$\text{b.hp.}_{\text{max}} = 156 \times 0.85 = 133$$

The originally assumed brake horsepower was based on a cylinder displacement of 82.5 cu. in. (Par. 2-4), but the torque was based on a cylinder displacement of $4.5^2 \times 0.785 \times 5.375 = 85.5$ cu. in. Hence the horsepower based on the original displacement will be about $(82.5/85.5) \times 133 = 128$. This is within less than 2.5 per cent of the originally assumed value of 125 b.hp., and therefore indicates that no serious errors have been made in the calculations.

4-12. Torque Variation with Number of Cylinders and Cylinder Arrangement.—In a one-cylinder engine, the torque is negative, *i.e.*, against rotation, for a large portion of the cycle. Hence to keep the engine turning, it is necessary to use a relatively heavy flywheel. This is obviously not practical for airplane engines, and, although the propeller acts to a considerable extent as a flywheel, because of its mass, it is desirable to use several cylinders to reduce the torque variation as well as increase the total power output. Figure 4-16 shows the effect of number of cylinders and arrangement on torque variation and ratio of maximum to mean torque.

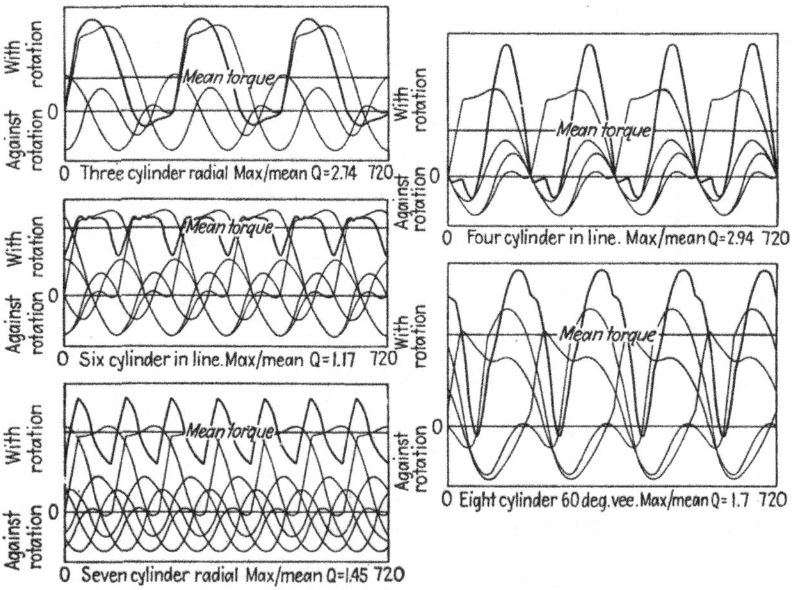

Fig. 4-16.—Effect of number of cylinders and arrangement on torque variation. (*From Angle, "Engine Dynamics and Crankshaft Design."*)

Suggested Design Procedure

Important. Make all constructions and curves to a large enough scale to permit accurate readings of values. Size B or larger drawing paper is recommended. Keep a record of the man-hours required on each item.

1. For the engine selected for your design, construct curves of piston travel, velocity, and acceleration through 360 deg. of crankshaft travel (for one cylinder).

2. Estimate the weight of reciprocating parts, and construct a curve of reciprocating inertia force vs. crank angle (through 720 deg. of crankshaft travel).

3. Superimpose the gas-force curve (see Suggested Design Procedure, page 35; item 4) on the reciprocating inertia-force-curve (see item 2 above) coordinates, and plot a curve of resultant force parallel to the cylinder axis.

4. Construct a single-cylinder torque curve (through 720 deg. of crankshaft travel), and draw a line on the diagram representing the mean torque. Determine the ratio of maximum to mean torque, and place the value found on the diagram.

5. Construct a curve of piston side thrust vs. crank angle (through 720 deg. of crankshaft travel).

6. Determine the firing order to be used, plot a curve of total engine torque vs. crank angle (through 720 deg. of crankshaft travel), and draw a line on the diagram representing the mean engine torque. Determine the ratio of maximum to mean torque, and place the value found on the diagram.

7. By using the mean engine torque value found in item 6, determine the indicated horsepower and brake horsepower for your engine. If the brake horsepower thus determined does not agree within 5 per cent of the originally assumed value of brake horsepower (Suggested Design Procedure, page 24, item 2), recheck the work for errors.

8. When items 1 to 7 have been completed and put in proper form, submit for checking and approval.

References

1. Angle: "Engine Dynamics and Crankshaft Design."
2. Huebotter: "Mechanics of the Gasoline Engine."
3. Cousins: "Analytical Design of High Speed Internal Combustion Engines."

CHAPTER 5

ANALYSIS OF BEARING LOADS

5-1. Crankshaft-bearing Loads.—Before the necessary sizes of the various crankshaft bearings can be determined, it is essential to know the loads to which they will be subjected, and before all these loads can be determined, it is necessary to know approximately the dimensions, *i.e.*, the mass of the moving parts. Obviously, this knowledge necessitates the making of assumptions based on the past experience of the designer or, in the

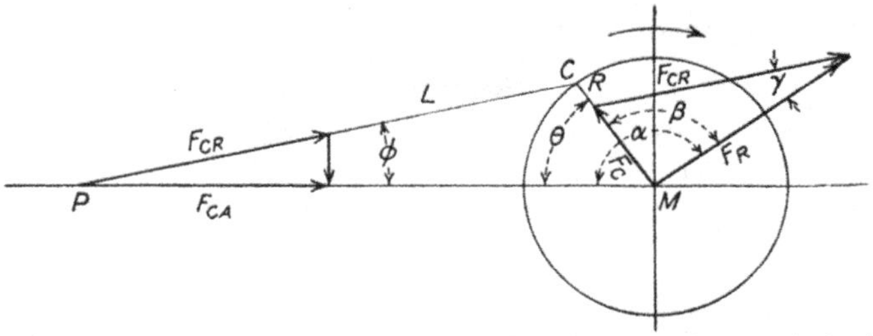

FIG. 5-1.—Method of determining the resultant force on the crankpin for an engine having one cylinder per crankpin.

case of students, of other designers. Only by utilizing the results of previous successful experience can an exhorbitant amount of trial-and-error effort be avoided.

5-2. Resultant Force on the Crankpin.—The resultant force on the crankpin may be found most readily by combining graphically the resultant forces along the connecting-rod axes with the centrifugal forces due to the weight of the lower end of the connecting rod.

Referring to Fig. 5-1, F_{CA} is the resultant force along the cylinder axis and F_C is the centrifugal force due to the rotating weight of the connecting rod.

The acceleration toward the axis of a rotating body necessary to keep the body moving in a circle is v^2/r; hence the centrifugal force on the body is

$$F_C = MA = M\frac{v^2}{r} = \frac{W_C v^2}{gr}$$

where F_C = centrifugal force, lb.
M = mass.
A = acceleration.
v = linear velocity, ft. per sec.
r = radius (crank arm), ft.
W_C = centrifugal weight, lb. (= the rotating weight *on* the crankpin for the case under consideration).
g = acceleration of gravity (= 32.2 ft. per sec.^{2}).

But

$$v = 2\pi rn = \frac{2\pi RN}{12 \times 60}$$

where n = r.p.s.
N = r.p.m.
R = crank arm, in.

Hence, by combining and reducing

$$F_C = 0.0000284W_C N^2 R \tag{5-1}$$

The centrifugal force is laid off to scale along the crank arm from the crankshaft axis M. F_{CR} is the component of F_{CA} along the connecting rod axis. From the diagram,

$$F_{CR} = \frac{F_{CA}}{\cos\phi}$$

but from Par. 4-2,

$$\cos\phi = \sqrt{1 - \left(\frac{R}{L}\right)^2 \sin^2\theta}$$

Therefore

$$F_{CR} = \frac{F_{CA}}{[1 - (R/L)^2 \sin^2\theta]^{1/2}} \tag{5-2}$$

Vector F_{CR} is laid off from the end of vector F_C and parallel to the axis of the connecting rod. The resultant F_R closes the force triangle.

Angle α represents the direction of the resultant force with respect to the cylinder axis, β with respect to the crank arm, and γ with respect to the connecting rod. Resultant forces on the crankpin are usually plotted as polar diagrams. Figure 5-2*a* shows polar diagrams for (*A*) an automotive engine and (*B*) an aircraft engine each plotted with respect to the cylinder axis. Figure 5-2*b* shows the data of Fig. 5-2*a* (diagram *B*) plotted with respect to the crank-arm axis. The effect of engine speed and

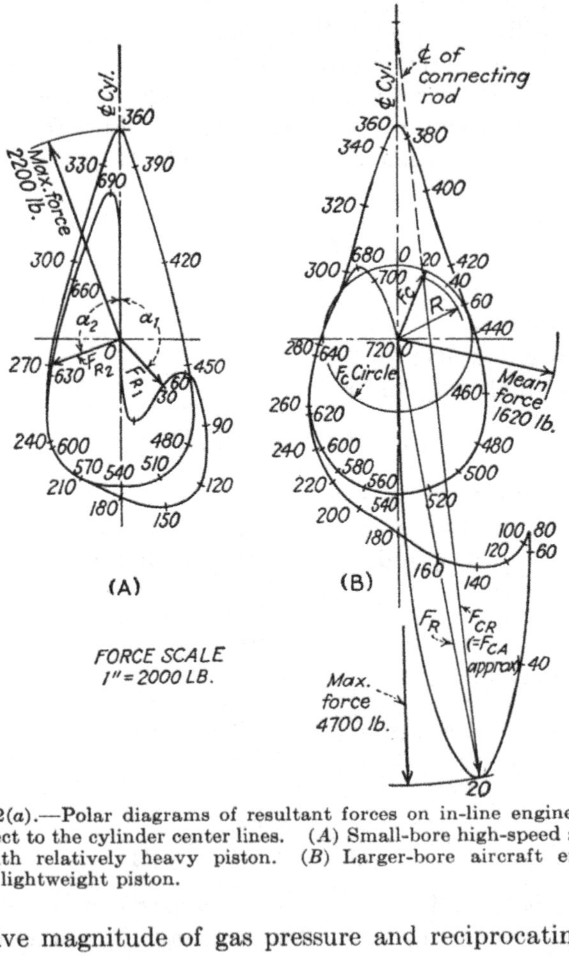

FIG. 5-2(*a*).—Polar diagrams of resultant forces on in-line engine crankpins with respect to the cylinder center lines. (*A*) Small-bore high-speed automotive engine with relatively heavy piston. (*B*) Larger-bore aircraft engine with relatively lightweight piston.

of relative magnitude of gas pressure and reciprocating inertia forces is readily apparent.

When more than one connecting rod is attached to a given crankpin, the vector F_{CR} must be included for each. Figure 5-3 shows the method of finding the resultant force on the crankpin for a V-type engine. Figure 5-4 shows polar diagrams with respect to (*a*) the engine axis, (*b*) the crank arm, and (*c*) the left connecting rod for the engine in Fig. 5-3.

When many connecting rods are attached to one crankpin through link pins, the determination of crankpin bearing loads

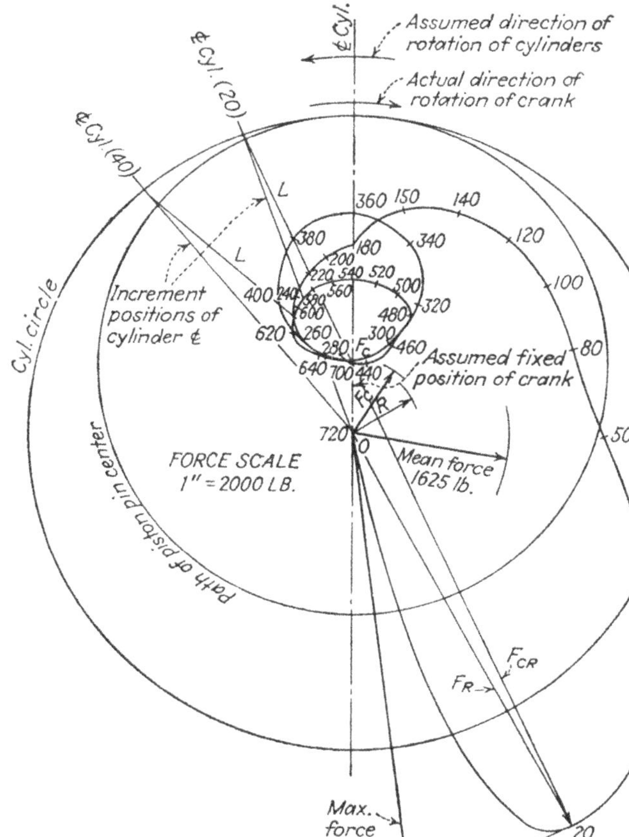

FIG. 5-2(*b*).—Polar diagram of resultant forces on an in-line engine crankpin with respect to the crank arm. This diagram corresponds to (*B*) in Fig. 5-2 (*a*). The diagram is most easily constructed by assuming that the crank arm remains fixed and the cylinders rotate backward at increment angles.

may be simplified somewhat by assuming that the forces in the articulated rods pass through the center of the crankpin. This assumption is not strictly correct, but it gives results sufficiently accurate for preliminary design purposes.

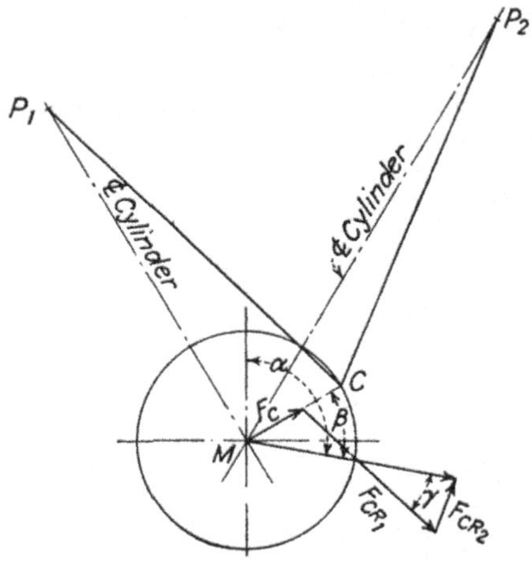

FIG. 5-3.—Method of finding the resultant force on the crankpin of a V-type engine. (*From Angle, "Engine Dynamics and Crankshaft Design."*)

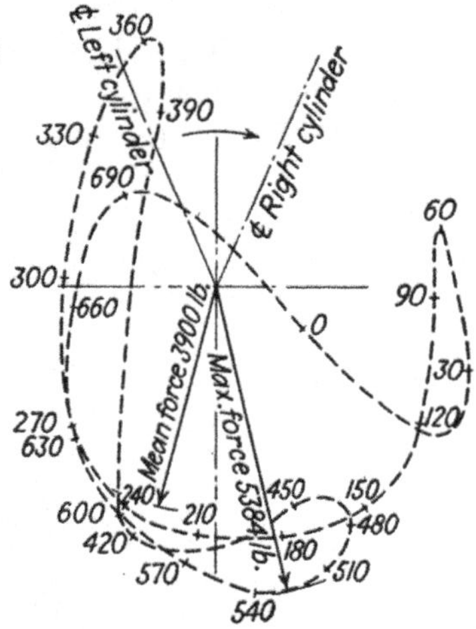

FIG. 5-4.(*a*)—Polar diagram of resultant force on V-type engine crankpin with respect to the engine axis. (*From Angle, "Engine Dynamics and Crankshaft Design."*)

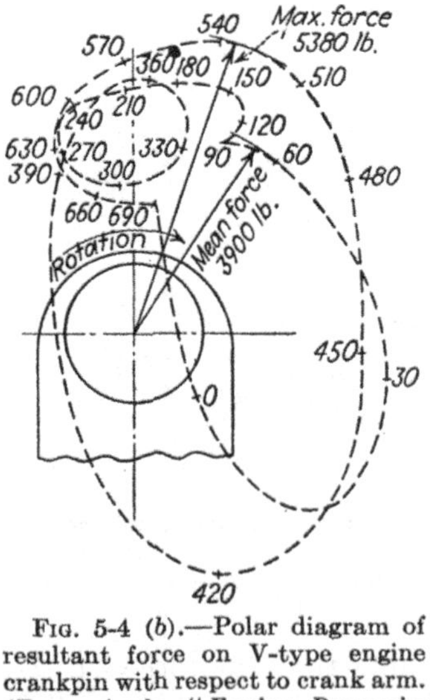

FIG. 5-4 (*b*).—Polar diagram of resultant force on V-type engine crankpin with respect to crank arm. (*From Angle, "Engine Dynamics and Crankshaft Design."*)

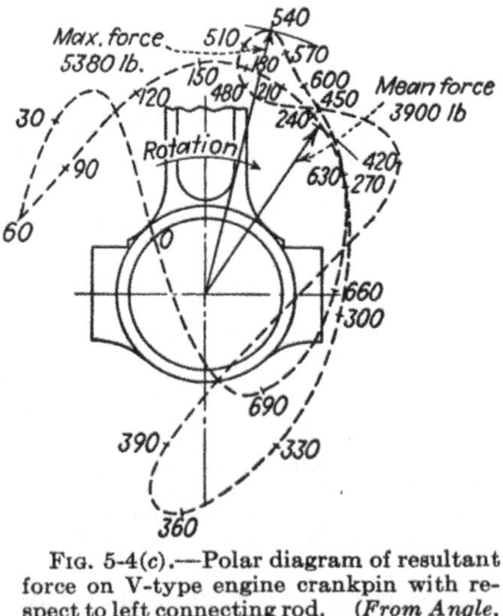

FIG. 5-4(*c*).—Polar diagram of resultant force on V-type engine crankpin with respect to left connecting rod. (*From Angle, "Engine Dynamics and Crankshaft Design."*)

5-3. Example.—Construct a polar diagram of crankpin bearing loads for the engine selected in Example 1, Table 2-1.

Procedure.—From Fig. A1-5, the probable rotating weight per crankpin for a 125-hp. engine of five cylinders will be about 10 lb. For a speed of 2,000 r.p.m. and a crank radius of 2.6875 in., the centrifugal force will be [from Eq. (5-1)]

$$F_C = 0.0000284 \times 10 \times \overline{2{,}000}^2 \times 2.6875 = 3{,}050 \text{ lb.}$$

The component of force along the connecting-rod axis F_{CR}, varies with the value of F_{CA} and θ. Table 5-1 shows, for the example, values of F_{CR} at increment angles. It is to be seen from this table that values of F_{CR} do not differ greatly from values of F_{CA}. For many purposes, these differences are small enough to be neglected, and values of F_{CA} may be used directly in plotting the polar diagram.

TABLE 5-1

1	2	3	4	5	6	7	8	9	10	11 (= 7)	12
θ	F_{CA}	$\sin\theta$	$\sin^2\theta$	$(R/L)^2\sin^2\theta$	$1-(R/L)^2\sin^2\theta$	$\sqrt{1-(R/L)^2\sin^2\theta}$	F_{CR}	θ	F_{CA}	$\sqrt{1-(R/L)^2\sin^2\theta}$	F_{CR}
0	800	0	0	0	1.0	1.0	800				
20	5,300	0.342	0.117	0.0072	0.9928	0.996	5,320	380	−1,400	0.996	−1,405
40	4,000	0.6428	0.413	0.0254	0.9746	0.987	4,050	400	−1,000	0.987	−1,013
60	2,650	0.866	0.751	0.0462	0.9538	0.977	2,710	420	−450	0.977	−461
80	2,200	0.9848	0.969	0.0591	0.9409	0.971	2,265	440	50	0.971	51.5
100	2,050	0.9848	0.969	0.0591	0.9409	0.971	2,110	460	450	0.971	464
120	2,000	0.866	0.751	0.0462	0.9538	0.977	2,045	480	750	0.977	768
140	1,800	0.6428	0.413	0.0254	0.9746	0.987	1,825	500	800	0.987	811
160	1,600	0.342	0.117	0.0072	0.9928	0.996	1,607	520	825	0.996	828
180	1,200	0	0	0	1.0	1.0	1,200	540	850	1.0	850
200	1,050	0.342	0.117	0.0072	0.9928	0.996	1,054	560	850	0.996	854
220	900	0.6428	0.413	0.0254	0.9746	0.987	912	580	850	0.987	861
240	800	0.866	0.751	0.0462	0.9538	0.977	819	600	825	0.977	845
260	600	0.9848	0.969	0.0591	0.9409	0.971	617	620	650	0.971	670
280	200	0.9848	0.969	0.0591	0.9409	0.971	206	640	300	0.971	309
300	−350	0.866	0.751	0.0462	0.9538	0.977	−358	660	−75	0.977	−77
320	−850	0.6428	0.413	0.0254	0.9746	0.987	−862	680	−200	0.987	−202
340	−1,300	0.342	0.117	0.0072	0.9928	0.996	−1,305	700	50	0.996	52
360	−1,500	0	0	0	1.0	1.0	−1,500	720	800	1.0	800

For example, five connecting rods are attached to the crankpin, and although the variation of resultant force (F_{CA}) per cycle (Fig. 4-6) is assumed to be the same in each cylinder, for any given angular position of the crank arm, the force along each individual connecting rod will be different owing to the fact that the various cylinders are operating on different parts of the cycle. Further, some of the connecting rods will be under compression and others will be under tension. An effort to combine graphically these forces is likely to result in some confusion if a system of keeping things straight is not used. Figure 5-5 shows a system that has been prepared for the example, and a similar figure may easily be arranged for any other engine.

In this figure, the force in the connecting rod F_{CR} is plotted against crank angle. Forces downward along the cylinder axis are taken as positive; upward forces are considered negative. The angular spacing of the cylinder axes is $^{360}\!/_5 = 72$ deg.; hence with No. 1 cylinder just starting expansion,

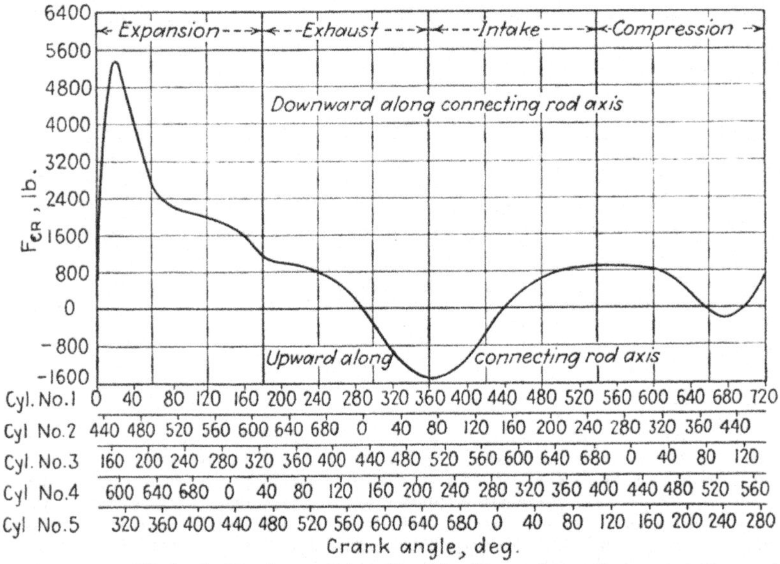

FIG. 5-5.—Methods for determining F_{cr} (the force along the connecting-rod center line) for any cylinder of Example 1 at any position of the crank arm with respect to the center line of No. 1 cylinder.

No. 3 (the next to fire) will be on the first part of the compression stroke. Number 3 is 144 deg. ahead of No. 1; hence its angular position for No. 1 at 0 will be $180 - 144 = 36$ deg. past the start of compression (Fig. 5-5). Call this 36-deg. point 0 for No. 3 cylinder, and lay off the crank-angle scale as indicated. Similarly, the 0 point may be found for the other cylinders and scales laid off as shown.

To illustrate the use of the scale, suppose No. 1 cylinder is at the 80-deg. point and it is desired to find the forces in each of the connecting rods. The solution consists in reading up to the curve from the 80-deg. point for each cylinder and taking the forces directly from the ordinate scale.

In plotting the polar diagram (Fig. 5-6), the cylinder axes are laid off from M at their proper angular relation. Then, with M (the crankshaft axis) as a center, the crank circle (radius R) and the centrifugal-force F_C circles are constructed to the dimension and force scales, respectively. These circles are divided into the desired angular segments (20 deg. in this instance).

To determine the resultant force on the crankpin at any given angular position of the crank arm, say 100 deg. in the direction of rotation past the beginning of expansion in No. 1 cylinder, the procedure is as follows. (*a*) Set the compass to the length of the master connecting rod (dimension scale), and with C (the 100-deg. position on the crank circle) as a center,

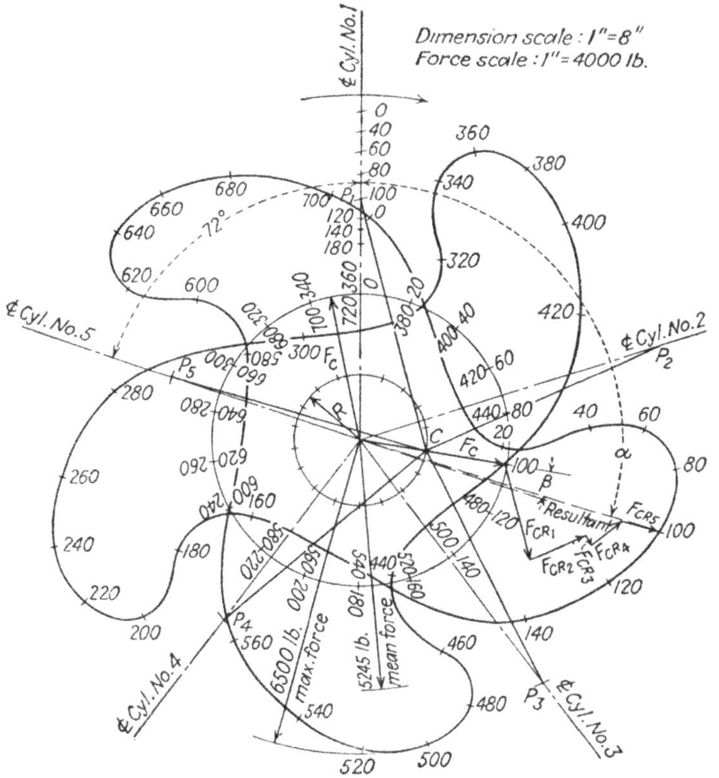

FIG. 5-6.—Polar diagram of resultant force on a five-cylinder radial-engine crankpin (Example 1).

strike arcs intercepting all the cylinder center lines (designated P_1, P_2, P_3, etc., in Fig. 5-6). Lines connecting these intercepts with C represent the center lines of the various connecting rods. (*b*) For a crank angle of 100 deg., read the values of F_{CR} for the various cylinders from Fig. 5-5. (*c*) With the 100-deg. intercept on the F_C circle as a starting point, lay off force vector F_{CR_1} parallel to No. 1 cylinder connecting rod P_1C and in the direction consistent with Fig. 5-5. From the end of F_{CR_1}, lay off F_{CR_2} parallel to P_2C and in the proper direction. Continue this procedure until all connecting-rod forces are laid off. The resultant force on the crankpin is represented by a vector connecting the crankshaft center (M) and the end of the last connecting-rod vector. The direction of this resultant with respect to the center line of cylinder No. 1 is α and with respect to the crank arm is β.

By connecting the ends of resultant force vectors determined at increment crank angles through 720 deg., the polar diagram (Fig. 5-6) was obtained. When constructed to a large enough scale to permit close accuracy, polar diagrams for radial-engine crankpins are symmetrical.

It should be noted that the maximum force from the diagram is considerably greater than the maximum value of F_{CR} (Fig. 5-5).

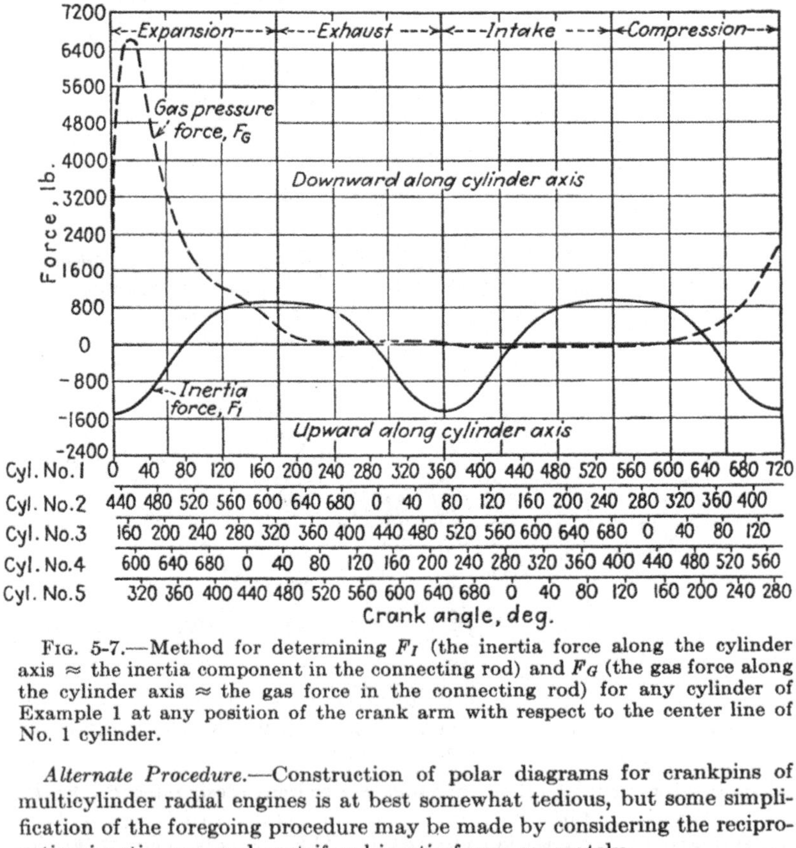

FIG. 5-7.—Method for determining F_I (the inertia force along the cylinder axis $\approx$ the inertia component in the connecting rod) and F_G (the gas force along the cylinder axis $\approx$ the gas force in the connecting rod) for any cylinder of Example 1 at any position of the crank arm with respect to the center line of No. 1 cylinder.

Alternate Procedure.—Construction of polar diagrams for crankpins of multicylinder radial engines is at best somewhat tedious, but some simplification of the foregoing procedure may be made by considering the reciprocating inertia, gas, and centrifugal inertia forces separately.

It has already been observed that the force on the crankpin due to the weight of the rotating part of the connecting rods is constant (for any given engine speed) and that it always acts along the crank-arm center line. By reproducing Fig. 4-6 in Fig. 5-7 and locating the positions of all the cylinders by the same method used for Fig. 5-5, the resultant forces F_{GR}, due to gas-pressure forces, and the resultant forces F_{IR}, due to inertia of the reciprocating parts, may be found separately. Figure 5-8 shows the method of determining the F_{IR} forces. The procedure in making this construction is the same as that for Fig. 5-6 except that inertia forces along the cylinder axes are used (obtained from Fig. 5-7) instead of the total force, *i.e.*, gas force + inertia force. It is seen from Fig. 5-8 that the resultant force on the crankpin due to the inertia of reciprocating parts (F_{IR}) is a constant*

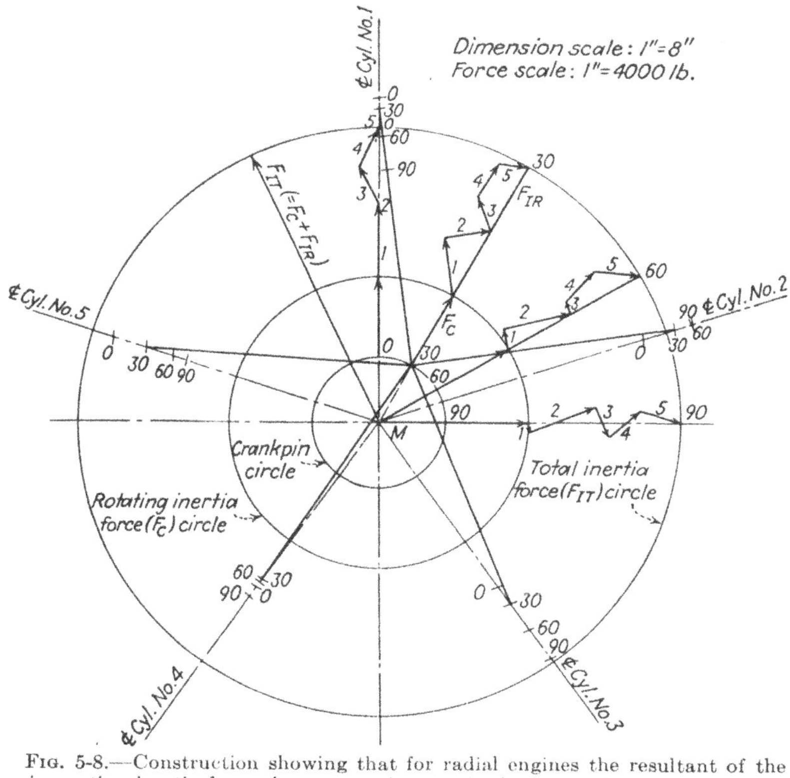

FIG. 5-8.—Construction showing that for radial engines the resultant of the reciprocating inertia forces is constant in magnitude (for a given engine speed) and that it always acts along the crank-arm center line. (Values apply only to Example 1.)

(for a given engine speed) and that it always acts parallel to the crank arm. Hence the construction need include only one determination of F_{IR} to obtain the force due to inertia of reciprocating parts for any angular position of the crank arm. In Fig. 5-8, the forces due to reciprocating inertia are added graphically to F_C, *i.e.*, construction is started from the end of the F_C vector. Hence the end of the F_{IR} vector is distant from M an amount equal to the total inertia force F_{IT}.

* Except for three-cylinder radials.

The resultant force on the crankpin due to gas-pressure forces is found by the construction shown in Fig. 5-9. In this figure, the gas forces in the various cylinders at any crank position are obtained from Fig. 5-7 and summed vectorially in the same way as for Figs. 5-6 and 5-8. The gas-force vectors are started from the end of the total inertia-force vector F_{IT}; hence a line connecting M and the end of the F_{GR} (resultant gas force) vector will give the magnitude and direction of the total resultant force F_R on the

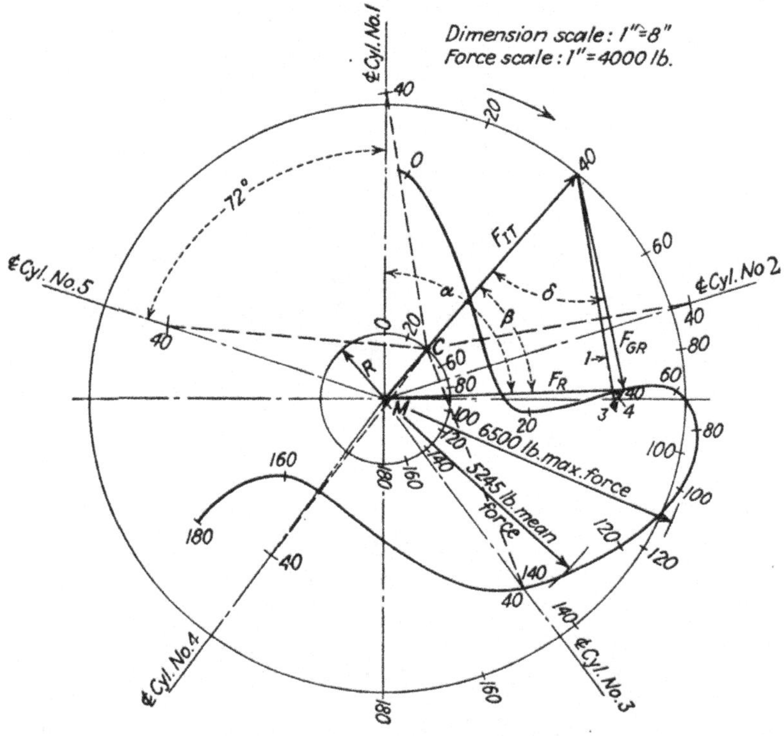

FIG. 5-9.—Alternate method of constructing a polar diagram of resultant force on a five-cylinder radial-engine crankpin (Example 1).

crankpin. Determination of F_R by the procedures illustrated in Figs. 5-8 and 5-9 is simpler than by that used in Fig. 5-6, because F_{IT} need be found for only one crank-arm position, and because the gas forces during most of the exhaust, intake, and the first part of the compression strokes are negligible. This last reduces the number of vectors to be summed in finding F_{GR}.

If carried through 720 deg. of increment-angle construction, Fig. 5-9 would give the same polar diagram as was obtained in Fig. 5-6. However, for obtaining the usual information desired, *i.e.*, the maximum and mean forces on the crankpin, this procedure is unnecessary as the cycle is repeated $n/2$ times during each revolution, n being the number of cylinders. For the five-cylinder engine in Example 1, construction at increment angles through $360/2.5 = 144$ deg. is all that is necessary to determine the maximum and mean resultant forces on the crankpin. However, for beginners in polar-diagram construction, it is advisable to carry the construction through at least twice this number of degrees of crank travel [*i.e.*, $360/(n/4)$] to provide more of a check on the work and to understand more completely the details of construction. Polar diagrams of forces on crankpins are frequently constructed with respect to the crank-arm axis (Fig. 5-4*b*). For radial engines, the construction is illustrated in Fig. 5-10. In this construction, the total inertia-force vector F_{IT} is laid off to scale as indicated in Fig. 5-10, the starting point being considered as the crankpin center C (corresponding to M in Fig. 5-9). From the end of the F_{IT} vector, resultant gas force vectors F_{GR} (determined by the construction illustrated in Fig. 5-9) are laid off at angles δ to the F_{IT} vector (δ also equals the angle of the gas-force resultant

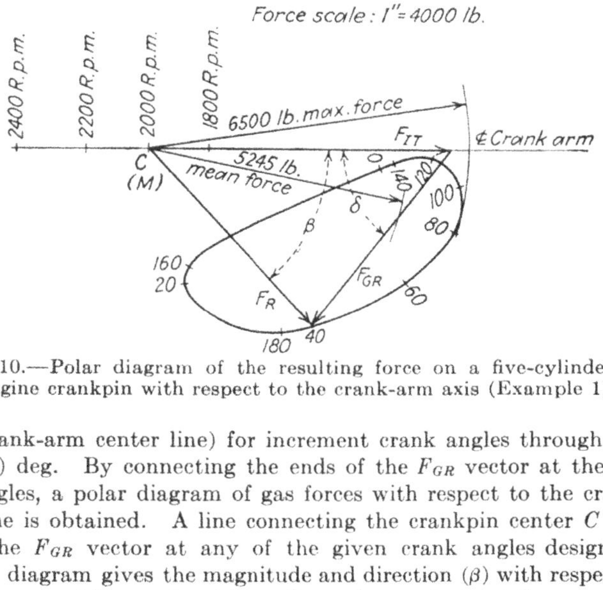

FIG. 5-10.—Polar diagram of the resulting force on a five-cylinder radial-engine crankpin with respect to the crank-arm axis (Example 1).

to the crank-arm center line) for increment crank angles through at least $360/(n/2)$ deg. By connecting the ends of the F_{GR} vector at the various crank angles, a polar diagram of gas forces with respect to the crank-arm center line is obtained. A line connecting the crankpin center C and the end of the F_{GR} vector at any of the given crank angles designated on the polar diagram gives the magnitude and direction (β) with respect to the crank-arm axis of the resultant force F_R on the crankpin at that crank angle. As the F_{GR} vector will retrace the polar diagram loop $n/2$ times per revolution, values and directions of F_R will also repeat $n/2$ times per revolution. Hence, one loop is sufficient for determining the maximum and mean forces on the crankpin.

The forces due to reciprocating and rotating weights, F_{IR} and F_C, vary as the square of the engine speed [Eqs. (4-8) and (5-1)]. Hence, the effect of engine speed on crankpin loadings may readily be found from Fig. 5-10. For instance, the maximum force F_R on the crankpin is 6,500 lb. for an engine speed of 2,000 r.p.m., and it is desired to know the maximum force at 2,400 r.p.m. Assuming that the gas forces remain the same,

$$F_R(2{,}400 \text{ r.p.m.}) = 6{,}500\,\frac{\overline{2{,}400}^2}{\overline{2{,}000}^2} = 9{,}300 \text{ lb. (approximately)*}$$

From this, it is apparent that very high crankpin loadings are likely to occur in power dives, and even in closed-throttle dives when the gas-force vector is negligible, the inertia load F_{IT} will rise to high values. Location of engine r.p.m. points along the crank-arm axis (Fig. 5-10) adds materially to the information conveyed by the diagram.

5-4. Crankpin Bearing Loads.—The unit loadings on bearings are based upon the force per square inch of projected bearing area, *i.e.*, the diameter of the crankpin (or journal) times its length. Numerous factors affect the allowable loadings such as distortion of the journal or connecting rod, condition of the lubricant, relative characteristics of the journal and bearing metal, and rubbing velocity. On the assumption of sufficient rigidity to the shaft and adequate lubrication, usual mean bearing pressures range from 750 to 2,000 lb. per sq. in. or more of projected area, and maximum pressures range up to 5,000 lb. per sq. in. (Tables A1-5 and A1-8).†

Rubbing velocity is the relative speed with which a point on the crankpin or journal moves by a point on the inner surface of the bearing. It may be calculated as follows:

$$V = \frac{\pi D}{12} \times \frac{N}{60} \tag{5-3}$$

where V = rubbing velocity, f.p.s.
D = crankpin or journal diameter, in.
N = engine speed, r.p.m.

Rubbing velocities (Table A1-5) range from 15 to 25 or more feet per second, but in more recent high-powered engines, 30 to 50 or more feet per second is proving quite satisfactory.

Rubbing factor, or *PV factor* as it is sometimes called, is the product of *mean* bearing load in pounds per square inch of projected bearing area and rubbing velocity in feet per second. *PV* factors are usually considered to be an indication of bearing capacity. Values range up to 50,000 or more with 20,000 to

* The maximum force is slightly less owing to the changed angularity (β) of vector F_R at 2,400 r.p.m. F_R may be found more accurately by first finding F_{IT} at the desired speed and then scaling F_R from the diagram.

† See reference 10 for additional data.

35,000 (depending on the size of the engine) being a limit recommended by some authorities. Actually, many details of design contribute in determining allowable values.

The ratio of rubbing velocity to unit bearing load is sometimes used as a still further criterion in determining allowable bearing loads. Lubrication engineers frequently express the conditions in a bearing by the relation[7] ZN/P where Z is the absolute viscosity[6] of the lubricant in centipoises, N is the speed of the journal in revolutions per minute, and P is the bearing load in pounds per square inch of projected area. Since V is a function of N [Eq. (5-3)], some engineers consider V/P as a better expression of bearing conditions than PV. Meyer[8] recommends using a value of $V/P > 0.016$.

5-5. Example.—For the engine selected in Example 1, determine the projected crankpin area, diameter, and length if the allowable maximum bearing load is not to exceed 1,200 lb. per sq. in. of projected area, the PV factor (based on mean loads) is not to exceed 20,000, $V/P > 0.016$, the rubbing velocity is not to exceed 20 f.p.s., and the ratio of length to diameter of the crankpin is to fall within the usual range for radial engines of 1.2 to 1.6 (Table A1-5).

Solution.—From Par. 5-3, the maximum force on the crankpin was found to be 6,500 lb. and the mean force was 5,245 lb. The smallest permissible projected crankpin area is 6,500/1,200 = 5.4 sq. in., and the corresponding mean pressure is 5245/5.4 = 970 lb. per sq. in. For the allowable rubbing velocity of 20 f.p.s., $PV = 970 \times 20 = 19{,}400 < 20{,}000$, the allowable rubbing factor, and $V/P = {}^{20}\!/_{970} = 0.0206 > 0.016$. From Eq. (5-3),

$$D = \frac{720V}{\pi N} = \frac{720 \times 20}{\pi \times 2{,}000} = 2.3 \text{ in.},$$

$$L = \frac{A}{D} = \frac{5.4}{2.3} = 2.35 \text{ in.},$$

and

$$\frac{L}{D} = \frac{2.35}{2.3} = 1.02$$

This L/D ratio would very likely be entirely satisfactory, but it is below the desired range. It is an indication that little difficulty will be had in meeting the bearing requirements, however. Assume $L/D = 1.25$, and to reduce V below its allowable limit, let $D = 2.25$ in. Then

$$L = 1.25 \times 2.25 = 2.82 \text{ in.},$$
$$A = 2.25 \times 2.82 = 6.35 \text{ in.},$$

$$\text{the mean pressure} = \frac{5{,}245}{6.35} = 825 \text{ lb. per sq. in.}$$

$$V = \frac{\pi \times 2.25 \times 2{,}000}{12 \times 60} = 19.65$$

$$\text{the maximum pressure} = \frac{6{,}500}{6.35} = 1{,}025 \text{ lb. per sq. in.}$$

$$PV = 825 \times 19.65 = 16{,}200$$

$$\frac{V}{P} = \frac{19.65}{825} = 0.0238$$

As far as bearing requirements are concerned, for the example, a crankpin diameter of 2.25 in. and an effective bearing length of 2.82 in. ± should be easily satisfactory.

5-6. Crankshaft Dimensions.—To perform its functions properly, an engine crankshaft must (*a*) be strong enough to withstand the forces to which it is subjected, (*b*) be rigid enough to prevent appreciable distortion, (*c*) have sufficient mass properly distributed so that it will not vibrate critically at the usual speeds at which it is operated, (*d*) have sufficient bearings of adequate size to handle the loads with available lubricants, and (*e*) for aircraft engines, have the shaft as light as possible. Obviously, some of these requirements are more severe than others, and in meeting the difficult requirements, usually the less difficult will be taken care of automatically. For instance, to meet the rigidity and vibration requirements, it is usually necessary to make the shaft much heavier and stronger than would be necessary for (*a*). Hence, a tedious stress analysis is generally unnecessary and seldom made on modern high-speed-engine shafts.

The present purpose (*d*) is to investigate main bearing loads and determine the necessary main bearing sizes. To do this, it is necessary to know the crankpin bearing loads, the position of the main bearings with respect to the crankpins, and the inertia loads due to unbalanced parts of the crankshaft. These last two items necessitate resort to past experience, if a great deal of trial-and-error effort is to be avoided. Unfortunately, all too few data have been assembled on crankshaft details, but some assistance may be found in Tables A1-5, A1-6, A1-7, A1-8, and A1-9.

5-7. In-line and V-engine Crankshafts.—Typical examples of in-line and opposed-engine crankshafts are shown in Figs. 5-11, 5-12, and 5-13. For air-cooled engines, crankshafts will usually have to be proportionally longer than for water-cooled engines because of the greater over-all diameter (including cooling fins) of the cylinders. This may necessitate increasing the shaft sections to provide sufficient stiffness and rigidity. To avoid excessive weight, this in turn sometimes necessitates the use of main bearings between all crank arms (Fig. 5-12). To

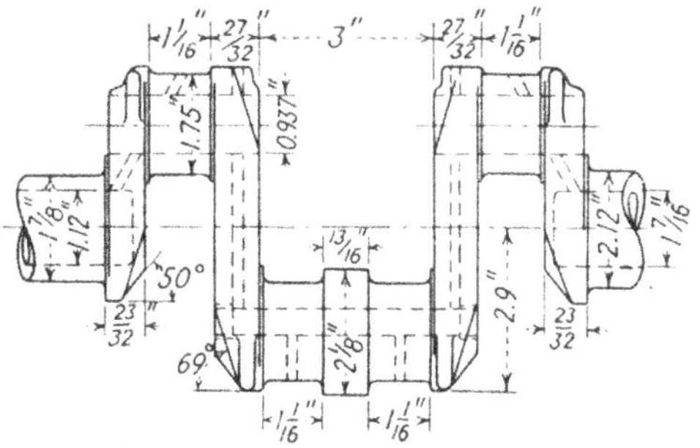

FIG. 5-11.—Continental A-40 crankshaft—four-cylinder opposed, two main bearings.

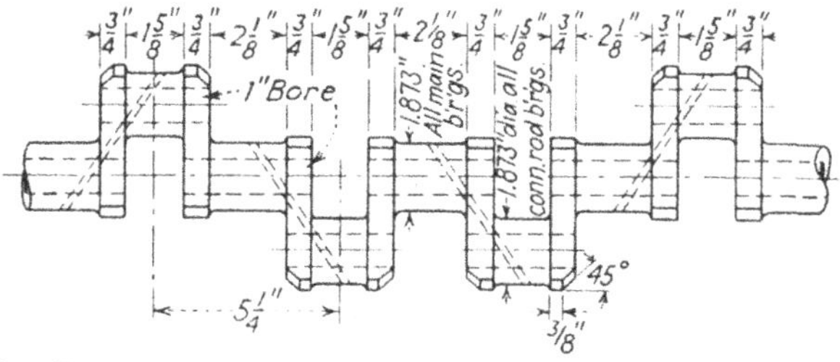

FIG. 5-12.—Four-cylinder in-line crankshaft. Tank-70, five main bearings.

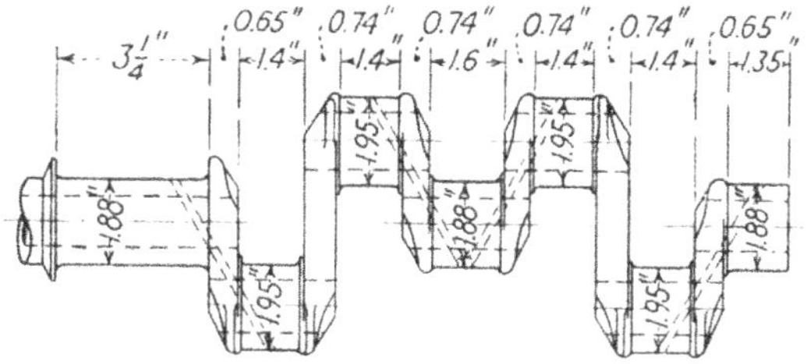

FIG. 5-13.—Continental A-50 crankshaft—four-cylinder opposed, three main bearings.

permit the necessary axial spacing of crankpin centers, the L/D ratio of bearings may be increased or the crank arms may be set at an angle. To keep down weight, it is obviously desirable to space the cylinder center lines as closely as cooling fins and other requirements will permit. Also for the same reason, for small in-line and V-type aircraft-engine crankshafts, counterweights are sometimes omitted. This last increases the main bearing loads due to inertia forces from unbalanced crank arms and crankpins and is a questionable saving.

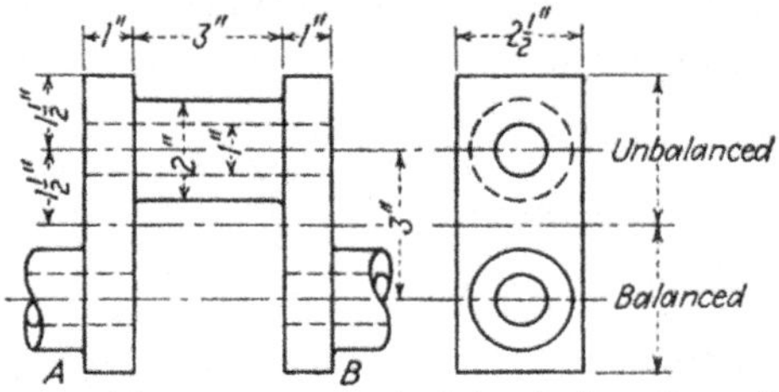

Fig. 5-14.—Diagram illustrating a method for finding the unbalanced weight to apply in determining main bearing loads in in-line and V-engines.

5-8. Example.—With reference to Fig. 5-14, determine the added load on main bearings A and B due to the unbalanced weight of the crank. Assume 2,000 r.p.m. and a density of shaft material of 0.28 lb. per cu. in.

Solution.

Volume of crankpin = 0.785 (4-1) 3 = 7.06 cu. in.
Weight of crankpin = 7.06 × 0.28 = 1.98 lb.
Distance to center of gravity = 3 in.
Volume of unbalanced part of crank arms
= 2 × 1 × 3 × 2.5 − 0.785 × 1 × 1 = 14.215 cu. in.
Weight of unbalanced part of crank arms = 14.215 × 0.28 = 3.98 lb.
Distance to center of gravity = 3 in.
Total weight of unbalanced parts = 1.98 + 3.98 = 5.96 lb.
Distance to center of gravity of crankpin and unbalanced parts of crank arms = 3 in.

The centrifugal force on the unbalanced crank is [from Eq. (5-1)]

$$F_C = 0.0000284 \times 5.96 \times \overline{2{,}000}^2 \times 3 = 1{,}605 \text{ lb.}$$

This load may be assumed to be divided equally between the two adjacent main bearings; hence each bearing will be subjected to a load of 802.5 lb. This load is, of course, in addition to that imposed by the forces acting *on* the crankpin.

The use of chamfered and rounded crank arms (Fig. 5-15) aids materially in reducing main bearing loads by reducing the unbalanced weight and distance to the center of gravity. These refinements are usually possible without sacrifice of crankshaft strength or stiffness. Chamfering at the main bearing end of the crank arm serves principally to reduce the weight of the shaft.

5-9. Radial-engine Crankshafts.—Although the added weight is undesirable, it is always necessary to counterbalance radial-engine crankshafts to attain static balance and to reduce the unbalanced loads on the main bearings. These loads are much greater in a radial engine due to the greater number of connecting rods attached to the crankpin. The determination of the size of counterweights will be considered later; their only effect in connection with the present consideration of main bearing loads is in permitting the assumption that the inertia forces are completely balanced by the counterweights.[4] Hence, the sum of the main bearing loads at any crank angle is represented by the gas-load vector F_{GR}, (Figs. 5-9 and 5-10) at that angle.

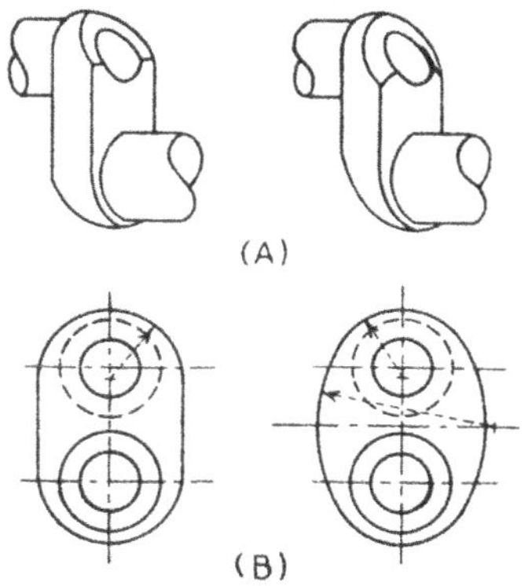

Fig. 5-15.—Crank arm details for noncounterweighted in-line and V-type aircraft-engine crankshafts. (*A*) Methods used in chamfering engine crankshaft arms. (*B*) Typical crank-arm contours. (*From Angle, "Engine Dynamics and Crankshaft Design."*)

The principal dimensions of several radial-engine crankshafts are given in Table A1-9. General arrangement of details is indicated in Fig. 5-16.

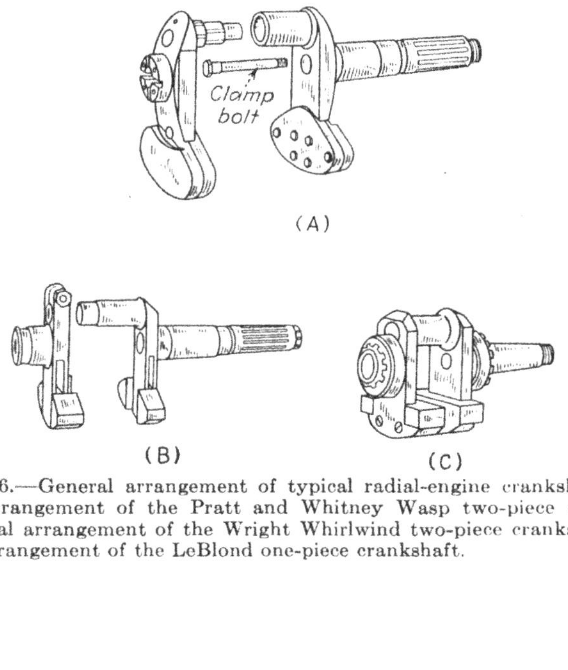

Fig. 5-16.—General arrangement of typical radial-engine crankshafts. (*A*) General arrangement of the Pratt and Whitney Wasp two-piece crankshaft. (*B*) General arrangement of the Wright Whirlwind two-piece crankshaft. (*C*) General arrangement of the LeBlond one-piece crankshaft.

5-10. Resultant Forces on Main Bearings.—Load distribution on crankshaft main bearings cannot be determined exactly because of uncertainty as to the effects of crankshaft and crankcase distortion, misalignment of bearings, bearing clearances, etc., but the following procedure is in common use and has proven satisfactory for conventional designs.

In in-line and V-engines where there is a main bearing on each side of the crankpin, the forces acting on the crankshaft bearings are obtained by considering the force at the crankpin, together with the centrifugal force due to the unbalanced part of the crank arms and crankpin (when counterweights are not used), to be equally divided between the two crankshaft bearings at each side of the crankpin. The load on end main bearings may be taken as one-half of the load on the adjacent crankpin bearing plus (vectorially) one-half of the centrifugal load due to the crank (when the shaft is not counterbalanced). The loads on intermediate and center main bearings may be taken as the vector sum of one-half the loads due to each adjacent crank, *i.e.*, one-half of each adjacent crankpin load plus (vectorially) one-half of each of the unbalanced adjacent crank-arm loads.

To illustrate the procedure, assume that Fig. 5-2*a* (*A*) represents a crankpin polar diagram for a conventional six-cylinder in-line engine having crank arms arranged as in Fig. 4-13 and a firing order of 1-5-3-6-2-4. Figure 5-17 shows a method of finding the resultant forces on the main bearings. In this figure, F_{R_1} is the resultant force on crankpin 1 at θ_1 (= 30 deg. of crank angle from the beginning of the power stroke). The magnitude and direction (α_1) of F_{R_1} is found in Fig. 5-2*a* (*A*). F_{CS_1} (Fig. 5-17) is the centrifugal force due to the unbalanced weight of crank 1. (This force is constant for a given engine speed and always acts along the crank-arm axis.) The resultant force on main bearing 1 is $F_{RM_1} = F_{RC_1}/2$ (Fig. 5-17), and its direction with respect to the engine axis is σ_1.

The resultant force on main bearing 2 is determined by taking one-half the vector sum of the resultant forces at cranks 1 and 2. Referring again to Figs. 5-17 and 5-2*a* (*A*), F_{R_2} is the resultant force on crankpin 2 when crank 1 is at θ_1 deg. from the beginning of the power stroke in No. 1 cylinder. F_{CS_2} is the centrifugal force due to the unbalanced part of crank 2, F_{RC_2} is the vector resultant of F_{R_2} and F_{CS_2}. $F_{RC_{1,2}}$ is the vector resultant of F_{RC_1}

and F_{RC_2}. The resultant force F_{RM_2} on main bearing 2 is equal to one-half of $F_{RC_{1,2}}$, and its direction with respect to the engine axis is σ_2.

An alternate method, which in some cases simplifies the construction of main-bearing polars, is to divide each vector

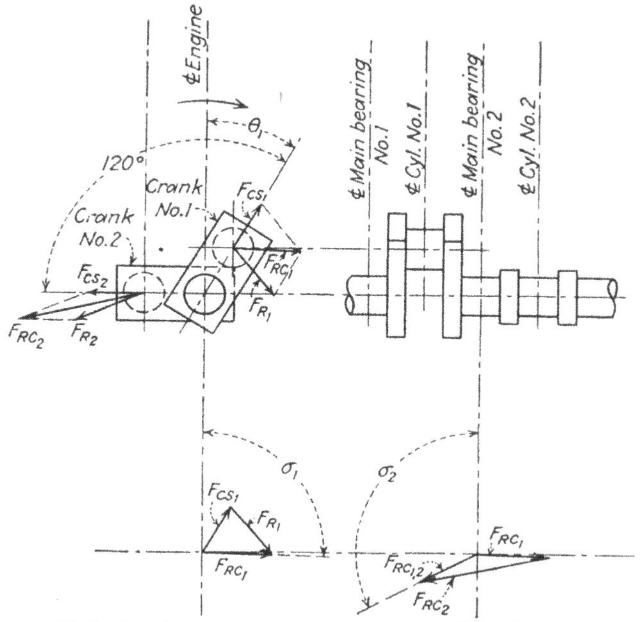

Main Bearing No. 1
Resultant force is

$$F_{RM1} = \frac{F_{RC1}}{2} \text{ at } \sigma_1 \text{ deg.}$$

to ℄ of engine

Main Bearing No. 2
Resultant force is

$$F_{RM2} = \frac{F_{RC1,2}}{2} \text{ at } \sigma_2 \text{ deg.}$$

to ℄ of engine.

FIG. 5-17.—Method of finding resultant forces on the main bearings of in-line engines. (Six-cylinder crankshaft illustrated.)

component by 2 before applying it in the construction. The resultant vectors are then F_{RM_1}, F_{RM_2}, etc., directly.

In finding F_{R_1} and F_{R_2}, it is necessary to take account of the firing order as well as the angular relation of the crank arms. To reduce the confusion in doing this, the system illustrated in Fig. 5-18 is useful. For instance, suppose crank 1 is 30 deg. ($= \theta_1$) past the beginning of the power stroke and it is desired to know the value and direction of F_{R_2}. From Fig. 5-18, crank 2

Cylinder Numbers

No. 1		No. 2		No. 3		No. 4		No. 5		No. 6	
∠ in terms of No. 1 cyl.	Event in the cycle	∠ in terms of No. 2 cyl.	Event in the cycle	∠ in terms of No. 3 cyl.	Event in the cycle	∠ in terms of No. 4 cyl.	Event in the cycle	∠ in terms of No. 5 cyl.	Event in the cycle	∠ in terms of No. 6 cyl.	Event in the cycle
0		240		480		120		600		360	
—	*P*	—	*E*	—	*I*	—	*P*	—	*C*	—	*I*
60		300		540		180		660		420	
—	*P*	—	*E*	—	*C*	—	*E*	—	*C*	—	*I*
120		360		600		240		0		480	
—	*P*	—	*I*	—	*C*	—	*E*	—	*P*	—	*I*
180		420		660		300		60		540	
—	*E*	—	*I*	—	*C*	—	*E*	—	*P*	—	*C*
240		480		0		360		120		600	
—	*E*	—	*I*	—	*P*	—	*I*	—	*P*	—	*C*
300		540		60		420		180		660	
—	*E*	—	*C*	—	*P*	—	*I*	—	*E*	—	*C*
360		600		120		480		240		0	
—	*I*	—	*C*	—	*P*	—	*I*	—	*E*	—	*P*
420		660		180		540		300		60	
—	*I*	—	*C*	—	*E*	—	*C*	—	*E*	—	*P*
480		0		240		600		360		120	
—	*I*	—	*P*	—	*E*	—	*C*	—	*I*	—	*P*
540		60		300		660		420		180	
—	*C*	—	*P*	—	*E*	—	*C*	—	*I*	—	*E*
600		120		360		0		480		240	
—	*C*	—	*P*	—	*I*	—	*P*	—	*I*	—	*E*
660		180		420		60		540		300	
—	*C*	—	*E*	—	*I*	—	*P*	—	*C*	—	*E*
720		240		480		120		600		360	

Firing order 1-5-3-6-2-4

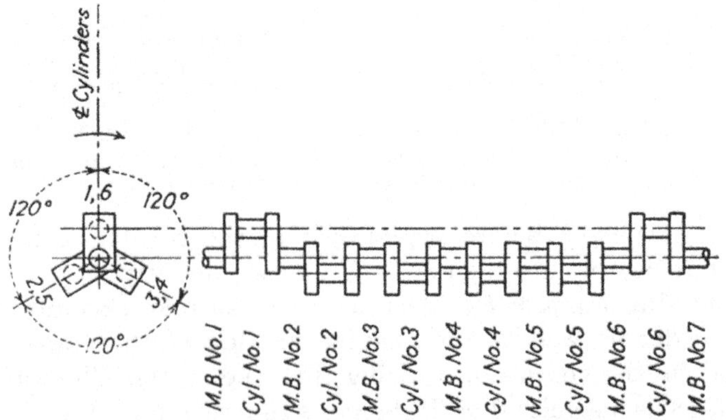

FIG. 5-18.—Method of determining the effect of crank angularity and firing order in constructing polar diagrams for main bearings. (Six-cylinder in-line engine illustrated.)

will be on the exhaust stroke and at the 270-deg. position in terms of crank 1. Hence, from Fig. 5-2a (A), F_{R_2} is the resultant force at the 270-deg. position, and its direction is α_2 deg. with respect to the center line of the cylinders. Similarly, F_{R_5} may be found for any position of crank 4, etc. Figure 5-18 applies only to the crankshaft and firing order given, but a similar chart may be readily constructed for any other engine.

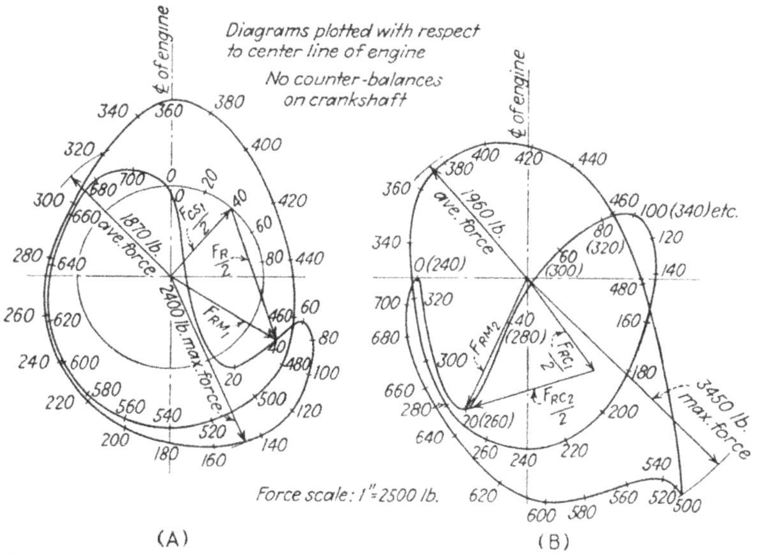

FIG. 5-19.—(A) End-main-bearing and (B) intermediate-main-bearing polar diagrams for a 4½-in. by 5⅜-in. six-cylinder in-line aircraft engine having a crankshaft and firing order as in Fig. 5-18. Angles in parentheses () are relative positions of crank arm on opposite end of main bearing under consideration.

Figures 5-19 and 5-20 show end-, intermediate-, and center-main-bearing polar diagrams for a six-cylinder in-line aircraft engine. The data are based on crankpin loadings as shown in Fig. 5-2a (B) in which the cylinder dimensions, gas-pressure forces, and reciprocating weights are the same as for Example 1. The rotating weight has been taken as 2.5 lb. per crankpin, and the speed is 2,000 r.p.m. The unbalanced weight of each crank arm is assumed to be 6.835 lb. at crank radius distance from the center of the crankshaft.

In single-bank radial engines, it is usually assumed that the inertia forces are completely balanced by the crankshaft counterweights and that the sum of the main bearing loads at any angular position of the crankshaft is represented by the gas-load vector F_{GR} (Figs. 5-9 and 5-10) at that angle.[4] In keeping with the preceding method for in-line and V-engine shafts, it would be logical to assume that each main bearing took one-half of this load. However, experimental evidence indicates that the loads are more nearly distributed as 40 per cent to the rear main

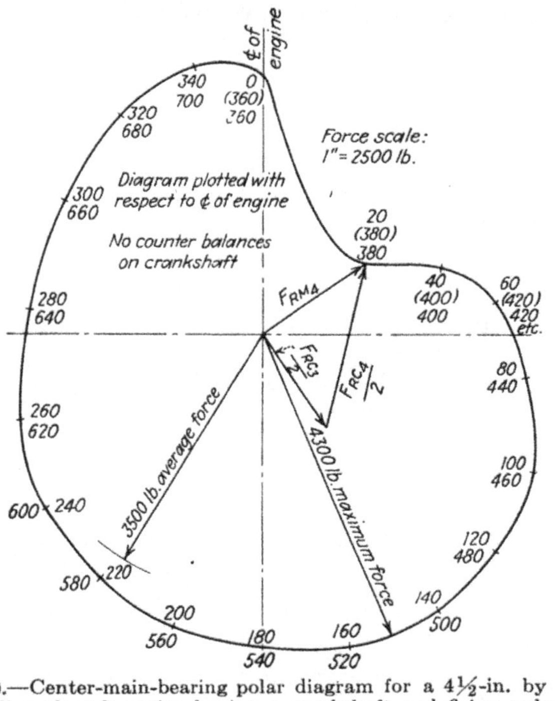

FIG. 5-20.—Center-main-bearing polar diagram for a 4½-in. by 5⅜-in. six-cylinder in-line aircraft engine having a crankshaft and firing order as in Fig. 5-18. Angles in parentheses () are relative positions of crank arm on opposite end of main bearing under consideration.

bearing, 75 per cent to the front main bearing, and 15 per cent radial load in the opposite direction on the thrust bearing in the nose of the crankcase (Table A1-10).

5-11. Example.—1. Construct polar diagrams for the main bearings of the engine in Example 1, using a load distribution of 40 per cent to the rear main bearing and 75 per cent to the front main bearing.

2. By assuming plain bearings, a maximum unit bearing load of 1,000 lb. per sq. in. of projected area, an allowable rubbing velocity of 20 f.p.s., and a maximum allowable rubbing (PV) factor of 15,000, determine the diameter and length of main bearings necessary.

3. By assuming ball or roller bearings, determine the sizes necessary.

Procedure 1.—On the assumption that the inertia forces are completely balanced, the gas loads are most conveniently obtained from either Fig. 5-9 or 5-10. In Fig. 5-10, δ represents the direction of the gas-force resultant F_{GR} with respect to the center line of the crank arm. If θ_1 represents the crank-arm position with respect to No. 1 cylinder center line, the direction of the F_{GR} vector is $\theta_1 + (180 - \delta)$ with respect to No. 1 cylinder. Values of F_{GR} and δ for various values of θ have been scaled directly from Fig. 5-10 and arranged for convenience in Table 5-2.

TABLE 5-2

θ	δ	$(180 - \delta)$	$\theta + (180 - \delta)$	F_{GR}	$0.4F_{GR}$	$0.75F_{GR}$
0	16	164	164	1,500	600	1,128
20	28	152	172	6,100	2,440	4,580
40	52.5	127.5	167.5	4,600	1,840	3,450
60	71	109	169	2,900	1,160	2,180
80	92.5	87.5	167.5	1,700	680	1,278
100	104	76	176	750	300	562
120	27	153	273	450	180	338
140	15	165	305	1,250	500	938
160	25	155	315	5,850	2,340	4,390
180	48	132	312	5,200	2,080	3,900

Values of $0.75F_{GR}$ (*A*) and $0.4F_{GR}$ (*B*) for the various values of θ are plotted with respect to the center line of No. 1 cylinder as shown in Fig. 5-21.

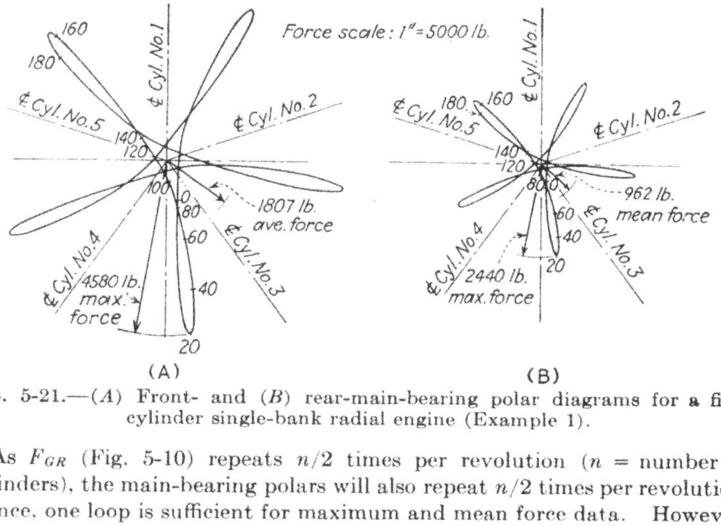

FIG. 5-21.—(*A*) Front- and (*B*) rear-main-bearing polar diagrams for a five-cylinder single-bank radial engine (Example 1).

As F_{GR} (Fig. 5-10) repeats $n/2$ times per revolution (n = number of cylinders), the main-bearing polars will also repeat $n/2$ times per revolution. Hence, one loop is sufficient for maximum and mean force data. However, the completed diagram may be quickly drawn by shifting the tracing paper angularly the proper amount for each loop. The complete polar is useful in showing the positions and directions of critical loadings on main-bearing supports and for relative wear studies. The relatively high ratio of maximum to mean force in radial-engine main bearings as compared with in-line-engine main bearings and the high rate of change of force (shock loadings) should be noted.

Procedure 2.—For the front main bearing, the projected bearing area will be

$$\frac{4{,}580}{1{,}000} = 4.58 \text{ sq. in.}$$

From Eq. (5-3),

$$D = \frac{720 \times V}{\pi \times N} = \frac{720 \times 20}{\pi \times 2{,}000} = 2.3 \text{ in.}$$

To reduce V below the allowable limit, let $D = 2.25$ in. Then

$$L = \frac{4.58}{2.25} = 2.04 \text{ in.}$$

$$V = \frac{2.25 \times \pi \times 2{,}000}{720} = 19.6 \text{ f.p.s.}$$

The mean force is 1,807 lb., hence

$$PV = \frac{1{,}807}{4.58} \times 19.6 = 7{,}750$$

For the rear main bearing, the projected area will be

$$\frac{2{,}440}{1{,}000} = 2.44 \text{ sq. in.}$$

and by assuming the same diameter as for the front main

$$L = \frac{2.44}{2.25} = 1.085 \text{ in.}$$

Procedure 3.—For the front main bearing [Fig. 5-21 (*A*)], the average force is 1,807 lb., the maximum force is 4,580 lb., and the speed is 2,000 r.p.m. On the assumption that ball bearings are to be used (Table A1-22), $L = 1{,}807$ lb., $Z = 0.88$ for an assumed bearing life of 2,500 hr. and $K = 2.0$, or 2.5 say 2.25. Then

$$C = 1{,}807 \times 0.88 \times 2.25 = 3{,}580 \text{ lb.}$$

The diameter of the main bearings should not be less than the diameter of the crankpin, and the front main bearing for direct drive will have to be greater in diameter than the largest diameter of the S.A.E. standard shaft end that is to be used. The crankpin diameter (Par. 5-5) is 2.25 in., and from Fig. A1-6, the logical propeller-shaft end will be S.A.E. taper type

No. 1. From Table A1-20, the maximum diameter of a taper type S.A.E. No. 1 shaft end is 2.05 in. Hence a front main bearing base diameter of 2.25 in. should be about adequate. From Table A1-22D, S.A.E. bearings 212, 312, and 412 are adequate in bore diameter, but from Table A1-22E, it is seen that at 2,000 r.p.m. the ratings are too low. From Tables A1-22F and A1-22G, it is evident that S.A.E. bearing 413 having a bore of 2.5591 in. or bearing 314 having a bore of 2.7559 in. could be used. Use of a shock factor of $K = 2$ or of a bearing life of 2,000 hr. would permit the use of S.A.E. bearing 412, but at the expense of a reduction in the factor of safety or life of the engine.

For the rear main bearing $L = 962$ lb., $Z = 0.88$, and $K = 2.25$. Then

$$C = 962 \times 0.88 \times 2.25 = 1{,}900 \text{ lb.}$$

From Tables A1-22D and A1-22E, S.A.E. bearing 212 having a bore diameter of 2.3622 in. could be used, or if it was desirable to have the same front and rear bearing bore diameters, bearings 213 or 214 could be used.

If roller bearings are desired (Table A1-23) for the front main, $L = 1{,}807$ lb., $Z = 0.64$, and $K \approx 1.5$. Then

$$C = 1{,}807 \times 0.64 \times 1.5 = 1{,}735 \text{ lb.}$$

From Tables A1-23B and A1-23C, bearings RLS-16-L or RLS-16-LL would be adequate. For the rear main, $L = 962$ lb., $Z = 0.64$, and $K \approx 1.5$. Then

$$C = 962 \times 0.64 \times 1.5 = 925 \text{ lb.}$$

From Table A1-23D, bearing RXLS − 2.25 would be adequate.

5-12. Relative Wear Diagrams.—For the purpose of determining the best location of the oilhole for crankpin bearings having force-feed lubrication, relative wear diagrams are useful. A method of constructing such diagrams follows:[3,2]

The bearing pressure is assumed to be evenly distributed over an arc of 180 deg. on the crankpin. The magnitude and direction of the force with respect to the crank arm is obtained at equal intervals throughout a complete cycle from a polar diagram of resultant forces on the crankpin. These forces are plotted as a series of half rings having their radial thicknesses proportional to the magnitude of the force and their mid-points falling on a line through the center of the crankpin in the direction of the application of the force considered. The summation of these rings produces an area which is termed the comparative wear on the crankpin. The best location for the oilhole is that point on the crankpin where the radial thickness of this resulting area is a minimum.

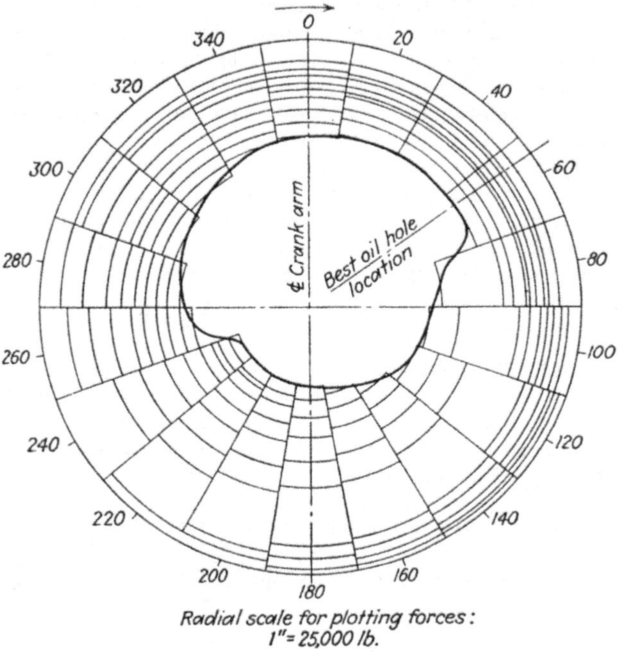

FIG. 5-22.—Diagram illustrating a method of constructing relative wear diagrams for crankpins. [*Data from Fig.* 5-2 (*a*) (*B*).]

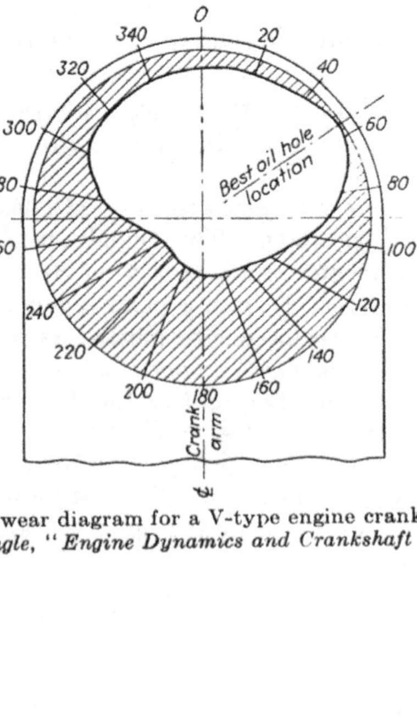

FIG. 5-23.—Relative wear diagram for a V-type engine crankpin (*From A.S.I.C.* 421 *and Angle, "Engine Dynamics and Crankshaft Design."*)

As an example of the construction, let Fig. 5-2 (*a*) (*B*) represent a polar diagram for a crankpin in which the best location for the oilhole is desired. By measuring to scale radially out to the curves [Fig. 5-2 (*a*) (*B*)] at increment crank angles, values of force in the direction of the crank-arm axis are found (Table 5-3). Plotting these forces as explained above gives Fig. 5-22.

TABLE 5-3

θ	Forces in direction of crank-arm axis	Total force in direction of crank-arm axis
0	2,250	2,250
20	1,150	1,150
40	900	900
60	780	780
80	790	790
100	870	870
120	1,040	1,040
140	1,320	1,320
160	1,550, 2,500, 3,700	7,750
180	1,600, 1,950	3,550
200	1,540, 1,680	3,220
220	1,300, 1,370	2,670
240	1,000, 1,010	2,010
260	800, 840	1,640
280	710, 720	1,430
300	730, 730	1,460
320	800, 840	1,640
340	500, 1,240	1,740
360	2,250	2,250

A relative wear diagram for a V-type engine is shown in Fig. 5-23.

For radial engines, the polar diagram with respect to the crank-arm axis (Fig. 5-10) should be used to construct the relative wear diagram for the crankpin.

This method of determining the oilhole location has been criticized by some sources on the grounds that it does not take into account the effect of centrifugal force on the oil in the crankpin. Thus they maintain that the oilhole should be located on the outer side of the crankpin regardless of what the wear diagram might show. Such a location would undoubtedly be satisfactory in most engines and would save considerable tedious construction.

Suggested Design Procedure

Important. Make all constructions and diagrams to a large enough scale to permit accurate work. Size B or larger drawing paper is recommended. Keep a record of the man-hours required on each item.

1. For the engine selected for your design, construct polar diagrams of forces on the crankpin (*a*) with respect to the engine axis and (*b*) with respect to the crank-arm center line.

For in-line and V-engines, construct the diagrams through 720 deg. of crank travel. For radials, construct the diagrams through a sufficient number of degrees of crank travel to accurately define the shape and spacing of the lobes. Then, for part *a*, complete the diagram through 720 deg. by shifting the tracing paper.

For part *b* of radial-engine polars, locate two or more higher r.p.m. points on the crank-arm axis.

2. Determine the maximum and mean forces, and locate values found on the diagrams constructed in item 1.

3. By using bearing loads, rubbing factors, etc., within the ranges given in Appendix 1, determine crankpin dimensions that will be adequate for bearing purposes.

4. Lay out to scale the general arrangement of crankshaft desired. Estimate the unbalanced weight per crank arm and the distance to the center of gravity.*

Refer to available sectional blueprints, specimen crankshafts, etc., for assistance in making the layout. Do not try to include details other than those necessary to the determination of unbalanced-weight data. For radials, this item is unnecessary at this point.

5. Construct main-bearing polar diagrams for all differently loaded main bearings.

For in-line and V-engines, construct the diagrams through 720 deg. of crankshaft travel.

For radials, construct the diagrams through a sufficient number of degrees of crankshaft travel to define accurately the shape and spacing of the lobes. Then complete the diagrams through 720 deg. by shifting the tracing paper.

6. Determine the maximum and mean forces, and locate values found on the diagrams constructed in item 5.

7. By using bearing loads, rubbing factors, etc., within the ranges given in Appendix 1, determine main-bearing dimensions that will be adequate for bearing purposes.

8. Construct a relative wear diagram for the crankpin of your engine, and show the best oilhole location, or locate hole on outside of crankpin in plane of crank arms.

* When this distance is not equal to the crank radius, as is usually the case, it is frequently the custom to use an *equivalent* unbalanced weight that is considered to act at crank radius from the center of rotation.

9. When items 1 to 8 have been completed and put in proper form, submit for checking and approval.

References

1. Heldt: "Automotive Engines."
2. Angle: "Engine Dynamics and Crankshaft Design."
3. *A.S.I.C.* 421.
4. *S.A.E. Jour.*, Vol. 28, No. 4, April, 1931.
5. *S.A.E. Jour.*, Vol. 29, Nos. 4 and 5, October, November, 1931.
6. Mark's "Handbook," 2d ed., p. 279.
7. *S.A.E. Jour.*, Vol. 35, No. 6, December, 1934.
8. Unpublished design notes of A. J. Meyer.
9. Design of Engine Bearings, *Automotive Ind.*, Aug. 1, 1939.
10. Willi: Engine Bearings from Design to Maintenance, *S.A.E. Jour.*, Vol. 45, No. 6, December, 1939.

CHAPTER 6

DESIGN OF RECIPROCATING PARTS

6-1. Design Requirements and Limitations.—The design of any machine element can be of reasonably certain effectiveness only when the designer (*a*) is fully aware of and properly considers the functions that the element must perform, and (*b*) is cognizant of the capabilities and limitations of the materials that can be used for the element. Hence, in proceeding with the design of individual parts of the engine, it is advisable to consider briefly the requirements, possibilities, and limitations of these parts in somewhat the same way that was done with the unit as a whole (Chap. 1).

6-2. Functions of the Piston.—Aircraft-engine pistons are called upon to satisfy a rather formidable list of requirements. Most, but not necessarily all, of these requirements are listed as follows:

The piston must

1. Take the gas-force load without appreciable distortion.

2. Fit closely enough in the cylinder to prevent piston slap, excessive blow by, or oil pumping.

3. Be capable of conducting away a large portion of the heat generated in the combustion chamber.

4. Have a coefficient of expansion such that the piston will not be too loose in the cylinder when cold or too tight when hot.

5. Have cross sections and a coefficient of heat flow sufficient to conduct away the heat absorbed by the head at a rate that will prevent hot spots and a resulting increased tendency of the fuel to detonate.

6. Have skirt dimensions sufficient to conduct a considerable portion of the heat absorbed by the head to the cylinder walls and to provide adequate bearing area to take the side thrust.

7. Be capable of giving up some of the heat absorbed to the lubricating oil without raising part of the oil to a temperature that might impair its lubricating qualities.

8. Provide adequate support for the piston rings.
9. Have adequate bearing area for the piston pin and supporting-pin bosses rigid enough to prevent excessive localized pin-bearing pressures.
10. Be as light in weight as possible.
11. Have adequate resistance to wear.

At best, some of these items are directly conflicting, and the designer is faced with the ever-present problem of judging where to strike a proper compromise. If he strives to reduce reciprocating inertia forces by reducing the piston weight to a very low value, he usually will have to sacrifice section thicknesses to a point where heat-flow characteristics will be impaired. Quite often the attainment of close piston fits in the cylinder necessitates the use of a denser metal to get the proper coefficient of expansion characteristics, and this is apt to mean a heavier piston. Many other conflicting problems requiring compromise solutions will occur to the student.

6-3. Piston Materials.—In keeping with the preceding requirements, an aircraft-engine piston should have

1. Adequate mechanical strength at working temperatures.
2. A low coefficient of linear expansion.
3. A high coefficient of thermal conductivity.
4. A low density.
5. A high resistance to abrasion.

By far the most common metals used for pistons are aluminum and cast iron. Important properties of these two metals are given in Table 6-1. Obviously neither of these metals is superior from every standpoint. However, owing partly to improved characteristics imparted by small quantities of other metals and partly to improved fabrication technique, almost all modern aircraft engines use aluminum-alloy pistons. Usually they are cast in permanent molds or forged.

The most commonly used aluminum alloys for pistons are S.A.E. 34 (Aluminum Co. designation 122), S.A.E. 321 (Aluminum Co. designation A132), and the so-called Y-alloy (Aluminum Co. designation 142). Important properties of these alloys as given by the Aluminum Company of America are listed in Tables A2-1 to A2-5. Alloy A132 is recommended by the Aluminum Co. as being particularly desirable for aircraft-engine pistons because of its low coefficient of expansion and low specific gravity.

6-4. Piston Dimensions.—Detailed dimensions of pistons are to a very considerable extent a matter of engineering judgment. The functions of the piston are so numerous and the heat flow, stresses, etc., are so involved that a rational approach is too complex to be of practical value. However, useful aids may be had from a study of previous designs (Table A1-14) and from empirical rules.

TABLE 6-1.—PROPERTIES OF PISTON METALS*

Metal	Specific gravity	Tensile strength, lb. per sq. in.	Coefficient of linear expansion, per deg. F.	Heat conductivity, B.t.u./(min.) (sq. ft.) (in.) (deg. F.)	Brinell hardness		
					At 32°F.	At 200°F.	At 400°F.
Aluminum..	2.7	15,000	0.0000124	24.0	150	138	120
Gray iron...	7.1	20,000	0.00000556	5.5	165	165	165
Magnesium..	1.74	.....	0.0000145	18.2	66	60	35

* Pure metals, not the alloys.

Huebotter and Young,[2] following extensive tests on automotive pistons, have drawn the following conclusions relative to piston design:

1. A deep section at the center of the head is very effective in lowering the maximum temperature. For this reason a liberal center-hole boss is recommended.
2. If ribs are used to reinforce the piston-pin bosses, they should extend to the center of the head.
3. The aluminum alloy piston has a wide margin of safety over the gray iron piston on a temperature basis.
4. The ring belt dissipates about 60 per cent as much heat from a given initial temperature as that . . . from the piston skirt.

Temperature gradients for a typical aluminum-alloy and a gray-iron piston are shown in Figs. 6-1 and 6-2. Although not geometrically identical, these two pistons are sufficiently similar in section to show the decided thermal advantage of aluminum. The relative weights of the two pistons are also of interest.

Typical aircraft-engine pistons are shown in Fig. 6-3. Current practice indicates the desirability of three or four rings above the pin bosses, frequently one ring near the bottom of the skirt,

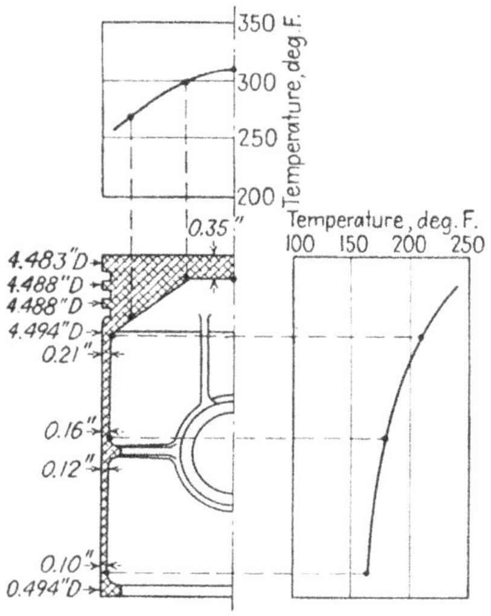

Fig. 6-1.—Temperature gradients in an aluminum-alloy piston. Weight = 2.635 lb. Cylinder diameter = 4.5 in. (*From Huebotter and Young, Flow of Heat in Pistons. Purdue University Engr. Exp. Sta. Bull.* 25).

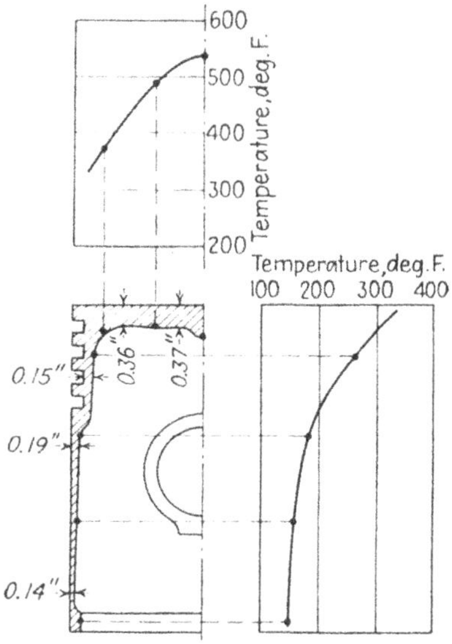

Fig. 6-2.—Temperature gradients in a gray-iron piston. Weight = 5.853 lb. Cylinder diameter = 4.5 in. (*From Huebotter and Young, Flow of Heat in Pistons. Purdue University Engr. Exp. Sta. Bull.* 25.)

ribs under the head for rigidity and better cooling, and amply supported pin bosses.

Piston clearance must be adequate to prevent "hot seizure" and small enough to prevent "cold slap." Customary practice for automotive engines has been

$$\text{For gray iron, clearance} = 0.001 \times \text{bore in inches.}$$
$$\text{For solid-skirt aluminum, clearance} = 0.0006 \times (\text{bore})^2 \text{ in inches.}$$

These rules should not be applied to special types such as Invar or steel-strut and flexible-skirt cam-ground pistons.

Aircraft pistons may be fitted with greater clearance as they operate nearer rated load most of the time (see Table A2-3 for coefficients of expansion). Clearance and other data on current automotive pistons will be found in Table A1-3. Swan[4] sug-

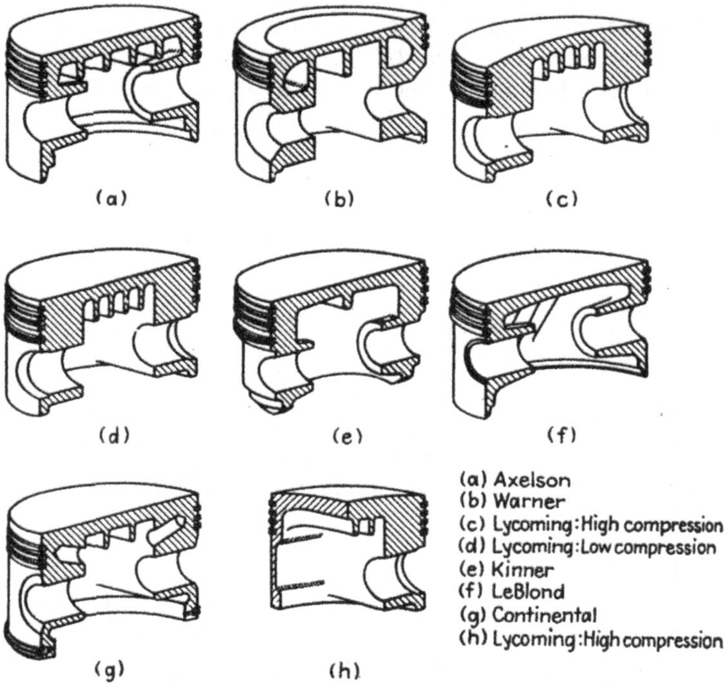

Fig. 6-3.—Typical aircraft-engine pistons.

gests the following piston clearances for Y and similar aluminum alloys:

	Inch per Inch of Diameter
Top of head	0.006
Bottom of head	0.004
Top of skirt	0.0025
Bottom of skirt	0.0015

Piston-ring and groove dimensions have been standardized by the Society of Automotive Engineers (see Table A1-16), but many aircraft engines are equipped with rings that do not conform to these standards.

Locate the piston pin approximately halfway between the lower ring groove above piston bosses and the end of the piston skirt. Current practice on relative diameter-length ratios of pistons and other details may be observed from Table A1-14. The piston skirt below the upper ring belt is usually considered to take the side thrust, *i.e.*, act as the bearing area between the piston and cylinder wall. This area is the equivalent of the crosshead bearing area in engines using that type of construction. Angle suggests using about 1 sq. in. of bearing surface for each 50 lb. of average side pressure. From the piston side thrust (see Par. 4-9 and Fig. 4-10), the average side thrust may be determined. Then the necessary length of piston skirt will be

$$P_L = \frac{T_{SA}}{D \times 50} \tag{6-1}$$

where T_{SA} = average side thrust against the cylinder wall, lb.
D = cylinder diameter, in.
P_L = length of the piston skirt.

The total length of the piston will be P_L plus the width of the upper ring belt. However, if the value of P_L as found from Eq. (6-1) is appreciably greater or less than current practice, the length of the skirt should be altered to fall within the range of values for similar engine pistons.

As there is very little side pressure on the piston in the direction of the piston-pin axis, a reduction in piston weight may be made by cutting away the skirt below the ends of the pin. However, it is doubtful if the gain in reduced inertia forces offsets the added complexity of construction and probable increased lubrication or oil-pumping problems except in very high-speed racing engines.

6-5. Piston Rings.—Piston rings should be (*a*) sufficiently elastic to exert the necessary side pressure against the cylinder walls and to permit insertion of the ring in its groove by sliding it over the piston, and (*b*) soft enough to prevent excessive wear on the cylinder walls. Close-grained cast gray iron is almost universally used for piston rings.

Special shapes and types of piston rings are sometimes used to permit closer control of the lubricant, more rapid seating, and to reduce blow by of the combustion gases (see reference 3). However, as it is quite common practice to purchase piston rings from companies specializing in their manufacture, the engine designer will probably do well to merely specify over-all

S.A.E. standard dimensions (Table A1-16) and follow the specific recommendations of the ring specialists on details.

End clearance on rings should be great enough to prevent any possible binding from heat expansion but small enough to prevent excessive gas leakage. A gap clearance of 0.003 in. per inch of bore is commonly specified[3] by automotive-engine manufacturers. Butt, diagonal, and lap-joint ring ends (Fig. 6-4) are most common. Side clearance of rings in piston grooves should be about 0.001 in. to minimize leakage through the piston groove behind the ring (Table A1-16).

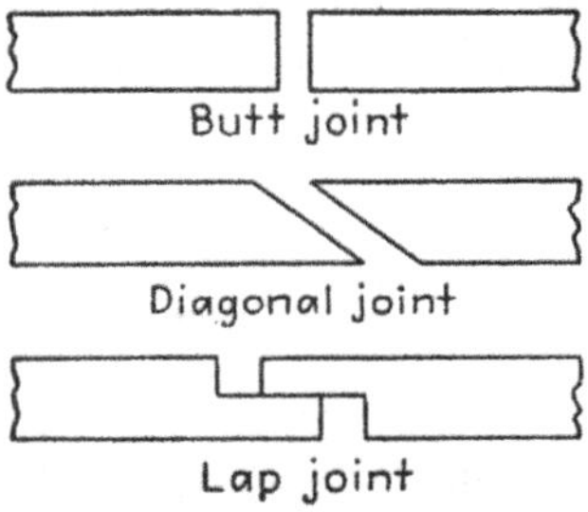

FIG. 6-4.—Common types of piston-ring joints.

6-6. Piston or Wrist Pins.—Piston pins, the connecting links between the piston and connecting rods, may be either clamped in the piston, clamped in the connecting rod, or full floating. This last method permits the pin to turn gradually so that wear is more evenly distributed, but it requires some form of snap ring or soft metal button to prevent the end of the hard pin from coming in contact with and scoring the cylinder wall.

The average distance between the pin bosses for the pistons in Table A1-14 is about 48 per cent of the piston diameter, and by allowing for end clearance on the small end of the connecting rod, the length of the piston-pin bearing in the connecting rod will be about 45 per cent of the piston diameter. For full-floating pins (after allowing for end buttons), this will permit about equal bearing areas in the upper end of the connecting rod and in the piston. Hence the piston-boss length may be made approximately one-fourth of the piston diameter, and the length of the pin bearing in the connecting rod may be made 45 per cent of the diameter.

The diameter of the piston pin will be determined by the maximum allowable bending moment and the allowable bearing pressure. Maximum stress in the pin will occur at full-throttle-low-speed (low inertia) conditions, and the pin may be assumed to take the full force of the explosion pressure in the combustion chamber. Maximum gas pressures were assumed to be about 75 per cent of the theoretical pressures (Par. 3-3). Hence the maximum force on the piston pin in pounds is

$$P_{max} = \frac{0.75\pi D^2 P_c}{4} = 0.59D^2P_c$$

where D = cylinder diameter, in.
P_c = calculated theoretical maximum pressure, lb. per sq. in. abs. [Eq. (3-3)].

The projected bearing area in the upper end of the connecting rod is

$$S = dL = 0.45Dd$$

where d = diameter of the piston pin, in.
L = effective length of the piston pin, in.
D = cylinder diameter, in.

Because of the low rubbing velocities, much higher bearing pressures may be used for piston pins than for crankpins provided suitable bearing metals such as phosphor bronze (Table A2-8) are used for connecting-rod bushings and the crankpins are casehardened.

Heldt[3] suggests an average piston-pin pressure of 3,200 lb. per sq. in. as representative of automotive practice, but values of 10,000 to 15,000 are not uncommon in high-powered aircraft engines. Thus

$$\frac{0.59D^2P_c}{0.45Dd} = B_p$$

where B_p = 3,000 to 15,000, with 5,000 to 10,000 probably being a safe range for small aircraft engines of good design. Hence the piston-pin diameter may be found from

$$d = \frac{DP_c}{K} \qquad (6\text{-}2)$$

where K = 4,000 to 8,000. Data in Table A1-14 indicate that piston-pin diameters are usually about 25 per cent of the piston diameters.

The piston-pin diameter as determined by Eq. (6-2) is that necessary for adequate bearing area. However, the piston pin must also be strong enough to withstand the stresses involved and as light in weight as possible. Reduction in weight may be made by using a hollow piston pin, the size of the hole in the pin being determined by the maximum bending moment and the allowable stress.

For determining the diameter of the hole in the piston pin d_I (Fig. 6-5), it may be assumed that the gas-force load P_{max} is equally divided between the pin bosses and acts as a concentrated load at the mid-point of their lengths. The reaction load in the connecting-rod bearing may be assumed as equally distributed over the length of the bearing.[3,4] Then from Fig. 6-5 (*B*), the maximum bending moment (at the mid-point along the piston pin) will be

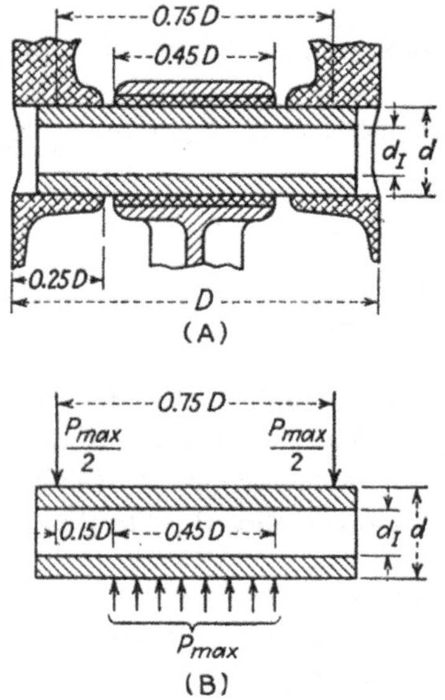

FIG. 6-5.—Average location and magnitude of forces on aircraft-engine piston pins.

$$M = \frac{P_{max}}{2} \times \frac{0.75D}{2} - \frac{P_{max}}{2} \times \frac{0.45D}{3} = 0.1375DP_{max} \quad (6\text{-}3)$$

where M = maximum bending moment, in.-lb.
P_{max} = maximum gas force on the piston, lb.
D = diameter of the piston, in.

For equilibrium conditions, this moment [Eq. (6-3)] must equal the internal moment.[5]

$$M = S\frac{I}{C} \quad (6\text{-}4)$$

where S = maximum allowable stress, lb. per sq. in.
I = moment of inertia of the piston-pin cross section.
C = one-half the diameter of the piston pin.

For hollow piston pins, the section modulus is

$$\frac{I}{C} = \frac{\pi}{32d}(d^4 - d_I^4) \quad (6\text{-}5)$$

where d = diameter of the piston pin, in.

$$d_I = \left(d^4 - 1.40\frac{dDP_{max}}{S}\right)^{0.25} \quad (6\text{-}6)$$

Expressing P_{max} in terms of P_c,

$$d_i = \left(d^4 - 0.825 \frac{dD^3P_c}{S}\right)^{0.25} \tag{6-7}$$

where the symbols are the same as above.

Piston pins may be made of plain carbon steel casehardened (S.A.E. 1020), nickel steel (S.A.E. 2315, 2320, or 2515), or chrome nickel steel (S.A.E. 3120, 3215, or 3220).* An allowable stress of 25,000 lb. per sq. in. may be used with the carbon steel, and 35,000 lb. per sq. in. with the alloy steels. S.A.E. 2315 steel is one of the most commonly used materials for aircraft-engine piston pins.

6-7. Knuckle or Link Pins.—Dimensions of knuckle or link pins for attaching the articulated rods to the master rod may be calculated in much the same way as those of piston pins. Owing to the greater mass of inertia-producing parts between the link pins and the gas force on the piston, link pins may be made somewhat smaller than piston pins. Probably the easiest way to determine the size is to use the same fundamental formulas that were used for the piston pins and assume higher allowable bearing loads and bending stresses. Meyer[6] suggests as an allowable bending stress 30,000 to 50,000 lb. per sq. in. Unit bearing loads may be somewhat higher than for piston pins because link pins have more positive force-feed lubrication. Link pins may be made of the same materials that are used for piston pins. For severe service, nickel chromium steel (S.A.E. 3125) may be used. For very severe service, such as very highly supercharged racing engines or Diesels, nickel molybdenum steel (S.A.E. 4615) may be advisable.

Link pins should be locked securely in place to prevent any endwise movement and resulting damage from contact with adjacent parts. A rather common method of securing link pins in one-piece master rods is to use small locking plates that are bolted to the outside of the master-rod flange between the link pins [Fig. 6-6 (*A*)]. The ends of the link pins are flared and either cut away or beveled, and the locking plates extend over their edges to prevent movement of the pins. When two-piece master rods are used, the cap bolts may be so located that

* For an explanation of the S.A.E. steel-numbering system, see Table A2-6. For detailed data on the various S.A.E. steels, see reference 8.

they pass through milled slots* in the sides of the link pins; thus securing them positively (see Fig. 6-14).

Link pins are commonly lubricated by pressure feed through holes in the master-rod flanges and a passageway inside the pin [Fig. 6-6 (*B*)].

Link pins should be located as close to the crankpin center as clearance and structural dimensions will permit (Par. 4-4). This

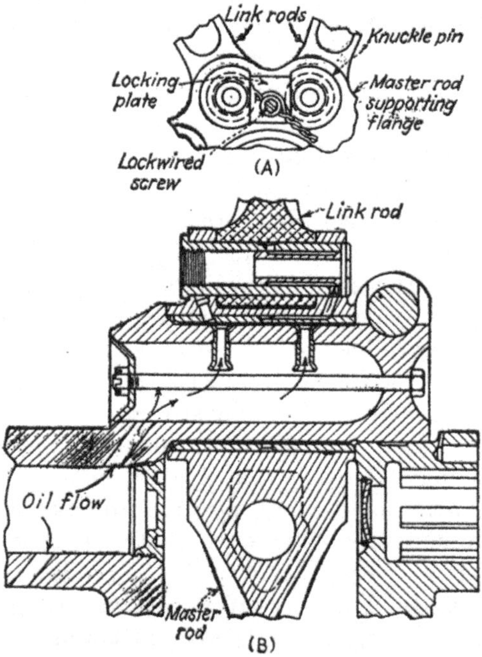

FIG. 6-6.—(*A*) Method of holding link pins in place, and (*B*) section through the crankpin and a link pin of a Lycoming Type R-680 nine-cylinder radial engine showing the means provided to lubricate the link-pin bearing.

makes it desirable to keep the diameter of the pins as small as bearing loads and strength requirements will allow. When six or eight articulated rods are attached to one master rod, care must be observed in providing adequate clearance between adjoining rods.

6-8. Connecting-rod Shank Stresses.—Connecting rods are subjected to

1. Compression stresses due to combined gas and inertia forces.

* Angle patent owned by Pratt and Whitney.

2. Tension stresses due to inertia forces.

3. Tension and compression stresses due to "whipping" or lateral acceleration of the rod.

4. Master rods in articulated systems are subjected to an additional bending stress owing to the axes of the articulated rods not passing through the center of the crankpin.

Considering these conditions in order:

1. Compression stresses are most severe at full-throttle low-speed (low-inertia) conditions, and as the connecting rod is of intermediate length in proportion to its cross-sectional area, the slenderness ratio L/k (= center-to-center length of the connecting rod divided by the least radius of gyration) usually falls

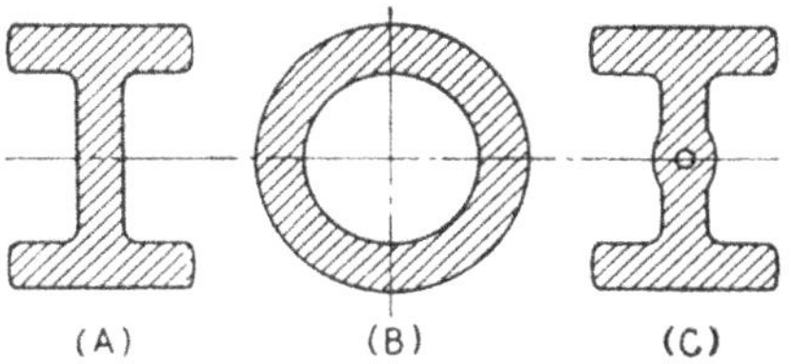

FIG. 6-7.—Connecting-rod shank sections.

within the range in which Rankine's column formula is most applicable. Hence critical compressive stresses may be found from

$$\frac{P_{max}}{A} = \frac{S_c}{1 + q(L/k)^2} \tag{6-8}$$

where P_{max} = maximum gas force on the piston, lb.
A = cross-sectional area of the connecting rod at the mid-point in its length, sq. in.
S_c = allowable stress, lb. per sq. in.
L = center-to-center length of the connecting rod, in.
k = least radius of gyration of the mid-section.
q = coefficient depending upon the arrangement of the column ends.

Connecting-rod shank sections used in aircraft engines are most often a modified form of I or H section, but tubular sections are frequently used in articulated rods, and sometimes, when oil is supplied under pressure to the piston-pin bearing, a hollow I section [Fig. 6-7 (*C*)] is used. In any event, the desired end is a rod of adequate strength and stiffness with a minimum weight.

In determining the shank dimensions, account should be taken of the fact that the end supports for the connecting rods are essentially free in the plane of rotation but fixed in the plane containing the crankpin and piston-pin axes. With free ends, the deflection of the rod under load will be as in Fig. 6-8 (*A*), whereas with fixed ends the deflection of the rod will be as in Fig. 6-8 (*B*). As the distance between the two inflection points *m* and *n*, Fig. 6-8 (*B*), is one-half of *L* and since the stress in the rod varies as L^2, the rod must be four times as strong in the plane of rotation as in the plane of the crankpin and piston-pin axes.

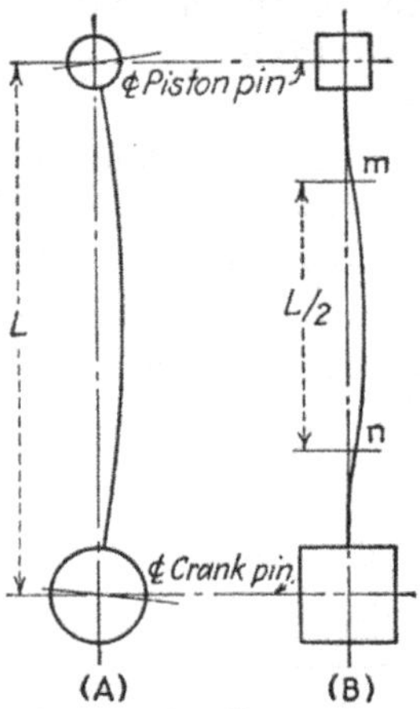

FIG. 6-8.—Connecting-rod deflection (exaggerated). (*A*) In the plane of rotation, (*B*) in the plane of the piston and crankpin axes.

For carbon-steel rods, S_c should not exceed 25,000 lb. per sq. in.; for alloy-steel rods, S_c should be held to less than 35,000 lb. per sq. in. For aluminum-alloy rods, S_c should be about 12,000 lb. per sq. in.

Values of $q = 1/10{,}000$ (for free ends) and $q = 1/40{,}000$ (for fixed ends) may be used. Values for moment of inertia and radius of gyration for several useful geometric shapes will be found in Table A3-1.

For I and H sections, a small draft angle (7 to 10 deg.) must be provided to permit forging, although this will be removed when the rods are machined all over, and usually all corners are rounded with fillets. These details make it difficult to determine the moment of inertia of the section, and to simplify the procedure, an equivalent section without draft or fillets is frequently used for determining over-all dimensions. Relative proportions of equivalent shank sections useful for this purpose are shown in Fig. 6-9.

2. Greatest tension in the connecting rod will occur (for normal operation) at highest speed at the beginning of the suction stroke. Still more severe conditions can exist in high-speed closed-throttle dives. Maximum tension may be found by means of Eq. (4-8). By using conservative values of stress, *i.e.*, values the same as for column effect in the rod (case 1 above) and investi-

gating for rated speed, usually the strength will be sufficient for any diving condition encountered. Ordinarily, a rod strong enough as a column is adequately strong in tension.

3. Whipping stresses are obviously greatest at highest speeds. These stresses are due to centrifugal force on the body of the connecting rod, and, as the forces act parallel to the crank arm, they tend to bend the connecting-rod shank. Magnitude of the maximum bending moment may be found from methods outlined in references 4, 9, and 11, but ordinarily, whipping stresses need not be investigated in aircraft engines as they are well below the maximum stresses due to the gas force on the piston.[3] Hence, a connecting-rod shank section adequate for column conditions (case 1 above) will be strong enough to withstand the maximum whipping stresses. This may not be true for very high-speed automobile racing engines, however.

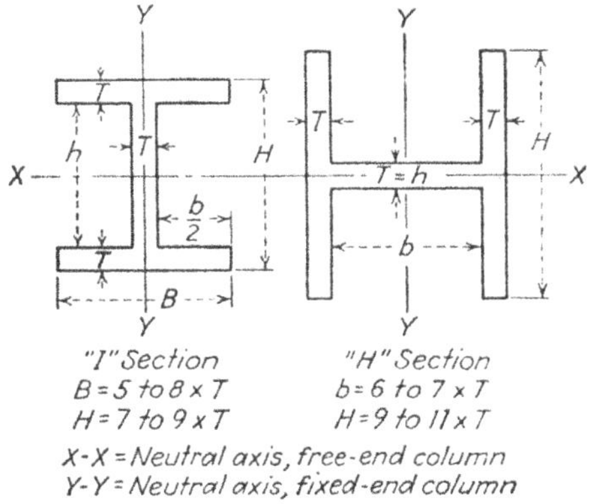

FIG. 6-9.—Equivalent mid-length connecting-rod shank sections representing usual proportions in aircraft-engine practice.

4. Bending stresses in the master rod of an articulated-rod system due to the forces in the link rods not passing through the center of the crankpin are ordinarily not critical because of the small distance between the line of these forces and the crankpin axis. The conventional practice of tapering the master-rod shank, *i.e.*, increasing the cross section toward the crankpin, usually provides an adequate safeguard against critical stresses in the master rod owing to forces in the articulated rods. In cases where extremely light weight is desired, it may be advisable to investigate these bending stresses, however. The method suggested in reference 4 may be used for this purpose.

In general, for conventional aircraft engines, a connecting-rod shank section adequate for case 1 above will be amply strong in tension, whipping, and bending due to articulated rods. In tapering the shank section, care should be taken to avoid reducing the section near the piston pin to a point where it becomes critical.

6-9. Connecting-rod Cap Bolts.—When a one-piece crankshaft is used (Figs. 5-11, 5-12, 5-13, and 5-16*c*), the big end of the connecting rod must be made in two pieces in order to get it onto the crankpin. In place, the two parts are held together by two or four bolts usually called *cap bolts.* These bolts are subjected to tensile stresses when the rod is under tension. The maximum tension occurs at maximum speed at the start of the suction stroke. This tensile force is due to the inertia of the reciprocating parts plus the centrifugal force due to the rotating parts, since these forces act in the same direction at the start of the suction stroke. The reciprocating inertia force may be found by means of Eq. (4-8), and the centrifugal force may be calculated by Eq. (5-1). The reciprocating weight may be taken as the sum of the weights of the piston, piston rings, piston pin, and one-third of the weight of the connecting rod, or the value found in item 2, Suggested Design Procedure, p. 56, (from Figs. A1-3 and A1-4) may be used. The centrifugal weight may be taken as two-thirds of the weight of the connecting rod minus the weight of the cap, or (for radials) the rotating weight may be the value you have used in calculating bearing loads (Fig. A1-5).

The diameter of the cap bolts may be found by using an allowable tensile stress of about 20,000 lb. per sq. in. Connecting-rod bolts should conform to S.A.E. standard dimensions and materials whenever possible (see Table A1-17). S.A.E. 2330 steel is one of the most commonly used aircraft-engine connecting-rod cap-bolt materials.

6-10. Connecting-rod Ends.—Connecting-rod ends provide the necessary backing for the bearing metal and transmit the loads to the bearing pins. In addition, they conduct away some of the heat generated in the bearing.

To avoid excessive localized bearing pressures, rod ends should be as free from distortion as possible. To provide this necessary rigidity and at the same time keep the weight to a minimum, designers frequently incorporate stiffener ribs in the bearing cap and flare the rod shank where it joins the rod end. These ends and caps are somewhat similar to curved beams, and for much the same reasons as with more conventional beams, they should have a high section modulus (I/C). In view of the shapes and loadings, the exact bending moments and stresses at any given section are difficult to determine, but a knowledge of beam characteristics will aid greatly in deciding upon detail arrangements. The somewhat vaguely defined ability known as engineering good judgment is of great value in detail design such as this, and, although, like personality, it is inherent in widely varying degrees in different individuals, it can be developed to a considerable extent by alert observation and clear thinking. Specifically, in the design of many machine parts, exact stresses either cannot be calculated or the calculations are so involved as to be impractical, but a correlation of the problem with simpler structures of similar characteristics usually aids in the intelligent selection of detail dimensions.

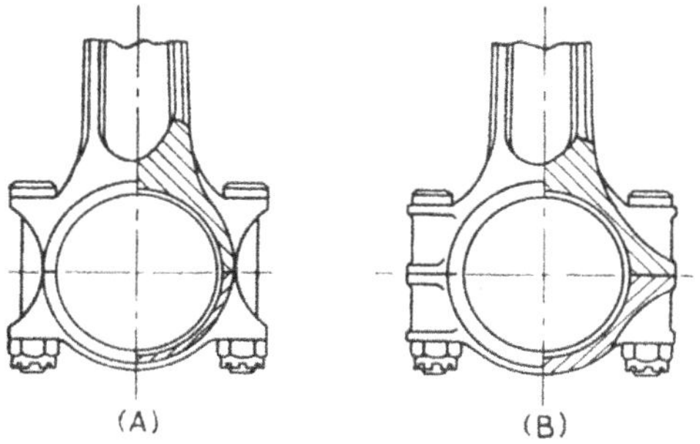

Fig. 6-10.—Aircraft-engine connecting-rod big ends. (*A*) Poor design that was subject to frequent crankpin-bearing failures. (*B*) Later design that eliminated the failures. (*From Ricardo, "High Speed Internal Combustion Engines."*).

Figure 6-10 is a case in point. Connecting rod (*A*) was found to cause frequent bearing trouble, and, although the exact stresses in the cap and rod end resulting from tightening the cap bolts and from inertia forces would be very difficult to determine, it is quite evident that such tightening and inertia forces would cause the cap and rod ends to buckle inward at the joint between them. Bearing failures very often start from high localized pressures, hence the logical solution to the problem (*B*) is quite apparent without even an approximate knowledge of the stresses in the parts. Intelligent avoidance of critical situations is just as important in the design of aircraft engines as in the flying of them.

Reentrant corners and abrupt changes of cross section almost invariably cause high localized stresses. In complex machine parts, it is usually impossible or impractical to calculate these stresses, but they can be avoided by using large fillets and gradual changes of section. The result of flaring the end of the connecting-rod shank where it joins the big end of the rod cannot readily be expressed mathematically, but even casual thought on the matter will show that it will reduce rod-end deflection (*i.e.*, tendency to high localized bearing pressures) in much the same way that distributing a concentrated load will reduce the deflection of a simple beam.

Innumerable other instances of this sort will occur as the design proceeds, and as the habit is formed of referring complex and formally insoluble problems to simple cases that are similar, in time that invaluable asset, good engineering judgment, will develop.

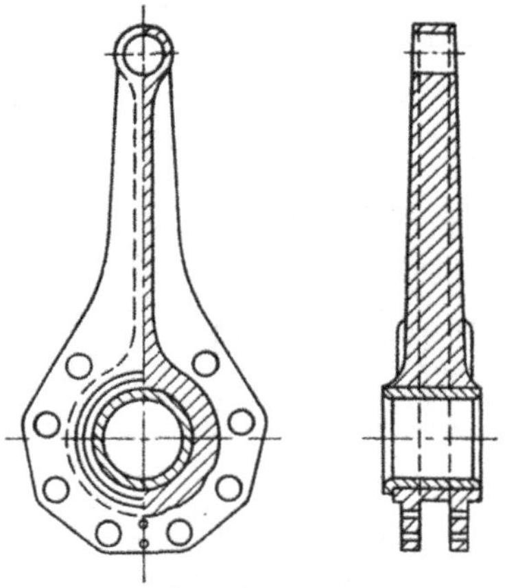

Fig. 6-11.—H-section master connecting rod. (*Wright.*)

To reduce weight and to some extent to avoid the reentrant corners at the cap-bolt heads, studs are sometimes forged integral with the connecting rod. However, this may increase the cost of the connecting rod as it tends to complicate the forging.

Typical connecting rods for different arrangements of cylinders are shown in Figs. 6-11 to 6-15. All these rods are from successful engines and will merit careful study.

Many radial engines use two-piece crankshafts, and this permits the use of one-piece connecting rods (Fig. 6-15). The advantage of a one-piece master rod lies mainly in the avoidance of highly stressed cap bolts and, to some extent, interference with the location of the link pins. This last, of course, becomes increasingly important as the number of cylinders per crankpin is increased. The obvious disadvantage is a more complex crankshaft.

6-11. Articulated Rods.—Articulated, or link, rods are subjected to the same general types of stresses as master connecting rods except that they are not subjected to bending stresses due to forces in other rods, item 4, Par. 6-8. Link-rod shanks should be designed according to the same general procedure as used in master rods. They usually are not tapered, however. Detailed data on articulated rods from several well-known makes of engines are given in Table A1-18.

6-12. Connecting-rod Materials.—Aircraft-engine connecting rods may be made of the following materials:

S.A.E. steels: 1040, 2315, 2340, 3140, 3240, 3250, 6135, and 6140.

Aluminum alloy: 25S (Tables A2-1 to A2-5).

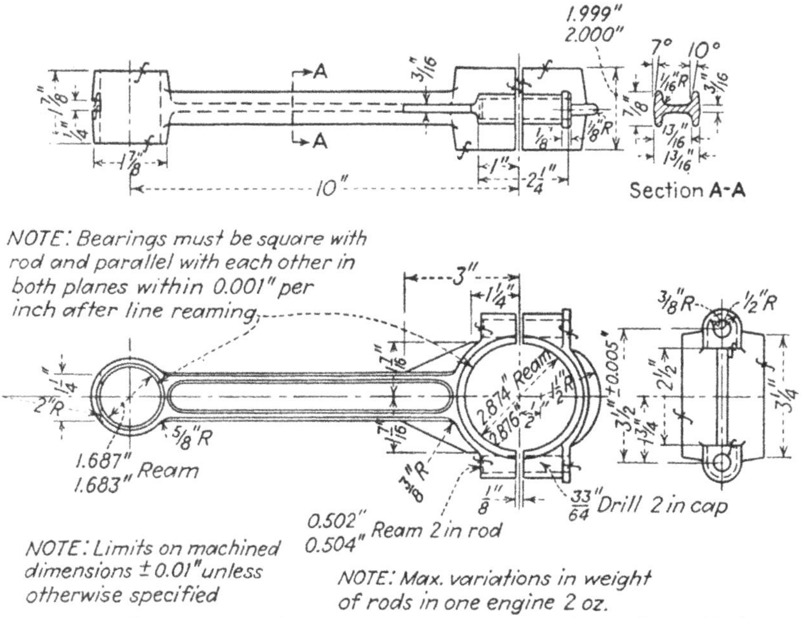

FIG. 6-12.—Connecting rod for a 4- by 5-in. in-line engine. (*From Huebotter, "Mechanics of the Gasoline Engine."*)

Chrome nickel steel (S.A.E. 3140) is one of the most commonly used materials for aircraft-engine connecting rods. Duralumin (25S) properly heat-treated, may be used for in-line engine rods and radial-engine articulated, or link, rods. For light or medium loadings, this aluminum alloy may be used without rod end bushings provided the piston and link pins are sufficiently hard. Nitralloy pins are satisfactory for this purpose, according to the Aluminum Company of America.

6-13. Bearings and Bearing Metals.—In general, there should be as great a difference as possible between the hardness of the bearing metal and the pin or journal. So-called white bearing metals or babbitts (Table A2-7) are very commonly used where the loads are not too great. These babbitts may be either in direct contact with the supporting metal of the connecting rod

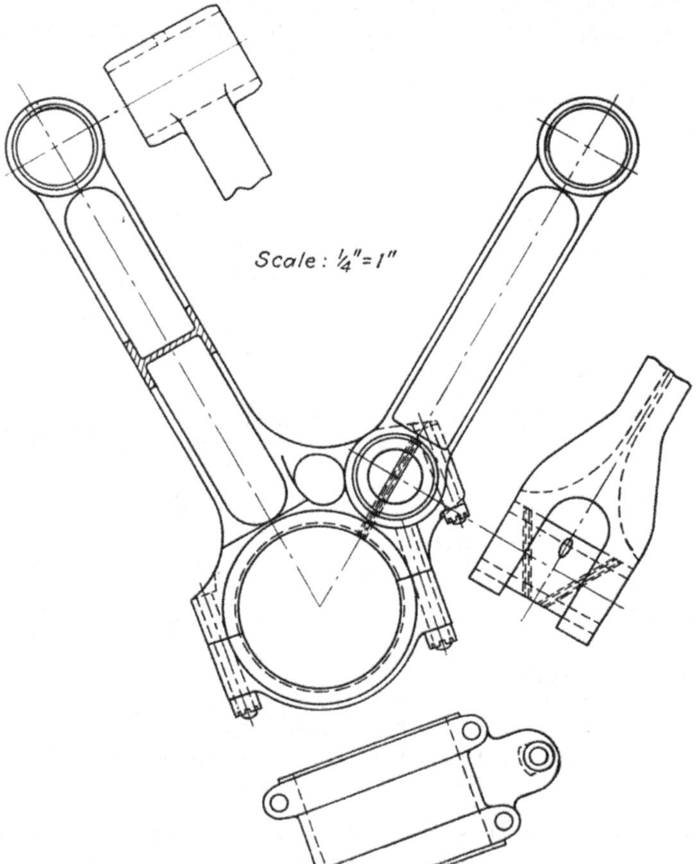

FIG. 6-13.—Master and articulated connecting rod used in the Packard 2500 airplane engine.

or, more commonly, bronze or steel backed. This last permits replacement of the bushings without replacing or rebabbitting the connecting rod, but the path for heat flow from the bearing may be less positive. Very thin (1/16 in. or less) steel-backed babbitt-lined replaceable bushings are in quite general use. For higher bearing pressures, copper-lead (Table A2-7) or cadmium alloys may be used. For piston and link-pin bushings, bronze bearing metals (Table A2-8) are commonly used.

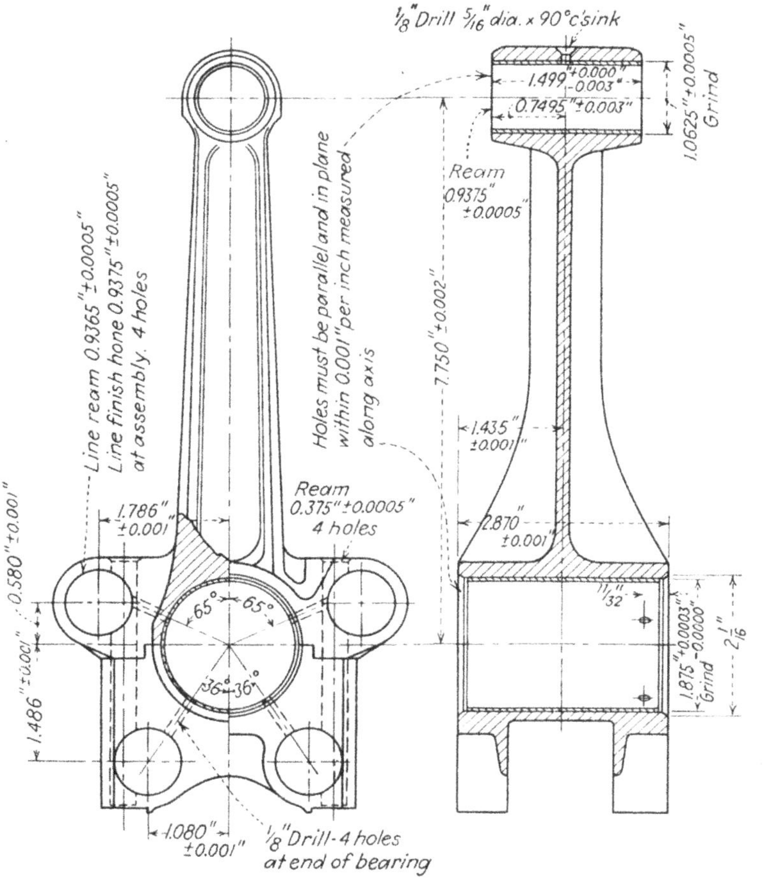

FIG. 6-14.—LeBlond five-cylinder radial-engine two-piece master-connecting-rod assembly.

Aluminum alloys have fair bearing qualities, so that the common practice of allowing the piston pin to bear directly on the inner surfaces of the piston bosses is satisfactory. Bronze bearing shells in the small ends of the connecting rods and in the link-pin ends of articulated rods may be designed for a wall thickness of $\frac{1}{16}$ to $\frac{1}{4}$ in.

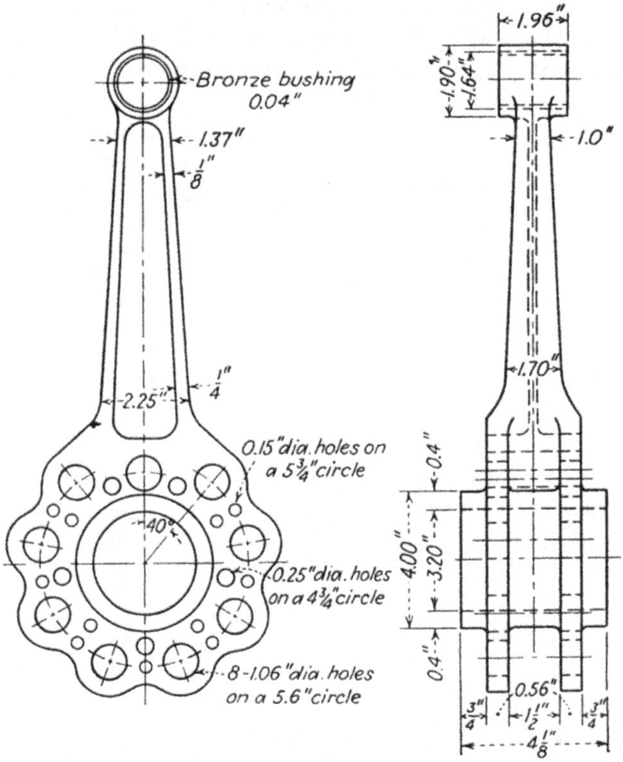

FIG. 6-15.—Master connecting rod used in the Wright 1820 Cyclone engine.

Suggested Design Procedure

Important. Give references for all formulas and empirical factors used. All drawings should be on standard-size paper, and *complete in all details* including dimensions, clearances, material specifications, and number required. Drawings (except as noted) should be blueprinted and properly folded (Fig. 2-4) for insertion in the design notebook. Keep a record of the man-hours required on each item.

1. Select materials and make all necessary calculations for the piston and piston pin.

2. Make a detailed drawing of the piston (at least two sectional views).

3. Make a detailed drawing of the piston pin and end buttons (or equivalent parts used to hold the pin in place).

4. Determine all necessary dimensions for the piston rings, and specify the S.A.E. standard size or sizes to be used.

5. *a*. Select materials, and make all necessary calculations for the master connecting rod and end bushings.

b. Same procedure for articulated rod, bushings, and link pin when used.

6. Check selected dimensions of connecting rod or rods with layout drawings (Suggested Design Procedure, page 24, item 4) to make certain of adequate clearance at all points. Alter layout drawings as necessary.

7. *a*. Make a detailed drawing of the master connecting rod (at least two views). Show bushings in place.

b. Same procedure for articulated rod and link pin when used.

8. Determine the weight of each of the reciprocating parts, and show in tabular form (*a*) name of part or item, (*b*) actual weight of parts, (*c*) weight of parts assumed in calculating bearing loads (Chap. V), (*d*) percentage increase or decrease of actual weights over assumed weights, (*e*) estimated change in maximum bearing loads due to *d*, (*f*) estimated change in mean bearing loads due to *d*.

Check the new bearing loads with values given in the tables of Appendix 1. If these loads are very far outside the usual ranges of loadings, etc., alterations in the design should be made to bring them back within the ranges of proven values. These changes may be made either by altering the weights of the parts or (when possible) by using bearing materials that will withstand the increased loads. If the loads are well below the assumed values, the specific weight of the engine is apt to be unnecessarily high.

9. Make an assembly drawing of the reciprocating parts on the layout drawings of Suggested Design Procedure, page 24, item 4. Show parts in section whenever such sectioning increases the clarity or legibility of the drawing. Include only principal over-all dimensions. Identify each part of the assembly drawing by a reference number corresponding to the detailed drawing or reference number of that part. When the detailed drawing contains more than one part, identify each part by the detailed drawing number and a letter. Do not blueprint the assembly drawings at this stage.

10. When items 1 to 9 have been completed and put in proper form, submit for checking and approval.

References

1. Alcoa Aluminum and Its Alloys, *Bulletin* of the Aluminum Company of America.
2. Huebotter and Young: Flow of Heat in Pistons, *Purdue Eng. Expt. Sta. Bull.* 25.
3. Heldt: "Automotive Engines."
4. Swan: "Handbook of Aeronautics," Vol. 2, Aero Engines, Design and Practice.
5. Poorman: "Strength of Materials."
6. Unpublished design notes of A. J. Meyer.
7. Ricardo: "High Speed Internal Combustion Engines."
8. S.A.E. "Handbook."
9. Huebotter: "Mechanics of the Gasoline Engine."
10. *S.A.E. Jour.*, Vol. 31, No. 6, December, 1932.
11. Lichty: "Internal Combustion Engines," 5th ed.
12. S.A.E. "Aircraft Engine Drafting Room Practice."

CHAPTER 7

CRANKSHAFT VIBRATION AND BALANCE

7-1. Fundamental Nature of Vibration.—When an elastic body is subjected to a force, it deflects, and if the force is suddenly removed, the restoring forces in the body return it to its original or neutral position. But the momentum acquired during the restoring action is such that the body passes beyond its neutral position and is deflected in the opposite direction. This in turn sets up restoring forces in the original direction of deflection, etc., so that the body oscillates or vibrates. If the original deflecting force is applied but once, the vibration gradually diminishes and finally ceases because of internal and external friction. However, if the deflecting force is applied repeatedly, the body will continue to vibrate, but the magnitude of the vibration will vary widely, depending upon the frequency of application of the deflecting force and the natural vibrating frequency of the body.

To illustrate, in Fig. 7-1 (*A*), *R* represents an elastic rod rigidly supported at one end. If the free end of the rod is struck a sudden blow with the hammer *H*, the rod will deflect from its neutral position *C* to a distorted position *D*. But restoring forces are set up in the rod when it is deflected so that as soon as the force of the hammer blow is spent the rod will spring back toward its neutral position. Its inertia of motion carries it beyond *C* to *E*, and then opposite restoring forces start it back toward *D*. This cycle or series of events will continue with gradually diminishing amplitude until the continuously opposing forces of internal and external friction bring the rod to rest at its neutral position.

If cam *S*, Fig. 7-1 (*A*), is rotated, the hammer blows will be repeated at a frequency depending upon the rate of rotation and the number of lobes on the cam. If the hammer blows occur when the vibrating rod is at *C* and moving in direction *M*, the vibration will obviously be damped, but if the blows are timed to occur at N, the force of the blows will aid in continuing and amplifying the vibration. Such a case is sometimes called *synchronous vibration*, or *resonance*. Figure 7-1 (B) illustrates the case for torsional vibration.

If the hammer blows occur at intermediate points between M and N, the resulting vibration of the rod will be intermediate. If the blows occur at one-half the frequency of the rod, vibration will be damped or aided as before, but the effect will be less

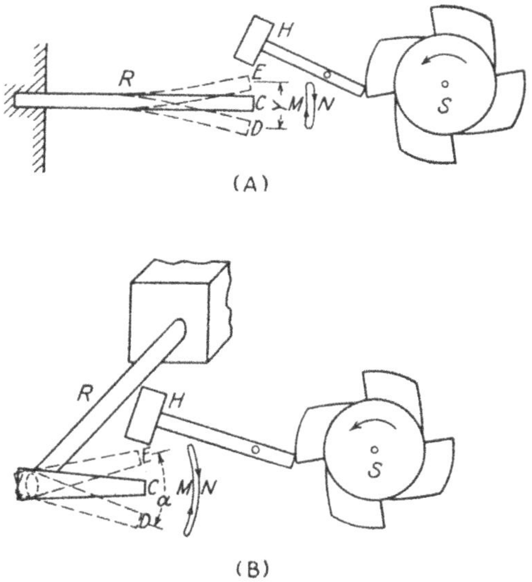

Fig. 7-1.—Fundamental idea of vibration in elastic machine parts. (A) Lateral vibration and (B) torsional vibration.

pronounced, and for one-fourth the frequency, the blows will have still less effect. Hence, as the speed of rotation of S is increased from a very low to a very high value, there will be several speeds at which vibration will be aided to a greater or less degree.

If the intensity of the hammer blows is varied, the vibration of the rod will be affected accordingly. Every elastic body has a natural period of vibration, *i.e.*, time per cycle of movement which depends upon its mass, moment of inertia (*i.e.*, dimensions), and stiffness. If these characteristics are changed, in the case of the rods in Fig. 7-1, the speeds of S at which synchronous vibration will occur will also be changed. However, blows occurring at frequencies other than those producing synchronous vibration will produce "forced" vibrations, but these forced vibrations are small compared with the ones produced at the "critical" speeds.

If the support for the vibrating member is not absolutely fixed and rigid, it will also vibrate. Thus the rods in Fig. 7-1 can produce vibrations in their supports.

By applying the foregoing principles to airplane engines, the hammer blows correspond to the varying gas and inertia forces and the crankshaft corresponds to the vibrating rod. The crankshaft is more complex than a simple rod, and its natural periods of vibration are harder to predict, but the basic idea is the same. In other words, at certain speeds, the varying forces occur at such a rate that vibration is greatly amplified. It is at these speeds that the engine is said to be "rough," and if they occur at the desired or usual speeds of operation, the engine is unsatisfactory. If the engine is operated for prolonged periods at a speed at which synchronous vibration occurs, some of the parts, usually the crankshaft, may fail owing to the increasing amplitude of the vibrations deflecting the parts beyond their fatigue strength. Thus a crankshaft can fail structurally even though it is many times stronger than the elementary formulas of mechanics would indicate as necessary.

7-2. Engine Balance.—In approaching the problem of deciding upon proper dimensions for the crankshaft, it is advisable first to consider the major causes of vibration. They are as follows:

1. Variation in engine torque.
2. Flexibility of the crankshaft in torsion.
3. Unbalanced rotating parts.*
4. Unbalanced reciprocating parts.*

In considering these items, it is important to distinguish between (*a*) vibration of the engine structure as a whole and (*b*) vibration of individual parts of the structure.

7-3. Variation in Engine Torque.—Variation in engine torque (Chap. 4) causes a corresponding variation in torque reaction, *i.e.*, piston side thrust (Fig. 4-10). This is an example of case *a*, Par. 7-2, and tends to rock the engine in the plane of rotation of the crank arms, the magnitude of the effect being largely

* See footnote ‡ Table 7-1.

dependent upon the ratio of maximum to mean torque in the engine (Figs. 4-15 and 4-16). This ratio varies with cylinder arrangement, but in general, it decreases with increase in the number of cylinders. There is no practical way to counterbalance this reaction-torque vibration, but rubber mountings and other devices are frequently incorporated to reduce the transmission of the rocking or vibration to the vehicle in which the engine is mounted.

7-4. Flexibility of the Crankshaft in Torsion.—At the other end of the connecting rod, the variation in torque sets up torsional oscillations in the crankshaft (item 2, Par. 7-2), which, owing to the large moment of inertia of the propeller, act in a manner similar to that of the supported rod in Fig. 7-1 (*B*). When these varying torque impulses occur at a frequency corresponding to the natural torsional frequency of the crankshaft, serious torsional vibration can occur if means for damping the oscillations are not provided. Torsional-vibration dampers as used on automotive engines[3] consist usually of friction disks attached to the end of the crankshaft and so mounted that they rotate with the shaft, but their inertia effect is such that they slide when torsional vibration occurs. The resulting friction between the disks and their supporting collar on the shaft tends to damp the torsional vibration quickly.

In aircraft engines, where the useful speed range is much less than in automotive power plants, it is usually possible to avoid the most troublesome critical torsional speeds by designing the shaft so that it does not have a natural period of severe vibration within the desired operating range. From detailed dimensions and other data, it is possible to predict with reasonable accuracy the shaft speed at which serious torsional vibrations will occur, but the calculations are long and somewhat complicated.[15,20] However, a crankshaft has many characteristics in common with the torsional pendulum, and a knowledge of the factors contributing to the natural period of such a pendulum will aid in the intelligent selection of shaft dimensions.

The expression for the time for one complete oscillation or cycle of a torsion pendulum is[8]

$$t = 2\pi \sqrt{\frac{Wk^2\theta}{Tg}} \qquad (7\text{-}1)$$

where t = time per cycle, sec.

W = weight of the mass at the end of the pendulum (corresponds approximately* to the crank arm and counterweight masses in an airplane engine), lb.

k = radius of gyration (of the crank arm and counterweight masses approximately*).

θ = angular displacement of the weight W from the neutral or static position at which it is held by the rod.

T = torque exerted by the pendulum rod on the weight W when it is displaced through the angle θ.

g = acceleration of gravity.

The twisting moment or torque in the pendulum rod is

$$T = S_s \frac{J}{r}$$

where S_s = torsional stress in the rod.

J = polar moment of inertia of the rod.

r = radial distance to the outer fiber of the rod.

But

$$S_s = \frac{E_s r \theta}{L}$$

where E_s = modulus of rigidity or modulus of elasticity in torsion (= about 12,000,000 lb. per sq. in. for steel).

L = length of the pendulum rod.

r and θ are as above.

Hence

$$T = \frac{E_s J \theta}{L}$$

and substituting this in Eq. (7-1)

$$t = 2\pi \sqrt{\frac{Wk^2L}{E_s J g}} \tag{7-2}$$

The frequency of vibration is the reciprocal of the time, hence

* The propeller has such a high inertia that it is approximately the equivalent of the rigid support in Fig. 7-1 (*B*). This simplyfying assumption suffices for preliminary considerations only, however.

$$f = \frac{1}{t} = \frac{1}{2\pi}\sqrt{\frac{E_sJg}{Wk^2L}} \tag{7-3}$$

where f = number of vibrations per second and all other symbols are as above.

If the frequency is expressed in vibrations per minute,

$$N = 60f = \frac{30}{\pi}\sqrt{\frac{E_sJg}{Wk^2L}} \tag{7-4}$$

where N = number of vibrations per minute and all other symbols are as above.

For steel, if g is in inches per second per second, Eq. (7-4) may be expressed as

$$N = 647{,}500\sqrt{\frac{J}{Wk^2L}} \tag{7-5}$$

where N = number of vibrations per minute.
J = polar moment of inertia of the pendulum shaft or rod, in.^{4}
L = length of the shaft or rod, in.
W = weight of the mass corresponding to the crank arms and counterweights, lb.
k = radius of gyration of the weight W, in.

The crankshaft is considerably more complex than the torsional pendulum, but it acts in much the same way. Hence its natural frequency of torsional vibration varies (*a*) as the square root of the polar moment of inertia of the shaft section, (*b*) inversely as the square root of the weight of crank arms and counterweights, (*c*) inversely as the radius of gyration of its crank arms and counterweights, and (*d*) inversely as the square root of its length.

Referring again to Fig. 7-1 (*B*), if cam *S* was rotated at such a speed that a blow was struck by the hammer every time the rod arm passed point *C* in direction *N*, severe vibration would result. If the cam were slowed down to where the hammer struck a blow every other time the arm passed *C* in direction *N*, vibration would again occur, but it would be less severe because the natural damping forces would have more time to act between blows. Similarly, as the speed of cam *S* was further decreased, still less severe vibrating periods would be encountered. The net result would be that within a sufficiently large range of cam speed there would be a series of speeds at which vibration would occur, and at each succeeding lower critical speed the vibration would

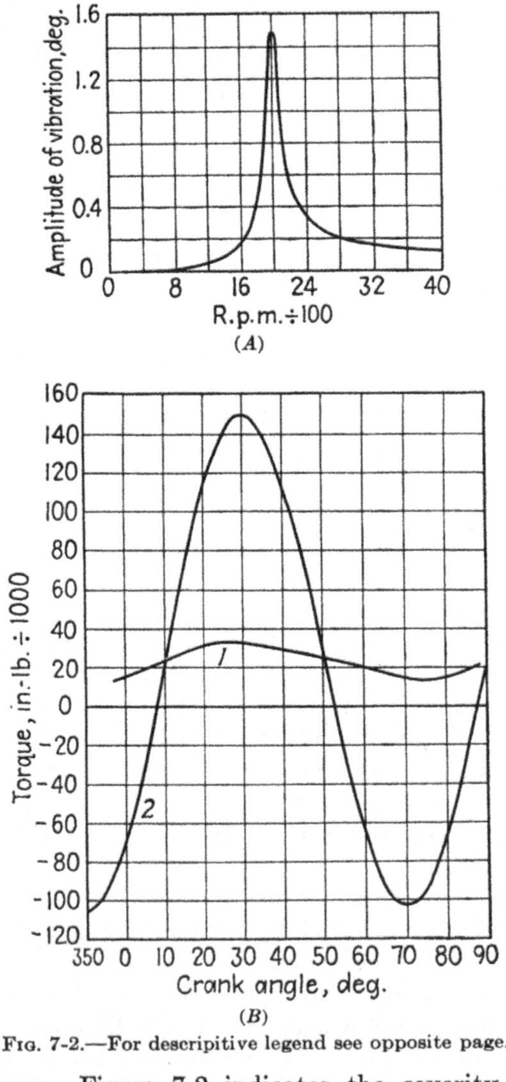

FIG. 7-2.—For descripitive legend see opposite page.

be less severe. Figure 7-2 indicates the severity of stresses (proportional to ratio of actual torque to gas torque) that have been found to exist in radial engines.

In crankshafts, study of these critical speeds is termed *harmonic analysis,* because it can be demonstrated mathematically that they can be represented by a constant mean value and a series of harmonics or sine-curve functions, *i.e.*, a Fourier series. Analysis by this means[3,10] is beyond consideration here, but it should be noted that the most severe vibrations occur at the higher speeds, and that by suitable design the worst critical speeds can usually be made to occur above the maximum speed at which

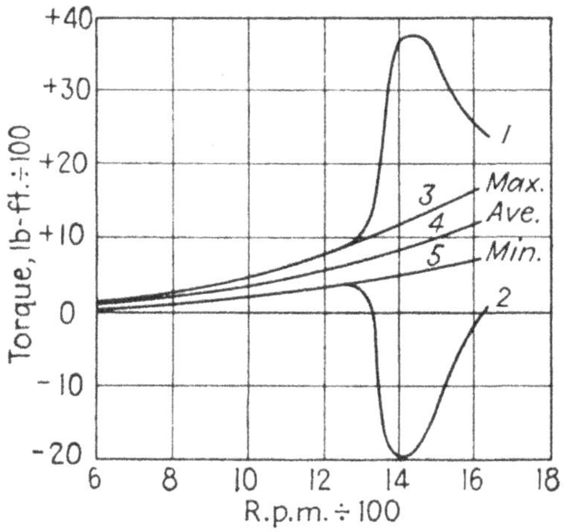

(*C*)

FIG. 7-2.—Effect on crankshaft torsional stresses due to resonant or critical speeds. (*A*) Effect of resonance on the amplitude of torsional vibration in a nine-cylinder radial engine. (*B*) (1) Torque acting on crankshaft and (2) torque in crankshaft at a resonant speed in a nine-cylinder radial engine. (*From S. A. E. Jour., Vol.* 38, *No.* 3). (*C*) (1), (2) Observed actual torque and (3), (4) and (5) torque due to gas pressure in a radial-engine crankshaft. (*From Judge, "Automobile and Aircraft Engines."*)

the engine is to be operated. Inspection of Eq. (7-5) indicates that the crankshaft may be designed against severe torsional vibration by increasing the polar moment of inertia of the shaft cross section, decreasing the weight of crank arms, eliminating or reducing the size of counterweights, reducing the radius of gyration of crank arms and counterweights, and using as short a crankshaft length as possible. Since J/L is a measure of the stiffness of the crankshaft and Wk^2 is the moment of inertia of crank arms and counterweights, it follows that the crankshaft should be as stiff as possible and have a low moment of inertia of its crank arms and counterweights. Stiffness may be increased without increase in weight by using hollow crankpins and journals. Chamfering and rounding of crank arms (Fig. 5-15) and elimination of counterweights wherever possible will contribute to reducing the moment of inertia.

An alternate method of reducing torsional vibration in radial engines that has been found to be very effective is the pendulum type of vibration absorber.[11] It can be demonstrated that by mounting a pendulous weight of suitable proportions opposite the crank arms, practically complete damping of torsional vibration can be had. Since radial engines require counterweights for proper balance, and since it is possible also to use these counterweights for the pendulum mass, practically complete elimination of torsional vibration may be attained without adding any dead weight to the engine. The device has so far found its greatest application in very high-powered engines where the crankshaft is already highly stressed and any additional vibration stresses become very critical.

7-5. Types of Crankshaft Balance.—Before considering the effects of unbalanced rotating and unbalanced reciprocating parts (items 3 and 4, Par. 7-2), it is advisable to fix clearly in mind the three types of crankshaft balance. They are[2]

1. Static balance.
2. Dynamic balance.
3. Deflection balance.

Considering these items in order, *static balance* is that condition in the crankshaft in which the algebraic sum of all moments of radial forces about the axis of support is zero. An example of this condition is illustrated in Fig. 7-3 (*A*) in which the shaft is supported by the bearings *M* and *N*. Obviously the shaft will remain in any position since $2W \times R = W \times 2R$, and the lever arms decrease in the same proportion for any angular position of the shaft to the condition shown in the figure.

The conditions for *dynamic balance* require that the algebraic sum of all moments of radial forces about an axis perpendicular to the axis of support must be equal to zero. Shaft (*A*), Fig. 7-3, will not meet this condition, for, during rotation, the centrifugal forces on the weights will produce a rotating couple that can be balanced only by reaction forces at the bearing supports *M* and *N*. In Fig. 7-3 (*B*), however, the shaft is in dynamic balance, as the taking of moments about either *M* or *N* will show. Thus taking moments about M,

$$5F_N = 2F(2 + 1) - (2F + 4F) = 0$$

To attain dynamic balance, a shaft must also be in static balance.

Deflection balance requires that there be no deflection of the shaft due to the centrifugal loads produced by the weights. Shaft (B), Fig. 7-3, when rotating, will be deflected as shown by the dashed curve (exaggerated). Obviously, deflection balance can be attained only when all forces are balanced by equal forces in their respective planes of rotation. Extreme deflection unbalance tends to produce high localized bearing pressures even though the shaft is dynamically balanced. This is due to distortion of the journals.

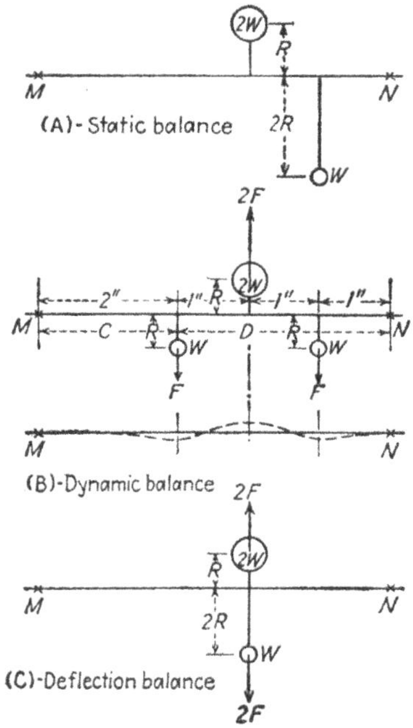

FIG. 7-3.—Three types of balance in engine crankshafts. (*Adapted from Huebotter, "Mechanics of the Gasoline Engine."*)

Static balance in an aircraft-engine crankshaft is necessary to satisfactory operation as, obviously, its absence would produce severe shaking (vibration) that could not be tolerated. Engines having only one crank throw, *i.e.*, single-bank radials, must have counterbalanced crankshafts to attain static balance. Multithrow crankshafts in practically all cases have the cranks symmetrically spaced about the crankshaft axis. Hence they are in inherent static balance if properly constructed.

Dynamic balance requires the use of counterweights on multi-throw as well as single-throw crankshafts. Nearly all modern automotive and aircraft engines are dynamically balanced, but counterweights on in-line and V-aircraft engines are not altogether desirable because (a) the added weight of counterbalancing is undesirable and (b) the added rotating mass contributes to lowering the severe torsional vibration periods into the useful speed range of the engine [Eq. (7-5)].

Complete deflection balance is impossible in conventional types of engines as the counterweights would have to be in the plane of the connecting rods and hence would not permit the necessary mechanical clearance. However, deflection balance can be closely approached by placing the counterweights as near the plane of the connecting rods as proper mechanical clearance will permit.

7-6. Unbalanced Rotating Parts.—The balance of rotating parts (item 3, Par. 7-2) is a relatively simple matter in conventional types of engines as it consists merely in providing counterweights on the opposite side of the axis of rotation to the unbalanced parts. In the case of an unbalanced crank throw, the weight or weights cannot be placed directly opposite the center of the crankpin, but a weight can be attached to each crank arm and so near the plane of rotation of the crankpin center that deflection unbalance is quite small.

The centrifugal force due to an unbalanced rotating mass [Eq. (5-1)], is a function of the weight of the unbalanced mass and the distance from the axis of rotation to its center of gravity. Obviously, a counterweight or counterweights located on the opposite side of the axis of rotation from the rotating mass to be balanced and having a weight and moment arm such that the product will equal the product of the unbalanced weight and its moment arm will balance the system. In aircraft engines, it is desirable to place the counterweights as far from the center of rotation as crankcase clearance and other limitations will permit as this will avoid the use of unnecessary dead weight in the engine.

7-7. Unbalanced Reciprocating Parts.—In the development of the expression for reciprocating inertia force [Eq. (4-8)], it was shown that

$$F_R = 0.0000284N^2WR(\cos \theta + Z \cos 2\theta) \qquad (7\text{-}6)$$

where F_R = reciprocating inertia force, lb.
N = r.p.m. of the crankshaft.
W = weight of reciprocating parts, lb.
R = crank radius, in.
θ = the angular displacement of the crankshaft from the dead center, deg.
Z = R/L.
L = the center-to-center length of the connecting rod, in.

This force acts along the cylinder center line, and its reaction tends to (a) distort the crankshaft and (b) move the supports for the crankshaft, *i.e.*, shake the engine. The amount of this shaking or vibration varies with the number and arrangement of the cylinders, and when favorable conditions exist, *i.e.*, these shaking forces vary at a frequency at or near the natural period of vibration of the engine or its supports, serious vibration can occur. Hence, it is advisable to investigate these forces with a view to balancing them by suitable counterweights or other means.

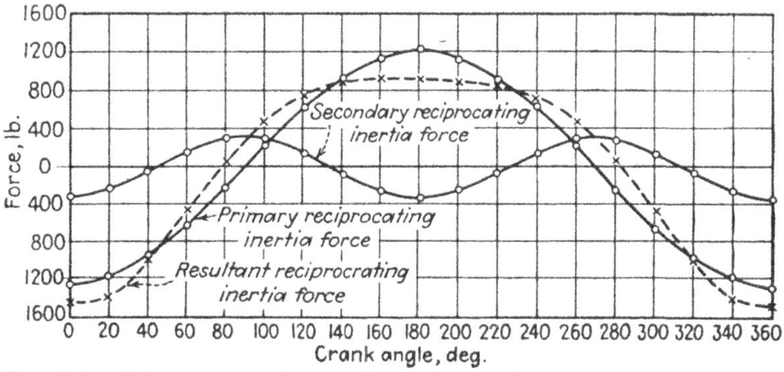

FIG. 7-4.—Primary, secondary, and resultant reciprocating inertia forces for a 4½- by 5⅜-in. 2,000-r.p.m., single-cylinder engine having a weight of reciprocating parts equal to 4 lb.

For this purpose, Eq. (7-6) may be divided into two terms, *i.e.*,

$$F_R = 0.0000284N^2WR \cos \theta + 0.0000284N^2WRZ \cos 2\theta \quad (7\text{-}7)$$

The first term in this expression is usually called the *primary* reciprocating inertia force, and the second is called the *secondary* reciprocating inertia force. They are also called the first and second harmonics since, if plotted against crank angle, they will form cosine curves* (Fig. 7-4).

Inspection of the primary force in Eq. (7-7) and comparison with Eq. (5-1) shows that the two are the same except for the value of W and the factor $\cos \theta$. This suggests a logical approach to balancing the primary reciprocating inertia force by attaching a counterweight opposite the crankpin such that its centrifugal force will balance the primary reciprocating inertia force. Such a procedure makes possible, for a single-cylinder engine, complete

* A cosine curve is the same as a sine curve displaced at an angle of 90 deg.

balance of the primary force at the 0- and 180-deg. positions where cos θ equals 1, but at the 90- and 270-deg. positions the primary reciprocating inertia force is zero

$$(\cos 90 \text{ and } \cos 270 = 0),$$

whereas the centrifugal force on the counterweight is the same as for the 0- and 180-deg. positions. Obviously, completely balancing the maximum primary reciprocating inertia force by a rotating counterweight on the crankshaft merely shifts the unbalance from the plane of the cylinder axis to a plane normal to this axis. This is illustrated in Fig. 7-5 where F_{RP} represents the primary reciprocating inertia force which always acts along the center line of the cylinder, F_c is the centrifugal force

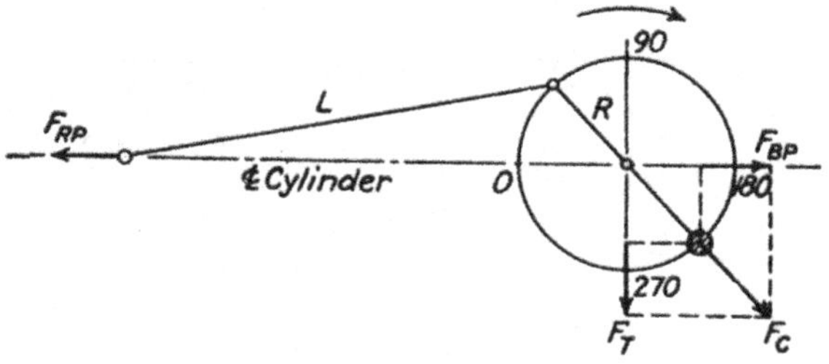

FIG. 7-5.—Counterweight method of balancing the primary reciprocating inertia force in a single-cylinder engine.

on the primary counterbalancing weight, F_{BP} is the component of F_c which opposes, *i.e.*, balances, F_{RP}, and F_T is the unbalanced transverse force component of F_c. Use of a counterweight that will balance one-half of the primary reciprocating inertia force is about the best compromise, as this will give forces parallel and normal to the cylinder axis of the single-cylinder engine which are each one-half of the magnitude of the initial primary reciprocating unbalance.

Several other means have been tried for balancing the primary force in a single-cylinder engine, but they require more or less modification of the simple and conventional crank chain and are not commonly used in aircraft engines. For a discussion of these methods, the student should consult references 1 and 9.

The secondary reciprocating inertia forces, as is indicated in Eq. (7-7) and Fig. 7-4, vary at *twice* crankshaft speed. Hence, they cannot be balanced by a counterweight rigidly attached to the crankshaft. Fortunately, the secondary unbalance is much less than the primary unbalance, the relative magnitude depending upon the L/R ratio. As a rule, no attempt is made to balance the secondary forces in a single-cylinder engine, but the Lanchester balancer[3] has been used to some extent on four-cylinder in-line automotive engines.

Equation (7-6) does not precisely represent the reciprocating inertia forces, as in its derivation (Pars. 4-2 and 4-5) the smaller

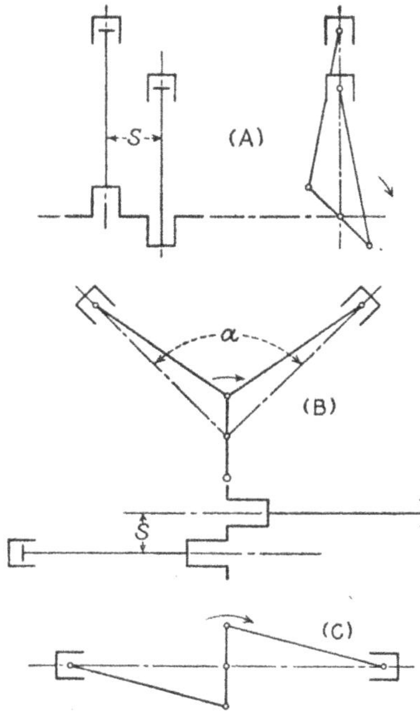

Fig. 7-6.—Various arrangements of two-cylinder engines.

terms were neglected. These smaller terms are functions of θ and R/L, and each succeeding term may be represented by a cosine curve of higher frequency and less magnitude. These higher harmonics produce minor shaking forces or vibrations, but due to their small magnitude, they are generally neglected.

7-8. Reciprocating Balance in Multicylinder Engines. 1. *The Two-cylinder In-line Engine.*—In a multicylinder engine, each cylinder when considered separately will produce shaking forces in the same way as in a single-cylinder engine, but by suitably arranging the different cylinders and the angular relations of the crank arms, part or all of the unbalance in the engine as a whole may be eliminated. In the two-cylinder in-line engine [Fig. 7-6 (*A*)], the crank arms are at an angular relation of 180 deg.; hence, when one piston is moving down the other is moving up. This displaces the primary reciprocating forces at 180 deg. and, as will be seen in Fig. 7-7, the resultant primary unbalance

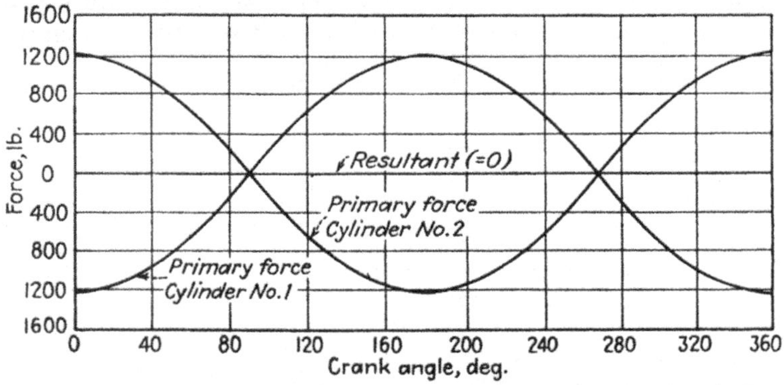

FIG. 7-7.—Construction showing that the primary reciprocating inertia forces are balanced in a two-cylinder engine having a crank arrangement as in Fig. 7-6 (*A*). 4½- by 5⅜-in. cylinder, 2,000 r.p.m., 4 lb. reciprocating weight.

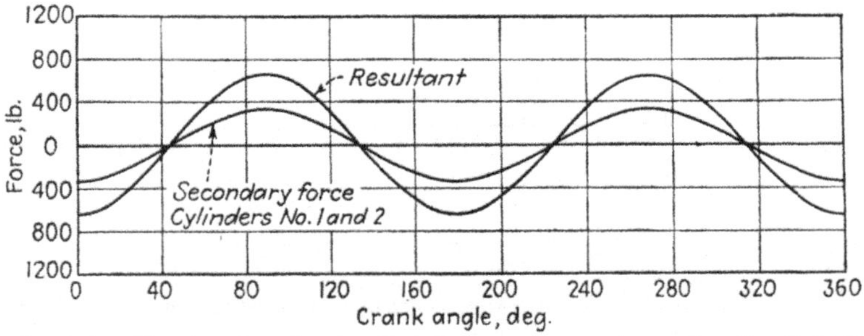

FIG. 7-8.—Construction showing that the secondary reciprocating inertia forces are not balanced in a two-cylinder engine having a crank arrangement as in Fig. 7-6 (*A*). 4½- by 5⅜-in. cylinder, 2,000 r.p.m., 4 lb. reciprocating weight. Amount of maximum unbalance is 2 × maximum unbalance for one cylinder.

for the engine is zero. This means that the engine will not tend to move up and down in the plane of the cylinders because of primary unbalance. However, each cylinder taken separately still has primary unbalance, and as these unbalanced forces in the different cylinders do not act along the same line, they will produce a rocking couple which tends to oscillate the engine about an axis normal to the plane of the cylinders. . The magnitude of this couple depends upon the magnitude of the primary unbalance in each individual cylinder and the distance between the center lines of the cylinders, *i.e.*, distance S, Fig. 7-6 (A). Obviously an engine of this type should have its cylinder center lines as close together as other limitations will permit.

The secondary reciprocating inertia forces of the engine in Fig. 7-6 (A) will also be displaced 180 deg., but upon combining them graphically (Fig. 7-8) it is seen that they do not cancel.

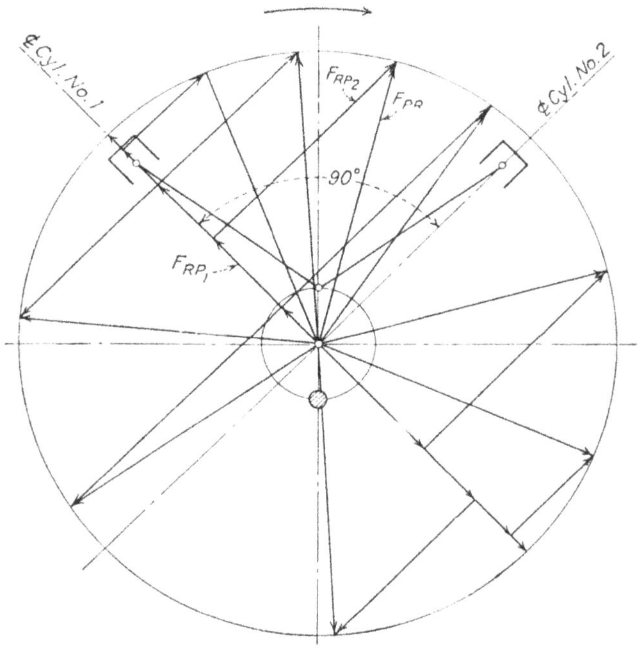

Fig. 7-9.—Construction showing that the primary reciprocating inertia forces give a constant radial rotating force in a 90-deg., single-crank, two-cylinder V-engine, Fig. 7-6 (B). $4\frac{1}{2}$- by $5\frac{3}{8}$-in. cylinder, 2,000 r.p.m., 4 lb. reciprocating weight per cylinder.

Hence, the secondary forces in this type of engine are not balanced, and as will be seen from the figure, the maximum secondary unbalance is twice that for one of the cylinders.

2. *The Two-cylinder Single-crank V-engine.*—By arranging the engine as in Fig. 7-6 (B), the unbalanced reciprocating forces due to each cylinder act at an angle α to one another. The resultant unbalance may be determined by adding them vectorially (Figs. 7-9 and 7-10). An example of primary unbalance determination is shown in Fig. 7-9 in which it is seen that the resultant

primary unbalance is a constant radial force that rotates with the crankshaft. For a 90-deg. V-engine, this primary unbalance is equal to the maximum primary unbalance for one of the cylinders taken separately. Obviously, since the primary unbalance of the engine in Fig. 7-6 (*B*) is constant and rotates with the crankshaft, it may be balanced by a suitable counterweight attached opposite the crank arm.

An example of secondary unbalance determination for the engine in Fig. 7-6 (*B*) is shown in Fig. 7-10. The resultant is a

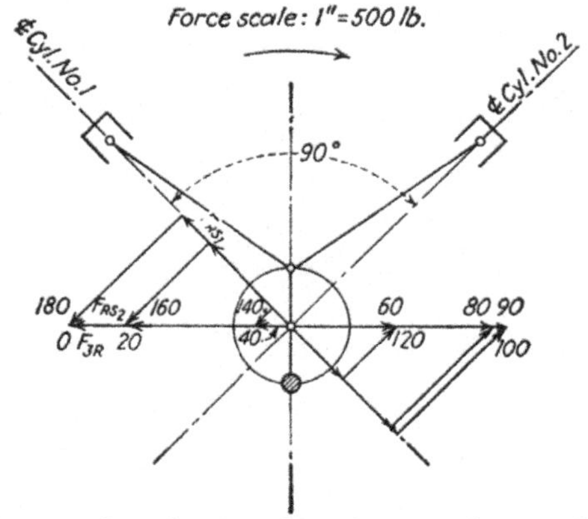

FIG. 7-10.—Construction showing that the secondary reciprocating inertia forces in a 90-deg, single-crank, two-cylinder V-engine, Fig. 7-6 (*B*) combine to form a transverse resultant force that varies at twice crankshaft speed. 4½- by 5⅜-in. cylinder, 2,000 r.p.m., 4 lb. reciprocating weight per cylinder.

transverse force which varies at twice crankshaft speed and has a value for a 90-deg. V-engine of

$$F_{SR} = \sqrt{F_{RS_1}^2 + F_{RS_2}^2}$$

where F_{SR} = secondary resultant force.

F_{RS_1} = secondary force in cylinder 1.

F_{RS_2} = secondary force in cylinder 2.

3. *The Two-cylinder Opposed Engine.*—In this type of engine, the pistons always move in opposite directions; hence the primary forces in one cylinder cancel these forces in the other cylinder. The secondary forces also cancel as will be seen from Fig. 7-11, but for the conventional crankshaft-cylinder arrangement shown in Fig. 7-6 (*C*), there will be a rocking couple which tends to oscillate the engine in the plane of the cylinder center lines. This

couple is proportional to the distance S between the cylinder center lines; hence, as in the case of the two-cylinder in-line engine [Fig. 7-6 (A)], it is desirable to have these center lines as close together as possible. When this distance S is small, the two-cylinder opposed engine is a rather satisfactory type for small inexpensive light planes.

4. *The Three-cylinder In-line Engine.*—In this type of engine (Fig. 7-12), the crank arms are at an angle of 120 deg. The

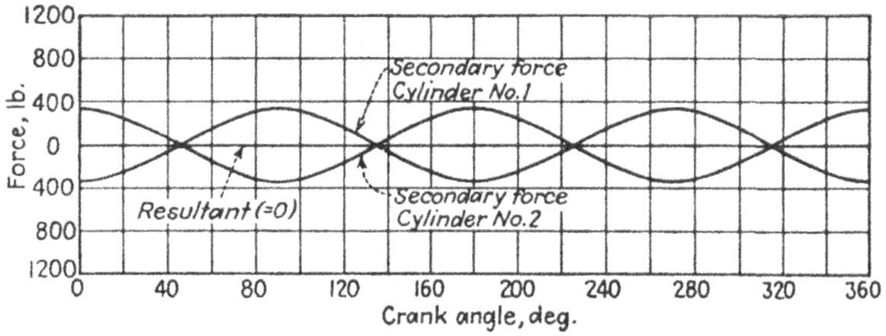

FIG. 7-11.—Construction showing that the secondary reciprocating inertia forces are balanced in a two-cylinder engine having a crank arrangement as in Fig. 7-6 (C). $4\frac{1}{2}$- by $5\frac{3}{8}$-in. cylinder, 2,000 r.p.m., 4 lb. reciprocating weight per cylinder.

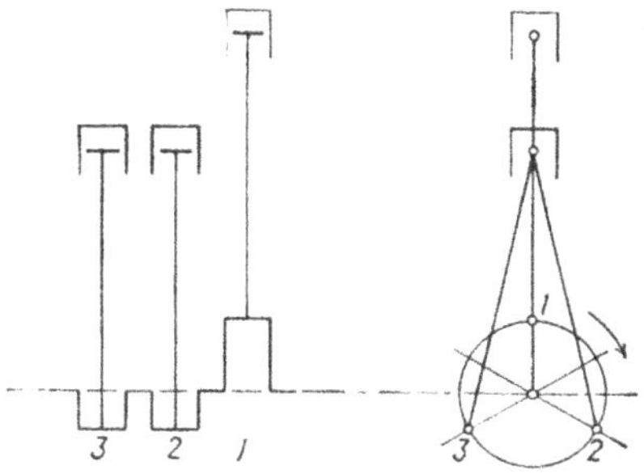

FIG. 7-12.—Crank arm arrangement for the three-cylinder in-line engine.

cylinder axes are all in the same plane, however, and a graphical determination of the degree of unbalance may be made by shifting the primary and secondary reciprocating inertia force curves through an angle of 120 deg. and then combining them. This has been done (Figs. 7-13 and 7-14), and from these diagrams it is seen that both the primary and secondary reciprocating inertia force resultants are zero. Hence, the three-cylinder in-line engine is balanced, but rocking couples still exist which tend to oscillate the engine about an axis normal to the plane of the cylinders.

The magnitude of these rocking couples can be reduced by placing the center lines of the cylinders as close together as possible.

5. *The Four-cylinder In-line Engine.*—This arrangement usually consists of two two-cylinder engines [Fig. 7-6 (*A*)], with the crank arms arranged so that the two inner cranks are parallel

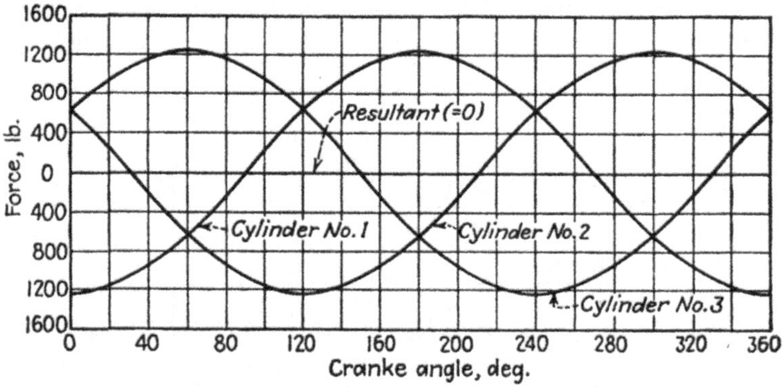

FIG. 7-13.—Construction showing that the primary reciprocating inertia forces in a three-cylinder in-line engine arranged as in Fig. 7-12 are balanced. 4½- by 5⅜-in. cylinder, 2,000 r.p.m., 4 lb. reciprocating weight per cylinder.

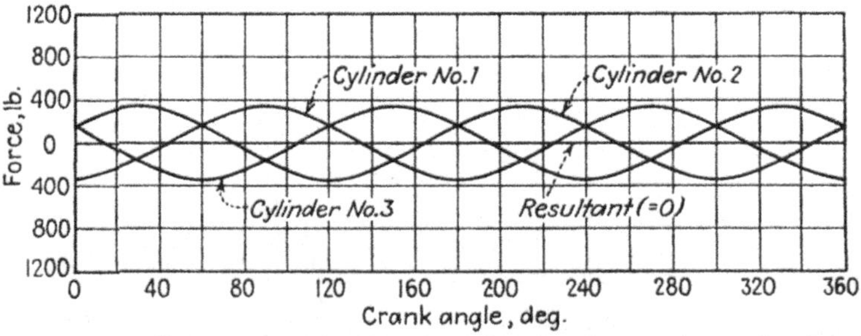

FIG. 7-14.—Construction showing that the secondary reciprocating inertia forces are balanced in a three-cylinder in-line engine arranged as in Fig. 7-12. 4½- by 5⅜-in. cylinder, 2,000 r.p.m., 4 lb. reciprocating weight per cylinder.

and extend in the same direction and the two outer cranks are parallel and at an angle of 180 deg. to the inner cranks. Such an arrangement will have the same balance characteristics as the two-cylinder in-line engine, and in addition the rocking couples will cancel. It should be noted, however, that the unbalanced secondary forces in the four-cylinder in-line engine will sum up to twice the magnitude of the two-cylinder engine of the same size cylinders, reciprocating weights, etc.

This unbalanced secondary force is of sufficient magnitude in the four-cylinder in-line engine to cause appreciable vibration, especially at synchronous speeds, and considerable attention has been directed toward effectively counteracting it. For this purpose, the Lanchester balancer or antivibrator has proved effective, but for aircraft engines, the added weight is objectionable.

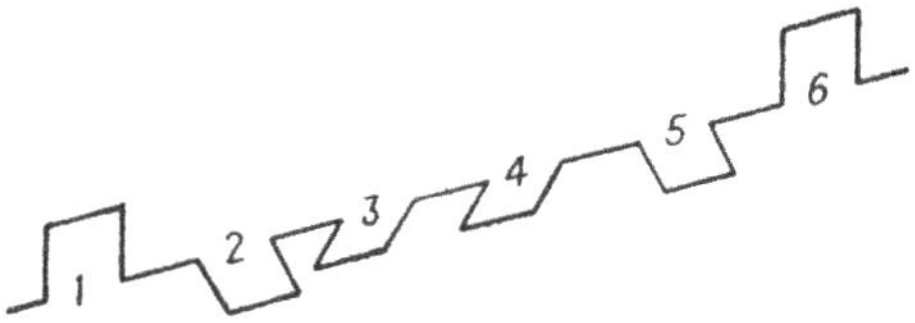

FIG. 7-15.—Conventional crank-arm arrangement in the six-cylinder in-line engine.

6. *The Six-cylinder In-line Engine.*—The usual crank arrangement for this type of engine (Fig. 7-15) is such that it consists of two three-cylinder engine crankshafts arranged so that crank arms 1 and 6, 2 and 5, and 3 and 4 are parallel and extend in the same directions, respectively. Hence, the primary and secondary forces are balanced and the rocking couples cancel.* The six-cylinder in-line engine is smooth and quite free from vibration due to reciprocating parts.†

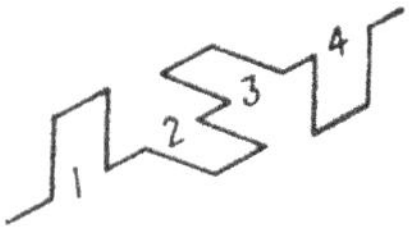

FIG. 7-16.—Conventional crank-arm arrangement in the eight-cylinder V-engine.

Since the twelve-cylinder V-engine is essentially two six-cylinder in-line engines attached to the same crankshaft, it is also inherently free from vibration due to reciprocating parts.

7. *The Eight-cylinder V-engine.*—Eight-cylinder V-engine crankshafts may be arranged either in one plane (the same as four-cylinder in-line engine shafts) or preferably in two planes (Fig. 7-16). With the one-plane arrangement, forging and other construction problems are simplified, but the secondary forces are not balanced. The magnitude of the unbalance depends upon the angle of the V,[3] the minimum severity being at an angle of 60 deg., but this gives unequal firing intervals.

It was noted in Par. 7-8, item 2, that the primary forces in a two-cylinder V-engine could be balanced by means of a rotating

* See footnote ‡ of Table 7-1.

† The sixth harmonics are not balanced, but the magnitude of the unbalance is so small that it is negligible, see reference 9.

weight attached opposite the crank arm, and since the eight-cylinder V-engine may be considered as four two-cylinder V-engines, it is apparent that primary reciprocating inertia forces may be balanced by means of suitable counterweights.

From Fig. 7-10, it was observed that the secondary forces in a 90-deg. two-cylinder V-engine combine to form a transverse

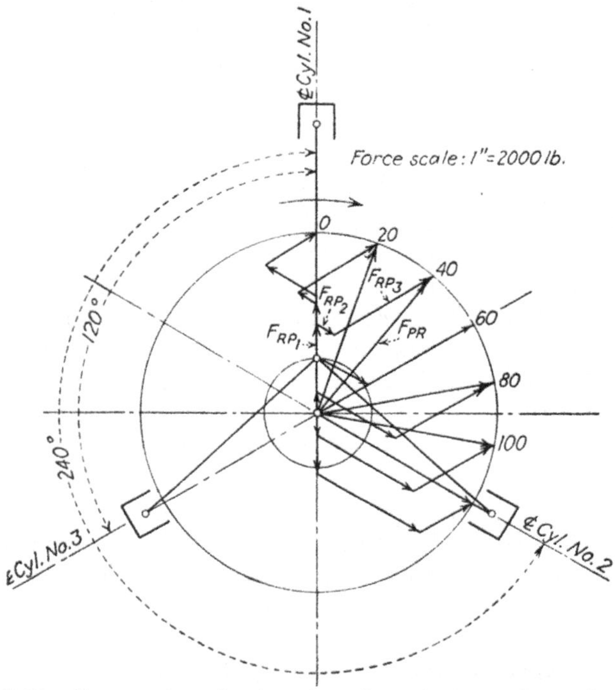

FIG. 7-17.—Construction showing that the primary reciprocating inertia forces in a three-cylinder single-crank radial engine have a constant resultant that acts outward along the crank-arm center line. 4½- by 5⅜-in. cylinders, 2,000 r.p.m., 4 lb. reciprocating weight per cylinder.

force having a period twice crankshaft speed. With the crank arrangement shown in Fig. 7-16, this force at crank 1 is opposite in direction to the secondary force at crank 4. Therefore the secondary forces at these two cranks will cancel. Similarly, the secondary forces at cranks 2 and 3 will cancel, and the engine will have inherent secondary reciprocating balance. Also the rocking couples will cancel; hence the eight-cylinder V-engine having a crank arrangement as in Fig. 7-16 may be made relatively free from vibration due to reciprocating inertia forces.

8. *Radial Engines.*—To permit even firing intervals, single-bank radial engines are built with an odd number of cylinders equally spaced around the crankshaft. The number of cylinders may be 3, 5, 7, or 9, with the latter three numbers much the more common.

Balance conditions in radial engines may be studied graphically by a procedure similar to that for the two-cylinder V-engine, but for engines of more than three cylinders, the construction becomes somewhat tedious, and analytical methods are preferable.

Figure 7-17 shows the construction procedure for a three-cylinder single-crank radial. In this figure, it is seen that the primary reciprocating inertia forces combine to form a constant rotating force that acts outwardly along the crank arm. Hence, the unbalanced primary force may be balanced by a suitable counterweight placed opposite the crank arm. Data for constructing Fig. 7-17 were taken from Fig. 7-4, and from the values shown, it is seen that the magnitude of the rotating primary unbalance is 1.5 times the maximum primary unbalance for one cylinder or one-half the maximum for all three cylinders.

This conclusion may also be reached by analytical methods,[1] and the unbalance expressed as

$$F_{PR} = \tfrac{3}{2}(0.0000284N^2WR) \qquad (7\text{-}8)$$

where F_{PR} = resultant primary unbalanced reciprocating inertia force, lb.
N = r.p.m. of the crankshaft.
W = reciprocating weight per cylinder, lb.
R = crank radius, in.

It may also be shown by analytical methods[1] that for any single-crank radial engine having an odd number of cylinders

$$F_{PR} = \frac{n}{2}(0.0000284N^2WR) \qquad (7\text{-}9)$$

where n is the number of cylinders and all the other terms are the same as in Eq. (7-8).

By procedure similar to that used in Fig. 7-17 or by analytical methods, it may be shown that the secondary reciprocating inertia forces in a three-cylinder radial engine are one-half balanced and that the resultant secondary force rotates in the opposite direction to the crankshaft and at twice crankshaft speed. Obviously, it is impractical to balance the secondary reciprocating inertia forces in a three-cylinder single-crank radial engine. The secondary forces in single-crank radials of more than three cylinders are balanced, however, and this already has been demonstrated for a five-cylinder radial (see Fig. 5-8).

9. *Summary of Reciprocating Balance.*—For convenient reference, and to check the reciprocating unbalance quickly, Table 7-1 may be used. It should be borne in mind, however, that even though a multicylinder engine is inherently balanced as a whole, its reciprocating parts may be individually unbalanced and produce stresses within the engine. For instance, a six-cylinder in-line engine is inherently balanced,* but individual cylinders are not, and the reciprocating parts in these cylinders cause stresses in the crankshaft and other parts even though there is no appreciable tendency for the engine as a whole to vibrate. Hence, counterweights greater than necessary for rotating unbalance help to reduce deflection unbalance even in an engine having inherent reciprocating balance. However, for aircraft engines, the added weight is objectionable.

7-9. Counterbalancing.—Counterweights are attached to engine crankshafts to

1. Attain static balance of rotating parts. In in-line and V-engines, the several crank arms are usually placed symmetrically about the shaft axis so that counterweights are unnecessary for static balance. However, in single-crank engines such as radials, counterweights are necessary to attain static balance.

2. Obtain dynamic balance. This condition is desirable as it aids in reducing bearing pressures, but, for aircraft engines, the added weight is undesirable.

3. Improve deflection balance. Placing counterweights as nearly opposite the crank-arm center of gravity as possible reduces the distortion of the shaft due to centrifugal forces on the crank arm, and this in turn reduces main bearing pressures.

4. Balance one-half of the primary reciprocating inertia force.

In aircraft engines, where weight is at a premium, it is desirable to have the counterweights as small as possible, and since centrifugal force is proportional to the product of weight and the distance from the center of gravity of the weight to the axis of rotation, it is desirable to place the weight as far as possible from

* See footnote ‡ of Table 7-1.

the center of the crankshaft [see Eq. (5-1)]. The limits on this radial distance are, of course, the clearances in the crankcase, and, as an increase in crankcase dimensions will increase the weight of that part, the proper length of counterweight arms

TABLE 7-1.—RECIPROCATING BALANCE IN CONVENTIONAL ENGINES

Cylinders		Angle between cranks	Primary forces	Secondary forces	Rocking couples	Feasible to balance
No.	Arrangement					
1	......	1 crank	Unbalanced	Unbalanced	None	One-half of primary force
2	In-line	180 deg.	Balanced	Unbalanced	Unbalanced	
2	Opposed	180 deg.	Balanced	Balanced	Unbalanced	
3	In-line	120 deg.	Balanced	Balanced	Unbalanced	
3	Radial*	1 crank	One-half balanced†	One-half balanced†	None	Other half of primary force
4	In-line	180 deg.	Balanced	Unbalanced	Balanced	
4	Opposed	180 deg.	Balanced	Balanced	Balanced	
5	Radial*	1 crank	One-half balanced†	Balanced	None	Other half of primary force
6	In-line	120 deg.	Balanced	Balanced	Balanced‡	
7	Radial*	1 crank	One-half balanced†	Balanced	None	Other half of primary force
8	V	180 deg. 4 cranks	Balanced	Unbalanced	Balanced	
8	V	90 deg.	Balanced	Balanced	Balanced	
9	Radial*	1 crank	One-half balanced†	Balanced	None	Other half of primary force

* Radial engines have some slight secondary unbalance owing to the articulated rod construction.

† Unbalance is one-half of maximum unbalance for one cylinder times number of cylinders.

‡ In reference 12 (now out of print), Prof. Sharp has demonstrated that if the unbalance of the connecting rods is taken into account, this statement is not strictly correct.

becomes a compromise. Usual procedure is to make the radial over-all length of the counterweights about equal to the radial distance from the crankshaft axis to the outermost part of the big end of the connecting rod. The shape of the weight is then made such that its center-of-gravity distance is as large as possible.

Figures 7-18 and 7-19 show some geometric shapes that give a high WR_G product for a minimum value of weight (W) and distance to outer fiber (R).

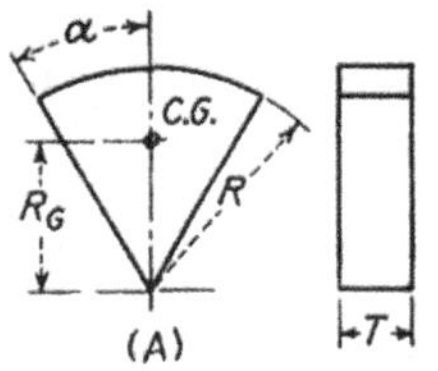

Circular Sector (A)

$$\text{Area} = 0.01745R^2\alpha, \text{ sq. in.}$$
$$\text{Weight} = 0.01745TR^2\alpha d, \text{ lb.}$$
$$R_G = 38.197\frac{R \sin \alpha}{\alpha}, \text{ in.}$$
d = density of material, lb. per cu. in.

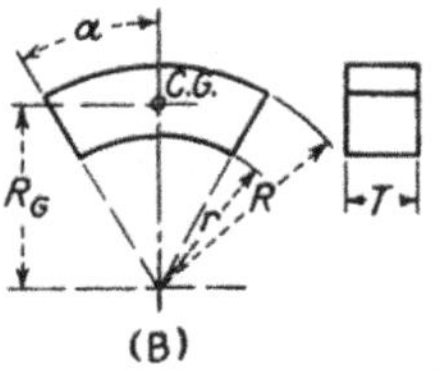

Circular Ring Sector (B)

$$\text{Area} = 0.01745(R^2 - r^2)\alpha, \text{ sq. in.}$$
$$\text{Weight} = 0.01745(R^2 - r^2)T\,d\alpha, \text{ lb.}$$
$$R_G = 38.197\frac{R^3 - r^3}{R^2 - r^2}\frac{\sin \alpha}{\alpha}, \text{ in.}$$
d = density of material, lb. per cu. in.

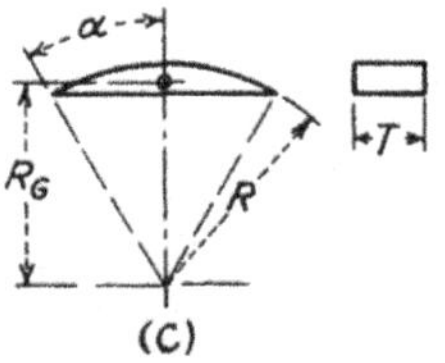

Circular Segment (C)

$$\text{Area} = R^2(0.01745\,\alpha - 0.5 \sin 2\alpha), \text{ sq. in.}$$
$$\text{Weight} = R^2(0.01745\,\alpha - 0.5 \sin 2\alpha), Td, \text{ lb.}$$
$$R_G = \frac{R \sin^3 \alpha}{0.02617\,\alpha - 0.75 \sin 2\alpha}, \text{ in.}$$
d = density of material, lb. per cu. in.

FIG. 7-18.—Counterweight shapes having a high value of weight × distance to center of gravity (WR_G) for given values of weight (W) and distance to outer fiber (R). (*From Huebotter, "Mechanics of the Gasoline Engine."*)

7-10. Example.—For an engine having the following characteristics, determine the size of counterweights necessary:

Five-cylinder single-bank radial, 4½- by 5-in. cylinders, 2,000 r.p.m., 4-lb. reciprocating weight per cylinder, 12-lb. rotating weight per crankpin, crank-arm details as shown in Fig. 7-20, ratio of connecting-rod length to crank radius = 4:1, allowable distance to outermost part of counterweight = 5.5 in.

Procedure.—From Par. 7-8, items 8 and 9, it is seen that one-half of the primary reciprocating inertia force is unbalanced and that it is feasible to balance this force by means of a rotating counterweight. From Eq. (7-9), the force acting along the crank arm due to reciprocating unbalance is

$$F_{PR} = \tfrac{5}{2}(0.0000284 \times \overline{2{,}000}^2 \times 4 \times 2.5) = 2{,}840 \text{ lb.}$$

From Eq. (5-1), the centrifugal force due to the big end of the connecting rods (rotating weight per crankpin) is

$$F_{CR} = 0.0000284 \times 12 \times \overline{2{,}000}^2 \times 2.5 = 3{,}408 \text{ lb.}$$

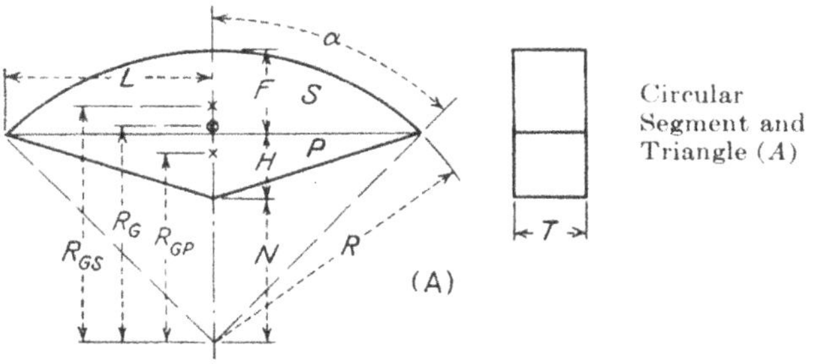

Area $= A_P + A_S = LH + R^2(0.01745\alpha - 0.5 \sin 2\alpha)$, sq. in.

$L = \sqrt{F(2R - F)}$ $\quad F = R(1 - \cos \alpha)$

Weight $= (A_P + A_S)Td$, lb. $\quad d$ = density of material, lb. per cu. in.

$$R_{GP} = N + \tfrac{2}{3}H$$

$$R_{GS} = \frac{R \sin^3 \alpha}{0.02617\alpha - 0.75 \sin 2\alpha} \qquad R_G = \frac{A_P R_{GP} + A_S R_{GS}}{A_P + A_S}$$

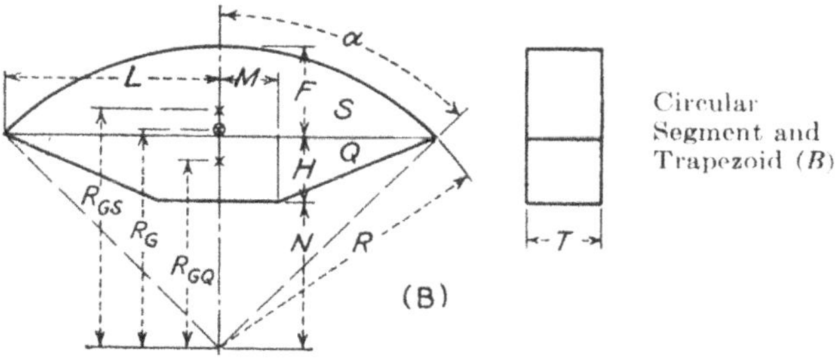

Area $= A_Q + A_S = (M + L)H + R^2(0.01745\alpha - 0.5 \sin 2\alpha)$

Weight $= (A_Q + A_S)Td$ $\quad d$ = density of material, lb. per cu. in.

$$R_{GQ} = N + \frac{H(M + 2L)}{3(M + L)} \qquad R_G = \frac{A_Q R_{GQ} + A_S R_{GS}}{A_Q + A_S}$$

$$R_{GS} = \frac{R \sin^3 \alpha}{0.02617\alpha - 0.75 \sin 2\alpha} \qquad L = \sqrt{F(2R - F)} \qquad F = R(1 - \cos \alpha)$$

FIG. 7-19.—Composite counterweight shapes.

The probable easiest way of finding the centrifugal force due to the unbalanced portion of the crank arms is to separate the arms into several simple geometric shapes, determine the values for each, and add the results.

Thus, for the crankpin the volume is

$$V_{CP} = 0.785(\overline{2.25}^2 - \overline{1.125}^2)3 = 8.93 \text{ cu. in.}$$

The density of steel may be taken as 0.28 lb. per cu. in.; hence the weight of the crankpin is

$$W_{CP} = 8.93 \times 0.28 = 2.5 \text{ lb.}$$

Since the crankpin is symmetrical about the crankpin center line, the distance from the axis of rotation to the center of gravity is 2.5 in.

From Eq. (5-1) the centrifugal force on the crankpin is

$$F_{CP} = 0.0000284 \times 2.5 \times \overline{2{,}000}^2 \times 2.5 = 710 \text{ lb.}$$

Calculation for weight and center of gravity of chamfered and rounded crank arms may be made by determining the weight and center of gravity of each geometric part.

A suggested procedure is to determine the weight or volume of the outer end of the crank arms beyond the crankpin center line and then take enough of the crank arms below the crankpin center line to place the center of gravity of the combined weights or volumes on the crankpin center line. The lower ends of the crank arms may be handled similarly with reference to the center line of main bearings, and then the mid-portion of the crank arms, which are usually more uniform in section, may be handled in the usual way.

Figure 7-21 is an enlarged detail sketch of the end of one of the crank arms of Fig. 7-20. In this detailed sketch, the various geometric shapes are designated by letters. For the parts above the crankpin center line, the following table can be used:

Name of part	Volume of part	Distance from ℄ to C. G. of part
Half cylinder....................	$V_B = \frac{\pi S^2 U}{2}$	$\bar{X}_B = \frac{4S}{3\pi}$
Ungula of a right circular cylinder.	$V_C = \frac{2}{3} R^2 Y$	$\bar{X}_C = \frac{3\pi R}{16}$
Ungula of a right circular cylinder.	$V_{C'} = \frac{2}{3} S^2 K = \frac{2}{3} S^3 \frac{Y}{R}$	$\bar{X}_{C'} = \frac{3\pi S}{16}$

Hence for part A

$$V_A = \frac{\pi R^2 U}{2} - \frac{\pi S^2 U}{2} - \left(\frac{2}{3} R^2 Y - \frac{2}{3} S^3 \frac{Y}{R}\right)$$

and since

$$V\bar{X} = \Sigma(v\bar{X})$$

$$\bar{X}_A =$$

$$\frac{\dfrac{\pi R^2 U}{2} \times \dfrac{4R}{3\pi} - \dfrac{\pi S^2 U}{2} \times \dfrac{4S}{3\pi} - \left(\dfrac{2}{3} R^2 Y \times \dfrac{3\pi R}{16} - \dfrac{2}{3} S^3 \dfrac{Y}{R} \times \dfrac{3\pi S}{16}\right)}{\dfrac{\pi R^2 U}{2} - \dfrac{\pi S^2 U}{2} - \left(\dfrac{2}{3} R^2 Y - \dfrac{2}{3} S^3 \dfrac{Y}{R}\right)}$$

or

$$\bar{X}_A = \frac{\dfrac{2R^3 U}{3} - \dfrac{2S^3 U}{3} - \left(\dfrac{\pi R^3 Y}{8} - \dfrac{\pi S^4 Y}{8R}\right)}{\dfrac{\pi R^2 U}{2} - \dfrac{\pi S^2 U}{2} - \dfrac{2}{3} Y \left(R^2 - \dfrac{S^3}{R}\right)}$$

Below the center line, the volume V_D is

$$V_D = VUS - \frac{\pi S^2 U}{2}$$

and the distance from the ℄ to the center of gravity is

$$\bar{X}_D = \frac{\dfrac{VUS^2}{2} - \dfrac{2S^3 U}{3}}{VUS - \dfrac{\pi S^2 U}{2}}$$

If we allow the remaining volume V_F to be such that

$$V_F \bar{X}_F = V_A \bar{X}_A - V_D \bar{X}_D$$

the center of gravity of $V_A + V_D + V_F$ will fall on the center line of the crankpin.

Using values from Fig. 7-20, the volume above the center of gravity of the crankpin is

$$V_A = \frac{\pi \times \overline{1.75}^2 \times 1.25}{2} - \frac{\pi \times \overline{0.5625}^2 \times 1.25}{2} - \frac{2 \times 0.75}{3}\left(\overline{1.75}^2 - \frac{\overline{0.5625}^3}{1.75}\right)$$

$$V_A = 3.93 \text{ cu. in.}$$

The distance to the center of gravity is

$$\bar{X}_A = \frac{\dfrac{2 \times 1.75^3 \times 1.25}{3} - \dfrac{2 \times 0.5625^3 \times 1.25}{3} - \dfrac{\pi(1.25 - 0.5)}{8}\left(1.75^3 - \dfrac{0.5625^4}{1.75}\right)}{3.93}$$

$$\bar{X}_A = 0.7 \text{ in.}$$

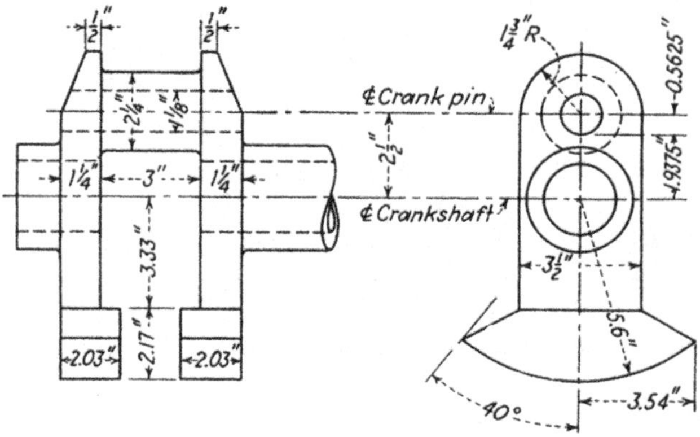

FIG. 7-20.—Crank-arm details for the example of Par. 7-10.

Below the crankpin center line,

$$V_D = 3.5 \times 1.25 \times 0.5625 - \frac{\pi \times 0.5625^2 \times 1.25}{2}$$

$$V_D = 1.839 \text{ cu. in.}$$

The distance to the center of gravity is

$$\bar{X}_D = \frac{\dfrac{3.5 \times 1.25 \times 0.5625^2}{2} - \dfrac{2 \times 0.5625^3 \times 1.25}{3}}{1.839}$$

$$\bar{X}_D = 0.296 \text{ in.}$$

To place the center of gravity of the combined volumes on the center line of the crankpin,

$$V_F\bar{X}_F = 3.93 \times 0.7 - (1.839 \times 0.296) = 2.207$$

From the detailed figure,

$$V_F = VUZ = 3.5 \times 1.25 \times Z = 4.375Z \text{ cu. in.}$$

and

$$\bar{X}_F = S + \frac{Z}{2} = 0.5625 + \frac{Z}{2}$$

hence

$$4.375Z\left(0.5625 + \frac{Z}{2}\right) = 2.207$$

or

$$Z = 0.6 \text{ in.}$$

Therefore,

$$V_F = 4.375 \times 0.6 = 2.63 \text{ cu. in.}$$

and

$$\bar{X}_F = 0.83 \text{ in.}$$

This still leaves $2.5 - (S + Z) = 2.5 - (0.5625 + 0.6) = 1.33$ in. of the crank arm and the part on the opposite side of the axis of rotation to be considered, but before calculating the effect of these parts, the centrifugal force on the upper part of the crank arms will be determined.

The weight of the several volumes found above is (for both crank arms)

$$W_{CA} = 2(V_A + V_D + V_F)d$$
$$= 2(3.93 + 1.839 + 2.63)0.28$$
$$= 4.7 \text{ lb.}$$

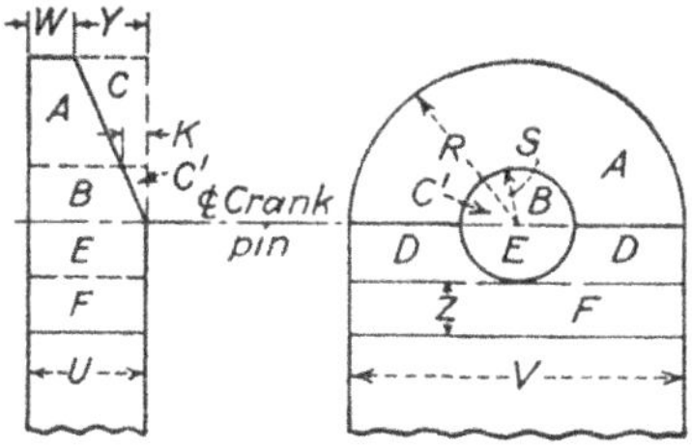

FIG. 7-21.—Enlarged detailed arrangement of the outer end of the crank arms in Fig. 7-20.

and the distance to the center of gravity from the axis of rotation is equal to crank radius, *i.e.*, 2.5 in. Hence, the centrifugal force on the upper end of the crank arms is

$$F_{CA} = 0.0000284 \times 4.7 \times \overline{2{,}000}^2 \times 2.5 = 1{,}335 \text{ lb.}$$

The remainder of the crank arms, *i.e.*, the parts extending from the center of rotation out a distance of 1.33 in. may be handled in much the same way as for other parts, but inasmuch as it will be necessary to extend the arms on the other side of the axis of rotation far enough to reach the counterweights, it is probable that most of the remaining part of the crank arms will be balanced. In fact, it is a convenient preliminary assumption (for the example being considered here) to let the point of attachment for the counterweights be 1.33 + J in. from the axis of rotation. This eliminates the inner 1.33 in. of the arms from the unbalance calculations, and the total force to be balanced is

$$F_{PR} + F_{CR} + F_{CP} + F_{CA} = 2{,}840 + 3{,}408 + 710 + 1{,}335 = 8{,}293 \text{ lb.}$$

The detailed shapes of counterweights to balance this force are largely a matter of individual preference; it should be kept in mind, of course, that minimum weight and adequate clearance are important items. Hence, for the purpose of the example, assume that counterweight shapes like Fig. 7-19 (B) are to be used. Further, assume that they are to be attached a distance (J) beyond the balanced portion of the crank arms of 2 in. The balancing effect of these 2 in. of extended crank arms is found as follows:

The volume for both arms is

$$V_{EX} = 3.5 \times 1.25 \times 2 \times 2 = 17.5 \text{ cu. in.}$$

The weight of the extended crank arms is

$$W_{EX} = 17.5 \times 0.28 = 4.9 \text{ lb.}$$

The distance from the axis of rotation to the center of gravity is

$$\bar{X}_{EX} = 1.33 + 1 = 2.33 \text{ in.}$$

and the centrifugal force is

$$F_{EX} = 0.0000284 \times 4.9 \times \overline{2{,}000}^2 \times 2.33 = 1{,}295 \text{ lb.}$$

Hence the net force to be balanced by counterweights is

$$F_N = 8{,}293 - 1{,}295 = 6{,}998 \text{ lb.}$$

Referring to Fig. 7-19 (B), let $\alpha = 40$ deg., $M = 3.5/2 = 1.75$ in., and $R = 5.5$ in. Then

$$F = R(1 - \cos \alpha) = 5.5(1 - 0.766) = 1.29 \text{ in.}$$
$$L = \sqrt{F(2R - F)} = \sqrt{1.29(2 \times 5.5 - 1.29)} \doteq 3.54 \text{ in.}$$
$$N = 1.33 + 2 = 3.33 \text{ in.}$$
$$H = R - (F + N) = 5.5 - (1.29 + 3.33) = 0.88 \text{ in.}$$

The area of the trapezoidal part of the counterweight is [from the data of Fig. 7-19 (B)]

$$A_Q = (M + L)H = (1.75 + 3.54)0.88 = 4.65 \text{ sq. in.}$$

The area of the circular segment is

$$A_S = R^2(0.01745\alpha - 0.5 \sin 2\alpha)$$
$$= 5.5^2(0.01745 \times 40 - 0.5 \sin 80) = 6.24 \text{ sq. in.}$$

The total area is

$$A_Q + A_S = 4.65 + 6.24 = 10.89 \text{ sq. in.}$$

The radial distance to the center of gravity of the trapezoid is

$$R_{GQ} = N + \frac{H(M + 2L)}{3(M + L)} = 3.33 + \frac{0.88(1.75 + 2 \times 3.54)}{3(1.75 + 3.54)}$$
$$= 3.82 \text{ in.}$$

The radial distance to the center of gravity of the circular segment is

$$R_{GS} = \frac{R \sin^3 \alpha}{0.02617\alpha - 0.75 \sin 2\alpha}$$
$$= \frac{5.5 \times 0.6428^3}{0.02617 \times 40 - 0.75 \times 0.9848} = 4.72 \text{ in.}$$

The radial distance to the center of gravity of the combined area is

$$R_G = \frac{A_Q R_{GQ} + A_S R_{GS}}{A_Q + A_S} = \frac{4.65 \times 3.82 + 6.24 \times 4.72}{4.65 + 6.24} = 4.345 \text{ in.}$$

The thickness of the counterweights may now be found as follows:

Let

$$F_{cw} = 6{,}998 \text{ lb. } (= F_N)$$

Then, the necessary weight of the counterweights is

$$W_{cw} = \frac{F_{cw}}{0.0000284 N^2 R_G} = \frac{6{,}998}{0.0000284 \times 2{,}000^2 \times 4.345} = 14.17 \text{ lb.}$$

Since this weight is divided equally between the two crank arms, for a density of the steel of 0.28 lb. per cu. in., the thickness of each counterweight is

$$T = \frac{W_{cw}}{2(A_Q + A_S)d} = \frac{14.17}{2 \times 10.89 \times 0.28} = 2.32 \text{ in.}$$

For this thickness of counterweights, from Fig. 7-20, it is seen that the space between the weights for passage of the connecting rods is

$$(2 \times 1.25 + 3) - (2 \times 2.32) = 0.86 \text{ in.}$$

This space probably would be inadequate for the size of connecting rods needed. The space could be increased (*a*) by shifting the counterweights to overhang the outside edges of the crank arms, (*b*) by increasing the radial distance to the center of gravity of the counterweights R_G, or (*c*) by using a counterweight metal of greater density. Shifting of the counterweights is objectionable because the main bearing supports will have to be moved out of line of the main bearings, and this will complicate the crankcase, although a slight shift sufficient for a small increase in space between the counterweights might be made without difficulty. Increase in R_G will require a larger diameter and hence an increased weight of crankcase. Increase in the density may be attained by using bronze ($d = 0.32$ lb. per cu. in.). Thus for bronze counterweights,

$$T = \frac{14.17}{2 \times 10.89 \times 0.32} = 2.03 \text{ in.}$$

and the space for passage of the connecting rods is

$$(2 \times 1.25 + 3) - (2 \times 2.03) = 1.44 \text{ in.}$$

As the width of the connecting rods for this size of engine would probably not exceed 1 or at most 1¼ in., this space should be adequate and the dimensions of the counterweights as determined above and shown in Fig. 7-20 should be satisfactory.

Suggested Design Procedure

Important. Include sample calculations of all items (as applies). Make layouts to a large enough scale to permit accuracy of measurements.

1. For the engine being designed, check arrangement of crank arms and rearrange if necessary to attain most suitable conditions for balance.

2. Determine necessary data and, for one cylinder through 360 deg. of crank travel, plot curves of

a. Primary reciprocating inertia force.

b. Secondary reciprocating inertia force.

3. Determine the centrifugal force that would have to be applied at the crankshaft to balance one-half of the maximum primary reciprocating inertia force in one cylinder.

4. Determine the centrifugal force on the crankpin due to unbalanced rotating weight.

5. By using the crankpin and main bearing dimensions determined in Suggested Design Procedure, page 86, data from Tables A1-8 and A1-9 (as apply), and data from available blueprints, etc., lay out the crankshaft (except the end portions) and determine the detail dimensions of the crank arms.

Make this crankshaft layout to scale, and leave sufficient room on the drawing for both an end view and a longitudinal view. Also leave room for later addition to the drawing of the shaft-end details, *i.e.*, propeller hub and auxiliary-drive connections.

NOTE: If the engine being designed is an in-line or V type, use the layout of Suggested Design Procedure, page 86 item 4 as a basis for the detail dimensioning.

6. Determine the centrifugal force on the crankshaft due to the unbalanced weight of the crank.

7. If more than one connecting rod is attached to each crankpin, determine the total centrifugal force that would have to be applied to the crankshaft to balance the part of the unbalanced portion of the reciprocating weight that it is feasible to balance.

8. Determine the total centrifugal force due to item 3 or 7 (as applies), item 4, and item 6.

9. Determine the dimensions of counterweights necessary to provide for item 8.

10. Lay out the counterweights (item 9 on the drawing of item 5).

Do not blueprint the crankshaft layout at this stage.

11. When items 1 to 10 have been completed and put in proper form, submit for checking and approval.

References

1. Angle: "Engine Dynamics and Crankshaft Design."
2. Huebotter: "Mechanics of the Gasoline Engine."
3. Heldt: "Automotive Engines."
4. Marks: "The Airplane Engine."
5. Marks: "Mechanical Engineers' Handbook."
6. Maleev: "Internal Combustion Engines."
7. *S.A.E. Jour.*, Vol. 28, No. 2, February, 1931.
8. Seely and Ensign: "Analytical Mechanics for Engineers."
9. Judge: "Automobile and Aircraft Engines."
10. Timoshenko: "Vibration Problems in Engineering."
11. *S.A.E. Jour.*, Vol. 38, No. 3, March, 1936.
12. Sharp: "Balancing of Engines."
13. Hazen and Montieth: Torsional Vibration of In-line Aircraft Engines, *S.A.E. Jour.*, Vol. 43, No. 2, August, 1938.
14. Malychevitch: Torsional Vibration of Diesel Engine Crankshafts, Part 1, *Automotive Ind.*, Mar. 18, 1939; Part 2, Mar. 25, 1939.
15. Wilson: "Practical Solution of Torsional Vibration Problems."

16. Biot: Vibration of Crankshaft and Propeller Systems. New Method of Calculation, *Jour. Inst. Aeronautical Sciences*, January, 1940.
17. Masi: Permissible Amplitudes of Torsional Vibration in Aircraft Engines, *S.A.E. Trans.*, Vol. 34 (*S.A.E. Jour.*, Vol. 45, No. 1, July, 1939).
18. Kearns: Vibration Characteristics of Aircraft Engine-propeller Systems, *S.A.E. Trans.*, Vol. 34 (*S.A.E. Jour.*, Vol. 45, No. 6, December, 1939).
19. Gerdan: Dampers and Suppressors, *Product Eng.*, Part 1, January, 1940; Part 2, February, 1940.
20. Tuplin: "Torsional Vibration."

CHAPTER 8

CRANKSHAFT DETAILS AND REDUCTION GEARING

8-1. Crankshaft Details.—The principal dimensions of the crankshaft having been determined, it is now possible to complete the details of parts. In doing this, it is important to give proper attention to the following points.

1. Provide adequate fillets for all reentrant corners. Such corners are points of high stress concentration and are apt to be critical if sharp. Large radius fillets[8] aid greatly in reducing this stress concentration.

2. Check the arrangement of parts for possible manufacturing difficulties. See if forging of the shaft can be simplified by some improvement in detail. Aircraft-engine crankshafts are usually machined all over. Check your details of design for any unnecessarily difficult or complicated machining operations. Skilled labor is expensive.

3. All main and connecting-rod bearings should be provided with pressure-feed lubrication. When plain bearings are used, as in in-line and V-engines, oil is usually supplied to the main bearings through passageways in the main-bearing supports. Drilled passageways in the shaft then lead some of this oil to the crankpins. In arranging these passageways, care should be exercised to avoid an excessive number of sharp turns as those turns increase the resistance to circulation of the oil. Avoid too small a size of passageway as the chances of solid particles obstructing the flow will be increased. Excessively large passageways may impair the strength or rigidity of the crank arms, especially if they pass close to highly stressed reentrant corners. Arrangement of oil passageways are shown in Figs. 5-11, 5-12, 5-13, and 6-6. Other arrangements may be studied from available blueprints and drawings.

Accessory drives such as magnetos, oil pumps, gun synchronizers, and superchargers are usually driven from the rear end of the crankshaft. The main accessory drive shaft for these parts is usually splined to the crankshaft, and the individual accessories are driven from this shaft through suitable gearing. Considerable leeway is available to the designer in arranging such drives, but for inexperienced designers, it is advisable to adhere rather closely to proven arrangements. Sectional cuts of current successful engines are of assistance in this respect, and examples of proven construction as shown on available blueprints and drawings should be studied.

Aircraft propeller shaft ends have been standardized by the S.A.E. Both taper and spline-type shaft ends are used, the latter being more common in large sizes of engines. Figure A1-6 and Tables A1-19 and A1-20 may be used for selecting the proper sizes and laying out the propeller end of the crankshaft. In general, the S.A.E. shaft number to select is that which has a maximum diameter nearest under the diameter of the crankshaft.

8-2. Reduction Gearing.—The brake horsepower that an engine can develop is a function of the speed, but in the case of aircraft engines, the rate of rotation is limited by the propeller efficiency so that it is usually inadvisable to operate direct-drive engines above 2,200 to 2,500 r.p.m. In many instances, the inherent tendency for the propeller efficiency to drop at high speeds can be offset by suitable reduction gearing.

Wood has shown[1] that for best performance the diameter of a two-bladed propeller necessary to absorb the sea-level rated power of an engine is

$$D = \frac{303}{\sqrt{\text{r.p.m.}}} \times \sqrt[4]{\frac{\text{b.hp.}}{\text{m.p.h.}}} \qquad (8\text{-}1)$$

where D = diameter of the propeller, ft.
r.p.m. = rated speed of the engine.
b.hp. = rated power of the engine.
m.p.h. = maximum speed of the plane in level flight.

With these data known, the corresponding propeller efficiency can be readily determined from Fig. 8-1. The effect of using reduction gearing in the engine may be demonstrated as follows:

Let R = the reduction gear ratio and e_m = the mechanical efficiency of the reduction gear. Then

$$R = \frac{\text{r.p.m.}_1}{\text{r.p.m.}_2}, \qquad e_m = \frac{\text{b.hp.}_2}{\text{b.hp.}_1}$$

where subscript 1 refers to the crankshaft and subscript 2 refers to the propeller shaft, and for the same air speed

$$D_2 = \frac{303}{\sqrt{\text{r.p.m.}_1/R}} \times \sqrt[4]{\frac{\text{b.hp.}_1 \times e_m}{\text{m.p.h.}}} \tag{8-2}$$

or

$$D_2 = D_1 \sqrt{R} \sqrt[4]{e_m} \tag{8-3}$$

The corresponding change in effective propeller pitch will be (for $V_1 = V_2$)

$$\frac{V_2/n_2D_2}{V_1/n_1D_1} = \frac{V_2/n_2D_2}{V_2/Rn_2(D_2/\sqrt{R}\sqrt[4]{e_m})} = \frac{\sqrt{R}}{\sqrt[4]{e_m}}$$

or

$$\frac{V_2}{n_2D_2} = \frac{V_1}{n_1D_1} \times \frac{\sqrt{R}}{\sqrt[4]{e_m}} \tag{8-4}$$

With the new value of effective pitch known, the gain in propeller efficiency may be readily found from Fig. 8-1.

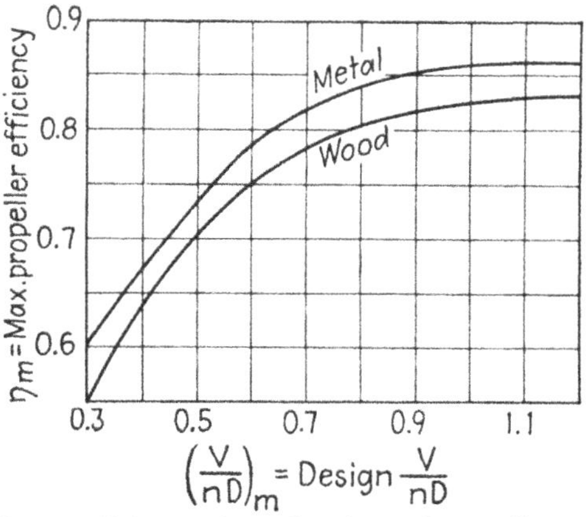

FIG. 8-1.—Maximum efficiency of wood and metal propellers as a function of design V/nD. (*From Wood, "Technical Aerodynamics."*)

Example.—An airplane attains 140 m.p.h. with an engine of 300 b.hp. operating at 2,700 r.p.m. (*a*) find the propeller diameter and maximum propeller efficiency. (*b*) What gain in propeller efficiency could be had if the engine was equipped with a 3:2 reduction gear having a mechanical efficiency of 90 per cent?

Solution.—From Eq. (8-1),

$$D_1 = \frac{303}{\sqrt{2{,}700}} \times \sqrt[4]{\frac{300}{140}} = 7.05 \text{ ft.} \quad (\textit{Ans. a})$$

and

$$\frac{V_1}{n_1D_1} = \frac{88 \times 140}{2{,}700 \times 7.05} = 0.649$$

From Fig. 8-1 (for metal propellers),

$$\eta_1 = 80\%$$

From Eq. (8-4),

$$\frac{V_2}{n_2D_2} = 0.649\,\frac{\sqrt{1.5}}{\sqrt[4]{0.9}} = 0.815$$

From Fig. 8-1,

$$\eta_2 = 84\%, \qquad \text{and} \qquad \eta_2 - \eta_1 = 84 - 80 = 4\% \quad (\textit{Ans. b})$$

Since for the same velocity, the thrust horsepower required will be the same, the b.hp. actually needed with the geared engine will be

$$\text{b.hp.}_2 = {}^{80}\!/_{84} \times 300 = 286$$

If the engine were slowed down enough to attain this gain in efficiency by direct drive, *i.e.*, maintain $V/nD = 0.815$, the power that it could develop would be approximately represented by the expression

$$\text{b.hp.} = \text{b.hp.}_R \times \frac{\text{r.p.m.}}{\text{r.p.m.}_R} \div \frac{dP}{dN} \qquad (8\text{-}5)$$

where the subscript R represents rated conditions and

$\frac{dP}{dN}$ = the slope of the full throttle curve (Fig. 8-2).

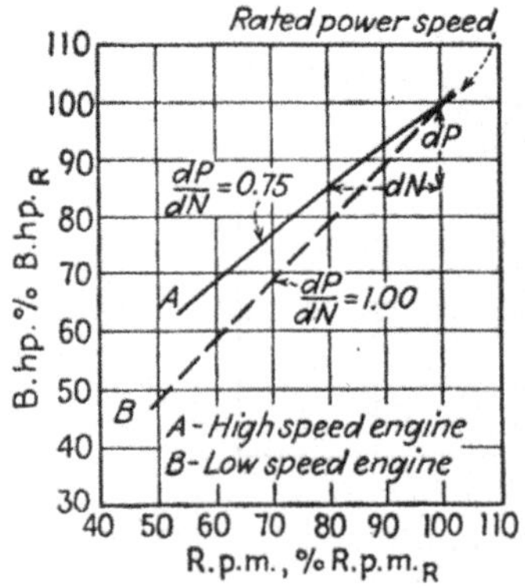

FIG. 8-2.—Variation of full-throttle brake horsepower with speed.

The value dP/dN is determined in part by many engine design factors, but the ratio of rated to maximum possible b.hp. is a major factor. The tendency for volumetric efficiency to drop at higher speeds is also a contributing factor to lower values of dP/dN.

For the example, dP/dN will probably not exceed 0.9; hence

$$\text{b.hp.} = 300 \times \frac{1{,}800}{2{,}700} \div 0.9 = 222$$

But maintenance of the original air speed requires 286 b.hp. This may be attained either by increasing the amount of supercharging, which will require a higher octane fuel, or by increasing the displacement of the engine.

If the increase in power from 222 to 286 b.hp. is attained by supercharging, the b.m.e.p. will have to be increased in direct proportion to the brake horsepower. If we assume that the 222 hp. is developed on a 73 octane number fuel, the b.m.e.p. for average conditions (Fig. 1-10) will be about 114 lb. per sq. in. To attain 286 b.hp., the b.m.e.p. will have to be about

$$114 \times {}^{286}\!/_{222} = 147 \text{ lb. per sq. in.}$$

To attain this m.e.p., the engine would have to be supercharged. From Fig. 1-10, it is seen that the fuel would have to have an octane number of upwards of 100. Such fuels are commercially available but rather expensive.

If the increase in power is attained by increasing the displacement, the weight will also be increased (Fig. 1-3) by upwards of

$$\frac{(1.85 \times 286) - (1.95 \times 222)}{1.95 \times 222} \times 100 = 22 \text{ per cent, and this is more}$$

than the increase due to adding reduction gearing.

To keep the bulk and weight of the engine as low as possible, it is essential that the reduction gearing be compact and of a material that will withstand extremely high allowable stresses. This calls for high grades of alloy steel, careful design, and precision workmanship in manufacture. Hence the advantages of a geared drive are partly offset by greater cost and complexity. The power output for which the engine is being designed is an important factor in the decision of whether or not reduction gearing shall be used. Very high power requirements usually dictate the use of reduction gearing, whereas small and medium-powered engines are usually direct drive. However, the continually increasing demand for small and medium-powered engines of lower specific weight (Fig. 1-3) may ultimately result in a greater use of geared drives in these sizes of power plants.

There are a large number of possible arrangements for reduction gearing, but experience has narrowed the field to three main types (Figs. 8-3 to 8-7).

The single reduction gear has the advantage of simplicity and somewhat lower cost, but it places the thrust line to one side of the axis of symmetry of the engine, and this may produce

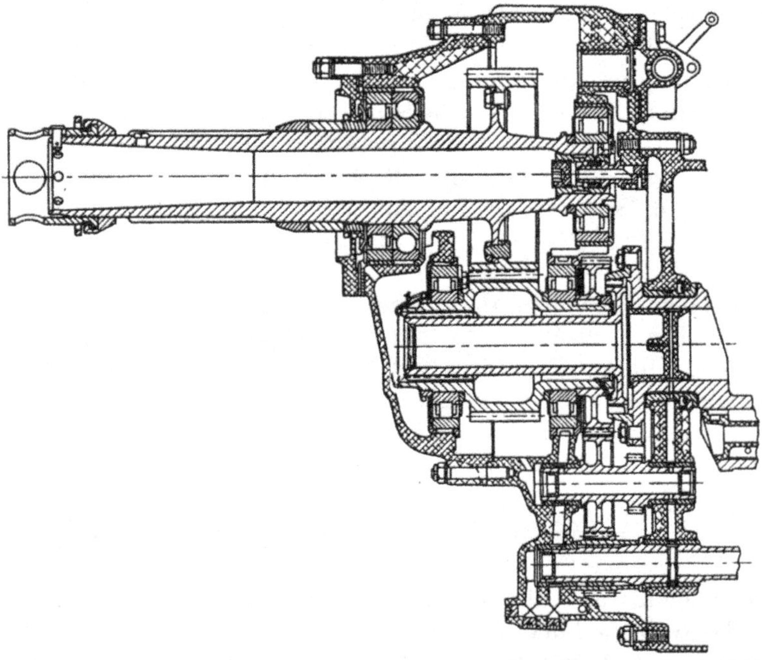

FIG. 8-3.—Single reduction gear used on geared models of ranger inverted V-12 engine. Gears are of the herringbone type.

complex stresses in the crankcase. However, when the propeller axis is placed above the crankshaft axis (as is usually the case with single-reduction gearing), the visibility forward in a single-engine tractor-type plane can be improved.

The planetary reduction type of gearing permits keeping the propeller and crankshaft axes concentric, and thereby reduces the complex and more or less indeterminate stresses in the nose of the crankcase. This gain is at the expense of some increase in complexity of the gearing. A particular advantage of planetary reduction gearing in high-powered engines is the ability to use more than one planetary gear. This permits dividing the load into several parts and reducing the strain on the gear teeth.

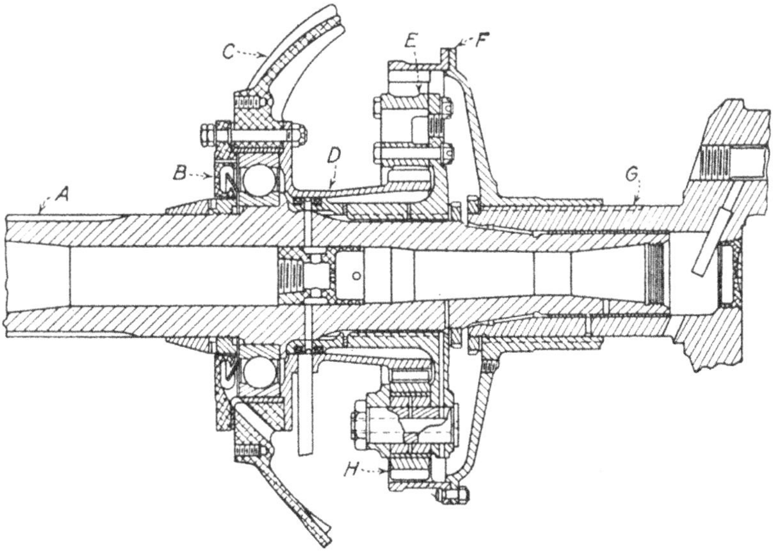

FIG. 8-4.—Arrangement of the Pratt and Whitney type planetary reduction gear. (A) Propeller shaft; (B) thrust-bearing assembly; (C) nose of crankcase; (D) fixed gear bolted to crankcase; (E) pinion cage splined to propeller shaft; (F) internal tooth-drive gear splined to crankshaft; (G) crankshaft; (H) planetary pinion.

FIG. 8-5.—Arrangement of the Wright Cyclone type of planetary reduction gear.

8-3. Gear Materials and Dimensions.—To provide adequate strength and minimum weight, aircraft-engine reduction gears are usually made of heat-treated and hardened alloy steels. Casehardened S.A.E. 2515 steel is used in reduction gears of Pratt and Whitney engines, and S.A.E. 4140 or Nitralloy (Table 8-1) is recommended by the Climax Molybdenum Company.

TABLE 8-1.—NITRALLOY STEELS SUITABLE FOR REDUCTION GEARS*

Steel No.	C	Si	Mn	Cr	Al	Mo
1	0.50–0.55	0.25–0.35	0.40–0.50	2.00–2.20	0.30–0.40	0.20–0.35
2	0.40–0.45	0.20–0.30	0.40–0.50	1.70–1.90	0.30–0.40	0.20–0.35
3	0.25–0.30	0.20–0.30	0.40–0.50	1.70–1.90	0.30–0.40	0.20–0.35

* From *The Moly Matrix*, Vol. 3, No. 8, July, 1936.

Allowable static stresses of 20,000 to 30,000 lb. per sq. in. or more may be used depending upon the nature and heat-treatment

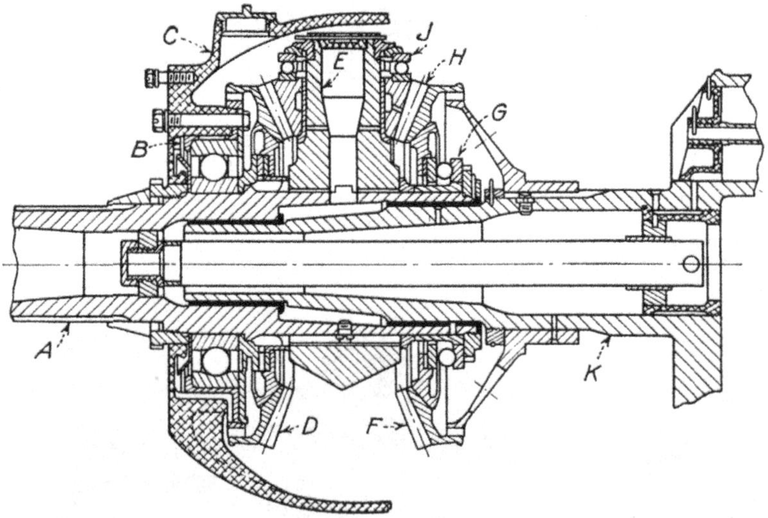

FIG. 8-6.—Arrangement of the Wright Cyclone 2 to 1 ratio bevel-type planetary reduction gear.

(*A*) Propeller shaft; (*B*) thrust-bearing assembly; (*C*) nose of crankcase; (*D*) fixed gear attached to crankcase; (*E*) pinion-gear supporting arm splined to propeller shaft; (*F*) drive gear splined to crankshaft; (*G*) drive-gear thrust bearing; (*H*) pinion gear; (*J*) pinion-gear thrust bearing; (*K*) crankshaft.

of the steel, and upon the accuracy of construction of the gear teeth. Pratt and Whitney engineers have found an allowable static stress of 22,000 lb. per sq. in. satisfactory. Buckingham recommends[2,3] the use of data in Tables 8-2 and 8-3 for determining the allowable static stress to be used.

Allowable stresses decrease with increase in pitch-line velocity. This necessary decrease is partly due to the increase in the effect

FIG. 8-7.—Arrangement of bevel type 1.58 to 1 planetary reduction gear from a Wright Cyclone.

of shock loads and inertia of parts at the higher speeds. However, a purely rational expression for variation of allowable stress with pitch-line velocity would be very difficult to obtain due to

TABLE 8-2.—FACTORS OF SAFETY FOR GEAR TEETH

For steady load on a single pair of gears	3
For suddenly applied loads on single pairs of gears	4
For steady loads on gears of a train beyond the first mesh	5
For suddenly applied loads on gears of a train beyond the first mesh	6

TABLE 8-3.—ULTIMATE STRENGTH OF GEAR MATERIALS

Material	Ultimate Strength, Lb. per Sq. In.
Cast iron	24,000
Semisteel	36,000
Bronze	36,000
Cast steel (S.A.E. 1235)	45,000
Forged steel (S.A.E. 1030)	60,000
Forged steel (S.A.E. 1045)	90,000
Forged steel (S.A.E. 3245)	120,000

the uncertainty of the numerous variables involved. Hence, empirical expressions are used, the most applicable to aircraft engines probably being the following:

For accurately cut gears, Buckingham suggests

$$S = S_0\left(\frac{1,200}{1,200 + V}\right) \tag{8-6}$$

and for pitch-line velocities above 4,000 f.p.m., the AGMA recommends the use of Eq. (8-7) for determining the allowable stresses

$$S = S_0\left(\frac{78}{78 + \sqrt{V}}\right) \tag{8-7}$$

where S = allowable stress at velocity V, lb. per sq. in.
S_0 = allowable static stress, lb. per sq. in.
V = pitch-line velocity, f.p.m.

$$V = \frac{\pi D n}{12} \tag{8-8}$$

where V is as in Eqs. (8-6) and (8-7).
D = pitch diameter, in.
n = speed of the gear, r.p.m.

The maximum safe tangential load that can be transmitted by the gear tooth (or the actual transmitted load at the pitch diameter) may be expressed as

$$W = \frac{SbY'}{P_d} = \frac{\pi SbY}{P_d} \tag{8-9}$$

where W = allowable or transmitted load at the pitch line, lb.
S = allowable or transmitted unit stress at velocity V, lb. per sq. in.
b = face width of the gear tooth, in.
Y' = Lewis outline factor (Table 8-4).
Y = Buckingham strength form factor (Table 8-4) ($Y' = \pi Y$).
P_d = diametral pitch.

TABLE 8-4.—GEAR-TOOTH FACTORS[2,3]
($Y' = \pi Y$)

Number of teeth	Lewis outline factor = Y'		Buckingham strength form factor = Y		
	14½ deg. involute and cycloidal	20 deg. full-depth involute	14½ deg. involute composite	20 deg. full-depth involute	20 deg. stub involute
12	0.210	0.245	0.067	0.078	0.099
13	0.220	0.261	0.071	0.083	0.103
14	0.226	0.276	0.075	0.088	0.108
15	0.236	0.289	0.078	0.092	0.111
16	0.242	0.295	0.081	0.094	0.115
17	0.251	0.302	0.084	0.096	0.117
18	0.261	0.308	0.086	0.098	0.120
19	0.273	0.314	0.088	0.100	0.123
20	0.283	0.320	0.090	0.102	0.125
21	0.289	0.327	0.092	0.104	0.127
23	0.295	0.333	0.094	0.106	0.130
25	0.305	0.339	0.097	0.108	0.133
27	0.314	0.349	0.099	0.111	0.136
30	0.320	0.358	0.101	0.114	0.139
34	0.327	0.371	0.104	0.118	0.142
38	0.336	0.383	0.106	0.122	0.145
43	0.346	0.396	0.108	0.126	0.147
50	0.352	0.408	0.110	0.130	0.151
60	0.358	0.421	0.113	0.134	0.154
75	0.364	0.434	0.115	0.138	0.158
100	0.371	0.446	0.117	0.142	0.161
150	0.377	0.459	0.119	0.146	0.165
300	0.383	0.471	0.122	0.150	0.170
Rack	0.390	0.484	0.124	0.154	0.175

However, the actual dynamic load on the gear tooth will be greater than W by the amount of an increment or impact load that results from acceleration and deceleration of the gear due to inaccuracy in the tooth profiles and tooth spacing. The AGMA

TABLE 8-5.—VALUES OF THE BUCKINGHAM DYNAMIC TOOTH LOAD CONSTANT, C

Material	Tooth form	Error in tooth action = e in inches			
		0.0005	0.001	0.002	0.003
Gray iron and gray iron	14½-deg. involute	400	800	1,600	2,400
Gray iron and gray iron	20-deg. full-depth involute	415	830	1,660	2,490
Gray iron and gray iron	20-deg. stub involute	430	860	1,720	2,580
Gray iron and steel	14½-deg. involute	550	1,100	2,200	3,300
Gray iron and steel	20-deg. full-depth involute	570	1,140	2,280	3,420
Gray iron and steel	20-deg. stub involute	590	1,180	2,360	3,540
Steel and steel	14½-deg. involute	800	1,600	3,200	4,800
Steel and steel	20-deg. full-depth involute	830	1,660	3,320	4,980
Steel and steel	20-deg. stub involute	860	1,720	3,440	5,160

recommends the use of Eq. (8-10) for determining the dynamic tooth load.

$$W_d = \frac{0.05V(bC + W)}{0.05V + \sqrt{bC + W}} + W \qquad (8\text{-}10)$$

where W_d = dynamic tooth load, lb.
V = pitch-line velocity,* f.p.m.
b = face width of the gear, in.
C = constant which depends on the material, tooth form, and accuracy of construction of the gear.
W is obtained from Eq. (8-9).

Values of C as determined by Buckingham are given in Table 8-5.

* For planetary (epicyclic) gear trains, the velocity of actual tooth engagement (Table A3-6) must be used to determine the dynamic loadings.[4]

To ensure against tooth breakage, the dynamic load should not produce a stress in the material greater than the flexural endurance limit S_t of the material. The stress produced may be checked by substituting W_d in Eq. (8-9) and solving for the dynamic tooth load stress S_d. The value thus found should not exceed the value of S_t (Fig. 8-8) corresponding to the Brinell hardness number of the gear material.* As an added precaution, Buckingham suggests

For steady loads, $S_t \geq 1.25S_d$.

For pulsating loads, $S_t \geq 1.35S_d$.

For shock loads, $S_t \geq 1.50S_d$.

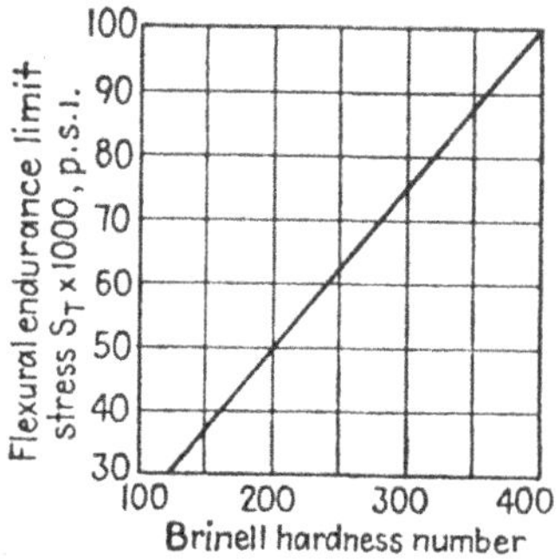

FIG. 8-8.—Relation between Brinell hardness number and flexural endurance-limit stress for steel. (*Data from Buckingham, "Manual of Gear Design."*)

Reduction gear teeth may be strong enough to transmit the desired horsepower and withstand the dynamic loading and yet be unable to resist rapid wear. This wear which is usually evidenced by a pitting of the tooth surfaces is generally conceded to be due to compressive fatigue stresses. An expression for equivalent static tooth load beyond which failure from pitting is likely to occur has been developed by Buckingham and adopted by the AGMA as follows:

$$W_w = DbKZ \tag{8-11}$$

where W_w = equivalent static tooth load beyond which pitting (wear) is likely to occur, lb.

D = pitch diameter of the pinion or smaller gear, in.

b = face width of the gears, in.

Z = ratio factor $= 2N_G/(N_P + N_G)$ for spur gears and $2N_G/(N_G - N_P)$ for internal gears.

N_P = number of teeth in the pinion.

N_G = number of teeth in the gear.

K = stress factor involving the maximum fatigue-limit compressive stress, the pressure angle of the gear teeth, and the moduli of elasticity of the material of the gears.[4]

* See Fig. A2-1 for estimating the Brinell hardness number of the gear steel being used

Values of K have been determined by Buckingham and are given in Fig. 8-9 and Table 8-6. Since the allowable static load W_w varies directly with K, it is readily apparent (Fig. 8-9) why casehardened or Nitralloy steels are highly desirable in reduction gearing. To avoid pitting, W_d should not exceed W_w, and for safety, $W_d \leq 0.75W_w$.

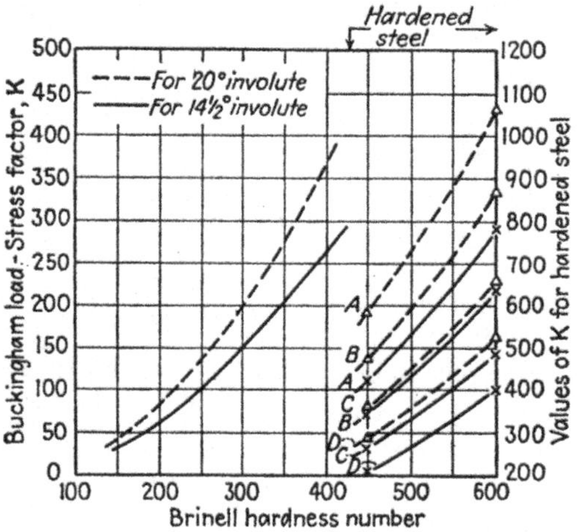

FIG. 8-9.—Values of the Buckingham gear-tooth fatigue constant or load-stress factor. Repetitions of stress in millions: A – 10, B – 20, C – 50, D – 100. (*Data from Buckingham, "Manual of Gear Design."*)

The torque on the gear when subjected to a tooth load W is

$$Q = \frac{WD}{2} = \frac{WN}{2P_d} \tag{8-12}$$

where Q = torque, lb.-in.
D = pitch diameter, in.
N = number of teeth in the gear.
W = allowable load at the pitch line, lb.
P_d = diametral pitch.

The face width b is usually made a function of the diametral pitch; the AGMA recommends as good practice

$$b = \frac{10}{P_d} \tag{8-13}$$

where b = face width, in.
P_d = diametral pitch.

TABLE 8-6.—VALUES OF THE BUCKINGHAM GEAR-TOOTH FATIGUE CONSTANT[2] OR STRESS-LOAD FACTOR

Material	Assumed maximum specific compressive stress, lb. per sq. in.	K for 14½-deg. tooth	K for 20-deg. tooth
Cast steel and cast steel	60,000	43	59
Forged steel and cast steel	65,000	50	68
Forged steel and forged steel	80,000	76	104
Hardened steel and cast steel	90,000	96	131
Forged steel and semisteel	80,000	114	156
Hardened steel and phosphor bronze	85,000	135	185
Hardened steel and semisteel	90,000	145	198
Heat-treated steel and heat-treated steel	120,000	171	234
Phenolic laminated and metal	32,000	189	259
Semisteel and semisteel	90,000	193	264
Hardened steel and heat-treated steel	130,000	201	275
Hardened steel and hardened steel	220,000	576	790

NOTE: Additional values of K in terms of Brinell hardness will be found in reference 4.

Hence the allowable torque may be expressed as

$$Q = \frac{5\pi SYN}{P_d^3} \tag{8-14}$$

where S = allowable unit stress at velocity V, lb. per sq. in.
Y = Buckingham strength-form factor (Table 8-4).
N = number of teeth in the gear.
P_d = diametral pitch.

The maximum safe horsepower corresponding to Eq. (8-14) is

$$\text{b.hp.} = \frac{2\pi n Q}{12 \times 33{,}000} \tag{8-15}$$

where n is the speed of the gear, r.p.m.
Q is the torque, lb.-in.

Combining Eqs. (8-14) and (8-15)

$$\text{b.hp.} = \frac{nSYN}{4{,}010 P_d^3} \tag{8-16}$$

or

$$P_d = \sqrt[3]{\frac{nSYN}{4{,}010 \text{ b.hp.}}} \tag{8-17}$$

8-4. Example of Single Reduction Gearing Calculation.—Determine suitable dimensions for a 3/2 single reduction gear to be used on a 150-b.hp. 2,700-r.p.m. aircraft engine.

Procedure.—Experience has shown (Par. 8-3) that casehardened S.A.E. 2515 is suitable for reduction gears. Using this material, the allowable static stress may be taken as $S_0 = 22{,}000$ lb. per sq. in. For strength (Table A3-4) a 20-deg. stub involute tooth should be used. The arrangement of the gearing will be as in Fig. 8-10.

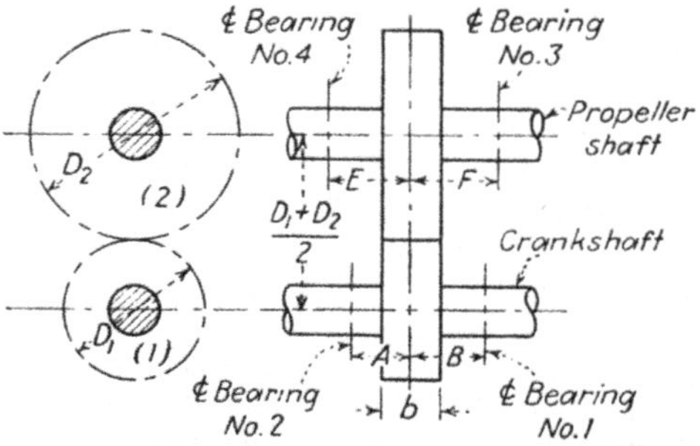

FIG. 8-10.—Arrangement for a single reduction gear.

To start, assume the drive gear No. 1 has 18 teeth, and to keep down weight and frontal area, let the distance between gear centers ≈ 6 in. Then $(D_1 + D_2)/2 = 6$, $D_2/D_1 = 3/2$. From which $D_1 = 4.8$ in., and

TABLE 8-7.—STANDARD DIAMETRAL PITCHES

1.00	2.50	6.00	12	24
1.25	3.00	7.00	14	28
1.50	3.50	8.00	16	32
1.75	4.00	9.00	18	36
2.00	5.00	10.00	20	40

$P_d = 18/4.8 = 3.75$. The nearest standard diametral pitch (Table 8-7) is either 3.5 or 4. Assume $P_d = 4$, then

$$D_1 = 18/4 = 4.5 \text{ in.}$$

and the distance between gear centers is

$$\frac{D_1 + D_2}{2} = \frac{4.5 + 1.5 \times 4.5}{2} = 5.625 \text{ in.}$$

The velocity of the pitch line [Eq. (8-8)] is

$$V = \frac{\pi}{12} \times 4.5 \times 2{,}700 = 3{,}180 \text{ f.p.m.}$$

and the allowable stress [Eq. (8-6)] is

$$S = 22{,}000 \left(\frac{1{,}200}{1{,}200 + 3{,}180}\right) = 6{,}020 \text{ lb. per sq. in.}$$

From Table 8-4, $Y = 0.12$, and from Eq. (8-16),

$$S = \frac{150 \times 4{,}010 \times 4^3}{2{,}700 \times 0.12 \times 18} = 6{,}600 \text{ lb. per sq. in.}$$

This is almost 10 per cent greater than the allowable stress, and although it would likely be taken care of in the factor of safety of the material, it is inadvisable to take unnecessary chances in so important an item.

A reduction in the stress may be made either by increasing the number of teeth or by reducing the diametral pitch, but for small changes, probably the easiest way is to widen the face of the gear. Thus the face width based on Eq. (8-13) is

$$b = \frac{10}{P_d} = \frac{10}{4} = 2.5 \text{ in.}$$

This width corresponds to a stress at full rated load of 6,600 lb. per sq. in. and the face width necessary to reduce this stress to 6,020 lb. per sq. in. is

$$b = 2.5 \times \frac{6{,}600}{6{,}020} \approx 2.75 \text{ in.}$$

Other dimensions of the gears now follow directly (Table A3-4) from the known dimensions, but before assuming that the preceding values of P_d and b will be satisfactory, it is necessary to investigate for dynamic loading and excessive wear.

For aircraft-engine reduction gears, the error in tooth action (Table 8-5) will probably not exceed 0.001 in., and for this error the dynamic tooth load constant for 20-deg. stub involute steel gears will be $C = 1{,}720$. From Eq. (8-9),

$$W = \frac{\pi \times 6{,}020 \times 2.75 \times 0.12}{4} = 1{,}560 \text{ lb.}$$

Hence, from Eq. (8-10)

$$W_d = \frac{0.05 \times 3{,}180(2.75 \times 1{,}720 + 1{,}560)}{0.05 \times 3{,}180 + \sqrt{2.75 \times 1{,}720 + 1{,}560}} + 1{,}560 = 5{,}750 \text{ lb.}$$

Substituting this value of W_d in Eq. (8-9) and solving for the stress,

$$S = \frac{W_d P_d}{\pi b Y} = \frac{5{,}750 \times 4}{\pi \times 2.75 \times 0.12} = 22{,}200 \text{ lb. per sq. in.}$$

From Table 8-2, the factor of safety for a single reduction gear should be about 4. Hence the tensile strength will be

$$22{,}000 \times 4 = 88{,}000 \text{ lb. per sq. in.}$$

and from Fig. A2-1, the corresponding Brinell hardness will be about 175.* From Fig. 8-8, the flexural endurance limit stress is

$$S_t = 42{,}000 \text{ lb. per sq. in.} > 1.5 \times 22{,}200 = 33{,}300 \text{ lb. per sq. in.}$$

Therefore, the gears are adequate to withstand the dynamic tooth loads.

* This represents the average hardness, not the surface conditions after casehardening.

For the check on probable wear resistance [Eq. (8-11)], $D = 4.5$ in., $b = 2.75$ in., and $Z = (2 \times 1.5 \times 18)/(18 + 1.5 \times 18) = 1.2$. By assuming that the steel gears will be casehardened to 500 Brinell hardness, a value of K (Fig. 8-9) of 350 to 750 may be used depending upon the desired life of the gears. For the operating life of an aircraft engine, a value of $K = 600$ would appear to be reasonable. Hence,

$$W_w = 4.5 \times 2.75 \times 600 \times 1.2 = 8{,}900 \text{ lb.}$$

and since $8{,}900 \times 0.75 = 6{,}700$ lb. > 5,750 lb., the gears are adequate for resistance to wear.

8-5. Example of Planetary-reduction Gearing Calculation.—Determine suitable dimensions for a 3/2 planetary reduction gear to be used on a 150-b.hp. 2,700-r.p.m. aircraft engine.

Procedure.—Since it will be of interest to compare this type of reduction gearing with the single reduction gearing (Par. 8-4), assume the same material, *i.e.*, S.A.E. 2515 casehardened, $S_0 = 22{,}000$ lb. per sq. in. Simple epicyclic spur gearing has been found to be suitable for reducing propeller shaft speed (Fig. 8-4); hence this system of gearing will be selected.

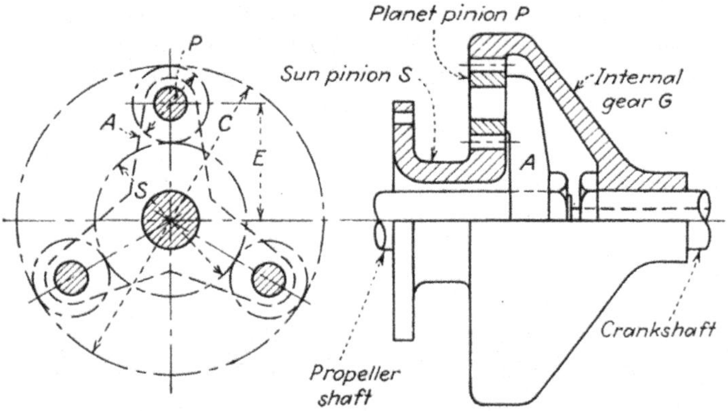

FIG. 8-11.—General arrangement for a three-planet pinion planetary reduction gear.

The arrangement of parts will be as in the figure of Table A3-6 except that in order to reduce and distribute the stress in the most critical parts more than one planet pinion will be used. For an engine of the size being considered, three* planet pinions will probably be adequate, and these may be supported by a Y-shaped arm (A, Table A3-6). Experience indicates that case 6 of Table A3-6 is a suitable arrangement; hence sun gear S will be fastened to the crankcase so that it cannot rotate, A will be splined to M_1 (the propeller shaft), and G will be splined to M_2 (the crankshaft).

* In simple planetary (epicyclic) drives,[4] the sum of the numbers of teeth in the internal gear and the sun gear must be divisible by the number of planet pinions used, *i.e.*, for three planet pinions, $(N_G + N_S) \div 3$ must equal a whole number, etc.

A preliminary sketch of the proposed arrangement of the gearing is shown in Fig. 8-11. Such layouts are useful in fixing more clearly in mind the necessary or desired arrangement o various parts even though later detailed study of means for lubrication, arrangement of bushings, methods of attaching the fixed gear to the crankcase, etc., may necessitate detailed changes.

This layout is sufficient to fix in mind the general plan of the reduction gear and to form a basis for the calculating of over-all dimensions that will be necessary. These sizes having then been determined, the layout may be altered in detail to fac litate manufacture, use of standard parts, arrangement of means for lubrication, etc. In this detail work, much assistance can be had from a study of the details of current successful designs (Figs. 8-4, and 8-5).

To keep the reduction unit as compact as possible, assume that the pitch diameter of the gear is 10 in., and since there will be three times as many contact points between gears as in the single reduction gearing (Par. 8-4), a larger value of diametral pitch may be used. Let $P_{dG} = 6$. Then

$$N_G = 6 \times 10 = 60 \text{ teeth,}$$

and from Table A3-6, case 6, for a $\frac{3}{2}$ reduction

$$\frac{60}{60 + N_S} = \frac{2}{3} \qquad \text{or} \qquad N_S = 30 \text{ teeth}$$

and the pitch diameter of sun pinion S is

$$D_S = {}^{30}\!/_{6} = 5 \text{ in.}$$

To fit the gears together, the pitch radius of S plus the pitch diameter of P will have to be equal to the pitch radius of G or

$$D_P = \frac{D_G}{2} - \frac{D_S}{2} = 5 - 2.5 = 2.5 \text{ in.}$$

The number of teeth on the pinion gear is

$$N_P = 2.5 \times 6 = 15 \text{ teeth}$$

From Table A3-6, case 6, the speed of the pinions around their own centers will be

$$n_P = 2{,}700 \times \frac{30}{15} \times \frac{60}{60 + 30} = 3{,}600 \text{ r.p.m.}$$

and the pitch-line velocity [Eq. (8-8)] is

$$V_P = \frac{\pi \times 2.5 \times 3{,}600}{12} = 2{,}355 \text{ f.p.m. } (= V_S = V_G, \text{ Table A3-6, case 6})$$

From Eq. (8-6), the allowable stress is

$$S = 22{,}000 \left(\frac{1{,}200}{1{,}200 + 2{,}355} \right) = 7{,}420 \text{ lb. per sq. in.}$$

The center distance of gears S and P is

$$E = \frac{D_S}{2} + \frac{D_P}{2} = 2.5 + 1.25 = 3.75 \text{ in.}$$

and the pitch-line velocity of the transmitted load (Table A3-6, case 6) is

$$V = 0.5236 \times 2{,}700 \times 3.75 \left(\frac{60}{60 - 15}\right) = 7{,}060 \text{ f.p.m.}$$

The transmitted load per planet pinion (Table A3-6, case 6) for a three-planet pinion reduction gear is

$$W = \frac{33{,}000 \times 150}{7{,}060 \times 3} = 234 \text{ lb. } (= W_S = W_G)$$

From Eq. (8-9), the face width of the pinion to produce a transmitted stress equal to the allowable stress is

$$b = \frac{234 \times 6}{\pi \times 7{,}420 \times 0.111} = 0.542 \approx \frac{9}{16} \text{ in.}$$

To increase the margin of safety, let $b = 5/8 = 0.625$ in.

To facilitate comparisons, assume the same accuracy of construction as in the single reduction gearing example of Par. 8-4, *i.e.*, an error in tooth action of 0.001 in. Then, from Table 8-5, the dynamic tooth load constant is $C = 1{,}720$ and the dynamic tooth load [Eq. (8-10)] is

$$W_d = \frac{0.05 \times 2{,}355(0.625 \times 1{,}720 + 234)}{0.05 \times 2{,}355 + \sqrt{0.625 \times 1{,}720 + 234}} + 234 = 1{,}234 \text{ lb.}$$

Substituting this value of W_d in Eq. (8-9) and solving for the stress

$$S_d = \frac{1{,}234 \times 6}{\pi \times 0.625 \times 0.111} = 34{,}000 \text{ lb. per sq. in.}$$

By using the same flexural endurance limit load as in the example of Par. 8-4, *i.e.*, $S_t = 42{,}000$ lb. per sq. in., it is apparent that the gear teeth would not break because of the shock loading. However, the factor of safety is $42{,}000/34{,}000 = 1.23 < 1.5$, the value recommended by Buckingham for shock loading.

To increase the factor of safety for shock or dynamic loading, let $b = 1.0$ in. Then

$$W_d = \frac{0.05 \times 2{,}355(1.0 \times 1{,}720 + 234)}{0.05 \times 2.355 \sqrt{1.0 \times 1{,}720 + 234}} + 234 = 1{,}654 \text{ lb.}$$

and

$$S_d = \frac{1{,}654 \times 6}{\pi \times 1.0 \times 0.111} = 28{,}500 \text{ lb. per sq. in.}$$

The factor of safety for shock loading is now $42{,}000/28{,}500 = 1.475 \approx 1.5$; hence the gears should be adequate to withstand the dynamic loading.

For the check on wear resistance, two conditions exist, *i.e.*, (*a*) wear between the pinion and sun gear and (*b*) wear between the pinion and internal gear. Considering case (*a*) first and referring to Eq. (8-11), $D_P = 2.5$ in., $b = 1$ in., and $Z = (2 \times 30)/(15 + 30) = 1.333$. Assuming the same degree of casehardening, *i.e.*, Brinell hardness = 500, as in the example of Par. 8-4, $K = 600$, and

$$W_w = 2.5 \times 1 \times 600 \times 1.333 = 2{,}000 \text{ lb.}$$

Since $W_d = 1{,}654 < 2{,}000 = W_w$, wear or pitting should not occur, but $W_d = 1{,}654 > 0.75 \times 2{,}000 = 1{,}500$ lb., hence the margin of safety is not large.

For case (*b*), $Z = (2 \times 60)/(60 - 15) = 2.67$ and

$$W_w = 2.5 \times 1 \times 600 \times 2.67 = 4{,}000 \text{ lb.}$$

Since $W_d = 1{,}654 < 0.75 \times 4{,}000 = 3{,}000$ lb., there is little likelihood of pitting or wear between the pinion and the internal gear.

8-6. Special Gears.—Since interchangeability is not important in reduction gears, it is possible to improve materially on conventional standard gears by resorting to special gearing. Of the possible variations, *variable-center-distance* gears are in considerable favor. With such gears, the pitch, center distance, addendum, and pressure angle may be varied to suit the tooth number of mating gears best. Reduction in wear, specific sliding, and greater ease of attainment of acceptable accuracy of tooth form are advantages claimed for variable-center-distance gears. Detailed data on these gears are to be found in references 3 and 12.

8-7. Reduction Gear Bearing Loads.—To ensure the selection of proper sizes of bearings for reduction gear shafts, bearing loads due to the gearing should be determined. In general, these loads are dependent upon the horsepower transmitted, the type of gearing, *i.e.*, whether single or double reduction, the tooth pressure angle, and the axial distance of the bearings from the center line of the gears. For helical or bevel gearing, additional thrust loads are also important. Methods of determining bearing loads due to gearing are given in Tables A3-9 to A3-12.

Example.—Determine bearing loads and select suitable sizes of antifriction bearings for the single-reduction gearing of Par. 8-4.

Procedure.—Referring to Fig. 8-10, the location of the bearings relative to the gears will be determined in part by the desired arrangement of the nose of the crankcase and by the location and arrangement of other adjacent parts. Assume for this example that these factors dictate a value of

$A = 2$ in., $B = 2$ in., $E = 2.5$ in., and $F = 2.5$ in. Referring to Table A3-9, the torque input is

$$Q = \frac{63{,}025 \times 150}{2{,}700} = 3{,}500 \text{ lb.-in.}$$

The tangential force is

$$P = \frac{3{,}500}{2.25} = 1{,}550 \text{ lb.}$$

The separating force for 20-deg. stub teeth is

$$S = 1{,}550 \tan 20° = 565 \text{ lb.}$$

The bearing loads calculated by the methods given in Table A3-9 are tabulated as follows:

Load due to	P	S
Bearing 1...	$P_1 = 1{,}550 \frac{2}{2+2} = 775$ lb.	$S_1 = 565 \frac{2}{2+2} = 282.5$ lb.
Bearing 2...	$P_2 = 1{,}550 \frac{2}{2+2} = 775$ lb.	$S_2 = 565 \frac{2}{2+2} = 282.5$ lb.
Bearing 3...	$P_3 = 1{,}550 \frac{2.5}{2.5+2.5} = 775$ lb.	$S_3 = 565 \frac{2.5}{2.5+2.5} = 282.5$ lb.
Bearing 4...	$P_4 = 1{,}550 \frac{2.5}{2.5+2.5} = 775$ lb.	$S_4 = 565 \frac{2.5}{2.5+2.5} = 282.5$ lb.

Total load

$$\text{On bearing 1} = L_1 = \sqrt{775^2 + 282.5^2} = 825 \text{ lb.}$$

By inspection, it is seen that $L_1 = L_2 = L_3 = L_4$; hence all bearings will be equally loaded.

Assume that ball bearings are to be used. Then, referring to Table A1-22, for bearings 1 and 2,

$L = 825$ lb.
$n = 2{,}700$ r.p.m.
$F = 1.0$ (assuming that the thrust will not exceed 10 per cent of L for straight spur gears).
$Z = 0.88$ (10 hr. per day $\times$ 300 days per year $= 3{,}000$ hr.; easily the life of the usual aircraft engine).
$K = 2.0$ (reduction gears are subjected to considerable shock due to torque variation and vibration).
$C = 825 \times 1 \times 0.88 \times 2.0 = 1{,}450$ lb.

By assuming that filling-notch type bearings are to be used (Table A1-22E), S.A.E. bearing 407 (heavy series), 308 (medium series), or possibly 210 (light series) could be used. The choice now largely rests with the bore most suitable to fit the crankshaft (see Table A1-22D), but other major dimensions may dictate in some cases.

To simplify the construction, the bearing adjacent to the crank arm might also be used as the end main bearing. In this case, the bearing load should be combined with the gear load to find the resultant load on the bearing.

For bearings 3 and 4, $L = 825$ lb., $n = 1{,}800$ r.p.m. (for a $\frac{3}{2}$ reduction gear), $F = 1.0$, $Z = 0.88$, $K = 2.0$, and $C = 1{,}450$ lb. as before. For filling-notch type bearings (Table A1-22E), S.A.E. bearings 406, 407, 308, or 209 should be strong enough. From Fig. A1-6, it appears that S.A.E. taper-type propeller shafts 1 or 2 should be adequate. The largest diameter of shaft end 1 is 2.05 in. and of shaft end 2 is 2.362 in. (Table A1-20). As the bearing on the propeller shaft end side of the reduction gear could not have a bore less than the maximum diameter of the S.A.E. shaft end, it would have to be at least an S.A.E. 211 bearing for the No. 1 shaft end and probably an S.A.E. 213 bearing if the No. 2 shaft end is used.

Again for the purpose of simplifying construction, it would be advisable to give some attention to the possibility of using the reduction gear bearing adjacent to the propeller for the dual purpose of taking the gear load and also the propeller thrust. Assume that the propeller efficiency at full throttle climb is 80 per cent, reduction-gear efficiency is 90 per cent, and the airplane speed under climb conditions is 60 m.p.h.

The thrust is

$$T = \frac{375 \times 150 \times 0.9 \times 0.8}{60} = 675 \text{ lb.}$$

The ratio of thrust to radial load is

$$\frac{T}{L} = \frac{675}{825} = 0.82$$

and from Table A1-22A, it is apparent that this value of T/L is beyond the recommended range for filling-notch-type bearings. For the nonfilling-notch type, $F = 1.3$ and

$$C = 825 \times 1.3 \times 0.88 \times 2.0 = 1{,}885 \text{ lb.}$$

From Table A1-22K, S.A.E. bearings 312, 313, or 314 should be adequate to handle the combined propeller thrust and reduction-gear load. As S.A.E. bearing 313 has a bore (= 2.5591 in.) nearest over the maximum diameter of the propeller shaft end, it is the more logical selection.

8-8. Reaction-torque Measurements.—In connection with the study of reduction gearing, a development by the Pratt and Whitney Company[6] for measuring brake horsepower in flight is of interest.

In this device (Fig. 8-12), the fixed gear (*D*, Fig. 8-4) is not bolted to the nose of the crankcase but instead is connected through suitable linkages to two pistons in small hydraulic cylinders. These hydraulic cylinders are connected to a high-pressure oil line, and one piston head is so shaped that it acts as a control valve. Hence the pistons are always kept floating near mid-travel position. A line from the hydraulic cylinders to a suitable pressure gage on the instrument board of the air-

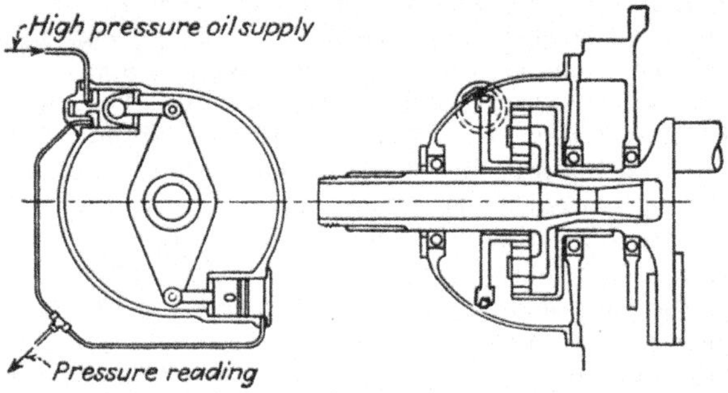

FIG. 8-12.—Schematic diagram of Pratt and Whitney torque indicator.

craft permits direct reading of the engine torque. The brake horsepower may readily be found from

$$\text{b.hp.} = K \times P_Q \times n \tag{8-18}$$

where K = a constant depending upon the reduction gear ratio and the dimensions of the torque indicator.
P_Q = torque-indicator gage reading, lb. per sq. in.
n = r.p.m. of the engine.

For an engine equipped with a constant-speed propeller, the gage could be calibrated to read in units of brake horsepower directly.

8-9. Thrust-bearing Details.—Thrust bearings are usually subjected to a combination of thrust and radial loads. The thrust load will usually be a maximum under full throttle climb conditions. The thrust in pounds may be determined from

$$T = \frac{375 \times \text{b.hp.} \times \eta \times e_{RG}}{V} \tag{8-19}$$

where T = thrust, lb.
b.hp. = brake horsepower.
η = propeller efficiency.
e_{RG} = mechanical efficiency of the reduction gearing (= 1.0 for direct drive).
V = climbing speed of the airplane, m.p.h.

The radial load on the thrust bearing depends upon the arrangement of shaft parts, the type of reduction gearing, and the arrangement of the engine, *i.e.*, in-line, radial, etc. An instance of procedure in combining reduction gear loads with the thrust load has been given in Par. 8-6. For direct-drive radial engines, Table A1-10 should be of assistance in determining the radial load on the thrust bearing.

With the thrust and radial loads known, the procedure in selecting a suitable thrust bearing is essentially the same as the

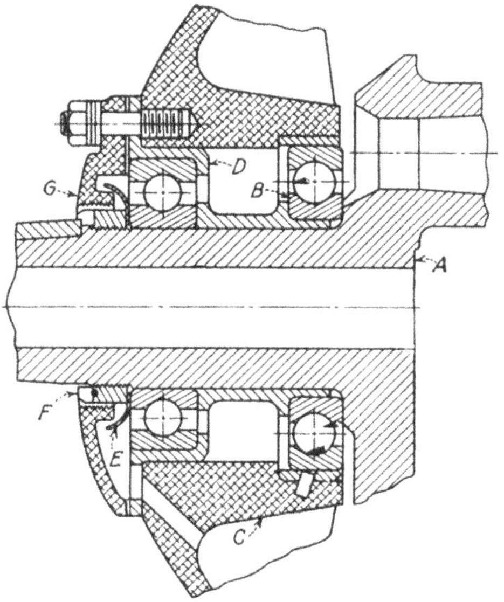

FIG. 8-13.—Front main and thrust bearing arrangement of Leblond, model 5-F. (*A*) Crankshaft; (*B*) front main bearing; (*C*) crankcase nose section; (*D*) thrust bearing; (*E*) oil slinger; (*F*) thrust bearing lock nut; (*G*) engine nose plate.

example of Par. 8-7. For most conventional arrangements, Table A1-22 gives adequate data for the preliminary selection. Before a design is given final approval for construction, it is advisable to have the manufacturer of the bearing to be used check the selection.

Detail arrangement of thrust-bearing installations vary considerably and probably can best be studied by reference to available cuts and drawings of typical engines. Figure 8-13 shows the detail arrangement of the thrust bearing used on the Leblond radial engine, model 5-F, and Fig. 8-14 shows the thrust bearing arrangement used on the Menasco six-cylinder in-line engine. Other arrangements are shown in Figs. 8-4 and 8-6.

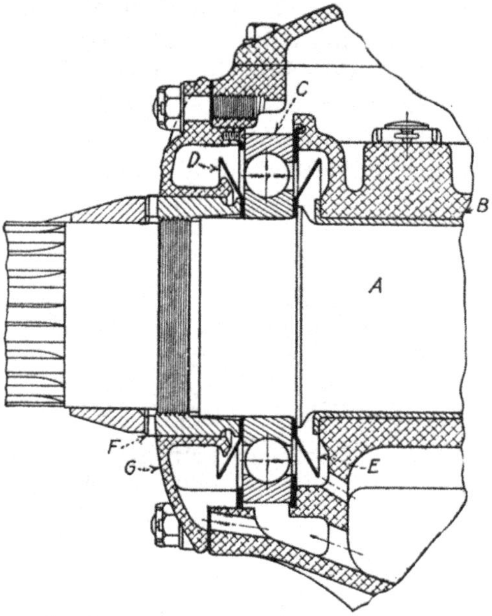

FIG. 8-14.—Thrust-bearing arrangement. Menasco six cylinder in-line engine. (*A*) Crankshaft; (*B*) front main bearing; (*C*) thrust bearing; (*D*) oil slinger; (*E*) excess oil slinger; (*F*) thrust bearing lock nut; (*G*) engine nose plate.

Suggested Design Procedure

Important. Give references for all formulas and empirical factors used. All drawings should be on standard-size paper and complete in all details, including dimensions, clearances, material specifications, number required, etc. Drawings (except as noted) should be blueprinted and properly folded (Fig. 2-4) for insertion in the design notebook. Keep a record of the man-hours required on each item.

1. Complete the remaining crank-arm and crankshaft details (see Suggested Design Procedure, page 142), and check for items suggested in Par. 8-1. Complete the layout and dimensioning of the crank arm and bearing portions of the crankshaft.

2. Select a suitable size of S.A.E. propeller shaft end.

3. Make a preliminary pencil sketch approximately to scale of the desired arrangement of the nose of the crankcase and the connecting parts between the crankshaft and the propeller shaft end. Locate the various parts of the crankshaft, reduction gearing (if used), supporting bearings, and supports for the bearings. Check the arrangement to be certain the parts can be fitted together. Arrange to use standard S.A.E. bearing sizes,

bearing lock nuts, splines, etc., wherever possible. (See tables in the appendix and/or the S.A.E. "Handbook.")

4. If reduction gears are to be used, select the desired arrangement and determine all necessary dimensions. Alter the sketch of item 3 above as necessary.

5. Determine bearing loads for reduction-gear supporting bearings, and select suitable bearings. Alter the sketch of item 3 above as necessary.

6. Determine thrust bearing loads, and select a suitable size of thrust bearing. Alter the sketch of item 3 above as necessary.

7. Complete the detail arrangement of crankshaft and propeller-shaft parts in the nose section. Use S.A.E. standard parts as applies. Complete the desired arrangement of detail parts in the nose section.

8. Make detail drawings of all parts that cannot be definitely identified by S.A.E. number. Include all dimensions, specify materials to be used (by S.A.E. number as applies), indicate heat-treatment and type of case-hardening as applies, check arrangement for lubricating of parts, and identify each part by number or letter in accordance with the part-numbering system being used.

9. Make an assembly drawing of the nose section parts on the layout drawing of Suggested Design Procedure, page 24, item 4. Show parts in section whenever such sectioning increases the clarity or legibility of the drawing. Include only principal over-all dimensions. Identify each part of the assembly drawing by a reference number corresponding to the detail drawing or reference number of that part. When the detailed drawing contains more than one part, identify each part by the detailed drawing number and a letter. Do not blueprint the assembly drawing at this stage.

10. When items 1 to 9 have been completed and put in proper form, submit for checking and approval.

Problems

1. An airplane attains 150 m.p.h. with an engine of 200 b.hp. operating at 3,200 r.p.m. direct drive.

a. Find the propeller diameter and maximum propeller efficiency.

b. What gain in propeller efficiency could be had if the engine was equipped with an $8/5$ reduction gear having a mechanical efficiency of 95 per cent?

c. What brake horsepower could the 3,200-r.p.m. direct-drive engine develop if it were slowed down to the geared-engine propeller speed of 2,000 r.p.m.?

d. By assuming 73 octane number fuel for the slowed-down engine in part *c*, what increase in octane number would be necessary to attain the power required by supercharging?

e. By assuming that the reduction gear increases the weight of the 3,200-r.p.m. engine by 12 per cent, what saving in weight will this represent over increasing the size of the engine at 2,000 r.p.m. to where it will develop 190 b.hp.?

2. An engine builder plans to use spur-type single reduction gearing for his in-line 150-b.hp. 3,000-r.p.m. engine. If the pitch diameter of the

pinion is 4 in. and the material is forged S.A.E. 1045 steel, what would be a safe allowable stress at pitch-line velocity?

3. If the pinion in Problem 2 has 16 teeth, a 20-deg. stub involute form, and a face width corresponding to AGMA recommendations, what maximum safe tangential load can be transmitted? Will this be adequate for rated brake horsepower?

4. The reduction-gear pinion in Problems 2 and 3 has a Brinell hardness number of 200 and an error in tooth action of 0.001 in. What is the dynamic tooth load stress? Will this dynamic or shock load stress the material beyond its flexural endurance limit?

5. The pinion in the preceding three problems is casehardened to 500 Brinell, and the reduction gear ratio is $3/2$. What stress factor K is necessary to prevent pitting and wear of the gear teeth? Explain briefly why a higher surface hardness would help reduce pitting and wear.

6. Select materials and determine suitable dimensions for a $4/3$ single reduction gear to be used on a 200-b.hp. 2,800-r.p.m. aircraft engine. Minimum desirable number of teeth on the pinion 15, minimum desirable diametral pitch 4. Center distance and frontal area to be as small as possible to keep down drag and shielding of cooling fins. Give reasons for each selection or assumption.

7. Select materials and determine suitable dimensions for a $4/3$ planetary reduction gear to be used on a 200-b.hp. 2,800-r.p.m. aircraft engine. Minimum desirable number of teeth on any gear 15, minimum desirable diametral pitch 6, minimum desirable number of pinion gears 3, maximum 6, internal gear to be splined to engine shaft, supporting arms for planet pinion gears to be splined to propeller shaft, sun pinion gear to be bolted to nose of crankcase. Make a sketch approximately to scale showing the arrangement of gearing, and tabulate calculated stresses, pitch diameters, numbers of teeth, diametral pitch, face width, and approximate over-all dimensions of assembled gearing. Give reasons for each selection or assumption.

References

1. Wood: "Technical Aerodynamics.'
2. Norman, Ault, and Zarobsky: "Fundamentals of Machine Design."
3. Buckingham: "Spur Gears."
4. Buckingham: "Manual of Gear Design."
5. "New Departure Handbook."
6. *S.A.E. Jour.*, Vol. 42, No. 2, February, 1938.
7. S.A.E. "Handbook."
8. Timoshenko: (*a*) "Strength of Materials" and (*b*) "Theory of Elasticity."
9. Heldt: Gear Steels, *Automotive Ind.*, Dec. 17, 1938.
10. Rasmussen: Gear Calculations Based on Dynamic Loading and Wear Resistance, *Product Eng.*, Part I, February, 1939; Part 2, March, 1939.
11. Larsen: Internal Gears, Formulas for Calculating Wear Resistance, *Product Eng.*, July, 1939.
12. Albert and Rogers: "Kinematics of Machinery."

CHAPTER 9

CYLINDERS AND VALVES

9-1. Functions of the Cylinder.—Aircraft-engine cylinders serve to

1. Provide space for confining accurately metered portions of the charge while it is passing through the cycle or process of having its chemical energy converted into heat and mechanical energy.
2. Guide the piston in its reciprocating motion.
3. Provide the thermal path for removal of a large portion of the heat energy liberated during combustion.
4. Support and guide the valves, and usually support part of the valve gear.
5. Support the spark plugs or, in the case of Diesels, the injector nozzles.
6. Support part of the inlet and exhaust passages for the charge.

These principal functions very largely dictate the details of design of the cylinder. For instance, requirement 1 necessitates a relatively gastight construction, an accurate combustion chamber space, sufficient structural rigidity to prevent distortion during combustion, fatigue resistance to rapidly varying forces, sufficient strength at operating temperatures, and resistance to corrosion by the products of combustion. Requirement 2 calls for a rigid cylinder to prevent binding of the piston, a smooth wall surface to keep friction to a minimum, a hard wall surface to keep wear low, and in conventional construction sufficient strength to transmit the axial force to the crankcase. Requirement 3 dictates the use of materials of high thermal conductivity. For air-cooled cylinders within the range of sizes used in aircraft engines, this calls for more or less elaborate cooling fins. For liquid-cooled cylinders, a means of confining the liquid closely around the cylinder must be provided. Requirement 4 requires the use of wear-resistant valve seats and guides and sufficient rigidity to prevent binding of parts or leakage of charge. Requirement (5) is evident, and requirement 6 necessitates a more or less complex construction.

In addition to these requirements, the cylinder must (7) be as light in weight as possible.

9-2. Types of Cylinder Construction.—As in the case of other aircraft-engine parts, compliance with the diverse requirements for cylinders results in numerous compromises in the selection and arrangement of parts. In the automotive. marine, and

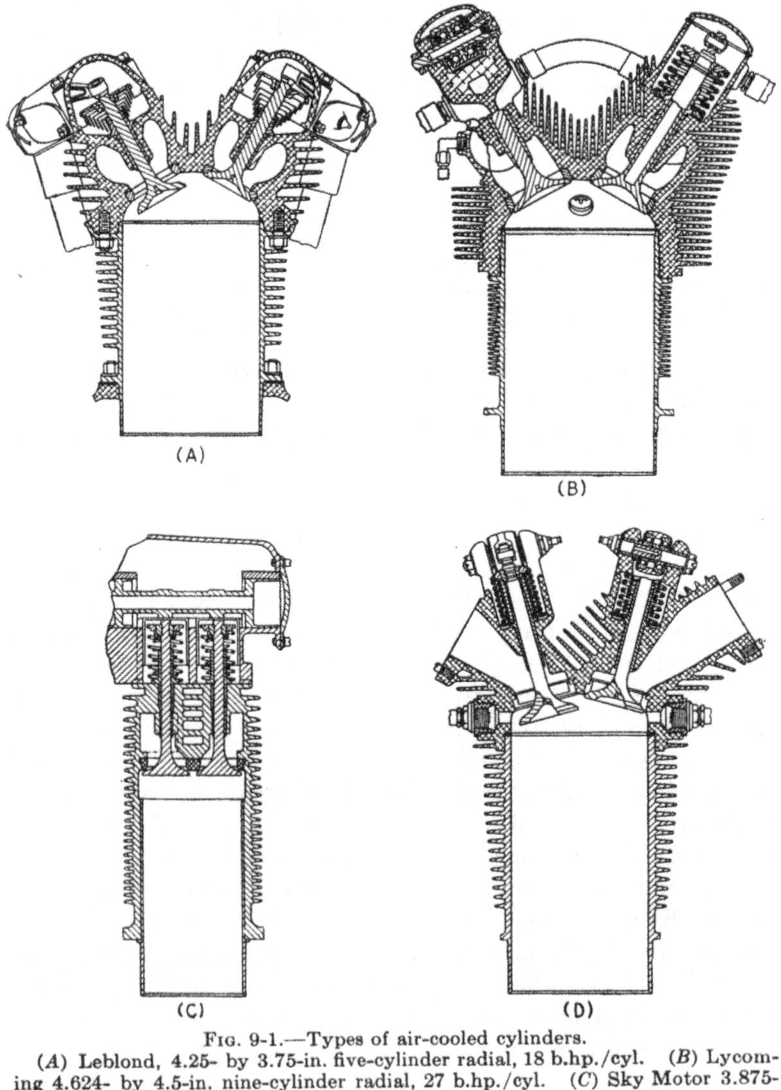

Fig. 9-1.—Types of air-cooled cylinders.

(*A*) Leblond, 4.25- by 3.75-in. five-cylinder radial, 18 b.hp./cyl. (*B*) Lycoming 4.624- by 4.5-in. nine-cylinder radial, 27 b.hp./cyl. (*C*) Sky Motor 3.875- by 4.25-in. four-cylinder in-line, 15 b.hp./cyl. (*D*) Menasco 4.5- by 5.125-in. six-cylinder in-line, 27 b.hp./cyl.

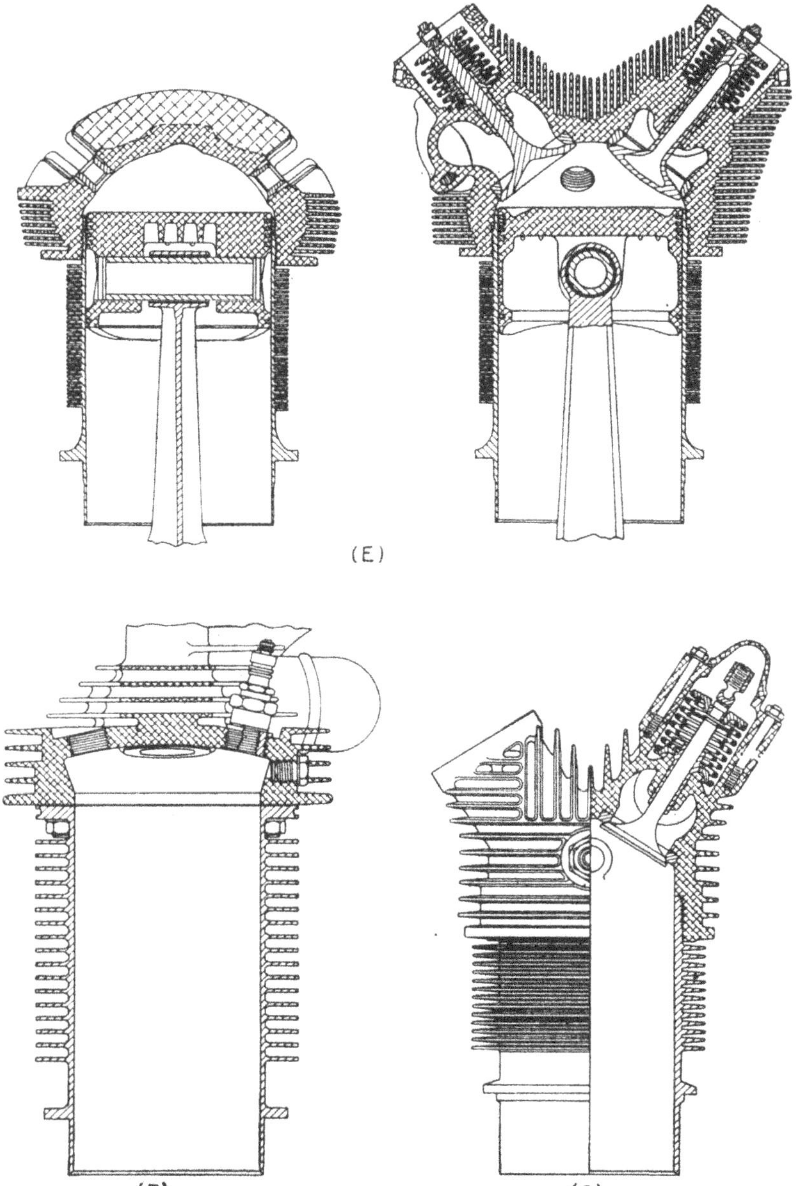

FIG. 9-1 (*Cont.*).—Types of air-cooled cylinders. (*E*) Wright 6.125- by 6.875-in. nine-cylinder radial, 90 b.hp./cyl. (*F*) Kinner 4.25- by 5.25-in. five-cylinder radial, 20 b.hp./cyl. (*G*) Lenape 4.125- by 4-in. three-cylinder radial, 17 b.hp./cyl.

industrial fields where weight is at less of a premium, cast iron is by far the most commonly used material for cylinders, but for aircraft-engine cylinders its strength-weight ratio is too low, and for air-cooled cylinder heads, its coefficient of thermal conductivity is inadequate for all but the smallest sizes of engines. In a few cases, cast iron is used because of its relatively low cost.

The majority of modern aircraft-engine cylinders are of composite construction, the barrel being machined from a steel forging bolted to or screwed and shrunk into a cast aluminum-alloy head. Aluminum has an excellent thermal conductivity, and the head sections may be made sufficiently thick, without excessive weight, to withstand the gas-force stresses, but it is not sufficiently wear resistant to serve as material for valve guides and valve seats, and usual practice is to use a more suitable alloy for these parts.

To resist wear, steel cylinder barrels are treated to a high surface hardness. In some types of engines, replaceable cylinder liners are used, but in aircraft engines this method is seldom used. Cylinder barrels are usually provided with a hold-down flange which permits their being bolted to the crankcase, but in some cases through bolts from the head have been used. In liquid-cooled engines, jackets of cast aluminum or of sheet steel have been used to confine the cooling medium. In air-cooled engines, fins are machined on the barrel and cast integral with the head. Figure 9-1 shows typical current arrangements of composite cylinder construction, and additional details may be studied from blueprints and drawings.

9-3. Cylinder Materials.—For cylinder barrels, Johnson[1] recommends S.A.E. 1050, S.A.E. 4140, or Nitralloy (Tables A2-9 and A2-11). Choice among the three will rest largely with the severity of the service, *i.e.*, the specific performance desired, and with the cost, the plain carbon steel being, of course, the least expensive.

For cylinder heads, aluminum alloys 142, 355, A355, or 195, (Tables A2-1, A2-2, A2-3, A2-4, and A2-5) are recommended. These alloys retain their mechanical properties well at elevated temperatures, a feature particularly desirable in cylinder-head materials.

Valve guides may be made of cast aluminum bronze,[1,2] S.A.E. Specification 68 (Table A2-12), wrought aluminum bronze,[1,2]

S.A.E. Specification 701 (Table A2-12), or hard cast bronze,[1,2] S.A.E. Specification 62 (Table A2-12). Valve seats may be made of cast aluminum bronze,[1,2] S.A.E. Specification 68 (Table A2-12), wrought aluminum bronze,[1,2] S.A.E. Specification 701 (Table A2-12), or NF-9, an alloy of copper, aluminum, iron, nickel, and manganese (Table A2-10). For severe service, valve seats may be faced with Stellite No. 6 (Table A2-10). Cylinder studs and nuts[1] may be made of S.A.E. 3140 or 6150 (Tables A2-9 and A2-11). Stellite No. 1 is recommended for valve stem tips.

9-4. The Cylinder Barrel.—To prevent rupture, the cylinder barrel must be strong enough to withstand the maximum gas force to which it is subjected. The greatest gas force occurs normally when the piston is near the top of the cylinder and hence shields the cylinder from direct action of the force, but under adverse conditions, such as preignition, the upper end of the cylinder barrel may be subjected to near maximum explosion pressure. The usual relation for the stress in thin-walled cylinders is obtained from

$$PR = St \tag{9-1}$$

where P = maximum pressure in the cylinder, lb. per sq. in.
R = radius of the cylinder (= bore/2), in.
S = stress, lb. per sq. in.
t = thickness of the cylinder wall, in.

For thin-walled cylinders with closed ends, the longitudinal stress in the walls is obtained from

$$PR = 2S't \tag{9-2}$$

where S' = longitudinal stress, lb. per sq. in.

Obviously the conditions of Eq. (9-1) are the more critical.

In applying Eq. (9-1) to engine cylinders, consideration must be given to manufacturing limitations. Thus, for cast gray iron automotive cylinders with integral jackets, the outer surface of the cylinder wall cannot be machined or closely checked for shift of the core during pouring. Hence, it is usual practice to add about ⅛ in. to the value of t as determined from Eq. (9-1). For an allowable stress of 6,000 lb. per sq. in. and an assumed maximum cylinder pressure of 500 lb. per sq. in., the cylinder wall thickness for cast iron can be

$$t = 0.0416 \times D + 0.125 \tag{9-3}$$

where D = cylinder diameter, in.
This agrees fairly well with the recommendations of Huebotter[3] and Heldt[4] for automotive engines.

For air-cooled cast-iron cylinders, both inside and outside surfaces can be machined, and an allowance for eccentricity need not be made. In addition, the cooling fins act as stiffener ribs, and t may be reduced somewhat to save weight. However, caution should be exercised as the longitudinal stress is not affected by the ribs, and due to the relatively small radii of fillets between the cooling fins, high local stresses may be set up.

For steel cylinder barrels, allowable stresses of 12,000 to 20,000 lb. per sq. in. may be used, depending upon the quality of the steel.

Cylinders are usually attached to the crankcase by means of hold-down studs which should be sufficient in number to distribute the stress in the cylinder flange and in the metal of the crankcase. S.A.E. coarse series threads[2] should be used for studs that are to be fitted into aluminum-alloy crankcases. The effective length of the threads in the soft metal should be two to three times the diameter of the stud. Some manufacturers use a reduced diameter of the stud (equal to or slightly less than the root diameter of the threaded portions) so that any misalignment or lack of parallelism will not cause a concentration of stress at the surface of the aluminum adjacent to the edge of the stud hole. Also this shifts the bolt "stretch" away from the threads. Ground threads are also used to increase accuracy and thereby reduce the possibility of part of the threads carrying all the load. The threads are ground after hardening.

Safe loads on studs and bolts are given in Table 9-1. The critical force tending to pull the cylinder off the crankcase is equal to the maximum gas pressure times the piston area. This force is also equal to the force tending to pull the cylinder head away from the barrel. For very high maximum cylinder pressures, as in Diesels, the force tending to pull the cylinders away from the crankcase may be sufficient to require an excessive number and size of studs and a rather massive supporting boss in the crankcase or the use of a stronger crankcase metal. To permit the use of a very light crankcase, the Packard Diesel used steel hoops around the crankcase on either side of the cylinders. In place, these hoops fitted over flanges at the base of the cylinders, and they were tightened until the crankcase was under considerable initial compression. This method of holding cylinders to the crankcase would not ordinarily be necessary in gasoline engines.

When the head is bolted to the barrel, bolt selection is much the same as for the hold-down studs. For screwed and shrunk heads, a sufficient number of threads should be used to ensure adequate resistance to shear of the weaker metal.

Cooling fins on steel cylinder barrels are usually machined into the steel forging. These fins serve primarily to conduct away excess heat, but they also serve to strengthen and increase the rigidity of the barrel.

TABLE 9-1.—SAFE LOADS FOR S.A.E. COARSE (N.C.) THREAD SERIES BOLTS*

Nominal bolt diameter, in.	Number of threads per inch	Ultimate strength, lb. per sq. in.		
		65,000 (carbon or nickel steel)	80,000 (nickel steel)	95,000 (nickel steel heat-treated)
1/4	20	186	229	272
5/16	18	322	396	470
3/8	16	488	601	714
7/16	14	675	830	986
1/2	13	915	1,125	1,340
9/16	12	1,186	1,460	1,730
5/8	11	1,480	1,820	2,170
3/4	10	2,240	2,760	3,280
7/8	9	3,140	3,860	4,580
1	8	4,120	5,060	6,010

From Marks: "Mechanical Engineers' Handbook."

* For aircraft-quality steels under steady loads where fatique is not critical or where the uncertainty of uniform load distribution is not a factor, these allowable loads may be safely doubled or possibly even tripled.

To prevent rapid wear, cylinder walls are usually hardened, and frequently the rubbing surfaces of pistons are made more wear resistant by anodic treatment, *i.e.*, chemically changing the surface to a harder compound. These processes reduce the rate of wear, but the principal cause of initial wear remains. To get at this cause of wear, it is necessary to consider briefly the usual methods of finishing metal surfaces. In cutting with a tool (Fig. 9-2*A*) the metal at the point of separation *P* is torn loose because the point of the tool is too blunt to reach the bottom of the separating space. This process generates much heat, and as the cutting oil does not effectively penetrate this space, most of the heat is absorbed by the metal. When the rate of cutting

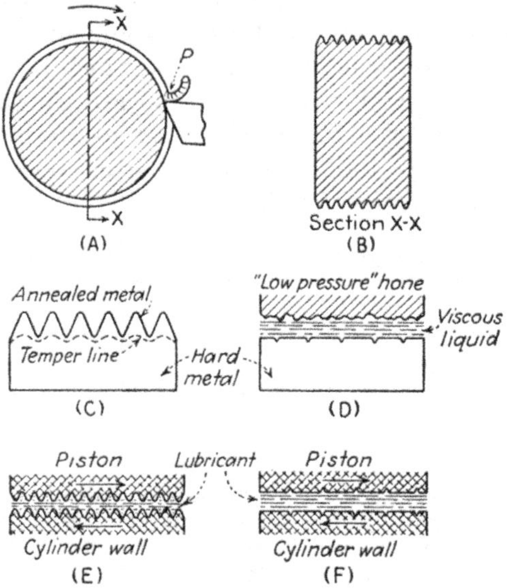

FIG. 9-2.—Types of bearing finishes: (*A*) tearing action of a cutting tool; (*B*) rough surface from conventional cutting or grinding; (*C*) effect of tearing action; (*D*) principle of the "superfinish" process; (*E*) conventional bearing; (*F*) "superfinished" bearing. [Sections (*C*) (*D*) (*E*) and (*F*) greatly enlarged.]

is sufficient, the thin sections between the grooves made by the tool point (Fig. 9-2*B* and *C*) will be annealed and softened. Subsequent casehardening will restore the temper, but the final finish grinding usually produces a recurrence of the annealing, for the cutting grains of the stone tear and heat the metal in much the same way as does the tool. Thus, by the conventional methods of finishing, bearing surfaces are far from smooth and if highly magnified would appear in section much the same as Fig. 9-2*C* and *E*.

In operation, a bearing such as Fig. 9-2*E* would be satisfactory as long as the viscosity of the lubricant was sufficient to prevent contact of the high points on the surfaces, but under the high load and temperature conditions encountered in aircraft engines, this ideal condition is unlikely to long exist. Once metallic contact occurs, the greatly increased friction and resulting heat thins the lubricant so that further metallic contact follows rapidly. The result is either fusion of the metal and failure, or if the process is deliberate and less severe, *i.e.*, the so-called "running in" of a new engine, the high points will be gradually removed. Unfortunately, in this latter case, the clearance will be increased, with resulting greater troubles from oil pumping, tendency to blow by, etc.

By removing the high points of the bearing surface (Fig. 9-2*D*), the possibility of metal-to-metal contact will be very much reduced, the running in time will be lessened or eliminated, and proper clearance will be maintained much longer. With conventional finishing, this smoothing is difficult because as the hone removes the high points it also digs deeper into the base metal to form new scratches. To get around this tendency to form new scratches, a new process of finishing (called *superfinish* as developed by the Chrysler Corporation) uses a very light pressure on the hone and a liquid of suitable viscosity, which allows the cutting edges of the hone to reach the high points and remove them but prevents the cutting edges from digging into the base metal and forming new scratches. Thus, the annealed high points are removed to form an exceedingly smooth surface having a hardness practically equal to the original case value. Such finishing greatly reduces the possibility of metal-to-metal contact, and even if it does occur, wear will be less rapid because the area of contact will be much greater.

9-5. Cooling Fins and Baffles.—To prevent an excessive temperature rise in the cylinder with resulting troubles from detonation, structural failure of working parts, etc., it is necessary to provide a good thermal path for heat flow from the combustion chamber to the cooling medium. For direct air-cooled engines,* this necessitates the use of fins on the head and around the upper part of the barrel. The transfer of heat from the cylinder to the cooling air consists of conduction of the heat through the fins to the fin surfaces and of the convection of the heat from the fin surfaces to the cooling air. For constant temperature

* A discussion on liquid cooling is to be found in reference 15.

conditions, the heat removed from the head and barrel must be equal to the heat absorbed.

The rate at which heat is given up to the air may be expressed[5] for the cylinder head as

$$H = a_h U_h (t_h - t_a) \tag{9-4}$$

and for the cylinder barrel as

$$H = a_b U_b (t_b - t_a) \tag{9-5}$$

where H is in B.t.u. per hr.

a_h and a_b = respective base areas of the head and barrel covered by cooling fins, sq. in.

U_h and U_b = respective over-all heat transfer coefficients for the head and barrel in B.t.u. per hr. per sq. in. of (head or barrel) *base* area per deg. F. difference between the average (head or barrel) temperature and the cooling air.

t_h and t_b = respective average temperatures of the head and barrel, deg. F.

t_a = temperature of the cooling air, deg. F.

The desirability of having a high over-all heat transfer coefficient is apparent.

Biermann and Pinkel[6] have shown that the over-all heat transfer coefficient may be expressed as

$$U = \frac{q}{S + T}\left[\frac{2}{a}\left(1 + \frac{W}{2R_b}\right)\tanh aW' + S_b\right] \tag{9-6}$$

where U is as in Eqs. (9-4) and (9-5).

q = surface heat transfer coefficient, B.t.u. per sq. in. *total* surface area per hr. per deg. F. temperature difference between the surface and the entering cooling air.

T = average fin thickness, in.

S = average space between fins, in.

W = fin width, in.

$W' = W + \frac{T_t}{2}$, the effective fin width.

T_t = fin-tip thickness, in.

R_b = radius from the center of the cylinder to the fin root, in.

$$a = \sqrt{\frac{2q}{KT}}.$$

K = thermal conductivity of the metal, B.t.u. per sq. in. per deg. F. through 1 in. per hr. ($K = 2.17$ for steel and 7.66 for aluminum Y alloy).

S_b = distance between adjacent fin surfaces at the fin root, in.

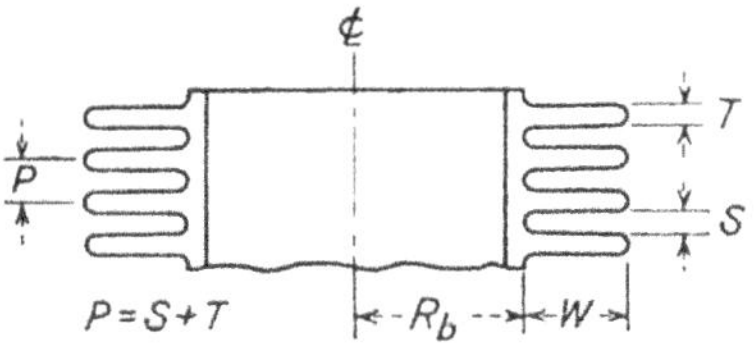

FIG. 9-3.—Fin nomenclature:
T = thickness
S = space
P = pitch
W = width
R_b = fin root radius

The relation of the various fin dimensions is shown in Fig. 9-3.

The surface heat-transfer coefficient[5] q of a body is a function of the gas velocity, density, conductivity, viscosity, specific heat, and dimensions of the body. These factors may be expressed by the relation

$$q = C_P \rho V \times f\left(\frac{\rho VS}{\mu}, \frac{\mu C_P}{K}, r_1, r_2, r_3, \ldots\right) \tag{9-7}$$

where C_P = specific heat at constant pressure.
ρ = gas density.
V = gas velocity.
μ = gas viscosity.
S = characteristic dimension.
$\frac{\rho VS}{\mu}$ = Reynolds number.
$\frac{\mu C_P}{K}$ = Prandtl number.
r_1, r_2, etc., = dimensionless ratios of other important dimensions of the body to S.

and f is read "function of."

Experimental studies[6] by the NACA to determine the effect of these various factors on q for finned cylinders indicate that the surface heat-transfer coefficient is principally affected by the fin spacing and air velocity.* Figure 9-4 shows results of these tests. Cross plotting against velocity at constant fin spacing showed q to vary about as the 0.796 power of the velocity.

* Gibson[16] found that a steel surface gave 5 to 10 per cent greater heat dissipation than aluminum and a coating of stove enamel increased the heat dissipation from cast aluminum fins about 10 per cent.

The values of q in Fig. 9-4 apply only to a cylinder of 4.66-in. diameter and atmospheric conditions of 29.92 in. Hg. and 80°F. However, by use of the theory of similitude it can be shown[6] that the data may be applied to other sizes of cylinders and other atmospheric conditions. Thus, if we let

$$J = \frac{D_x}{D_T} \tag{9-8}$$

where D_x = outer cylinder wall diameter in inches for the cylinder under investigation.

D_T = outer cylinder wall diameter in inches for the cylinder upon which Fig. 9-4 is based (= 4.66 in.),

and relate the other dimensions and the velocities by

$$D_T = \frac{D_x}{J}; \qquad T_T = \frac{T_x}{J}; \qquad W_T = \frac{W_x}{J}; \qquad S_T = \frac{S_x}{J};$$

$$V_T = JV_x \tag{9-9}$$

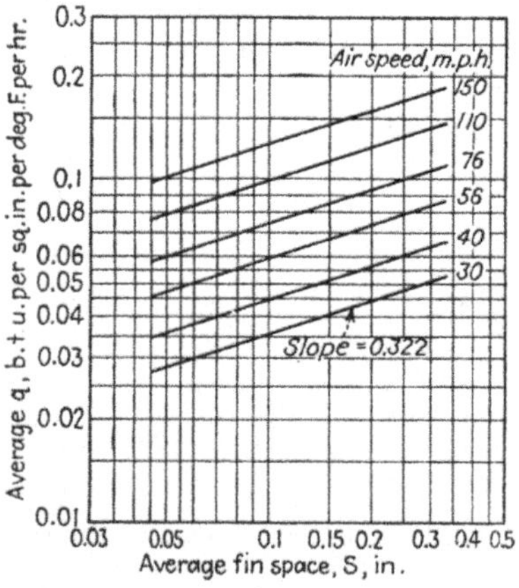

FIG. 9-4.—Variation of average q with fin spacing. (*From NACA Tech. Rept.* 488.)

where T = fin thickness, in.

W = fin width, in.

S = fin spacing, in.

V = velocity,

and the subscripts correspond to the subscripts for the diameters, the two cylinders may be regarded as dimensionally similar.[6] The cylinder of diameter D_x may now be converted to an equivalent cylinder of diameter D_T, and by making the conversion for the other factors [Eq. (9-9)], Fig. 9-4 may be entered to find q_T. The surface heat-transfer coefficient q_x may now be found from the relation[6]

$$q_x = \frac{q_T}{J} \tag{9-10}$$

The effect of altitude may be corrected by[6]

$$V_S = \frac{\rho V_a}{0.0734} \tag{9-11}$$

where V_a = velocity at altitude, m.p.h.
ρ = weight density at altitude, lb. per cu. ft.
V_S = equivalent velocity at sea level (*i.e.*, corresponding to Fig. 9-4), m.p.h.

Example 1.—Determine the surface heat-transfer coefficient for a cylinder barrel of 4-in. bore, a fin width of 0.7 in., a fin pitch of 0.25 in., and a fin thickness of 0.0625 in. The velocity past the cylinder (which corresponds to velocity of best climb) is 60 m.p.h., and atmospheric conditions are standard.

Solution.—Assuming a cylinder wall thickness of ⅛ in., and using the symbols of Eqs. (9-8) and (9-9),

$$J = \frac{4 + 2 \times 0.125}{4.66} = \frac{4.25}{4.66} = 0.91$$

$$S_T = \frac{0.25 - 0.0625}{0.91} = 0.206 \text{ in.}$$

$$V_J = 60 \times 0.91 = 54.5 \text{ m.p.h.}$$

From Fig. 9-4, $q_T = 0.07$, and from Eq. (9-10), $q_x = \frac{0.07}{0.91}$

$= 0.077$ B.t.u./(sq. in.)(deg. F.)(hr.)

Example 2.—Determine the over-all heat transfer coefficient U, if the cylinder barrel in Example 1 is of steel, and the fins are rectangular in cross section.

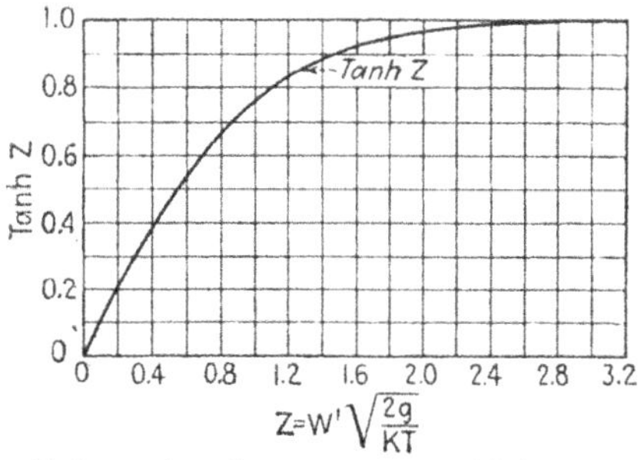

FIG. 9-5.—Values of tanh Z. (*From NACA Tech. Rept.* 488.)

Solution.—From Example 1, $q = 0.077$, $T = 0.0625$ in. ($= T_t$ for rectangular fins), $S = 0.1875$ in. ($= S_b$ for rectangular fins), $W' = 0.7$ in. ($= W'$ for rectangular fins). For steel, $K = 2.17$,

$$a = \sqrt{\frac{2q}{KT}} = \sqrt{\frac{2 \times 0.077}{2.17 \times 0.0625}} = 1.067, \qquad 2R_b = 4.25 \text{ in.}$$

Substituting these values in Eq. (9-6),

$$U = \frac{0.077}{0.1875 + 0.0625}\left[\frac{2}{1.067}\left(1 + \frac{0.7}{4.25}\right) \tanh 1.067 \times 0.7 + 0.1875\right]$$

$$U = 0.308(2.185 \tanh 0.745 + 0.1875)$$

From Fig. 9-5, tanh 0.745 = 0.64

$$U = 0.308(2.185 \times 0.64 + 0.1875)$$
$$U = 0.495 \text{ B.t.u./(hr.)(sq. in.)(deg. F.) difference}$$

Example 3.—Determine the over-all heat-transfer coefficient for the aluminum Y alloy head fins of the cylinder in Example 1 for an average fin width of 1 in., a pitch of 0.38 in., and an average fin thickness of 0.1 in. Assume that the fin-tip thickness is 0.08 in. and the fin-root thickness is 0.12 in. Assume an average head thickness of 0.375 in.

Solution.—The data of Fig. 9-4 are based on cylinder barrels, but it may be assumed that the air-flow characteristics around the head approximate the conditions of the barrel; hence

$$J = \frac{4 + 2 \times 0.375}{4.66} = 1.02$$
$$S_T = \frac{0.38 - 0.1}{1.02} = 0.274 \text{ in.}$$
$$V_J = 60 \times 1.02 = 61 \text{ m.p.h.}$$

From Fig. 9-4, $q_T = 0.09$, and from Eq. (9-10),

$$q_x = \frac{0.09}{1.02} = 0.0882 \text{ B.t.u./(sq. in.) (deg. F.) (hr.)}$$

For the over-all heat-transfer coefficient, $q = 0.0882$, $T = 0.1$ in., $T_t = 0.08$ in., $S = 0.28$ in., $S_b = 0.26$ in., $W = 1$ in., $W' = 1 + \frac{0.08}{2} = 1.04$ in., $2R_b = 4.75$ in., for aluminum Y alloy $K = 7.66$, $a = \sqrt{\frac{2 \times 0.0882}{7.66 \times 0.1}} = 0.48$

Substituting these values in Eq. (9-6),

$$U = \frac{0.0882}{0.28 + 0.1}\left[\frac{2}{0.48}\left(1 + \frac{1}{4.75}\right) (\tanh 0.48 \times 1.04) + 0.26\right]$$
$$U = 0.232[(5.04 \times \tanh 0.5) + 0.26]$$

From Fig. 9-5, tanh 0.5 = 0.475

$$U = 0.615 \text{ B.t.u./(hr.) (sq. in.) (deg. F.) difference}$$

Pinkel[5] has shown that for a Pratt and Whitney 1340-H cylinder, the relation among the average cylinder barrel and head temperatures, the indicated horsepower, and the over-all heat-transfer coefficient may be expressed as in Fig. 9-6. These curves are based on data from one size and design of cylinder, but they serve to show the approximate temperatures of other sizes of reasonably similar cylinders.

Example 4.—Determine the approximate cylinder wall and head temperatures for the cylinder in Examples 1, 2, and 3, if the engine is a five-cylinder radial rated at 70 b.hp.

Solution.—From Fig. A1-2, the mechanical efficiency will be about 85 per cent, hence the indicated horsepower per cylinder will be

$$\text{i.hp.} = \frac{70}{0.85 \times 5} = 16.5$$

For the cylinder wall,

$$\frac{U}{\text{i.hp.}^{0.64}} = \frac{0.495}{16.5^{0.64}} = 0.0825$$

From Figure 9-6, the approximate average cylinder-barrel temperature is 325°F.

For the cylinder head,

$$\frac{U}{\text{i.hp.}^{0.64}} = \frac{0.615}{16.5^{0.64}} = 0.1027$$

From Fig. 9-6, the approximate average cylinder-head temperature is 355°F.

Example 5.—Determine for the engine of the preceding four examples the portion of the heat supplied which is removed by the cooling fins. Air temperature 80°F., number of cooling fins on the cylinder barrel = 20.

Solution.—The area of the barrel covered by cooling fins is

$$a_b = 20 \times 0.25 \times \pi \times 4.25 = 66.7 \text{ sq. in.}$$

The heat removed per hour through the barrel fins is [from Eq. (9-5)]

$$H_b = 66.7 \times 0.495(325 - 80) = 8{,}100 \text{ B.t.u. per hr.}$$

The area of the head covered by cooling fins is somewhat irregular and the effective area is uncertain due to the complicated heat flow around the valves. However, the area may be approximated by assuming that

$$\frac{a_{h1}}{a_{b1}} = \frac{a_{h2}}{a_{b2}} \qquad (9\text{-}12)$$

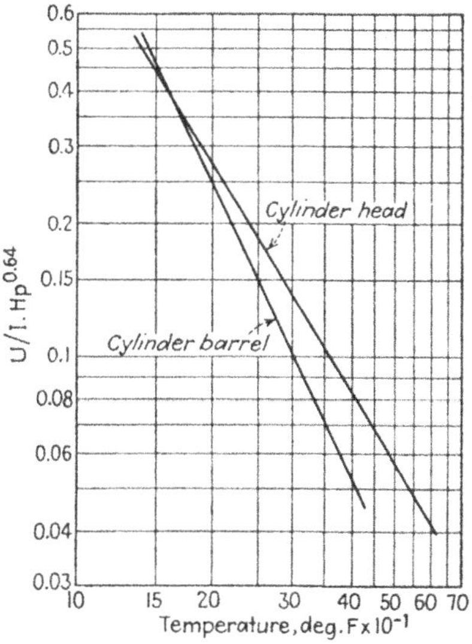

FIG. 9-6.—Approximate relation between cylinder-barrel and-head temperatures, indicated horsepower, and the over-all heat-transfer coefficient. (*From NACA Tech. Rept.* 488.)

where a_{h1} and a_{b1} are the respective base areas of the head and barrel of the cylinder under consideration. a_{h2} and a_{b2} are the respective base areas of the head and barrel of a similar cylinder that has been measured.

According to Pinkel,[5] the Pratt and Whitney 1340-H cylinder has a base area of barrel covered by fins of $a_{b2} = 68.7$ sq. in., and a base area of head of $a_{h2} = 142$ sq. in. Hence, if we assume that the cylinder under consideration is similar to the Pratt and Whitney 1340-H cylinder, the base area of the head may be taken as

$$a_{h1} = \frac{66.7 \times 142}{68.7} = 138 \text{ sq. in.}$$

When possible, of course, it is much more advisable actually to measure the areas for the cylinder under consideration.

The heat removed per hour through the head fins is [From (Eq. 9-4)]

$$H_h = 138 \times 0.615(355 - 80) = 23{,}400 \text{ B.t.u. per hr.}$$

At rated load, an engine of this power should have a brake thermal efficiency of at least 25 per cent. Hence, the heat supplied per cylinder per hour is

$$H_s = \frac{2545 \times 70}{0.25 \times 5} = 142{,}300 \text{ B.t.u. per hr.}$$

The percentage of the heat supplied that passes out through the cooling fins is

$$\frac{8{,}100 + 23{,}400}{142{,}300} = 0.221, \text{ or } 22.1\%$$

According to Swan,[15] for adequate cooling, the heat dissipated from the cooling fins should be about equal to 50 to 60 per cent of the heat equivalent of the brake horsepower. On this basis, for adequate cooling of the engine in Example 5, it would be necessary for the fins to dissipate not more than

$$\frac{2{,}545 \times 70 \times 0.6}{5} = 21{,}378 \text{ B.t.u. per cyl. per hr.}$$

Since the fins are capable of dissipating

$$8{,}100 + 23{,}400 = 31{,}500 \text{ B.t.u. per cyl. per hr.}$$

under the assumed conditions, it is evident that the assumed fin dimensions are easily adequate.

As the specific power output of an engine is increased by supercharging, a limit is quickly reached at which ordinary methods of air cooling are inadequate. To extend this limit, controlled cooling by means of deflectors or baffles is used. These baffles are designed to deflect the air into the fin spaces where it would not flow normally, such as the downstream or rear side of the cylinders. The effect of a well-designed close-fitting baffle in reducing cylinder temperature, particularly at the rear side of the cylinder, is shown in Fig. 9-7.

The NACA has investigated the effect of several types of baffles and deflectors on temperature reduction and heat transfer,[7] and the findings indicate that the surface coefficient q is increased about 30 per cent by the use of a good shell baffle. The shape of the baffle is also of importance, the conclusions being that (*a*) the shell should fit tightly around the ends of the fins, (*b*) the entrance angle (α in Fig. 9-7) should be about 145 deg., (*c*) the rearward extension of the baffle behind the cylinder should

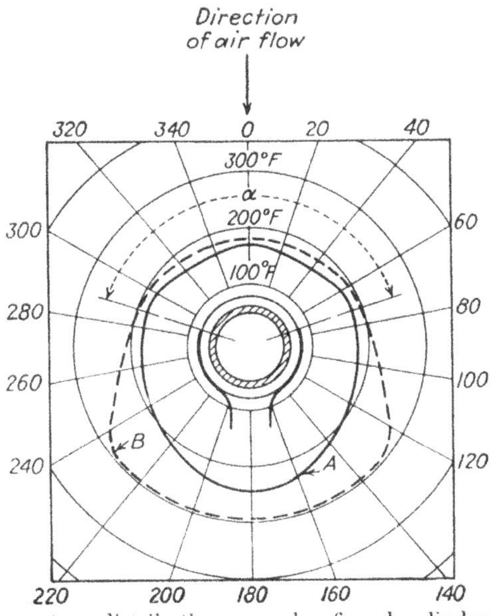

FIG. 9-7.—Temperature distribution around a finned cylinder: (*A*) with shell baffle; (*B*) without baffles. (*From S.A.E. Jour., Vol. 35, No. 4.*)

be about 3 in., and (*d*) the ratio of exit area to free-flow area between the fins should be between 1.6 and 2.3. It was also observed that both with and without baffles the surface heat-transfer coefficient q varied as the 0.85 power of the air speed. This is slightly greater than the 0.796 reported in reference 6.

To force the cooling air between the fins and through the baffles, it is necessary to provide a pressure drop or difference in pressure between the inlet and exit sides. This available pressure drop can be provided by suitable engine cowling such as the NACA or equivalent cowling. The quantity of air that must be forced between the fins is dependent upon the specific weight, specific heat, and temperature rise of the air, or

$$Q = \frac{H}{3{,}600wc_P(t_o - t_i)} \tag{9-13}$$

where Q = flow, cu. ft. per sec.
H = heat to cooling, B.t.u. per hr. [corresponding to the H in Eqs. (9-4) and (9-5)].
w = specific weight of the air, lb. per cu. ft.
c_P = specific heat of the air = 0.24 B.t.u. per lb.
t_o = outlet air temperature at the baffle exit, deg. F.
t_i = inlet air temperature = atmospheric temperature, deg. F.

Since the temperature rise $(t_o - t_i)$ is relatively small, Q will have to be quite large, but since

$$Q = AV = CA\sqrt{2gh} \tag{9-14}$$

where A = cross-sectional area of the space between the fins, sq. ft.
V = mean velocity of the air between the fins, ft. per sec.
C = a coefficient relating theoretical and actual velocity.
g = acceleration of gravity, ft. per sec.^{2}
h = head or pressure drop causing flow, feet of air, and

since air drag is proportional to h, it is apparent that increase in heat transfer by increasing the velocity will be at the expense of a rapid increase in drag of the engine. The air-drag horsepower necessary to provide the necessary Q may be expressed as

$$HP_{AD} = \frac{Q \times P_d}{550} \tag{9-15}$$

where Q = air flow between the fins, cu. ft. per sec.
P_d = pressure drop, lb. per sq. ft.

But $P_d = hw$ where w is the specific weight of the air, lb. per cu. ft.

Therefore,

$$HP_{AD} = \frac{C'wAh^{3/2}}{550} \tag{9-16}$$

where $C' = C \times \sqrt{2g}$ and the other symbols are the same as in the preceding equations.

From this, it is évident that it is more economical from a power-loss standpoint to increase the heat transfer by increasing the area

A, and this may be best attained by increasing the width of the fins. Unfortunately, manufacturing limitations have prevented the use of cast fins having a width much greater than about 1.5 to 2.0 in., a thickness less than about 1/16 in., and a space

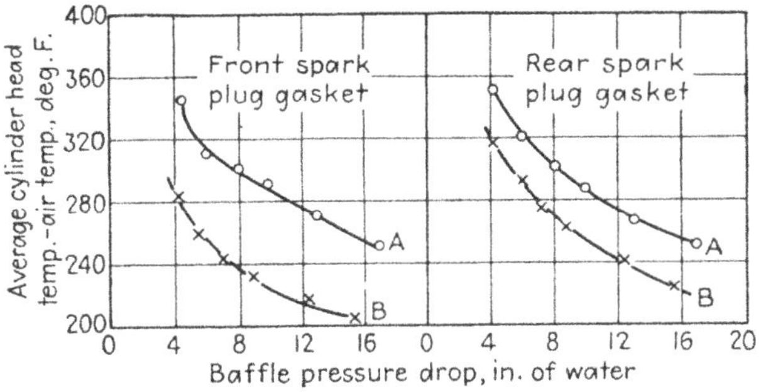

FIG. 9-8.—Effect of turbulence and pressure drop on cylinder-head temperature for *A* Wright Cyclone cylinder. Curves *B* approximate the turbulence of flight conditions. (*From Campbell, Cylinder Cooling and Drag of Radial Engine Installations, S.A.E. Jour., Vol. 43, No. 6, December,* 1938.)

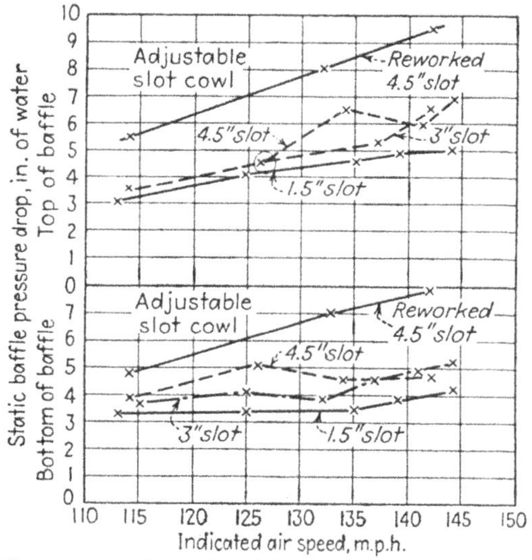

FIG. 9-9.—Baffle pressure drops for different types of cowling and air speeds. (*From S.A.E. Jour., Vol.* 43, *No.* 6.)

between fins of much less than 5/32 in., and even these dimensions are attained at considerable expense and foundry troubles. To attain greater heat transfer than these dimensions will permit at present necessitates either increasing the pressure drop *h*, by improved cowling[10] or for extreme cases by blower cooling,[8] or increasing the effectiveness of heat transfer, *i.e.*, controlling the turbulence of the air.[10] An example of the effect of turbulence and pressure drop on cylinder temperature is shown in Fig. 9-8. For recent data on cowl and baffle design for very high-

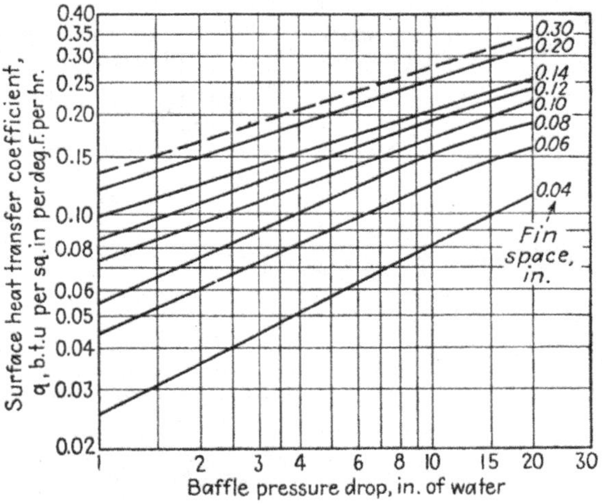

FIG. 9-10.—Effect of baffle pressure drop and fin spacing on surface heat transfer coefficient. (*From S.A.E. Jour., Vol.* 41, *No.* 3.)

performance engines, the student should consult references 8, 9, and 10.

Attainable pressure drops across baffled cylinders vary so widely with the design details of the cowling that it is necessary to know in detail the arrangement of the parts of the cowling in order to decide just what pressure drop can be obtained. Very few data of this sort are available, but reference to Fig. 9-9 indicates that baffle pressure drops of 4 to 6 in. of water at maximum speed and 2 to 4 in. of water at climbing speed would be reasonable assumptions for engine-design purposes. The effect of pressure drop on the surface heat transfer coefficient is shown in Fig. 9-10. These data are based on a cylinder of 4.66 in. diameter enclosed in a special type of baffle designed for blower cooling (Fig. 9-11). However, in the absence of more complete

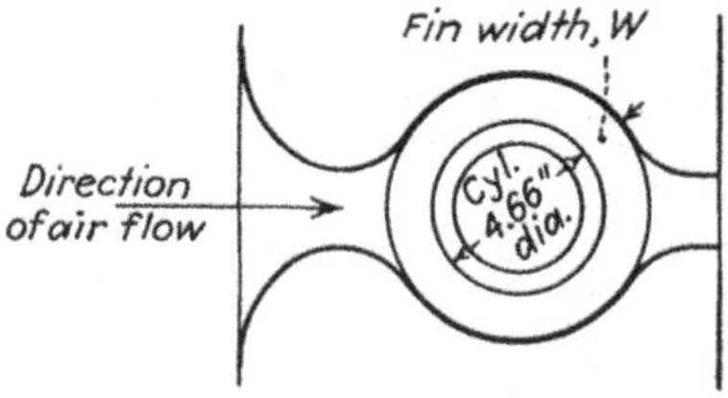

FIG. 9-11.—Type of baffle used to determine the data of Fig. 9-10. (*From S.A.E. Jour., Vol.* 41, *No.* 3.)

data, they may be approximately applied to the general case by use of Eqs. (9-8), (9-9), (9-10), and (9-11).

Example 6.—What increase in the proportion of heat to cooling could be attained for the engine in Example 5 if the cylinders were baffled and a baffle pressure drop of 2 in. of water was available?

Solution.—From Example 1, $J = 0.91$, $S_T = 0.206$ in., and

$$VJ = 54.5 \text{ m.p.h.}$$

From Fig. 9-10, $q_T = 0.151$, and from Eq. (9-10),

$$q_x = \frac{0.151}{0.91} = 0.166 \text{ B.t.u./(sq. in.)(deg. F.)(hr.)}$$

For the over-all heat transfer coefficient of the cylinder barrel,

$$a = \sqrt{\frac{2q}{KT}} = \sqrt{\frac{2 \times 0.166}{2.17 \times 0.0625}} = 1.565$$

and

$$U_b = \frac{0.166}{0.1875 + 0.0625}\left[\frac{2}{1.565}\left(1 + \frac{0.7}{4.25}\right)(\tanh 1.565 \times 0.7) + 0.1875\right]$$

$$U_b = 0.91 \text{ B.t.u./(hr.)(sq. in.)(deg. F. difference)}$$

From Example 3, $J = 1.02$, $S_T = 0.274$, and $VJ = 61$ m.p.h. From Fig. 9-10, $q_T = 0.164$, and from Eq. (9-10),

$$q_x = \frac{0.164}{1.02} = 0.161 \text{ B.t.u./(sq. in.)(deg. F.)(hr.)}$$

For the over-all heat transfer coefficient for the cylinder head,

$$a = \sqrt{\frac{2q}{KT}} = \sqrt{\frac{2 \times 0.161}{7.66 \times 0.1}} = 0.649$$

and

$$U_h = \frac{0.161}{0.28 + 0.1}\left[\frac{2}{0.649}\left(1 + \frac{1}{4.75}\right)(\tanh 0.649 \times 1.04) + 0.26\right]$$

$$U_h = 1.042 \text{ B.t.u./(hr.)(sq. in.)(deg. F. difference)}$$

For the cylinder barrel,

$$\frac{U}{\text{i.hp.}^{0.64}} = \frac{0.91}{16.5^{0.64}} = 0.1515$$

From Fig. 9-6, the approximate average cylinder-barrel temperature is 240°F.

For the cylinder head,

$$\frac{U}{\text{i.hp.}^{0.64}} = \frac{1.042}{16.5^{0.64}} = 0.1735$$

From Fig. 9-6, the approximate average cylinder-head temperature is 250°F.

The heat removed per hour through the barrel fins is [from Eq. (9-5)]

$$H_b = 66.7 \times 0.91(240 - 80) = 9{,}200 \text{ B.t.u. per hr.}$$

The heat removed per hour through the head fins is [from Eq. (9-4)]

$$H_h = 138 \times 1.042(250 - 80) = 24{,}450 \text{ B.t.u. per hr.}$$

The percentage of the heat supplied that passes out through the cooling fins is

$$H_{CF} = \frac{9{,}200 + 24{,}450}{142{,}300} = 0.2365, \text{ or } 23.65\%$$

The increase in heat to cooling is

$$23.65 - 22.1 = 1.55\%$$

The percentage increase is

$$\frac{1.55}{22.1} \times 100 = 7\%$$

Example 7.—Assuming an octane number such that the cylinder temperatures without baffles, *i.e.*, $t_b = 325°$F. and $t_h = 355°$F. as found in Example 4, are satisfactory, what increase in indicated horsepower per cylinder would be possible with baffles and a baffle pressure drop of 2 in. of water.

Solution.—For the cylinder barrel, for $t_b = 325°$F. (from Fig. 9-6),

$$\frac{U_b}{\text{i.hp.}^{0.64}} = 0.0825$$

and for $U_b = 0.91$ (Example 6)

$$\text{i.hp.} = \left(\frac{0.91}{0.0825}\right)^{1/0.64} = 43$$

For the cylinder head for $t_b = 355°$F. (from Fig. 9-6),

$$\frac{U_h}{\text{i.hp.}^{0.64}} = 0.1027$$

and for $U_h = 1.042$ (Example 6)

$$\text{i.hp.} = \left(\frac{1.042}{0.0127}\right)^{1/0.64} = 37.2$$

The limiting part is the cylinder head, and for $t_h = 355°$F., the percentage increase in horsepower by using baffles and a 2-in. pressure drop is

$$\frac{37.2 - 16.5}{16.5} \times 100 = 125\%$$

The value of baffles and cowling is readily apparent.

9-6. Valve Requirements and Materials.—During operation, aircraft-engine valves are subjected to or must withstand

1. Combustion temperatures ranging up to 3,000°F. or more.
2. Exhaust temperatures of the order of 1,200 to 1,500°F.
3. Pressures of 500 or more pounds per square inch without leaking.
4. Rapid hammering of the valve face against its seat and of the tappet against the end of the stem.
5. Wear due to friction in the valve guides.
6. Corrosion or oxidation by various constituents in the charge and in the products of combustion.

To meet these conditions a valve must have

1. A high strength at unusually high working temperatures.
2. Maximum resistance to distortion or warping.
3. Sufficient hardness and resistance to impact to prevent rapid wear.
4. Resistance to corrosion and oxidation.
5. No tendency to air-harden when cooled rapidly.

These requirements are difficult to meet and some of them are conflicting. Hence, few materials are entirely satisfactory, and valves, especially exhaust valves, have long been regarded as limitations to further increases in performance. However, gradual progress in design details, particularly in regard to more effective cooling, and now developments in special steels are continually pushing back the limitations. Suitable valve materials[4,11,12] are S.A.E. 3140, chrome nickel steel, silcrome steel, and colbalt-chromium steel (see also Tables A1-3 and A2-11). The advantage in physical properties of austenitic steels over hardenable steels for exhaust valves at high-performance operating temperatures is indicated by Fig. 9-12.

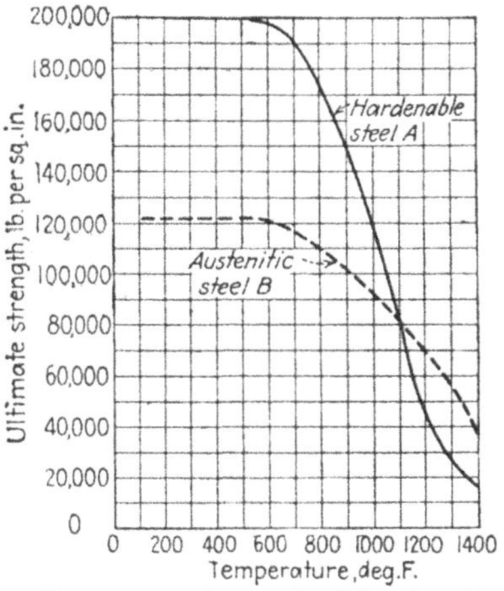

FIG. 9-12.—Strength of hardenable vs. austenitic steel. Note superiority of austenitic steel at high temperatures. (*From Wil-Rich Forum, Vol.* 11, *No.* 3.)

Austenitic steels also have high corrosion resistance. For severe service, Stellite-faced seats, special hardening, and internal cooling by salts or sodium may be used. Salts usually used are lithium and potassium nitrate. These salts or sodium, when sealed in the hollowed-out stem of the exhaust valve, melt at operating temperatures and are thrown back and forth in the space due to the reciprocating motion of the valve. In this way, heat is mechanically carried from the hot head of the valve to the relatively cooler stem adjacent to the valve guide. The hollow stem is usually filled a little more than half full of the coolant.

9-7. Breathing Capacity and Valve Size.—The area of a valve head exposed to the combustion gases increases as the square of the diameter, but the area through which heat can escape to the valve seat increases only as the first power of the diameter (for a given seat width). Hence, large valves are more difficult to cool. Large valves are also heavier, and this means greater load on the valve gear during the rapid acceleration of opening

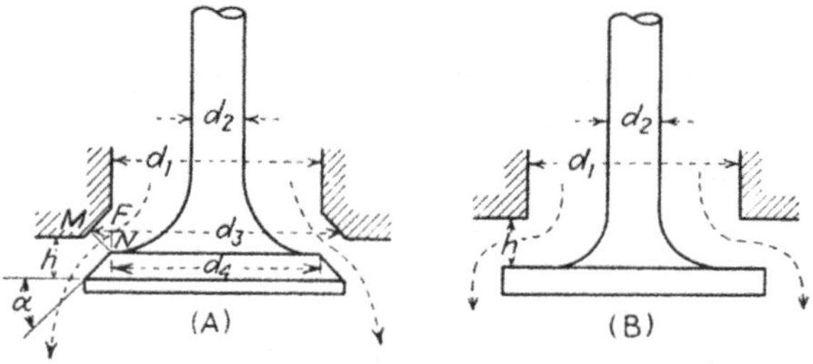

FIG. 9-13.—(*A*) Conventional conical-seated valve. (*B*) Flat-seated valve.

and on the valve springs during closing. However, too small a valve tends to restrict the flow of charge and thereby reduce the volumetric efficiency and power of the engine (Fig. 9-15). Valve size is also limited by the available area of combustion-chamber wall. Multiple valves, *i.e.*, two intake and two exhaust valves per cylinder, permit a greater area of opening for the charge and increase the ratio of area of valve seats to head but at the expense of increased complexity of valve gear and cylinder head. At present, multiple valves are used principally on large liquid-cooled engines. Adequate area of opening is obtained in radial engines with one intake and one exhaust valve by inclining the valves at an angle to the cylinder axis (Fig. 9-1).

Valve seats (Fig. 9-13) may be flat, *i.e.*, normal to the valve-stem axis, or inclined (conical), the latter being the practice in aircraft engines. Flat-seated valves give a somewhat greater area of opening for a given valve lift, but gas in passing through the opening has to make two approximately 90-deg. turns which tend to increase the turbulence and the resistance to passage. Gas in passing a conical-seated valve can assume a more nearly streamlined flow with resulting less friction. Hence, although the conical-seated valve has the disadvantage of a less area of opening for a given valve lift, it is generally conceded to give higher volumetric efficiency than the flat-seated valve. In addition, the conical seat helps to keep the valve head aligned with the stem. Both 30-deg. and 45-deg. seats are used, but the latter is by far the most common.

Theoretically, the area of opening through the valve seat should be about the same as the minimum net area of the port. For the flat-seated valve (Fig. 9-13), the area through the seat is

$$A_{FS} = \pi d_1 h \tag{9-17}$$

where d_1 = minimum port diameter, in.
h = lift of the valve, in.

The area of the port is

$$A_v = 0.785(d_1^2 - d_2^2) \tag{9-18}$$

where d_2 = diameter of the valve stem.

Combining Eqs. (9-17) and (9-18), the lift necessary to avoid pinching the flow is

$$h = \frac{(d_1^2 - d_2^2)}{4d_1} \tag{9-19}$$

Inlet valve stem diameters are usually about 25 per cent of the valve port diameters.

On this basis, Eq. (9-19) becomes

$$h = \frac{d_1^2 - (0.25d_1)^2}{4d_1} = 0.234d_1 \tag{9-20}$$

The area of opening through a conical-seated valve (wide valve face) is the lateral area of the frustum of a right circular cone. Referring to Fig. 9-13 A, the diameters of the frustum of the cone are d_1 and d_3 and the slant height is

$$MN = S = h \cos \alpha$$

where h = lift, in.,
α = angle of the valve seat,

also

$$d_3 = d_1 + 2MF = d_1 + 2S \sin \alpha = d_1 + 2h \cos \alpha \sin \alpha$$

Since the lateral area of the frustum of a right circular cone is

$$A = \frac{\pi}{2} S(d_1 + d_3)$$

where S is the slant height and d_1 and d_3 are the diameters of the bases, the area of opening through the conical valve seat may be expressed as

$$A_{CS} = \pi h \cos \alpha (d_1 + h \cos \alpha \sin \alpha) \qquad (9\text{-}21)$$

For the most commonly used conical angle of $\alpha = 45$ deg., Eq. (9-21) reduces to

$$A_{CS} = 1.11h^2 + 2.22d_1h \qquad (9\text{-}22)$$

Equating this value of A_{CS} to the net area of the valve port and letting $d_2 = 0.25d_1$.

$$h^2 + 2d_1h = 0.663d_1^2$$

From which

$$h = 0.29d_1 \qquad (9\text{-}23)$$

For the line MN (Fig. 9-13) to pass through the valve face with this amount of lift, d_4 would have to be considerably less than d_1. If d_4 is only slightly less than d_1 (the usual case), the area of opening will approximate the lateral area of a frustum of a right cone plus the lateral area of a cylinder of diameter d_1, *i.e.*, the lift will be intermediate between that for a conical seat and that for a flat-seated valve. For very narrow valve seats,* the lift for a conical-seated valve will approach that for a flat-seated valve. Hence, for minimum restriction to flow, $h = 0.25d_1$.

* Swan[15] suggests the following exhaust valve seat widths:

Port diameter, in.	7/8–1 1/8	1 1/4–1 7/8	2–2 1/4	2 3/8–2 7/8	3–3 1/4
Projected seat width, in.	5/64	3/32	1/8	5/32	7/32

See also Lichty, "Internal Combustion Engines," 5th ed., p. 309.

Lewis and Nutting[13] have found that velocities through the valve opening more nearly approach the theoretical velocity $V\ (= \sqrt{2gH})$ when the lift is a smaller percentage of the diameter of the valve [= f (diameter valve port)]. This is explained as probably being due to the "jet action" at low ratios of lift to diameter. Figure 9-14 shows the results of some tests on flow through poppet valves. In these results, the *coefficient of efflux* is defined as the ratio of the observed mean velocity through the valve to the mean velocity that would theoretically result from a pressure drop equal to that across the valve. Thus valve lifts somewhat less than $h = 0.25d_1$ may be used without serious impairment of the flow $Q\ (= AV)$, because as the area of opening through the valve is decreased by decreasing the lift, the coefficient of efflux and hence the actual velocity is increased. In practice, it may be desirable to make the lift considerably less than $h = 0.25d_1$ in order to reduce noise and acceleration forces on the valve gear.

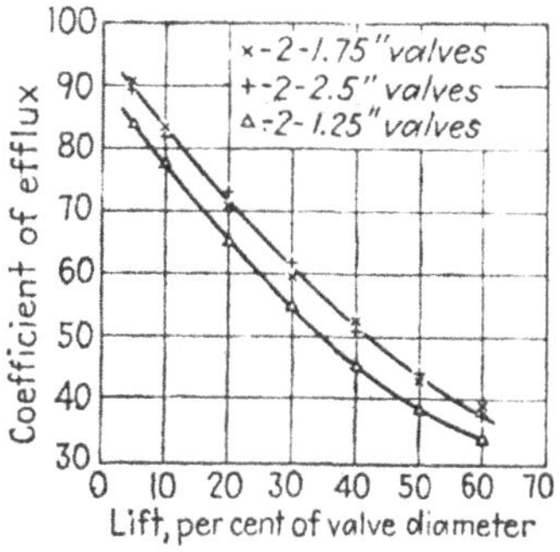

FIG. 9-14.—Effect of valve lift on coefficient of efflux. (*From NACA Tech. Rept. No. 24.*)

The mean velocity of the gas through the valve port may be calculated by the relation

$$V_v = V_p \frac{A_p}{A_v} \tag{9-24}$$

where V_v = mean velocity of the gas through the valve port, f.p.s.

V_p = mean piston speed, f.p.s. (= 2 × stroke in feet × r.p.s.).

A_v = net area of the valve port, sq. in.

A_p = area of the piston, sq. in.

Equation (9-24) is a convenient way of relating velocity and valve size, but it does not take into account the effect of numerous factors such as pressure surging, back pressure on the exhaust, effectiveness of scavenging, compression ratio, and supercharging. However, a knowledge of usual mean velocity values aids in approximating the probable valve size that should be used.

Figure 9-15 shows the effect of gas velocity and valve size on power output on a small engine. Usual mean gas velocities through inlet valves of nonsupercharged aircraft engines of conventional design and corresponding brake mean effective pressures are shown in Fig. A1-8.

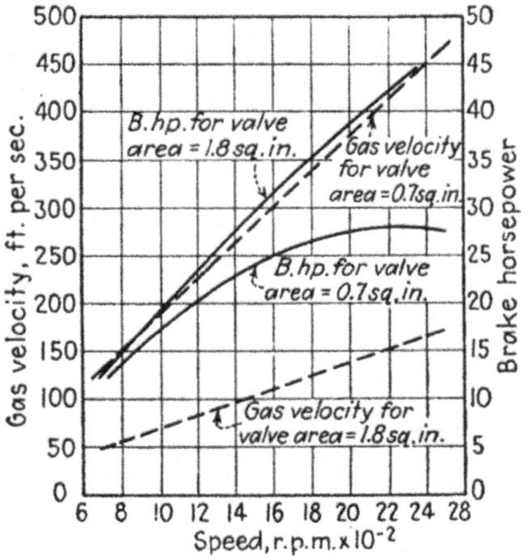

FIG. 9-15.—Effect of valve size on brake horsepower. (*From Pomeroy as reported in Judge, "Automobile and Aircraft Engines."*)

9-8. Valve Details.—Valves are usually formed from steel rod by upsetting. Flat- and spherical-head mushroom shapes may be used, but for aircraft engines, concave or so-called tulip heads are more common (Fig. 9-1). This dishing of the head serves to lighten the valve and possibly reduce its tendency to warp. However, the metal should be sufficient in section to provide the necessary thermal path to the stem and to provide sufficient strength against any tendency for the spring to pull the valve into the port. The valve should not appreciably overhang the seat, as the thin section at this point is exposed to the hot combustion gases on three sides and can easily be overheated.

The head should be joined to the stem by a large-radius fillet to avoid the excessive stress incident to the reentrant corner as well as to provide a more direct path for heat flow to the stem. Screw-driver slots and spanner holes for valve-grinding purposes are objectionable because they provide added area for heat absorption and act as obstructions to the heat flow in the metal. A rubber suction cup on the valve-grinding tool eliminates the need for slots and spanner holes in the head. In general, the shape of the valve head should be such that it will absorb the least heat both from the gases in the combustion chamber and from the feed back of heat from the gases in the exhaust pipe.

When tulip-shaped heads are used, the rim of the head should be thick enough to carry the heat circumferentially and dissipate it gradually in event of poor seating on one side. Exposed thin sections should be avoided as they are apt to be the source of fine failure cracks following overheating.

Valve stems should be sufficiently large in diameter to provide an adequate thermal path for the heat that is transferred to the valve guides. A valve-stem diameter of about one-fourth the valve-port diameter fairly well represents current practice, but in high-powered engines, this may be insufficient to prevent overheating and subsequent elongation of the stem. A larger diameter of valve stem tends to obstruct the flow of gas through the port, and the usual alternative is to use a hollow stem filled with a salt or metallic sodium. At operating temperatures, the fused salt or metal is thrown back and forth in the stem and transfers the heat by convection. Salt- or sodium-cooled valves are quite effective, but they are more expensive than solid-steel valves. Hence, they are generally used only in the more critical exhaust valves (Fig. 9-1).

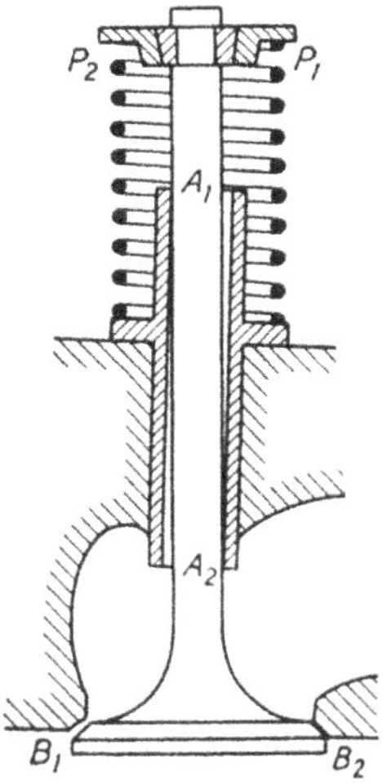

FIG. 9-16.—Effect of unequal spring pressure on retainer ($P_1 > P_2$). Excessive wear at A_1 and A_2. Unequal valve seating at B_1 and B_2 tends to bend stem or impair efficiency. (Diagram exaggerated.)

In aircraft-engine practice, valve-spring retainers are usually held in place by split taper collars which fit in grooves in the valve stem (Fig. 9-1). These grooves should be shallow and have fillets in reentrant corners to avoid undue stress in the stem. Even with squared ends, valve springs seldom exert the same pressure at all points around the spring retainer. This unequal pressure tends to bend the valve stem and cause binding and more rapid wear of the stem and guide (Fig. 9-16). Jardine and Jardine[12] suggest the use of a spherical seat between the valve-spring retainer and the split collar as a means of controlling this trouble. A shallow groove for a snap ring is sometimes machined in the valve stem between the top of the valve guide and the valve-spring retainer to keep the valve from falling into the cylinder and doing mechanical damage in event of spring breakage.

Aluminum alloys used for cylinder heads are not suitable for valve seats, and usual practice is to use inserts of cast or wrought aluminum bronze* (Par. 9-3). Inserts are sometimes cast in place, but the difficulty of holding them accurately in position during casting of the head is an objection. More often the inserts are screwed and shrunk or simply shrunk in place in the finished casting. Regardless of the method of putting them in, it is highly important that a good thermal contact be had at all times between the inserts and the surrounding head metal. To ensure this condition, the difference in temperature between the head and insert should be large during insertion so that difference in coefficient of expansion of the two metals will not cause the insert to get loose during operation. The insert may be cooled in liquid air just before insertion or the head can be heated. Swan[15] recommends a head temperature of 320 to 350°C. for insertion, a shrinkage interference of 0.0035 in. per inch diameter of the insert, and an insert thickness of about 0.4 in. Too thin a section of head metal around the insert may result in distortion and partial pulling away around part of the insert. Poor arrangement of fins for uniform heat flow from the insert can also contribute to warping and poor thermal contact. Slight peening of the metal around the insert helps remove the possibility of the insert getting loose under extremely adverse conditions and doing mechanical damage.

Aluminum-alloy head metals are also not suitable for valve guides chiefly because of poor wearing qualities. Adequate and effective lubrication between valve stems and guides is difficult to attain, and to reduce wear, hard alloys of aluminum bronze or other materials (Par. 9-3) are usually used. Good thermal contact between the guide and head metal is essential and may be attained by shrink fits, but the desirability of being able to replace worn guides makes this type of fit objectionable. Guides are usually held in place by means of a shoulder resting on the head metal and forming the support for one end of the valve spring. Valve guides should not extend into the port

* Stellite-faced steel seats are frequently used for extreme service.

appreciably beyond the head metal as unequal heating and resulting warpage may bind the valve stem. Split-valve guides are sometimes used to permit oversize valve tips that have a greater bearing area. Clearance between the valve guide and valve stem should be small to minimize leakage of gas due to difference in pressure at opposite ends of the guide.

9-9. The Combustion Chamber.—Since the clearance or combustion-chamber volume is related to the piston displacement and compression ratio by the expression

$$\frac{V_D + V_C}{V_C} = CR$$

the volume of the combustion chamber may be expressed as

$$V_C = \frac{V_D}{CR - 1} = \frac{\pi d^2 S}{4(CR - 1)} \qquad (9\text{-}25)$$

where V_C = combustion chamber volume, cu. in.
V_D = piston displacement, cu. in.
CR = compression ratio.
d = cylinder diameter, in.
S = stroke, in.

The shape of the combustion chamber has much to do with the effectiveness of a given design,[14] and in automotive engines, recent practice has been to use a more or less elongated chamber with the part farthest from the spark plug flattened to produce a narrow space between the piston and head (Fig. 9-17*F*). During burning, the charge ignites at the spark-plug points and the flame spreads through the mixture as an approximately spherical front of rapidly increasing radius proportional to the flame speed through the mixture. The flame front is not strictly spherical owing to the distorting effect of turbulence and the chilling of the part of the charge closest to the combustion chamber walls, but this does not alter the basic idea of combustion control.

At the beginning of combustion (Fig. 9-18*A*), the mixture is at compression pressure and temperature, but as the flame spreads through the charge, the burning portion expands and compresses the unburned portion ahead of the advancing flame front (Fig. 9-18*B*). Since this compression is very rapid, the temperature of the unburned portion also rises rapidly, and if it reaches its spontaneous ignition temperature before the advancing flame can reach it and burn it, the last portion of the charge will ignite spontaneously with a resulting very rapid liberation of heat. This extremely rapid heat liberation produces a pressure rise much more rapid than that resulting from normal combustion. This is the usual explanation of the phenomenon known as

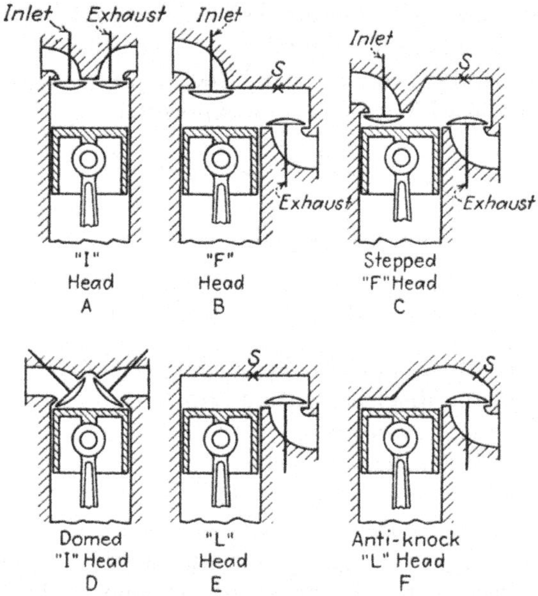

FIG. 9-17.—Combustion-chamber and valve arrangements.

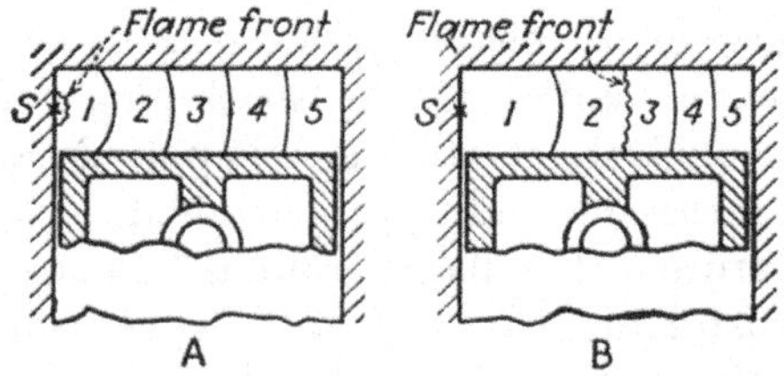

FIG. 9-18.—Basic idea of combustion in gasoline-engine cylinders (see text).

detonation, and its adverse effect on bearings, pistons, valves, and power output is well known.

Chilling the last portion of the charge helps to reduce detonation by reducing the rate of temperature rise, and this in turn reduces the amount of charge that will reach its spontaneous ignition temperature. Flattening the space occupied by the last portion of the charge (Fig. 9-17*F*) increases the surface-volume ratio of this part of the combustion chamber and thereby increases heat flow from this part of the mixture. Close proximity of the relatively cool intake valve (Fig. 9-17*C*) also helps to reduce the rate of temperature rise of the last portion of the charge. Thus, an elongated combustion chamber is desirable from the standpoint of controlling detonation.

Unfortunately, an elongated combustion chamber reduces the area of flame front during normal combustion, and this reduces the rate of heat liberation and pressure rise. Thus, the net area

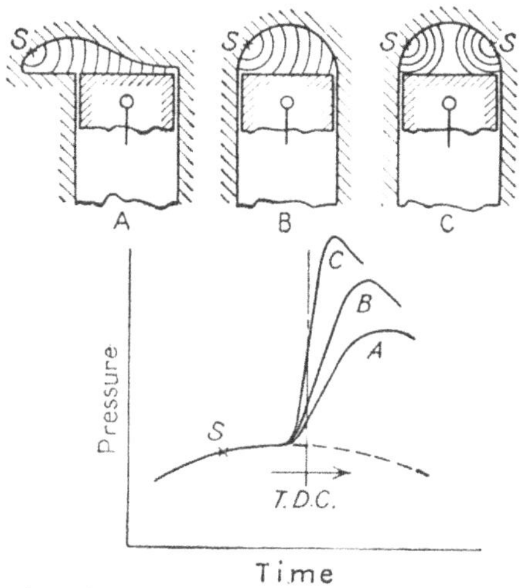

FIG. 9-19.—Effect of combustion-chamber shape and flame-front area on rate of pressure rise and maximum pressure.

of the indicator card is reduced (Fig. 9-19), and this lowers the specific power output and efficiency. The effect of combustion-chamber shape on rate of pressure rise dP/dT and on maximum pressure ($\propto$ to maximum temperature) is shown diagrammatically in Fig. 9-20. Case *C* of this figure is usually conceded to be better than case *A* or *B* because the start of pressure rise is less abrupt than case *A* (*i.e.*, less shock and roughness), and maximum pressure (and temperature) is not so high as in case *B* (*i.e.*, less tendency to detonate).

A compact combustion chamber shape (Fig. 9-19) will permit a greater flame front area and therefore is desirable for high-performance engines, but higher octane fuels must be used to offset the poorer inherent resistance to detonation. Still more rapid heat liberation is possible with dual ignition (Fig. 9-19), and the gain in power can be demonstrated by switching from one to both magnetos in flight.

The sphere is the most compact geometric shape, but for a combustion chamber, this would require a concave piston head. A spherical segment permits a flat piston head, and aircraft-engine combustion chambers approximating this shape are extensively used (Fig. 9-1) especially in engines using valves set at an angle to the cylinder axis.

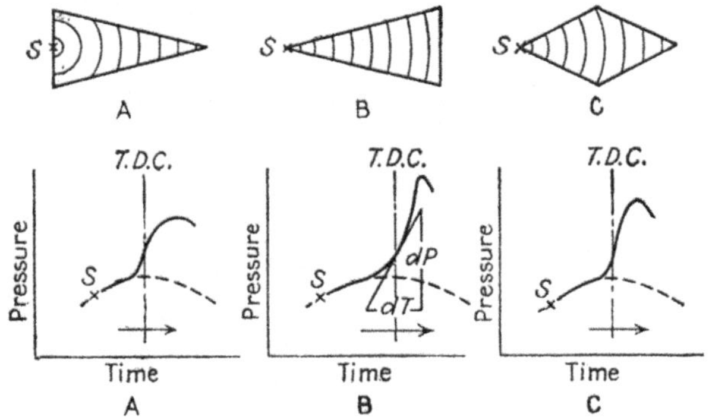

FIG. 9-20.—Effect of combustion-chamber shape and flame-front area on pressure variation during combustion.

The volume of a spherical segment having only one base is

$$V = \frac{\pi h}{6}(3r_2^2 + h^2) \qquad (9\text{-}26)$$

where h = altitude of the segment, in.
r_2 = radius of the base, in.

For combustion chambers, r_2 should be about equal to the radius of the cylinder barrel, and on this basis, the altitude h needed to provide the necessary combustion-chamber volume may be found by combining Eqs. (9-25) and (9-26).

Thus

$$\frac{\pi d^2 S}{4(CR - 1)} = \frac{\pi h}{6}\left(\frac{3d^2}{4} + h^2\right)$$

from which

$$h^3 + 0.75d^2h - \frac{1.5d^2S}{(CR - 1)} = 0$$

Let

$$a = 0.75d^2 \qquad \text{and} \qquad b = \frac{1.5d^2S}{CR - 1}$$

then

$$h = \sqrt[3]{\frac{b}{2} + \sqrt{\frac{b^2}{4} + \frac{a^3}{27}}} + \sqrt[3]{\frac{b}{2} - \sqrt{\frac{b^2}{4} + \frac{a^3}{27}}} \qquad (9\text{-}27)$$

Example.—An engine has a cylinder diameter of 4 in., a stroke of 4.5 in., and a compression ratio of 5:1. If the combustion chamber is in the shape of a spherical segment, what is the distance along the cylinder axis from the center of the piston head (at T.D.C.) to the inner surface of the cylinder head?

Solution

$$a = 0.75 \times 4^2 = 12$$

$$b = \frac{1.5 \times 4^2 \times 4.5}{(5 - 1)} = 27$$

From Eq. (9-27),

$$h = \sqrt[3]{\frac{27}{2} + \sqrt{\frac{27^2}{4} + \frac{12^3}{27}}} + \sqrt[3]{\frac{27}{2} - \sqrt{\frac{27^2}{4} + \frac{12^3}{27}}}$$

or

$$h = 1.78 \text{ in.}$$

Owing to the need for a flat plane for the valve seats, the shape of the valve heads, and the irregular shape of the exposed end of the spark plug, the actual combustion-chamber shape deviates somewhat from a spherical segment. If not taken into account, these detail parts will change the compression ratio, and it is desirable for the preliminary design to estimate the change in volume produced by these parts and correct the clearance volume dimensions accordingly. In any event, it is particularly important to have the compression ratio the same in each of the cylinders.

When the valve stems are parallel to the cylinder axis, the combustion chamber usually approximates a short cylinder, and to permit the use of large valves, the diameter of this cylinder is frequently made greater than the diameter of the cylinder barrel, but the height should not be so small as to reduce too greatly the area of the flame front. A check should also be made to see if mechanical clearance between the valves and piston head is adequate.

Cylinder heads should be sufficiently thick in section (*a*) to withstand the maximum bursting force at explosion pressure, (*b*) to be rigid enough to prevent distortion of valve seats and binding of the valves in the valve guides, and (*c*) to conduct away adequately the heat absorbed from the combustion gases. In conventional types of heads, rigidity and thermal conductivity are critical; hence, if the head thickness is sufficient for *b* and *c*, the strength will usually be adequate. The fins on air-cooled cylinder heads act as truss members and contribute greatly to the rigidity as well as to the strength. In spherical-segment types of combustion chambers, therefore, a head thickness sufficient for heat flow will also be adequate for strength and rigidity, but in short cylindrical chambers, the flat top may tend to bow outward under explosion pressure sufficient to distort the valve seats. The case is a close parallel to that of steam boilers wherein stay bolts are needed for flat heads, but not for hemispherical heads. However, since stay bolts cannot be used in cylinder heads, the best alternative is greater external trussing and the use of large-radius fillets at the juncture of the flat and cylindrical portions of the head. The stiffening effect of the manifolds is also quite useful.

Valve location is determined by the size of the valves, arrangement of the valve gear, and cooling factors. In in-line engines, valves set parallel to the cylinder axes can be operated by an overhead camshaft without rocker arms, a factor in favor of simplicity, but usually valve size is restricted to the point where there is a reduction in b.m.e.p. In addition, it is more difficult to provide uniform air cooling all around the valve seats. Multiple valves boost the b.m.e.p. but do not greatly help the cooling of the valve seats, and in addition they add to the complexity of the valve gear. Valves set at an angle appear to be about the best solution for small and medium-size engines and in many instances for large engines as well, since uniform air cooling of seats is improved, larger valves may be used, and push rods and rocker arms do not involve any greater, if as much, complexity as overhead camshafts. There are so many possible arrangements of valves and valve gear, however, that specific rules cannot be laid down, but the designer should be able to justify his particular selection.

Spark-plug location is highly important in elongated combustion chambers designed primarily for detonation control, but in compact chambers little choice is available. In general, dual spark plugs are located on opposite sides of the head and between the valves. Plug bosses should not be too close to the valve seats, and the seats should not be too close together as the narrow separating section of head metal may be shunted off from the cooling air, be overheated, and crack. Spark-plug points set too far back in elongated bosses may be exposed only to stratified burned charge and give poor ignition characteristics. Poor heat flow from plug bosses can cause overheating of plug points and preignition. Excessive heat flow may cause overcooling and fowling of plug points. S.A.E. standard spark-plug dimensions for use in determining plug-boss details are given in reference 2. Arrangement of cooling fins around the spark-plug boss should be such as to permit ready access to the plug without undue danger of breaking the fins.

Suggested Design Procedure

1. Decide upon the type of cylinder construction to be used, and make detailed sectional sketches (to scale) of the proposed arrangement of the parts. Include enough different sections to show the arrangement clearly.

A suggested way of making these sketches is to put them on tracing paper placed over ¼-in. cross-sectioned paper. These sketches should be considered as a plan of procedure and as such should be given careful study as they are prepared. Hasty assembly of a "picture" without regard to fitting of parts, logical dimensions, etc., is of little value. Be able to justify details of the arrangement by reference to current practice whenever possible, but do not try to make the sketches detailed finished drawings. They should be the preliminary bird's-eye views.

2. With the desired arrangement of the entire cylinder well fixed in mind, determine the cylinder barrel details, *i.e.*, material, wall thickness, hold-down flange dimensions, method of attaching to head, etc.

3. If air-cooled, select fin dimensions by reference to Fig. A1-7 or current practice. If liquid cooled, use data in reference 15 or equivalent.

4. For fin dimensions selected, determine percentage of heat to cooling and alter dimensions of fins or use baffles if inadequate.

If fins are more than adequate, a saving in cost of manufacture may be possible by using less effective but more easily constructed fins.

5. Make a detailed dimensioned sectional drawing of the cylinder barrel. Leave space on the drawing for adding the cylinder head.

6. Determine the valve dimensions necessary for adequate breathing capacity.

7. Determine the remaining detailed dimensions of the valves, and make a detailed dimensioned drawing of intake and exhaust valves.

Salt- or sodium-cooled exhaust valves will probably be advisable if the b.m.e.p. is much above 115 to 120 lb. per sq. in.

8. Determine the dimensions of the combustion chamber necessary to give the desired compression ratio.

9. Make detailed dimensioned drawings of the valve guides and valve seats.

10. Make a detailed dimensioned sectional drawing of the cylinder head except the supports and housing for the valve gear.

This drawing should be on the same sheet and a part of the cylinder barrel drawing.

11. Make an assembly drawing of the cylinder (except the supports and housing for the valve gear) on the layout drawing of Suggested Design Procedure, page 24, item 4. Show parts in section whenever such sectioning increases the clarity or legibility of the drawing. Include only principal over-all dimensions. Identify each part of the assembly drawing by a reference number corresponding to the detailed drawing or reference number of that part.

12. When items 1 to 11 have been completed and put in proper form, submit for checking and approval.

Problems

1. Determine the surface heat transfer coefficient for a cylinder barrel of 4.5-in. bore, a fin width of 0.5 in., a fin pitch of 0.3 in., and a fin thickness of 0.0625 in. Velocity of air past cylinders is 65 m.p.h.

2. Determine the over-all heat transfer coefficient U if the cylinder barrel in Problem 1 is of steel and the fins are rectangular in cross section.

3. Determine the over-all heat transfer coefficient for the aluminum Y alloy head fins of the cylinder in Problem 1 for an average fin width of 1.1 in., a pitch of 0.4 in., and an average fin thickness of 0.125 in. Assume fin-tip thickness of 0.1 in. and fin-root thickness of 0.15 in. Assume an average head thickness of 0.375 in.

4. Determine the approximate cylinder wall and head temperatures for the cylinder in Problems 1, 2, and 3 if the engine is a 75-b.hp. four-cylinder opposed type.

5. Determine for the engine of the preceding four problems the proportion of the heat supplied which is removed by the cooling fins. Air temperature 80°F., number of cooling fins on the cylinder barrel 14.

6. What increase in the proportion of heat to cooling could be attained for the engine in Problem 5 if the cylinders were baffled and a baffle pressure drop of 2.25 in. of water was available.

7. Assuming an octane number of fuel such that the cylinder temperatures without baffles as found in Problem 4 are satisfactory, what increase in indicated horsepower per cylinder would be possible with baffles and a baffle pressure drop of 2.25 in. of water? What brake horsepower could the engine develop?

8. Take necessary measurements and determine the probable increase in brake horsepower that could be attained in a Continental A-40 engine without increase in cylinder temperatures if pressure baffles giving a 2-in. pressure drop were used. Assume the most critical conditions will be had at full-throttle climb of 45 m.p.h. with 73 octane number fuel in each case.

Assume an atmospheric temperature of 80°F. The A-40 is rated 40 b.hp. at 2,575 r.p.m.

References

1. Johnson: Aircraft Engine Materials, *S.A.E. Jour.*, Vol. 40, No. 4, April, 1937.
2. S.A.E. "Handbook."
3. Huebotter: "Mechanics of the Gasoline Engine."
4. Heldt: "Automotive Engines."
5. Pinkel: The Heat Transfer Processes in Air-cooled Engines, *Jour. Aeronautical Sciences,* Vol. 4, No. 10, August, 1937.
6. Biermann and Pinkel: Heat Transfer from Finned Metal Cylinders in an Air Stream, *NACA Tech. Rept.* 488.
7. Schey and Rollin: The Effect of Baffles on the Temperature Distribution and Heat Transfer Coefficients of Finned Cylinders, *NACA Tech. Rept.* 511.
8. Schey and Ellerbrock: Blower Cooling of Finned Cylinders, *NACA Tech. Rept.* 587.
9. Beisel, MacClain, and Thomas: The Cowling and Cooling of Radial Air-cooled Aircraft Engines, *S.A.E. Jour.*, Vol. 34, No. 4, May, 1934.
10. Campbell: Cylinder Cooling and Drag of Radial Engine Installations, *S.A.E. Jour.*, Vol. 43, No. 6, December, 1938.
11. Blank: Valve Materials, *Automotive Ind.*, Nov. 19, 1938.
12. Jardine and Jardine: Designing Valves and Related Parts for Maximum Service, *S.A.E. Jour.*, Vol. 37, No. 1, July, 1935.
13. Lewis and Nutting: Air Flow through Poppet Valves, *N.A.C.A. Tech. Rept.* 24.
14. Taub: Combustion Chamber Progress Correlated, *S.A.E. Jour.*, Vol. 27, No. 4, October, 1930.
15. Swan: "Handbook of Aeronautics," Vol. 2, Aero-Engines, Design and Practice.
16. Judge: "Automobile and Aircraft Engines."
17. Colwell: The Trend in Poppet Valves, *S.A.E. Jour.*, Vol. 45, No. 1, July, 1939.
18. Wallace: Superfinish, *S.A.E. Jour.*, Vol. 46, No. 2, February, 1940.
19. Colwell: Modern Aircraft Valves, *S.A.E. Jour.*, Vol. 46, No. 4, April, 1940.
20. Coler: Enamels for Aircraft Engine Cylinders, *Aero Digest*, Vol. 36, No. 4, April, 1940.

CHAPTER 10

VALVE GEAR

10-1. Usual Valve Gear Arrangements.—In the conventional four-stroke-cycle engine, movable valves are necessary to allow induction of new charge and removal of burned charge from the cylinder. These valves are usually of the poppet type, although single sleeve valves[8] are being used to a considerable extent in England.[9]

The motions of the poppet valves are derived from cams on a shaft or shafts driven at one-half crankshaft speed in the case

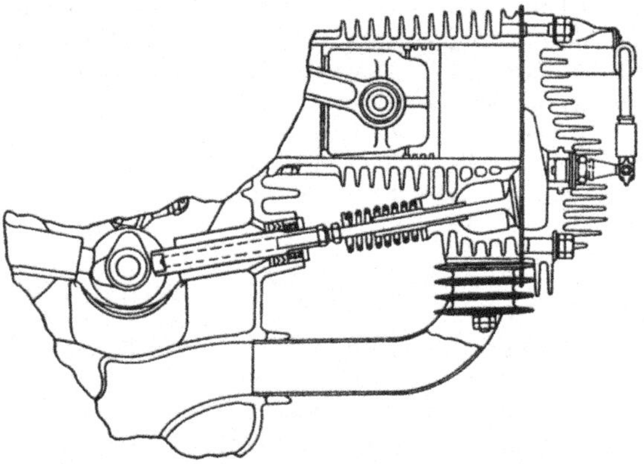

Fig. 10-1.—Arrangement of the valve gear used on the Continental A-40 L-head engine.

of in-line and V-engines and in some radial engines. In the majority of radial engines, however, the valve motions are derived from a cam ring or disk which rotates at much less than one-half of crankshaft speed. Where a camshaft is used, an individual cam is usually provided for each valve, but when a cam disk is used, several cams are provided in each of two races (one race for intake valves and one race for exhaust valves), and each cam operates in succession all valves connected to its race.

One camshaft operating all the intake and exhaust valves is usual practice in in-line and V-engines, but more than one camshaft is sometimes used. The camshaft may be located (*a*) in the crankcase or (*b*) over the top of the cylinder heads. In the first case (Figs. 10-1 and 10-2), an L-head arrangement may be used and the valve tips may ride directly on the cams or more often seat on cam-follower tappets, or overhead valves may be used and the motion of the valves transmitted through cam followers, push rods, and rocker arms. In the second case

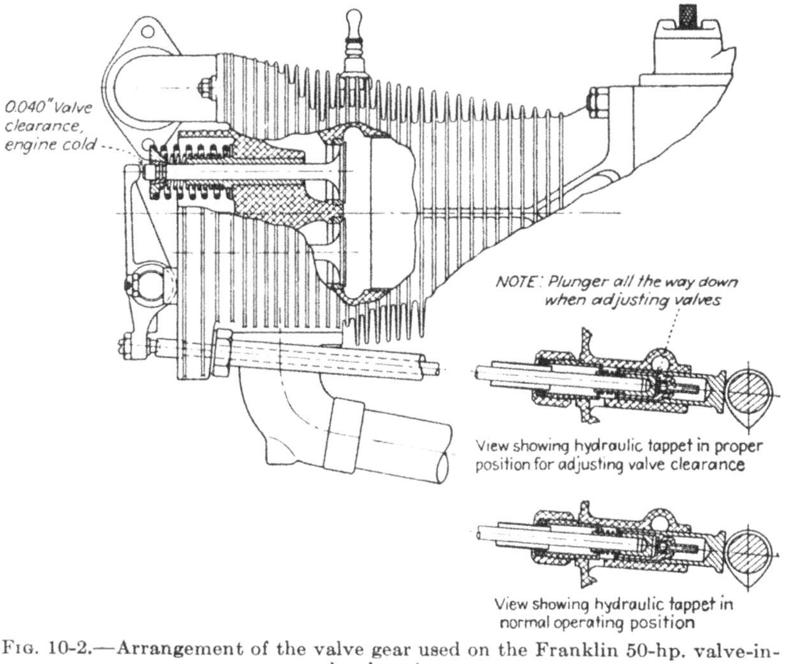

FIG. 10-2.—Arrangement of the valve gear used on the Franklin 50-hp. valve-in-head engine.

(Figs. 10-3 and 10-4), the overhead cams may act directly against the valve tips, or more often rocker arms transmit the motion. This last arrangement increases the complexity of the valve gear but permits the use of larger valves set at an angle to the center line of the cylinder. Usually, the overhead camshaft is driven from the crankshaft through bevel gearing and a torque tube, but in a few cases a positive chain drive is used. Overhead camshafts operating directly on the valve stems are particularly well adapted to high-speed operation because the weight of parts that must be returned by the valve spring is less than with rocker arms and push rods. With inclined valves and rocker arms

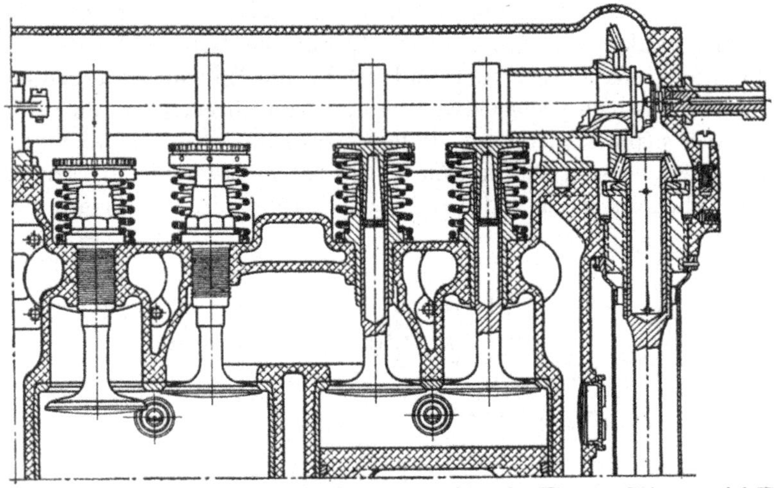

FIG. 10-3.—Arrangement of the valve gear used on the Hispano-Suiza, model-H engine.

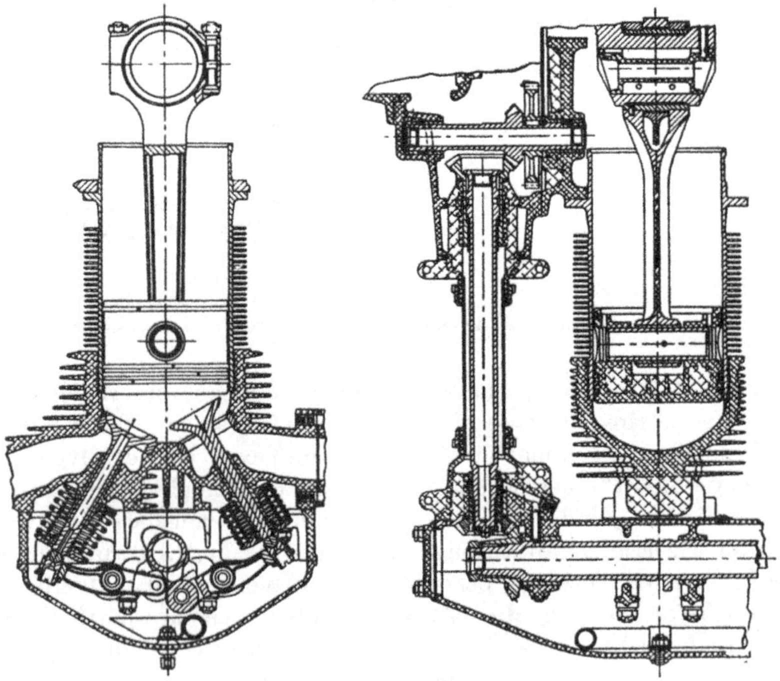

FIG. 10-4.—Arrangement of the valve gear used on the Ranger inverted air-cooled engines.

(Fig. 10-4), the weight is somewhat greater, but still low enough to permit high-speed operation, and the arrangement has the added advantage of permitting larger valves and higher volumetric efficiency. A disadvantage of overhead camshafts is the need for greater cylinder rigidity to keep the camshaft bearings aligned. Also, with air-cooled types, some difficulty may be encountered in designing the camshaft housing to permit adequate cooling-air flow around the cylinder head.

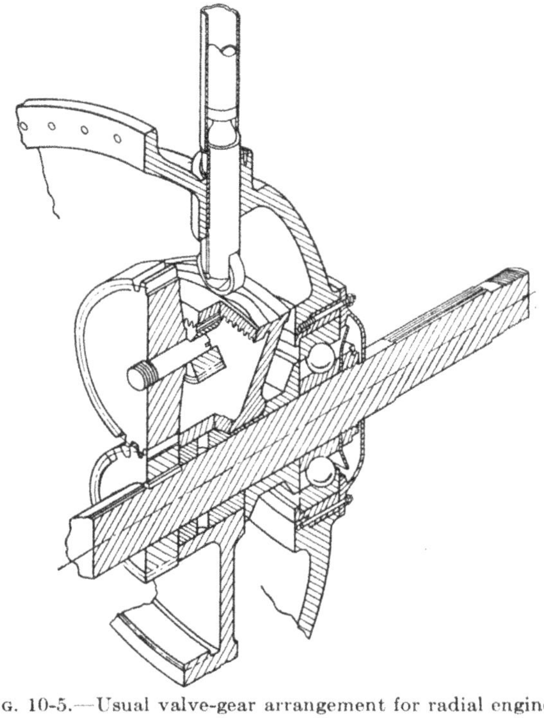

FIG. 10-5.—Usual valve-gear arrangement for radial engines.

In radial engines, push rods and rocker arms are about the only feasible way of transmitting the motion of the cam followers to the valves. Usually, all intake-valve followers ride on one cam race and all exhaust-valve followers ride on an adjacent race which is attached to or is a part of the same cam disk or ring (Fig. 10-5).

10-2. Valve Timing.—The opening and closing of the intake and exhaust valves at the proper point in the cycle has much to do with the effective performance of a four-stroke-cycle engine, and the proper point is usually not the dead-center position of the piston. Reasons for this may be shown diagrammatically by means of pumping-loop diagrams (Fig. 10-6).

Considering first the effect of exhaust-valve opening time (Fig. 10-6*A*, *B*, and *C*), if the exhaust valve is opened at bottom dead

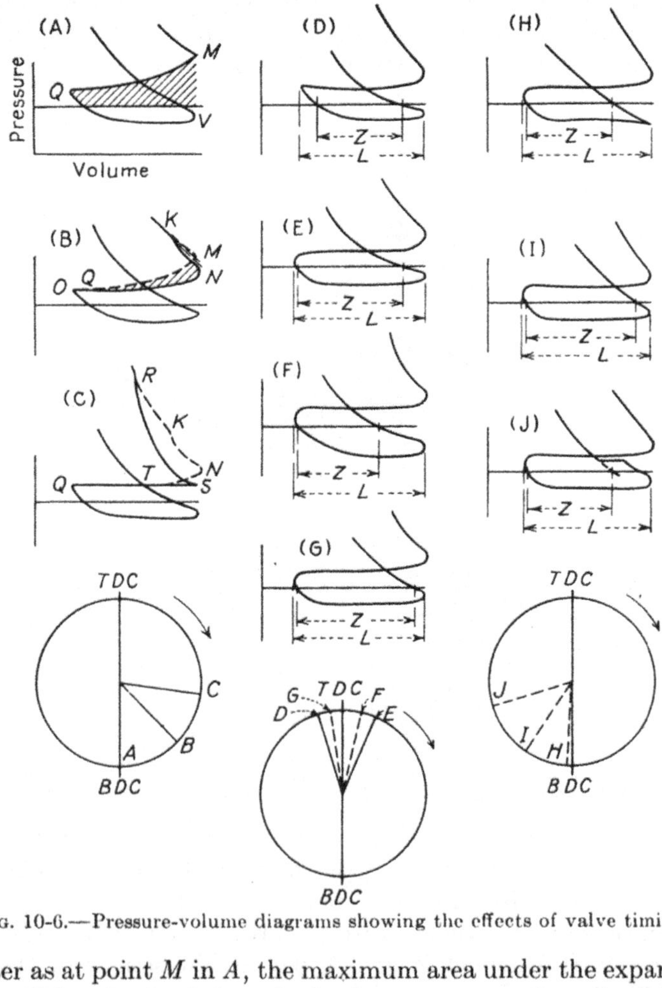

Fig. 10-6.—Pressure-volume diagrams showing the effects of valve timing.

center as at point *M* in *A*, the maximum area under the expansion line will be obtained, but the back pressure on the exhaust will be large and the net work of the cycle will be reduced. By opening the exhaust valve early as at point *K* in *B*, area *KMN* will be lost but the larger area *MQN* representing the reduction in the work of pushing the exhaust gases out of the cylinder will be saved, the net result being a gain in the net power output. However, if the exhaust valve is opened too early, as at R in C, the area $RKNS$ lost under the expansion line will be greater than the saving in work during exhaust as represented by the area TNS. Therefore, the condition B is most desirable.

The effect of varying the exhaust-valve closing time is shown diagrammatically in Fig. 10-6D and E. If the exhaust valve is closed before the end of the exhaust stroke, it will start to close even earlier with the result that there will be a pinching or throttling of the flow of burned gases out of the cylinder. This will cause a build up in pressure during the last part of the exhaust stroke and result in a higher pressure in the clearance space at the beginning of the intake stroke. Before any new charge can be drawn in, the burned charge in the clearance space must expand down at least to atmospheric pressure, and the higher the initial clearance pressure, the greater the portion of the suction displacement that will be required for the expanded clearance gases. Thus too early a closing of the exhaust valve tends to reduce the volumetric efficiency Z/L.

By closing the exhaust valve after the top dead-center position as in E, less pinching of the exhaust gas during the last part of the exhaust stroke will be had and, in addition, use may be made of the "carry-out" effect produced by the kinetic energy of the high-velocity gases passing to the exhaust manifold. This last results in a lowering of the clearance pressure and an increase in the volumetric efficiency.

The effect of varying the intake-valve opening time is illustrated in Fig. 10-6F and G. In F, the intake valve starts to open after the start of the intake stroke, and since it takes several degrees of crankshaft travel to open the valve completely, during a considerable portion of the intake stroke the incoming charge will be throttled. The result will be a lower pressure in the cylinder, and unless other factors, discussed later, more than offset this effect, compression will start on the following stroke at a pressure well below atmospheric, and this, as shown in F, will cause a low volumetric efficiency.

By starting to open the intake valve before top dead center as in G, the valve will be more nearly wide open during the suction stroke, and owing to the reduced amount of throttling, a higher volumetric efficiency will be had. Thus the exhaust valve should close after top dead center, and the intake valve should open before top dead center.

Obviously, with this timing both valves will be partly open at the same time, and this condition is called *valve overlap*. At first thought, it would seem that opening the intake valve before the exhaust valve was closed would result in a flow of burned gas back through the intake valve and into the intake manifold where it would have to be returned to the cylinder before any new charge could come in, or the other extreme, *i.e.*, new charge would flow into the cylinder and on out into the exhaust manifold without being burned. But this will not happen at high throttle settings if the valve overlap is not too great (*a*) because the inertia of the high-velocity burned gases will cause these gases to continue on out through the exhaust-valve opening even if another means of escape is provided and (*b*) because the valves are both so nearly closed during this overlap period that very little if any new charge will get into the cylinder far enough to be carried out by the escaping exhaust gases. The inertia of the new charge tends to hold it back, and this further contributes to preventing escape of new charge to the exhaust. A large valve overlap is not conducive to good economy at part-throttle operation, however, because under this condition, the pressure in the intake manifold is quite low and the tendency for exhaust gas to flow back through the intake valve is increased. A sudden rush of burned gas into the intake manifold is likely to push some new charge back out through the carburetor where it will be lost. Also the new charge which is later taken into the cylinder will be more highly diluted and therefore burn less efficiently.

Intake-valve closing time has a rather major effect on engine performance, and this is illustrated in Fig. 10-6*H*, *I*, and *J*. In case *H*, the intake valve closes very close to bottom dead center which means that it starts to close well before the end of the suction stroke. Hence the incoming charge will be throttled during the last part of the suction stroke and the pressure in the cylinder will be lowered, or at least there can be no use made of the so-called "ramming effect" which results from the kinetic energy of the high-velocity incoming charge. With the pressure at the beginning of compression well below atmospheric, the compression line will cross the atmospheric line farther from the end of the card and the volumetric efficiency will be low as in case *H*.

By leaving the intake valve open until well after bottom dead center as in case *I*, the kinetic energy of the incoming charge will tend to "ram" more charge into the cylinder and build up the pressure even though the cylinder volume is starting to decrease by the return of the piston. However, if the intake valve is held open too long after bottom dead center, the volumetric efficiency will be reduced because the returning piston will overcome the inertia of flow or ramming effect of the incoming charge and then start forcing the mixture back out through

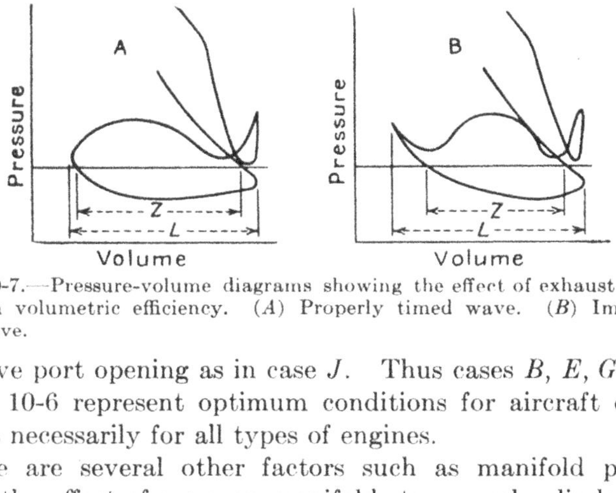

FIG. 10-7.—Pressure-volume diagrams showing the effect of exhaust pressure waves on volumetric efficiency. (*A*) Properly timed wave. (*B*) Improperly timed wave.

the valve port opening as in case *J*. Thus cases *B*, *E*, *G*, and *I* in Fig. 10-6 represent optimum conditions for aircraft engines but not necessarily for all types of engines.

There are several other factors such as manifold pressure waves, the effect of common manifolds to several cylinders, and engine speed, which may alter or even completely change the engine performance that would be expected from the above optimum cases. For instance (Fig. 10-7*A* and *B*), the sudden outrush of burned gas from the cylinder when the exhaust valve opens sets up pressure waves in the exhaust pipe that have a frequency and amplitude dependent upon the diameter, length, turns, and branches of the exhaust pipe and upon the speed of the engine. If these waves are traveling away from the cylinder at the time the exhaust valve is closing, they will tend to reduce the pressure in the clearance space, as in case *A*, and therefore contribute to high volumetric efficiency. But if the waves are traveling toward the cylinder, *i.e.*, building up pressure at the inner end of the manifold, they ram some of the escaped gases back into the cylinder and raise the pressure in the clearance space, as in case *B*. As far as volumetric efficiency is concerned, case *A* in Fig. 10-7 will produce results similar to case *E* in Fig. 10-6, and case *B* in Fig. 10-7 will produce results similar to case *D* in Fig. 10-6. Thus changing to the correct valve timing while retaining an improper manifold or engine speed can result in no improvement or even a reduction in volumetric efficiency. In a similar way, improper intake manifolding can offset good intake-valve timing.

In branched manifolds or manifolds leading to a common header, the pressure waves due to one cylinder can be properly timed for that cylinder and yet cause adverse affects in other cylinders.

A given optimum valve timing is in general correct for only one engine speed. The most pronounced effect of speed occurs in connection with intake-valve closing time. The principal reason for this lies in the fact that ramming effect varies with speed, whereas the intake-valve closing time remains fixed in conventional engines. At low speeds, the ramming effect is small and the returning piston can quickly overcome the inertia of inrushing gas. Hence, a valve timing such as case *I* in Fig. 10-6 may be too late a closing for the low-speed condition and the volumetric efficiency will be low as in case *J*.

A very high engine speed will in general give a correspondingly high ramming effect which will be overcome less quickly by the returning piston. Thus at high speed, case *I* (Fig. 10-6) will represent too early a closing time for the intake valve and the volumetric efficiency will be low as in case *H*.

Obviously it would be desirable to vary the intake-valve closing time with speed, but since this is impracticable except in experimental engines, the fixed valve timing used must be a compromise.

In automobile engines requiring good performance over a wide range of speed, a valve timing such as case *I*, Fig. 10-6, is the usual compromise because it gives reasonably good average volumetric efficiency throughout the speed range. In airplanes and in racing engines, however, high power at high engine speed is essential and high power at low engine speed is of secondary importance. Hence the intake-valve closing time should be farther after bottom dead center. Figure 10-8 shows the variation of volumetric efficiency with engine speed for a fixed valve timing. Table 10-1 gives the valve timing of 18 aircraft engines

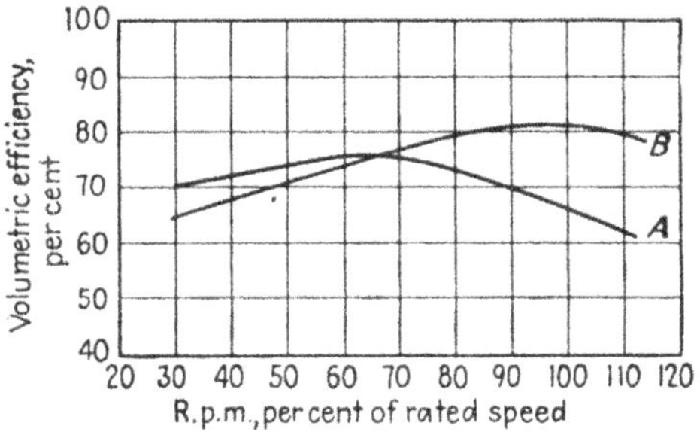

FIG. 10-8.—Volumetric efficiency vs. engine speed for (*A*) good power over a wide range of speed and (*B*) maximum power at high speed.

together with a summary of average, maximum, and minimum timings.

TABLE 10-1.—AIRCRAFT-ENGINE VALVE-TIMING DATA

Engine type	I.V.O. deg. B.T.C.	I.V.C. deg. A.B.C.	E.V.O. deg. B.B.C.	E.V.C. deg. A.T.C.
9 cyl. radial	26	76	71	31
14 cyl. 2-row radial	20	76	76	20
7 cyl. radial	5	55	55	5
7 cyl. radial	8	45	55	8
7 cyl. radial	8	21	63	20
7 cyl. radial	9	51	51	9
5 cyl. radial	35	90	80	45
7 cyl. radial	41	94	91	54
5 cyl. radial	10	77	69	33
9 cyl. radial	15	45	60	15
4 cyl. in-line	17	77	50	10
12 cyl. V	15	65	70	30
7 cyl. radial	10	60	60	10
3 cyl. radial	4	57	45	19
2 cyl. opposed	5	55	55	5
4 cyl. opposed	10	55	55	10
4 cyl. in-line	14	68	42	22
5 cyl. radial	0	60	60	0
Average	14	59	62	19
Maximum	41	94	91	54
Minimum	0	21	42	0

Maximum overlap = 95 deg. Minimum overlap = 0 deg. Average overlap = 33 deg.

10-3. Valve Cams and Followers.—The requirements for valve cams are somewhat conflicting in that for high volumetric efficiency the valves should be opened and closed quickly and held wide open for a large part of the open time, whereas to keep down acceleration forces and spring tension, the valves should be opened and closed gradually. In addition, the cam contours should be such that they are not too difficult or expensive to manufacture.

Many types of cam contours have been tried in the attempt to attain more fully the best compromise conditions, but the majority of cams now in current use fall into one of the following classifications:

1. Tangent cams with roller or round-nose followers.
2. Convex flank or "mushroom" cams with flat followers in sliding contact.
3. Concave-flank (hollow-faced) cams with roller followers.
4. Constant-acceleration cams, generated cams, etc., having component parts formed of curves other than straight lines or arcs of circles. Only the first three types will be analyzed here. For details of types 4, see references 1 and 2. Some manufacturers use composite curves made up to meet special requirements.

All cams have in common a base circle on which the follower rides during the time the valve is closed, an opening flank so shaped as to open the valve in the desired way, a cam nose which may include a period of dwell on which the follower rides during the time when the valve is wide open, and a closing flank which allows the valve to close properly. The problem of design is largely one of properly shaping the flank and nose portions of the cam.

10-4. Tangent Cams.—Tangent cams are frequently used in in-line and V-engines, usually with roller followers, although a round-nose sliding follower may be used. They have straight-line flanks tangent to the base and nose circles and are relatively easy to lay out and manufacture, but they require a stiff valve spring to ensure contact of the follower with the cam surface at all points. Figure 10-9 shows the general layout of a tangent cam with a roller follower.

For the tangent cam (Fig. 10-9), let

R_B = the radius of the base circle.
R_N = the radius of the nose circle.
R_F = the radius of the roller follower.

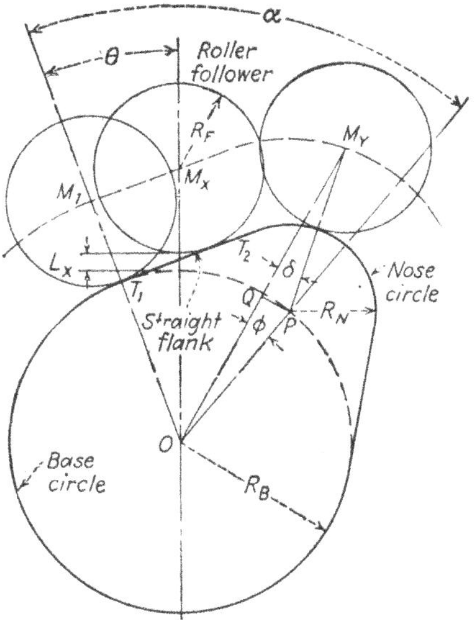

FIG. 10-9.—Tangent cam with roller follower on straight flank portion M_X and on nose circle M_Y.

α = one-half of the angular motion of the camshaft from initial opening to final closing of the valve (α = one-fourth of the angular travel of the crankshaft for a ratio of camshaft to crankshaft speed of 1:2).

θ = the angular travel of the camshaft from initial opening, *i.e.*, the point of tangency of the flank and base circle.

ϕ = the angle between a line through the centers of the base and follower circles and a line through the centers of the base and nose circles.

L_x = the lift of the follower corresponding to θ, in.

L_Y = the lift of the follower corresponding to ϕ, in.

L_M = the maximum lift of the cam follower.

δ = the angle between a line through the centers of the base and follower circles and a line through the centers of the nose and follower circles.

M_x = instantaneous position of follower center when follower is on the straight flank.

M_y = instantaneous position of follower center when follower is on the nose circle.

M_1 = position of follower center when follower is tangent to base circle and straight flank (*i.e.*, at T_1).

M_2 = position of follower center when follower is tangent to straight flank and nose circle (*ie.*, at T_2).

M_3 = position of follower center at maximum lift or when follower is tangent to nose circle and dwell circle.

Referring to Fig. 10-9,

$$\cos\theta = \frac{R_F + R_B}{R_F + R_B + L_x} = \frac{S}{S + L_x}$$

where $S = R_F + R_B$ or the lift is

$$L_x = \frac{S}{\cos\theta} - S = S\left(\frac{1}{\cos\theta} - 1\right), \text{ in.} \qquad (10\text{-}1)$$

The velocity is

$$V_x = \frac{dL_x}{dt} = \left(\frac{dL_x}{d\theta}\right)\left(\frac{d\theta}{dt}\right)$$

but for a given engine speed

$$\frac{d\theta}{dt} = \frac{\theta}{t} = \frac{2\pi N}{60}$$

where N = r.p.m. of the camshaft.

$$V_x = \frac{2\pi N}{12 \times 60}\left(\frac{dL_x}{d\theta}\right) = 0.00873N\frac{dL_x}{d\theta} = 0.00873NS\frac{d\left(\frac{1}{\cos\theta}\right)}{d\theta}$$

$$V_x = 0.00873NS\frac{\tan\theta}{\cos\theta}, \text{ ft. per sec.} \qquad (10\text{-}2)$$

The acceleration is

$$A_x = \frac{dV_x}{dt} = \frac{dV_x}{d\theta}\left(\frac{d\theta}{dt}\right) = \frac{2\pi N}{60}\left(\frac{dV_x}{d\theta}\right) = 0.000914N^2S\frac{d}{d\theta}\left(\frac{\tan\theta}{\cos\theta}\right)$$

$$A_x = 0.000914N^2S\frac{(1 + 2\tan^2\theta)}{\cos\theta}, \text{ ft. per sec.}^2 \qquad (10\text{-}3)$$

Inspection shows that the acceleration will increase as long as the follower remains on the straight flank because $\tan\theta$ increases and $\cos\theta$ decreases with increase of θ. Hence the maximum acceleration on the straight flank will occur at the point of tangency T_2 of the flank and nose circles.

On the nose circle at position M_Y, Fig. 10-9, the lift will be

$$L_Y = OM_Y - S$$

where $S = R_F + R_B$, but

$$OM_Y = h \cos \phi + i \cos \delta$$

where $h = OP$ and $i = R_N + R_F$. (NOTE: OP need not equal R_B as shown in Fig. 10-9.) But from the right triangle PQM_Y

$$(i^2 - PQ^2)^{1/2} = QM_Y = i \cos \delta$$

and

$$PQ = h \sin \phi$$

therefore

$$i \cos \delta = (i^2 - h^2 \sin^2 \phi)^{1/2}$$

$$L_Y = h \cos \phi + (i^2 - h^2 \sin^2 \phi)^{1/2} - S \qquad (10\text{-}4)$$

where L_Y = lift, in.

By the same procedure as above (since $d\phi = -d\theta$), the velocity is

$$V_Y = \frac{-2\pi N}{12 \times 60} \times \frac{dL_Y}{d\phi} = -0.00873N \frac{d}{d\phi} [h \cos \phi + (i^2 - h^2 \sin^2 \phi)^{1/2} - S]$$

$$= -0.00873N \left[-h \sin \phi - \frac{h^2 \sin \phi \cos \phi}{(i^2 - h^2 \sin^2 \phi)^{1/2}} \right]$$

$$= 0.00873\, Nh \sin \phi \left[1 + \frac{h \cos \phi}{(i^2 - h^2 \sin^2 \phi)^{1/2}} \right] \qquad (10\text{-}5)$$

where V_Y is in ft. per sec.

The acceleration on the nose circle is

$$A_Y = + \frac{dV_Y}{dt} = - \frac{2\pi N}{60} \times \frac{dV_Y}{d\phi} = -0.000914N^2h \frac{d}{d\phi} \left[\sin \phi \left(1 + \frac{h \cos \phi}{(i^2 - h^2 \sin^2 \phi)^{1/2}} \right) \right]$$

$$A_Y = -0.000914N^2h \left[\cos \phi + \frac{(i^2 - h^2 \sin^2 \phi)^{1/2} h(-\sin^2 \phi + \cos^2 \phi) - h \sin \phi \cos \phi \left(- \frac{h^2 \sin \phi \cos \phi}{(i^2 - h^2 \sin^2 \phi)^{1/2}} \right)}{(i^2 - h^2 \sin^2 \phi)} \right]$$

$$A_Y = -0.000914N^2h \left[\cos \phi + \frac{h \cos 2\phi}{(i^2 - h^2 \sin^2 \phi)^{1/2}} + \frac{h^3 \sin^2 2\phi}{4(i^2 - h^2 \sin^2 \phi)^{3/2}} \right], \text{ ft. per sec.}^2 \qquad (10\text{-}6)$$

When the valve is wide open, $L_Y = L_3$ = maximum lift, $\phi = 0$, and Eq. (10-6) reduces to

$$A_3 = -0.000914N^2h\left(1 + \frac{h}{i}\right) \qquad (10\text{-}7)$$

To ensure adequate clearance, the cam base circle is usually about ⅛ in. larger in diameter than the camshaft. The diameter of the camshaft is determined from stiffness requirements (see Par. 10-15), and the diameter of the cam disk in radial drives is a matter of adequate clearance for driving gears (see Pars. 10-9 and 10-10), but for the moment, if it is assumed that the base-

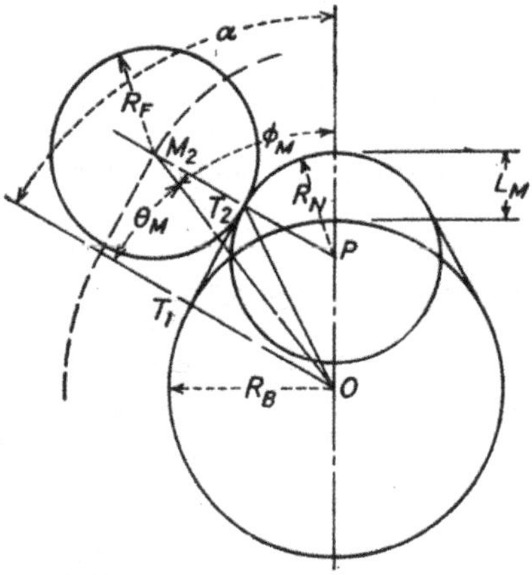

FIG. 10-10.—Relation of various parts in a tangential cam.

circle diameter is known, the other tangential cam dimensions may be found as follows:

Referring to Fig. 10-10 and using the same nomenclature as in Fig. 10-9,

$$OP = h, \qquad \text{and} \qquad h \cos \alpha = R_B - R_N$$

also

$$h + R_N = L_M + R_B$$

and solving simultaneously for R_N

$$(L_M + R_B - R_N) \cos \alpha = R_B - R_N$$

or

$$R_N = R_B - \left(\frac{L_M \cos \alpha}{1 - \cos \alpha}\right) \qquad (10\text{-}8)$$

and with R_N known, the position of the nose circle center may be found from

$$h = \frac{R_B - R_N}{\cos \alpha} \tag{10-9}$$

From Fig. 10-10, it is seen that

$$\theta_M = \text{arc tan} \frac{h \sin \alpha}{R_F + R_B} \tag{10-10}$$

When the angle of crankshaft travel from initial opening to final closing of a valve is quite large or when it is desired to hold the valve wide open for a part of the open time in order to reduce throttling of the charge through the valve port to a minimum and thereby raise the volumetric efficiency, a "period of dwell" is built into the cam (Fig. 10-11). This period of dwell is an arc of a circle concentric with the base circle and having a radius equal to the radius of the base circle plus the total lift, *i.e.*,

$$\text{Radius of dwell} = R_B + L_M$$

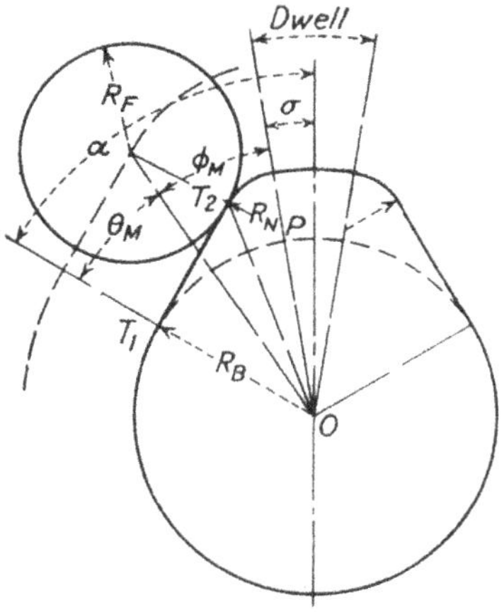

FIG. 10-11.—Tangential cam with a period of dwell.

The period of dwell is somewhat arbitrary with the designer, but it is limited by the capacity of the valve spring and to a lesser extent by the allowable stress at the line of contact between the cam and roller follower.

Referring to Fig. 10-11, the distance between the centers of the base and nose circles is

$$OP = h \qquad \text{and} \qquad h \cos(\alpha - \sigma) = R_B - R_N$$

also

$$L_M + R_B = h + R_N$$

and solving simultaneously for R_N

$$(L_M + R_B - R_N) \cos(\alpha - \sigma) = R_B - R_N$$

from which

$$R_N = R_B - \frac{L_M \cos(\alpha - \sigma)}{1 - \cos(\alpha - \sigma)} \tag{10-11}$$

With R_N known, the position of the nose circle center may be found from

$$h = \frac{R_B - R_N}{\cos(\alpha - \sigma)} \qquad (10\text{-}12)$$

From Fig. 10-11. it is seen that

$$\theta_M = \text{arc tan}\,\frac{h \sin(\alpha - \sigma)}{R_F + R_B} \qquad (10\text{-}13)$$

10-5. Example of Tangent-cam Calculations.—Determine dimensions and accelerations for a tangent cam to operate an inlet valve on an in-line engine, (*a*) for zero dwell, and (*b*) for a period of dwell angle of 20 deg. of camshaft travel. Available data are as follows: inlet-valve opening, 15 deg. before top center; inlet-valve closing, 65 deg. after bottom center; maximum lift, 0.5 in.; diameter of camshaft, 1.575 in.; diameter of the roller follower, 1 in.; speed of engine, 2,000 r.p.m.

Procedure a.—Using the nomenclature of the preceding article,

$$R_B = \frac{1.575}{2} + \frac{0.125}{2} = 0.85 \text{ in.}$$

$$L_M = 0.5 \text{ in.}$$

$$\alpha = \frac{15 + 180 + 65}{4} = 65 \text{ deg.}$$

From Eq. (10-8),

$$R_N = 0.85 - \frac{0.5 \times \cos 65}{1 - \cos 65} = 0.85 - \frac{0.5 \times 0.4226}{1 - 0.4226} = 0.484 \text{ in.}$$

From Eq. (10-9),

$$h = \frac{0.85 - 0.484}{0.4226} = 0.865 \text{ in.}$$

From Eq. (10-10),

$$\theta_M = \text{arc tan}\,\frac{0.865 \sin 65}{0.5 + 0.85} = \text{arc tan}\,0.581 = 30°9.5'$$

$$\phi_M = 65° - 30°9.5' = 34°50.5'$$

From Eq. (10-3), the maximum acceleration on the straight flank is

$$A_{xM} = A_2 = 0.000914 \times \left(\frac{2{,}000}{2}\right)^2 (0.5 + 0.85)\,\frac{[1 + 2(\tan 30°9.5')^2]}{\cos 30°9.5'}$$

$$A_2 = 2{,}395 \text{ ft. per sec.}^2$$

For the deceleration when $\phi = 0$, from Eq. (10-7),

$$i = R_N + R_F = 0.484 + 0.5 = 0.984 \text{ in.}$$

$$A_{YM} = A_3 = -0.000914 \times \left(\frac{2{,}000}{2}\right)^2 \times 0.865\left(1 + \frac{0.865}{0.984}\right) = -1{,}485 \text{ ft. per sec.}^2$$

Procedure b.—For the cam with the 20-deg. period of dwell,

$$R_B = 0.85 \text{ in.}$$
$$L_M = 0.5 \text{ in.}$$
$$\sigma = 10 \text{ deg.}$$
$$\alpha - \sigma = 65 - 10 = 55 \text{ deg.}$$

From Eq. (10-11),

$$R_N = 0.85 - \frac{0.5 \cos 55}{1 - \cos 55} = 0.85 - \frac{0.5 \times 0.5736}{1 - 0.5736} = 0.179 \text{ in.}$$

From Eq. (10-12),

$$h = \frac{0.85 - 0.179}{0.5736} = 1.17 \text{ in.}$$

From Eq. (10-13),

$$\theta_M = \text{arc tan} \frac{1.17 \sin 55}{0.5 + 0.85} = \text{arc tan } 0.71 = 35°22.5'$$
$$\phi_M = 55° - 35°22.5' = 19°37.5'$$

From Eq. (10-3), the maximum acceleration on the straight flank is

$$A_{xM} = A_2 = 0.000914 \left(\frac{2{,}000}{2}\right)^2 (0.5 + 0.85) \frac{[1 + 2(\tan 35°22.5')^2]}{\cos 35°22.5'}$$
$$A_2 = 3{,}040 \text{ ft. per sec.}^2$$

From Eq. (10-7) for $i = R_N + R_F = 0.179 + 0.5 = 0.679$

$$A_3 = -0.000914 \left(\frac{2000}{2}\right)^2 1.17 \left(1 + \frac{1.17}{0.679}\right)$$
$$A_3 = -2{,}915 \text{ ft. per sec.}^2$$

However, for rapid valve deceleration, Eq. (10-7) will not give the maximum deceleration and Eq. (10-6) should be used to find the maximum value. Thus substituting various values of ϕ in Eq. (10-6) and solving for A_Y, the deceleration is found to be a maximum when $\phi = \phi_M = 19°37.5'$, at which value $A_Y = 3{,}920$ ft. per sec.^{2}

From this, it is evident that increase in volumetric efficiency by using a period of dwell is attained at the expense of greater stress at the line of contact between the cam and roller during acceleration and greater spring loading during deceleration. For the example, the increase in stress during acceleration will be $[(3{,}040 - 2{,}395)/2{,}395] \times 100 = 27$ per cent, and at the full-open position, the increase in spring loading will be

$$[(2{,}915 - 1{,}485)/1{,}485] \times 100 = 96.2 \text{ per cent.}$$

However, since the maximum deceleration occurs in case *b* when the lift is 0.309 in., a fairer comparison would be to note the increase in loading over case *a* at a lift of 0.309 in. For case *a* at a lift of 0.309 in., from Eq. (10-4)

$$0.309 = 0.865 \cos \phi + (0.984^2 - 0.865^2 \sin^2 \phi)^{1/2} - 1.35$$

or

$$\phi = 28° \, 30'$$

and from Eq. (10-6), the deceleration on the nose circle is

$$A_Y = -0.000914 \left(\frac{2,000}{2}\right)^2 0.865 \left[\cos 28°30' + \frac{0.865 \cos 57°}{(0.984^2 - 0.865^2 \sin^2 28°30')^{1/2}} + \frac{0.865^3 \sin^2 57°}{4(0.984^2 - 0.865^2 \sin^2 28°30')^{3/2}}\right]$$

$$A_Y = -1,240 \text{ ft. per sec.}^2$$

The increase in spring loading is

$$\frac{2,915 - 1,240}{1,240} \times 100 = 135\%$$

Thus the use of a period of dwell is limited first by the ability of the spring to keep the follower on the cam.

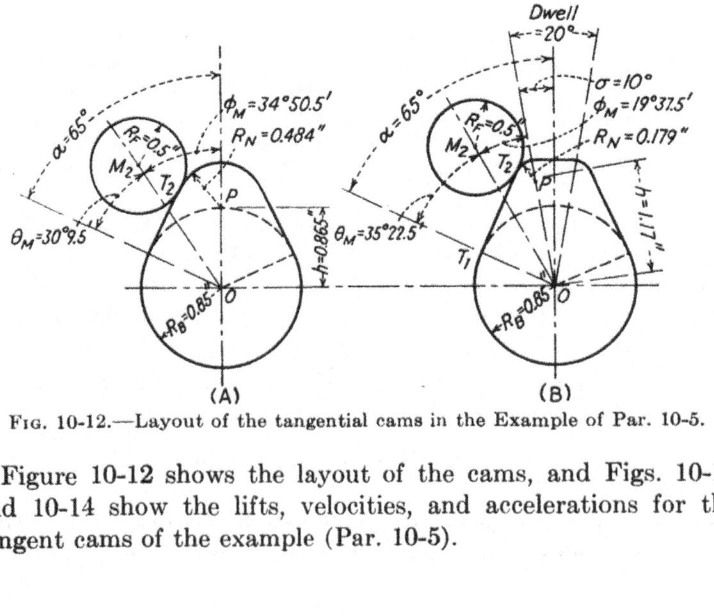

FIG. 10-12.—Layout of the tangential cams in the Example of Par. 10-5.

Figure 10-12 shows the layout of the cams, and Figs. 10-13 and 10-14 show the lifts, velocities, and accelerations for the tangent cams of the example (Par. 10-5).

10-6. Mushroom Cams.—Mushroom cams derive their name from the mushroom-shaped sliding follower which usually has a flat surface in contact with the cam. This type of cam, which is extensively used in automotive practice, differs from the

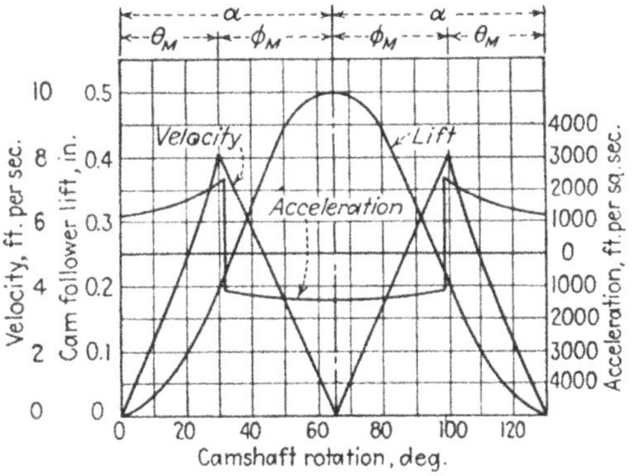

FIG. 10-13.—Lift, velocity, and acceleration curves for the tangent cam, Case (a), Par. 10-5.

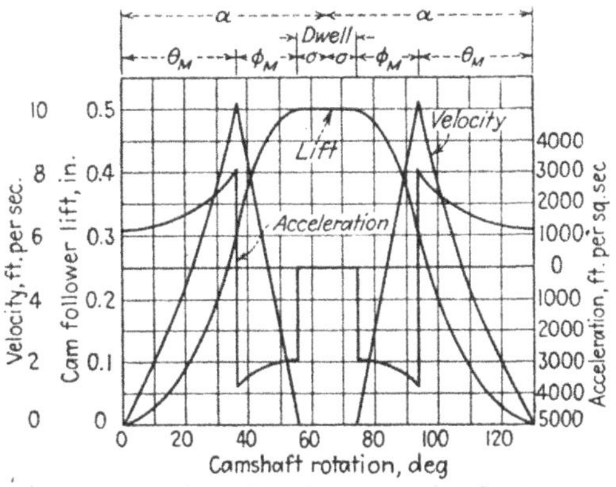

FIG. 10-14.—Lift, velocity, and acceleration curves for the tangent cam, Case (b), Par. 10-5.

tangential cam in that the flanks are arcs of circles tangent to the base and nose circles, respectively. Figure 10-15 shows the general arrangement of a mushroom cam with a flat-faced sliding follower. In the following analysis, symbols (as apply) are the same as in the analysis of tangential cams. In addition,

R_{FL} = radius of the flank circle.

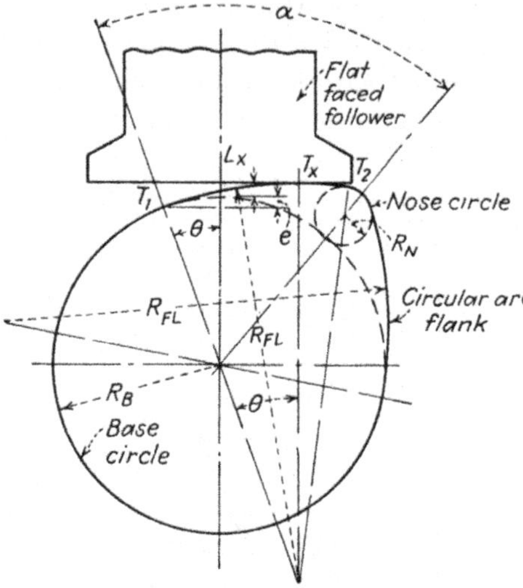

FIG. 10-15.—Mushroom cam with flat follower on convex circular-arc flank portion (T_X).

Referring to Fig. 10-15, for the lift on the flank circle

$$e + L_x = R_{FL} - R_{FL} \cos \theta$$
$$e = R_{FL}(1 - \cos \theta) - L_x$$

also

$$e = R_B - R_B \cos \theta = R_B(1 - \cos \theta)$$

eliminating e

$$R_B(1 - \cos \theta) = R_{FL}(1 - \cos \theta) - L_x$$

or

$$L_x = (R_{FL} - R_B)(1 - \cos \theta) \qquad (10\text{-}14)$$

For the velocity on the flank circle,

$$V_x = \frac{dL_x}{dt} = \frac{dL_x}{d\theta} \times \frac{d\theta}{dt} = \frac{2\pi N}{12 \times 60} \times \frac{dL_x}{d\theta} = 0.00873N \frac{dL_x}{d\theta}$$

$$V_x = 0.00873N(R_{FL} - R_B) \sin \theta \qquad (10\text{-}15)$$

For the acceleration on the flank circle,

$$A_x = \frac{dV_x}{dt} = \frac{dV_x}{d\theta} \times \frac{d\theta}{dt} = 0.000914N^2(R_{FL} - R_B) \cos \theta \qquad (10\text{-}16)$$

On the nose circle, the lift, velocity, and the acceleration would be similar to that on the nose circle of a tangential cam except for the change from a finite to an infinite radius of the cam follower. Referring to Fig. 10-16, the lift on the nose circle is

$$L_Y = R_N + h \cos \phi - R_B \tag{10-17}$$

where $h = OP$, and $\phi = 0$ when $L_Y = L_3$ ($= L_M$, the maximum lift). The velocity on the nose circle is

$$V_Y = \frac{dL_Y}{dt} = \frac{dL_Y}{d\phi} \times \frac{d\phi}{dt} = 0.00873N \frac{dL_Y}{d\phi}$$

$$V_Y = 0.00873Nh \sin \phi, \text{ ft. per sec.} \tag{10-18}$$

The acceleration on the nose circle is

$$A_Y = \frac{dV_Y}{dt} = \frac{dV_Y}{d\phi} \times \frac{d\phi}{dt} = -0.000914N^2h \cos \phi, \text{ ft. per sec.}^2 \tag{10-19}$$

at maximum lift, $\phi = 0$ and

$$A_Y = A_3 = -0.000914N^2h \tag{10-20}$$

As with tangential cams, there are definite relations between the various dimensions and angles of the mushroom cam.

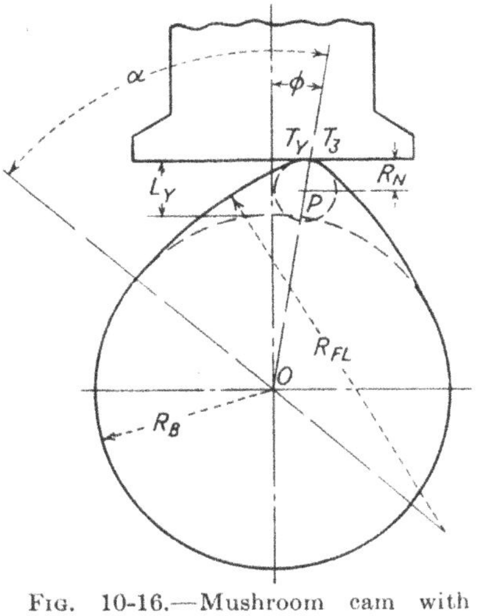

FIG. 10-16.—Mushroom cam with flat-faced follower on the nose-circle portion (T_Y).

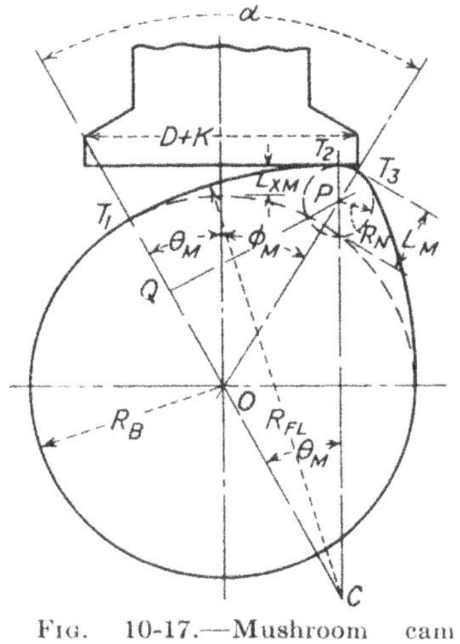

FIG. 10-17.—Mushroom cam with flat-faced follower in contact at the point of tangency of the flank and nose circles.

The radius of the base circle R_B is determined from shaft stiffness requirements (see Par. 10-15), the lift L_M is determined from valve and volumetric efficiency requirements, and the radius of the nose circle is assumed. Then, the distance between the base circle and nose circle centers, h ($= OP$, Fig. 10-17) is

$$h = R_B + L_M - R_N \tag{10-21}$$

For the radius of the flank circle (Fig. 10-17),

$$(R_{FL} - R_N)^2 = (h \sin \alpha)^2 + (R_{FL} - R_B + h \cos \alpha)^2$$

$$R_{FL}^2 + R_N^2 - 2R_N R_{FL} = h^2 \sin^2 \alpha + R_{FL}^2 + R_B^2 - 2R_B R_{FL} + 2R_{FL}h \cos \alpha - 2R_B h \cos \alpha + h^2 \cos^2 \alpha$$

$$R_{FL} = \frac{R_N^2 - R_B^2 - h^2 + 2R_B h \cos \alpha}{2(R_N - R_B + h \cos \alpha)} \tag{10-22}$$

Again from Fig. 10-17, the maximum angular travel of the cam while the follower is on the flank circle is

$$(R_{FL} - R_N) \sin \theta_M = h \sin \alpha$$

$$\theta_M = \text{arc sin} \left[\frac{h \sin \alpha}{(R_{FL} - R_N)} \right] \tag{10-23}$$

and the maximum angular travel of the cam from the point of tangency of the flank and nose circles to the point of maximum lift is

$$\phi_M = \alpha - \theta_M \tag{10-24}$$

10-7. Example of Mushroom-cam Calculations.—Determine dimensions and accelerations for a mushroom cam with flat-faced follower to operate an inlet valve on an in-line engine. Available data are as follows: inlet valve opening, 15 deg. before top center; inlet valve closing, 65 deg. after bottom center; maximum lift, 0.5 in.; diameter of camshaft, 1.575 in.; speed of engine, 2,000 r.p.m.

Procedure.—Using the nomenclature of Par. 10-6 and assuming the diameter of the base circle = diameter of camshaft + ⅛ in.,

$$R_B = \frac{1.575}{2} + \frac{0.125}{2} = 0.85 \text{ in.,} \qquad L_M = 0.5 \text{ in.,}$$

$$\alpha = \frac{15 + 180 + 65}{4} = 65 \text{ deg.}$$

Assume $R_N = 0.25$ in. From Eq. (10-21), $h = 0.85 + 0.5 - 0.25 = 1.1$ in. From Eq. (10-22),

$$R_{FL} = \frac{0.25^2 - 0.85^2 - 1.1^2 + (2 \times 0.85 \times 1.1 \times 0.4226)}{2[0.25 - 0.85 + (1.1 \times 0.4226)]} = 3.99 \text{ in.}$$

From Eq. (10-23),

$$\theta_M = \text{arc sin} \frac{1.1 \times 0.9063}{(3.99 - 0.25)} = \text{arc sin } 0.2665 = 15°8'$$

From Eq. (10-24),

$$\phi_M = 65° - 15°8' = 49°52'$$

The layout of the cam is shown in Fig. 10-18.

From Eq. (10-16), it is evident that the acceleration on the flank circle will be a maximum when $\theta = 0$.

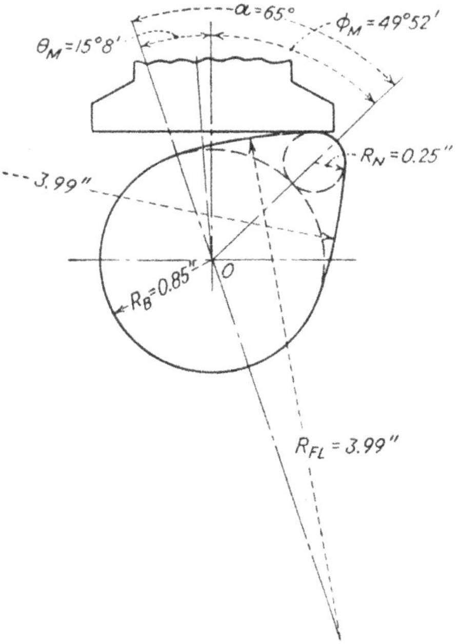

Fig. 10-18.—Layout of the mushroom cam in the Example of Par. 10-7.

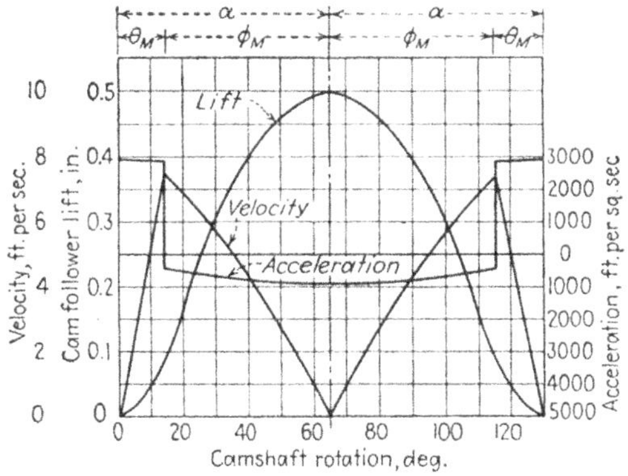

Fig. 10-19.—Lift, velocity, and acceleration curves for the mushroom cam, Par. 10-7.

For the data of the example,

$$A_{xM} = 0.000914(2{,}000/2)^2(3.99 - 0.85) \cos 0 = 2{,}870 \text{ ft. per sec.}^2$$

Other values of A_x together with values of L_x [calculated from Eq. (10-14)] and V_x [calculated from Eq. (10-15)] may be read from Fig. 10-19. From

Eq. (10-19), it is evident that the deceleration on the nose circle is a maximum when $\phi = 0$.

For the data of the example [using Eq. (10-20)],

$$A_{YM} = A_3 = -0.000914 \left(\frac{2,000}{2}\right)^2 \times 1.1 = -1,006 \text{ ft. per sec.}^2$$

Other values of A_Y [calculated from Eq. (10-19)] together with values of L_Y [calculated from (Eq. 10-17)] and V_Y [calculated from (Eq. (10-18)] may be read from Fig. 10-19.

Comparing Figs. 10-13, 10-14, and 10-19, it is seen that the initial acceleration for the mushroom cam is higher; hence the initial shock load on the valve gear will be higher. The deceleration loads are not greatly different for the tangent cam with zero dwell and the mushroom cam but with dwell deceleration loads rise rapidly. Mushroom cams are seldom built with a dwell period.

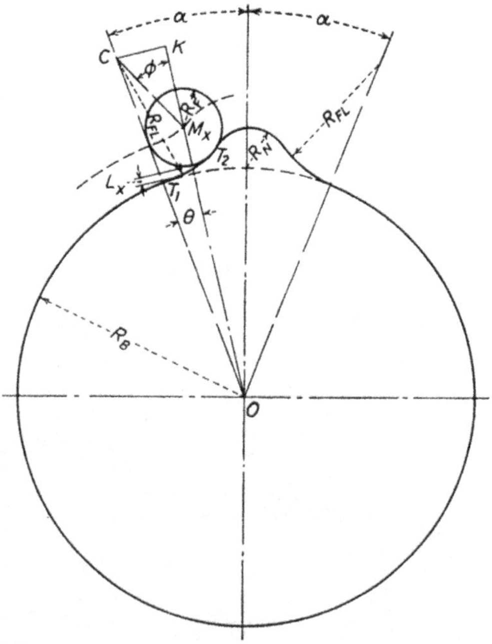

FIG. 10-20.—Hollow-faced cam with roller follower on concave circular-arc flank.

10-8. Hollow-faced Cams.—In the case of radial engines using a cam ring (Par. 10-1), the radius of the base circle R_B is quite large relative to the other dimensions. Also the ring rotates at less than one-half of crankshaft r.p.m., hence α is a relatively small angle.

Under these conditions, tangent and mushroom types of cam contours are not suitable with any appreciable lift because of the excessively small nose-circle radius that can be used; in many instances they are not even possible.

For radial engines using cam disks, therefore, other types of cam profiles should or must be used, and of the possible types, the so-called hollow-faced cam is probably most common. This type of cam (Fig. 10-20) differs from the mushroom cam basically in that it has a concave flank and uses a roller follower.

Referring to Fig. 10-20, from the geometry of the figure,

$$\frac{OK}{OC} = \cos\theta$$

$$OC = R_B + R_{FL}$$

$$OK = OM_x + M_xK$$

$$OM_x = R_B + L_x + R_F$$

$$M_xK = \sqrt{M_xC^2 - CK^2} = \sqrt{(R_{FL} - R_F)^2 - (R_B + R_{FL})^2 \sin^2\theta}$$

Hence

$$\frac{R_B + L_x + R_F + \sqrt{(R_{FL} - R_F)^2 - (R_B + R_{FL})^2 \sin^2\theta}}{R_B + R_{FL}} = \cos\theta$$

From which the lift on the flank circle is

$$L_x = (R_B + R_{FL})\cos\theta - \sqrt{(R_{FL} - R_F)^2 - (R_B + R_{FL})^2 \sin^2\theta} - (R_B + R_F) \quad (10\text{-}25)$$

For the velocity on the flank circle,

$$V_x = \frac{dL_x}{dt} = \frac{dL_x}{d\theta} \times \frac{d\theta}{dt} = \frac{2\pi N}{12 \times 60}\frac{dL_x}{d\theta} = 0.00873N\frac{dL_x}{d\theta}$$

$$V_x = 0.00873N\frac{d}{d\theta}[(R_B + R_{FL})^2 \cos\theta - \sqrt{(R_{FL} - R_F)^2 - (R_B + R_{FL})^2 \sin^2\theta} - (R_B + R_F)]$$

$$= 0.00873N\left\{-(R_B + R_{FL})^2 \cos\theta + \frac{(R_B + R_{FL})^2 \sin\theta\cos\theta}{[(R_{FL} - R_F)^2 - (R_B + R_{FL})^2 \sin^2\theta]^{1/2}}\right\}$$

$$V_x = 0.00873N(R_B + R_{FL})\sin\theta\left\{\frac{(R_B + R_{FL})\cos\theta}{[(R_{FL} - R_F)^2 - (R_B + R_{FL})^2 \sin^2\theta]^{1/2}} - 1\right\} \quad (10\text{-}26)$$

For the acceleration on the flank circle

$$A_x = \frac{dV_x}{dt} = \frac{dV_x}{d\theta} \times \frac{d\theta}{dt} = \frac{2\pi N}{60} \times \frac{dV_x}{d\theta}$$

$$= 0.000914N^2(R_B + R_{FL}) \frac{d}{d\theta}$$

$$\left(\sin\theta \left\{\frac{(R_B + R_{FL})\cos\theta}{[(R_{FL} - R_F)^2 - (R_B + R_{FL})^2 \sin^2\theta]^{1/2}} - 1\right\}\right)$$

$$A_x = 0.000914N^2(R_B + R_{FL})$$

$$\left\{\frac{(R_B + R_{FL})\cos 2\theta}{[(R_{FL} + R_F)^2 - (R_B + R_{FL})^2 \sin^2\theta]^{1/2}}\right.$$

$$\left. + \frac{(R_B + R_{FL})^3 \sin^2\theta \cos^2\theta}{[(R_{FL} - R_F)^2 - (R_B + R_{FL})^2 \sin^2\theta]^{3/2}} - \cos\theta \right\} \quad (10\text{-}27)$$

On the nose circle, the hollow-faced cam with roller follower is the same as the tangent cam with roller follower. Hence for the hollow-faced cam,

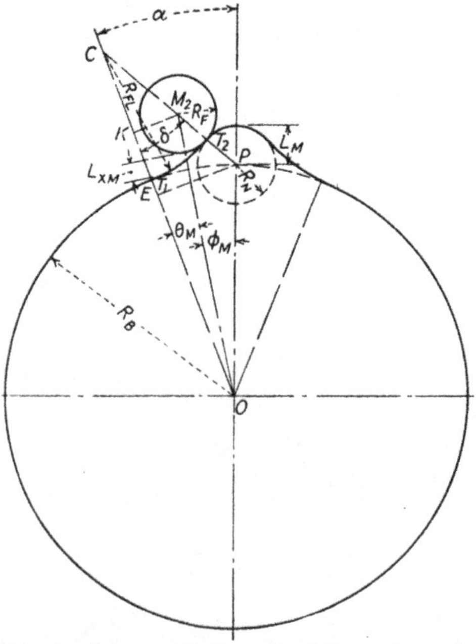

FIG. 10-21.—Hollow-faced cam with the roller follower in contact at the point of tangency of the flank and nose circles.

For L_Y, use Eq. (10-4).
For V_Y, use Eq. (10-5).
For A_Y, use Eq. (10-6).

As with tangent and mushroom cams, the parts of hollow-faced cams are definitely related. When applied to radial-engine cam rings, the radius of the base circle R_B is no longer determined from shaft-stiffness requirements but instead is based upon the gear-drive dimensions (see Par. 10-9). The lift L_M is determined from valve and volumetric efficiency requirements, and the radii of the nose circle R_N and the roller follower R_F are assumed. The angle α is no longer one-fourth of the total valve open time, but is

$$\alpha = \frac{\text{total valve open time, deg.}}{2R} \tag{10-28}$$

where $R = \dfrac{\text{crankshaft r.p.m.}}{\text{cam-ring r.p.m.}}$ [see Par. 10-9 and Eq. (10-34)].

From Fig. 10-21, the distance between the base and nose circle centers, h $(= OP)$, is

$$h = R_B + L_M - R_N \tag{10-29}$$

For the radius of the flank circle,

$$EP = h \sin \alpha$$

also

$$EP = (R_{FL} + R_N) \sin \delta$$

$$\sin \delta = \frac{h \sin \alpha}{(R_{FL} + R_N)}$$

$$\sin^2 \delta = \frac{h^2 \sin^2 \alpha}{(R_{FL} + R_N)^2} \tag{a}$$

$$R_B - T_1E = OE = h \cos \alpha$$

$$T_1E = R_B - h \cos \alpha \tag{b}$$

$$R_{FL} + T_1E = CE = (R_{FL} + R_N) \cos \delta$$

$$T_1E = (R_{FL} + R_N) \cos \delta - R_{FL} \tag{c}$$

Combining (*b*) and (*c*)

$$\cos \delta = \frac{R_B + R_{FL} - h \cos \alpha}{R_{FL} + R_N}$$

$$\cos^2 \delta = \frac{(R_B + R_{FL} - h \cos \alpha)^2}{(R_{FL} + R_N)^2} \tag{d}$$

adding (*a*) and (*d*)

$$\sin^2 \delta + \cos^2 \delta = \frac{h^2 \sin^2 \alpha}{(R_{FL} + R_N)^2} + \frac{(R_B + R_{FL} - h \cos \alpha)^2}{(R_{FL} + R_N)^2}$$

From which

$$(R_{FL} + R_N)^2 = h^2 \sin^2 \alpha + (R_B + R_{FL} - h \cos \alpha)^2$$

Solving this expression for the radius of the flank circle gives

$$R_{FL} = \frac{h^2 + R_B^2 - R_N^2 - 2R_B h \cos \alpha}{2(R_N - R_B + h \cos \alpha)} \tag{10-30}$$

Again referring to Fig. 10-21,

$$\frac{KM_2}{EP} = \frac{CM_2}{CP}$$

but

$$EP = h \sin \alpha$$
$$CM_2 = R_{FL} - R_F$$
$$CP = R_{FL} + R_N$$

Hence

$$KM_2 = \left(\frac{R_{FL} - R_F}{R_{FL} + R_N}\right) h \sin \alpha \tag{a}$$

Also

$$\tan \theta_M = \frac{KM_2}{OK} = \frac{KM_2}{(OC - CK)} \tag{b}$$

but

$$OC = R_B + R_{FL} \tag{c}$$
$$CK = CM_2 \cos \delta = (R_{FL} - R_F) \cos \delta = \sqrt{(R_{FL} - R_F)^2 \cos^2 \delta}$$
$$CK = \sqrt{(R_{FL} - R_F)^2 (1 - \sin^2 \delta)}$$

From step (*a*) in the development of Eq. (10-30),

$$\sin^2 \delta = \frac{h^2 \sin^2 \alpha}{(R_{FL} + R_N)^2}$$

Hence

$$CK = \sqrt{(R_{FL} - R_F)^2 \left(1 - \frac{h^2 \sin^2 \alpha}{(R_{FL} + R_N)^2}\right)} \tag{d}$$

Combining (*b*), (*c*), and (*d*),

$$KM_2 = \left\{(R_B + R_{FL}) - \sqrt{(R_{FL} - R_F)^2 \left[1 - \frac{h^2 \sin^2 \alpha}{(R_{FL} + R_N)^2}\right]}\right\} \tan \theta_M \tag{e}$$

combining (*a*) and (*e*)

$$\theta_M = \text{arc tan} \left(\frac{\left(\frac{R_{FL} - R_F}{R_{FL} + R_N}\right) h \sin \alpha}{\left\{(R_B + R_{FL}) - \sqrt{(R_{FL} - R_F)^2 \left[1 - \frac{h^2 \sin^2 \alpha}{(R_{FL} + R_N)^2}\right]}\right\}}\right) \tag{10-31}$$

$$\phi_M = \alpha - \theta_M \tag{10-32}$$

When a period of dwell is used with the hollow-faced cam, Eqs. (10-30), (10-31), and (10-32) may be used by replacing α by $(\alpha - \sigma)$ where σ = dwell angle/2. As in the case of tangential cams, the period of dwell is somewhat optional, but as the radius of the nose circle R_N is reduced, the spring load to keep the follower on the cam increases rapidly [Eq. 10-6]. If R_N is maintained large, then the radius of the flank circle R_{FL} is smaller [Eq. (10-30)] and the acceleration on the flank A_x increases. Thus the period of dwell is limited either by the allowable spring load or by the stress at the line of contact between the roller follower and the flank circle.

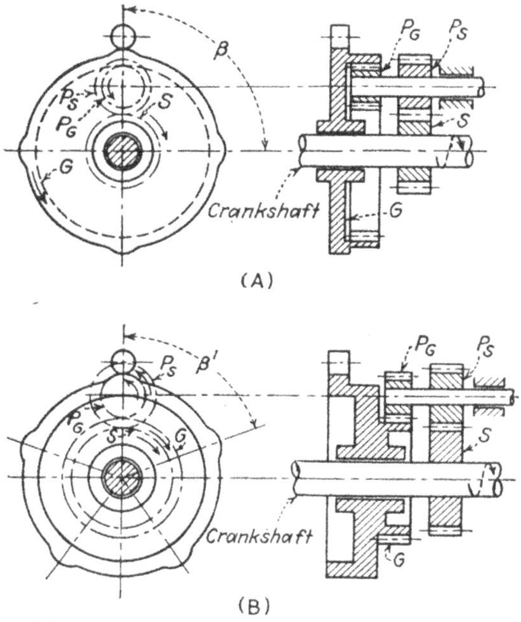

FIG. 10-22.—Schematic arrangement of usual cam-ring drives for radial engines. (*A*) For opposite rotation of cam ring and crankshaft. (*B*) For same direction of rotation of cam ring and crankshaft.

10-9. Radial-engine Cam Rings.—Four-stroke-cycle radial engines have an odd number of cylinders per bank to permit even firing, and for such engines, a cam ring containing more than one cam is generally used. Usually two rows of cams are used (one row for intake and one row for exhaust valves), both rows of cams are integral with the same cam ring, and this ring is concentric with and connected by suitable gearing to the crankshaft (Fig. 10-22).

Each intake cam operates the intake valves of all cylinders in succession, and the corresponding action occurs with each exhaust cam. Hence the cam ring must rotate at such a speed that X intake cams operate Y intake valves in two revolutions or 720 deg. of crankshaft travel.

For opposite rotation of cam ring and crankshaft (Fig. 10-23), cam A is just ready to operate the follower of No. 1 cylinder. Usual cylinder numbering is in succession in the direction of rotation; hence the next cylinder to start the event is No. 3, and

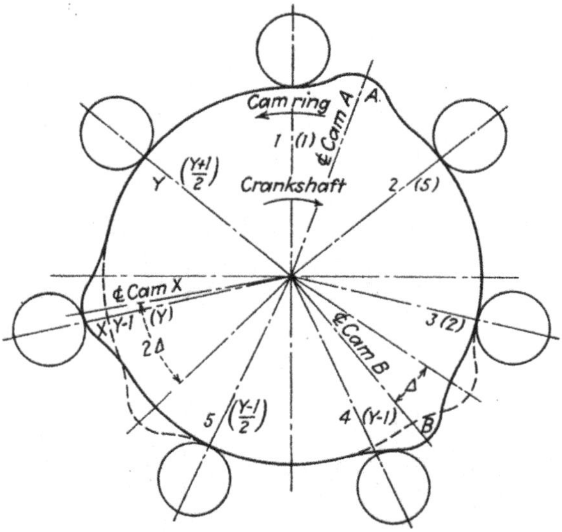

FIG. 10-23.—Radial-engine cam-ring analysis.

the next will be No. 5. But for purposes of this analysis, it is preferable to number the cylinders in the order of firing or cyclic sequence, and this method of numbering is indicated in parentheses adjacent to the usual numbering.

For an engine of Y cylinders (Fig. 10-23), No. (2) valve will start to open $720/Y$ deg. after the start of No. 1 cylinder, and to be in position to accomplish this, cam B will have to move through an angle of Δ deg. For X equally spaced cams,

$$\Delta = \frac{360}{X} - \frac{720}{Y}$$

If there are more than two cams, the third cam will have to move through an angle of 2Δ [$= (3 - 1)\Delta$] to operate valve (3), and cam X will have to move through an angle of $(X - 1)\Delta$ to operate valve $\left(\frac{Y-1}{2}\right)$. When cam X is ready to operate its valve, cam A will also have moved through $(X - 1)\Delta$ deg.; hence cam A can operate valve $\left(\frac{Y+1}{2}\right)$ by moving an additional angle Δ or a total motion of $(X - 1 + 1)\Delta = X\Delta$ deg. But this distance that cam A moves from the start of opening valve (1) to the start of opening valve $\left(\frac{Y+1}{2}\right)$ is $\frac{360}{Y}$ or

$$\frac{360}{Y} = X\Delta = X\left(\frac{360}{X} - \frac{720}{Y}\right)$$

from which

$$X = \frac{Y-1}{2} \tag{10-33}$$

When cam A has reached valve $\left(\frac{Y+1}{2}\right)$, all the other cams will be in the same respective position as when cam A started to operate valve (1); hence one row of X cams can operate all the intake valves for Y cylinders, and a parallel row of X cams attached to the same cam ring can operate all the exhaust valves.

In the preceding analysis, when the crankshaft turned through $720/Y$ deg., the cam ring turned through an angle of Δ deg. in the opposite direction. Thus the speed ratio is

$$R = \frac{720/Y}{\Delta} = \frac{720/Y}{(360/X) - (720/Y)} = \frac{1}{(360Y/720X) - 1}$$

but from Eq. (10-33)

$$X = \frac{Y-1}{2}$$

hence

$$R = \frac{1}{\dfrac{360Y}{720\left(\dfrac{Y-1}{2}\right)} - 1}$$

from which

$$R = Y - 1 \tag{10-34}$$

or the ratio of cam ring r.p.m. to crankshaft r.p.m. is

$$\frac{1}{R} = \frac{1}{Y - 1} \tag{10-35}$$

By a similar line of reasoning to the above, it can be demonstrated that for the *same* direction of rotation of the cam ring and crankshaft the number of cams required is

$$X' = \frac{Y + 1}{2} \tag{10-36}$$

and the ratio of crankshaft to cam ring r.p.m. is

$$R' = Y + 1 \tag{10-37}$$

In general, opposite rotation of the cam ring to crankshaft is desirable because fewer cams have to be accurately machined on the cam ring and because the relative velocity between the cam ring and follower rollers is usually less than for the same direction of rotation.

Because the drive gear attached to the crankshaft and the driven gear to which the cam ring attaches are concentric about the crankshaft axis, some limitations are placed on the possible gear-tooth combinations that will give the correct crankshaft/cam-ring ratio R or R'. It is usually not desirable for any gear to have less than 12 teeth, and to keep down weight, and relative velocity between the cam ring and roller follower, the largest gear should have a small over-all diameter. Also care should be exercised to see that the gears do not interfere with adjacent parts.

Referring to Fig. 10-22, it is seen that for opposite rotation of crankshaft and cam ring (case A) the gear train is a special case of compound epicyclic gearing in which the arm (A, Table A3-7) does not revolve. From case 1 of Table A3-7, the speed ratio of the driven gear to the drive gear is

$$\frac{1}{R} = \frac{N_S N_{PG}}{N_G N_{PS}} \tag{a}$$

where R = ratio of crankshaft to cam ring, r.p.m.
N_S = number of teeth on the drive gear S, Fig. 10-22A.
N_{PG} = number of teeth on the pinion gear P_G, Fig. 10-22A.
N_G = number of teeth on the driven gear G, Fig. 10-22A.
N_{PS} = number of teeth on the pinion gear P_S, Fig. 10-22A.

From Fig. 10-22*A* it is apparent that

$$D_S + D_{PS} = D_G - D_{PG} \qquad (10\text{-}38)$$

where D = the pitch diameter of the various gears designated by the subscripts. But

$$D = \frac{N}{P_d}$$

where N = number of teeth in the gear.
P_d = diametral pitch.
Hence, for all gears having the same diametral pitch, Eq. (10-38) becomes

$$N_S + N_{PS} = N_G - N_{PG} \qquad (10\text{-}39)$$

Suitable gear-tooth combinations must satisfy both Eqs. (10-37) and (10-39), and with so many unknowns, a trial solution is necessary. However, the problem may be simplified by assuming ratios of N_S/N_{PS}. Thus, for a seven-cylinder single-bank radial engine with opposite crankshaft and cam-ring rotation, from Eq. (10-34), $R = 6$. Assuming $N_{PG} = 12$ teeth and $N_S/N_{PS} = 1$, from Eq. (10-37)

$$\frac{1}{6} = 1 \times \frac{12}{N_G} \qquad \text{or} \qquad N_G = 72 \text{ teeth}$$

since from the foregoing assumption $N_S = N_{PS}$, Eq. (10-39) will be satisfied when

$$2N_S = 72 - 12$$

Hence

$$N_S = N_{PS} = 30 \text{ teeth}$$

Several other gear combinations are given in Table 10-2, some of which are more desirable than others. In general, the best combination is one which, for the desired minimum number of teeth on the smallest pinion, gives the lowest number of teeth on the internal gear without interference of parts anywhere. The minimum chance for interference between the crankshaft and pinion P_G will exist when N_{PG} is small numerically. Over-all dimensions will be least when N_G and N_{PS} are small numerically.

TABLE 10-2.—REDUCTION-GEAR TOOTH COMBINATIONS SUITABLE* FOR COMPOUND EPICYCLIC GEAR TRAINS AS APPLIED TO RADIAL-ENGINE CAM-RING DRIVES

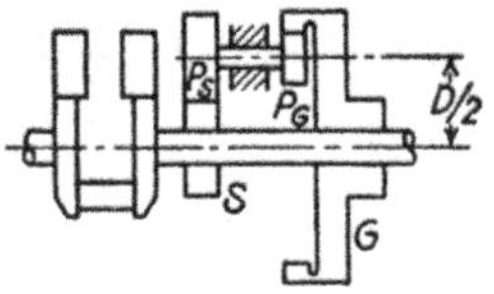

$$R = \frac{N_{PS}N_G}{N_S N_{PG}}$$

$$N_S + N_{PS} = N_G - N_{PG}$$

$$R = \frac{\text{crankshaft r.p.m.}}{\text{cam-ring r.p.m.}}$$

N = number of teeth

$\frac{N_S}{N_{PS}}$	5-cyl. radial, $R = 4$				7-cyl. radial, $R = 6$				9-cyl. radial, $R = 8$			
	N_{PG}	N_{PS}	N_S	N_G	N_{PG}	N_{PS}	N_S	N_G	N_{PG}	N_{PS}	N_S	N_G
1/1	12	18	18	48	12	30	30	72	12	42	42	96
	14	21	21	56	14	35	35	84	14	49	49	112
	16	24	24	64	16	40	40	96	16	56	56	128
	18	27	27	72	18	45	45	108	18	63	63	144
2/1	12	28	56	96	12	44	88	144	12	60	120	192
	15	35	70	120	15	55	110	180	16	80	160	256
1/2	36	24	12	72	18	24	12	54	12	24	12	48
	39	26	13	78	21	28	14	63	24	48	24	96
	42	28	14	84	24	32	16	72	36	72	36	144
3/2	15	30	45	90	15	48	72	135	15	66	99	180
	20	42	63	120	20	64	96	180	20	88	132	240
2/3	24	24	16	64	20	36	24	80	15	39	26	80
									30	78	52	160
3/4	28	32	24	84	28	56	42	126	14	40	30	84
									21	60	45	126
5/4	18	32	40	90	18	52	65	135				
7/8									15	48	42	105

* As far as satisfying Eqs. (10-37) and (10-39) is concerned. However, before definitely selecting any particular combination, a layout to scale should be made to check on interference of parts.

10-10. Example of Radial-engine Cam Calculations.—Determine dimensions and accelerations for a hollow-faced cam to operate the inlet valves on a seven-cylinder radial engine. Available data are as follows: inlet valve opening, 15 deg. before top center; inlet valve closing, 65 deg. after bottom center; maximum lift of roller follower, 0.5 in.; diameter of roller follower,

1 in.; diameter of engine crankshaft, 3 in.; opposite rotation of crankshaft and cam ring; speed of engine, 2,000 r.p.m.

Procedure.—From Eq. (10-33), the number of inlet cams required will be

$$X = \frac{7 - 1}{2} = 3$$

From Eq. (10-34), the ratio of crankshaft speed to cam-ring speed will be

$$R = 7 - 1 = 6$$

From Table 10-2, tentatively select

$$N_{PG} = 18, \qquad N_{PS} = 45, \qquad N_S = 45, \qquad N_G = 108$$

Since the loads to be transmitted will not be excessive, 14½-deg. involute gears should be satisfactory, and a diametral pitch of 12 (see Table 8-7) should be low enough. The pitch diameters corresponding will be

$$D_{PG} = {}^{18}\!/_{12} = 1.5 \text{ in.}, \qquad D_{PS} = {}^{45}\!/_{12} = 3.75 \text{ in.},$$
$$D_S = {}^{45}\!/_{12} = 3.75 \text{ in.}, \qquad D_G = {}^{108}\!/_{12} = 9 \text{ in.}$$

The diameter of gear S will have to be sufficient to permit a hole center large enough to slip over the crankshaft.

From Table A3-3, the minimum dedendum is

$$\frac{1.157}{P_d} = \frac{1.157}{12} = 0.0964 \text{ in.}$$

and $3.75 - 2 \times 0.0964 = 3.5572 > 3$ in., hence the sun gear S will be sufficient. The gear P_G will have to fit in place without touching the crankshaft, and from Table A3-3 the addendum is

$$\frac{1}{P_d} = \frac{1}{12} = 0.0833 \text{ in.}$$

The center distance between the crankshaft and P_G is

$$\frac{D_G}{2} - \frac{D_{PG}}{2} = \frac{9}{2} - \frac{1.5}{2} = 3.75 \text{ in.}$$

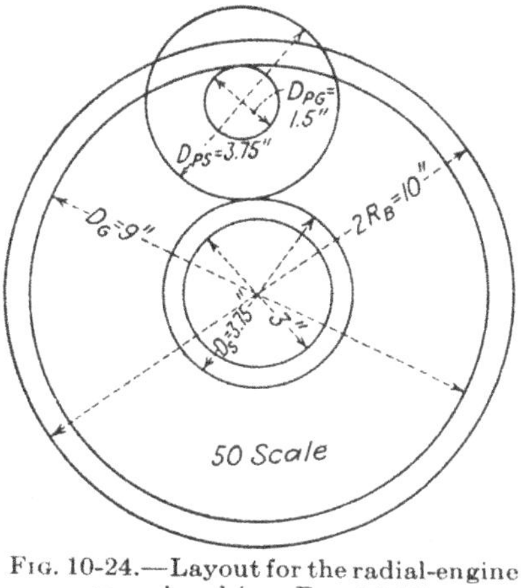

FIG. 10-24.—Layout for the radial-engine cam-ring drive, Par. 10-10.

and as $3.75 > (1.5/2) + 0.0833 + {}^{3}\!/_{2} = 2.3333$ in., the pinion gear P_G will fit in place without interference. The layout of the pitch circles for the various gears is shown in Fig. 10-24.

The cam ring is usually made a part of the internal gear G either integrally or by attachment; hence a logical value for the radius of the base circle (Fig. 10-24) is $R_B = 5$ in.

From Eq. (10-28),

$$\alpha = \frac{15 + 180 + 65}{2 \times 6} = 21.666° = 21°40'$$

Assume the radius of the nose circle is $R_N = 1.25$ in. Then from Eq. (10-29)

$$h = 5 + 0.5 - 1.25 = 4.25 \text{ in.}$$

From Eq. (10-30), the radius of the flank circle is

$$R_{FL} = \frac{4.25^2 + 5^2 - 1.25^2 - 2 \times 5 \times 4.25 \times 0.9293}{2(1.25 - 5 + 4.25 \times 0.9293)} = 5.05 \text{ in.}$$

From Eq. (10-31),

$$\theta_M = \text{arc tan}\left[\frac{\left(\dfrac{5.05 - 0.5}{5.05 + 1.25}\right) 4.25 \times 0.3692}{\left((5 + 5.05) - \left\{(5.05 - 0.5)^2\left[1 - \dfrac{4.25^2 \times 0.3692^2}{(5.05 + 1.25)^2}\right]\right\}^{1/2}\right)}\right]$$

from which

$$\theta_M = 11°19'$$
$$\phi_M = 21°40' - 11°19' = 10°21'$$

Cam-ring dimensions and the general arrangements are shown in Fig. (10-25).

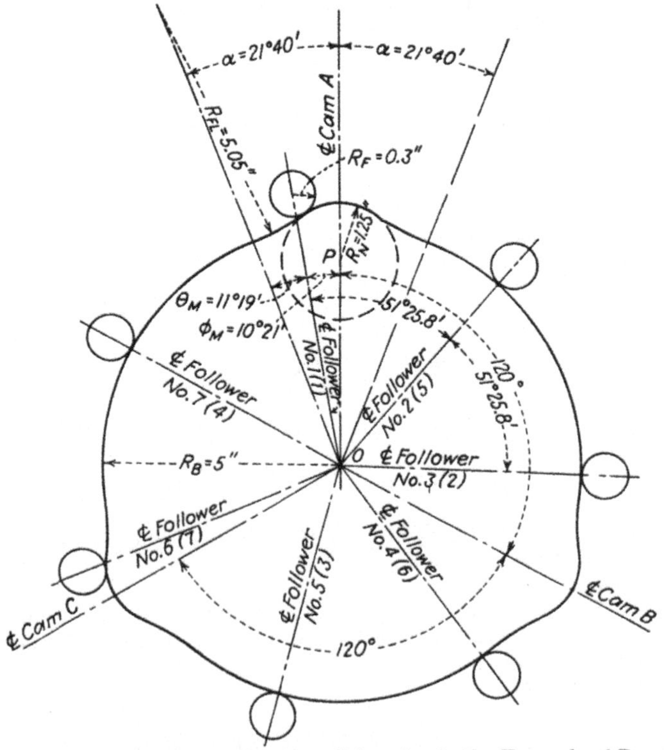

FIG. 10-25.—Cam-ring layout for the radial engine in the Example of Par. 10-10.

The acceleration on the flank circle may be found by means of Eq. (10-27). For example at $\theta = 11°19'$

$$A_x = 0.000914 \left(\frac{2{,}000}{6}\right)^2 (5 + 5.05) \left\{\frac{(5 + 5.05) \times 0.9230}{[(5.05 - 0.5)^2 - (5 + 5.05)^2 \times 0.1962^2]^{1/2}} + \frac{(5 + 5.05)^3 \times 0.1962^2 \times 0.9806^2}{[(5.05 - 0.5)^2 - (5 + 5.05)^2 \times 0.1962^2]^{3/2}} - 0.9806\right\}$$

$$A_x = 1{,}788 \text{ ft. per sec.}^2$$

The acceleration on the nose circle may be found from Eq. (10-6). For example at $\phi = 10°21'$,

$$A_Y = -0.000914 \left(\frac{2{,}000}{6}\right)^2 4.25 \left[0.98375 + \frac{4.25 \times 0.9354}{(1.75^2 - 4.25^2 \times 0.1797^2)^{1/2}} + \frac{4.25^3 \times 0.3535^2}{4(1.75^2 - 4.25^2 \times 0.1797^2)^{3/2}}\right]$$

$$A_Y = 1{,}780 \text{ ft. per sec.}^2$$

These values of A_x and A_Y together with the accelerations for other values of θ and ϕ are shown in Fig. 10-26. In addition, Fig. 10-26 shows the effect of reducing the radius of the nose circle from $R_N = 1.25$ to $R_N = 0.5$ in. In this latter case, it is seen that spring loads increase quite rapidly indicating that a period of dwell would be limited by the ability of the spring to keep the follower on the cam.

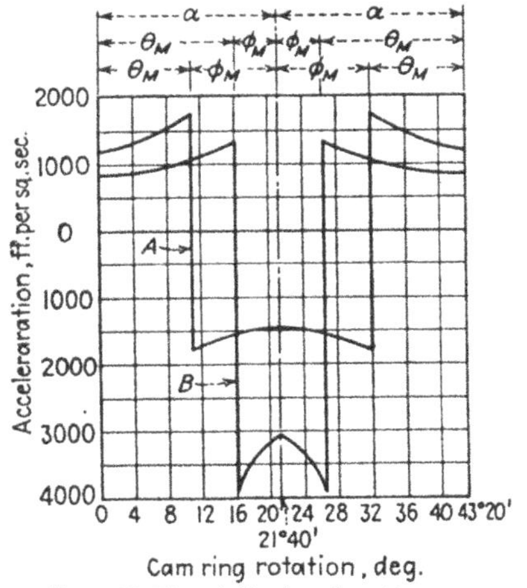

FIG. 10-26.—(*A*) Acceleration curve for the radial-engine cam in the Example of Par. 10-10. (*B*) Effect of reducing the radius of the nose circle from $R_N = 1.25$ to $R_N = 0.5$ in.

10-11. Cam Ramps.

—Owing to difference in temperature rise from cold, or idling, conditions to hot, or full-throttle, conditions of the various parts of the valve gear and the cylinder, a difference in expansion of the parts usually results which necessitates providing for valve clearance to prevent the valve being held slightly off its seat at some conditions of engine operation. This clearance is of the order of 0.005 to 0.015 in., depending upon the arrangement of parts, and it is sufficient to change the valve timing appreciably between the cold and hot conditions of operation. For engines in which quietness of operation is a factor as well as more accurate valve timing and reduced shock loading on the cam, cam ramps are used. These quieting ramps may be made in various ways,[1,2] and in general they consist of an incline occupying 15 to 30 deg. of camshaft travel and a rise equal to the clearance. They may be incorporated by reducing R_B by the amount of the clearance and then connecting this undercut base circle with the point $\theta = 0$ by means of the ramp.

Figure 10-27 shows the effect of a cam ramp without clearance, wherein the follower rides on the base circle R_B, but with clear-

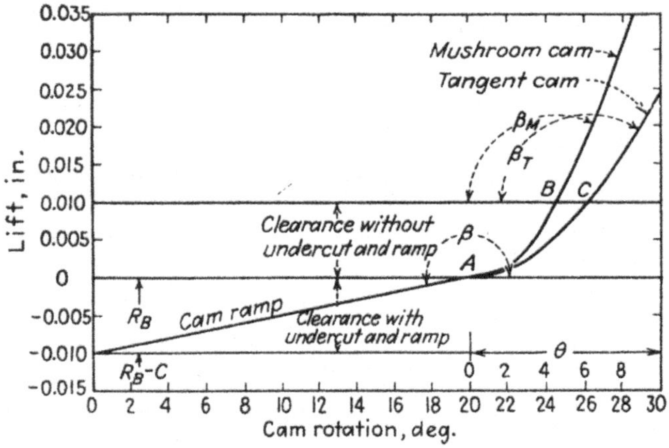

FIG. 10-27.—Effect of cam ramp in reducing shock and noise. (Note: See page 112 of Reference 10 for ramp formula.)

ance, the follower will not bear against the base circle. For instance, with 0.010-in. clearance the follower will strike the flank at B or C at a much sharper angle than it would at A without clearance. Hence the shock will be greater with clearance, and it is seen from the figure that the shock with mushroom cams is greater than with comparable tangent cams because $\beta_M < \beta_T$. With the cam ramp and anywhere from 0- to 0.010-in. clearance, the follower will strike the flank at A and the shock will be less severe because β is much greater than β_M or β_T. This, in turn will mean quieter operation.

In aircraft engines, cam ramps are probably less important than in automobile engines (*a*) because quiet idling is less essential, (*b*) because the noise of the cam cannot be heard above the roar of the unmuffled motor and propeller, (*c*) because the clearance can usually be set correctly for the much narrower operating range, and (*d*) because aircraft engines most commonly use tangent or hollow-faced cams that are inherently less noisy from this source.

Automatic clearance adjusters are of interest in this connection since they maintain the clearance at zero without danger of

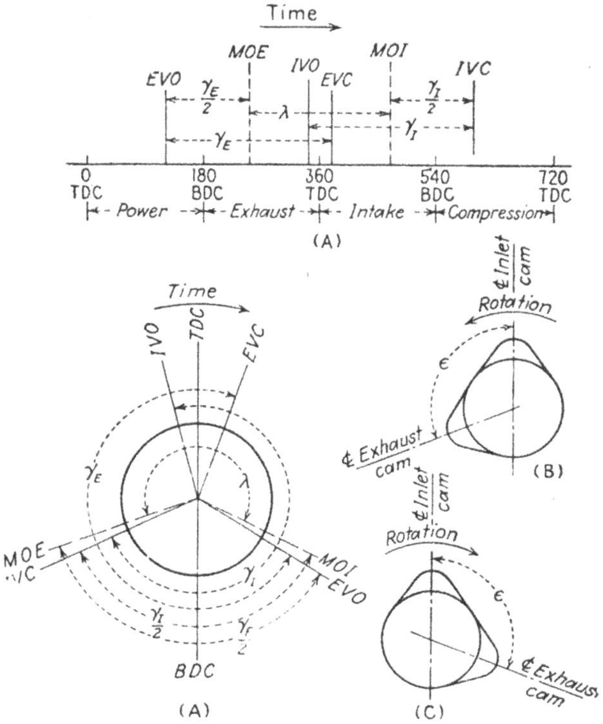

FIG. 10-28.—Method of determining inlet- and exhaust-cam spacing. (*A*) Valve-timing diagram referred to crankshaft. (*B*) Cam spacing for opposite rotation of crankshaft and camshaft, and (*C*) for same direction of rotation.

holding the valves off their seats and are therefore probably the best solution to the problem. Figure 10-2 shows one type of adjuster as used on the Franklin engine.

10-12. Cam Spacing.—The angular spacing of the inlet and exhaust cams on the camshaft or cam ring depends upon the valve timing, ratio R of crankshaft to camshaft or cam-ring speed, and upon the firing order. Referring to Fig. 10-28,

γ_I = the angular travel of the crankshaft from inlet-valve opening (I.V.O.) to inlet-valve closing (I.V.C.).

γ_E = the angular travel of the crankshaft from exhaust-valve opening (E.V.O.) to exhaust-valve closing (E.V.C.).

M.O.I. = mid-opening position of the inlet valve.

M.O.E. = mid-opening position of the exhaust valve.

λ = the angle between M.O.E. and M.O.I. measured according to cyclic sequence.

From the figure,

$$360 - \lambda = \frac{\gamma_I}{2} - \text{I.V.C.} + \frac{\gamma_E}{2} - \text{E.V.O.}$$

or

$$\lambda = \text{I.V.C.} + \text{E.V.O.} - \left(\frac{\gamma_I + \gamma_E}{2}\right) + 360 \qquad (10\text{-}40)$$

and

$$\epsilon = \frac{\lambda}{R} \qquad (10\text{-}41)$$

where ϵ = the angle between the exhaust and inlet cam measured according to cyclic sequence.

$$R = \frac{\text{crankshaft r.p.m.}}{\text{camshaft (or ring) r.p.m.}}$$

For example, in a four-cylinder opposed engine having a valve timing of inlet-valve opening, 15 deg. before top center; inlet-valve closing, 65 deg. after bottom center; exhaust-valve opening, 60 deg. before bottom center; and exhaust-valve closing, 20 deg. after top center,

$$\gamma_I = 15 + 180 + 65 = 260 \text{ deg.},$$
$$\gamma_E = 60 + 180 + 20 = 260 \text{ deg.}$$

and from Eq. (10-40)

$$\lambda = 65 + 60 - \left(\frac{260 + 260}{2}\right) + 360 = 225 \text{ deg.}$$

From Eq. (10-41),

$$\epsilon = 225/2 = 112.5 \text{ deg.}$$

For a seven-cylinder radial engine using the preceding valve timing and opposite rotation of the crankshaft and cam ring,

$R = 6$ [Eq. (10-34)], and

$$\epsilon = 225/6 = 37.5 \text{ deg.}$$

To permit closer axial spacing of the inlet- and exhaust-cam races and to place the center lines of the followers more nearly coin-

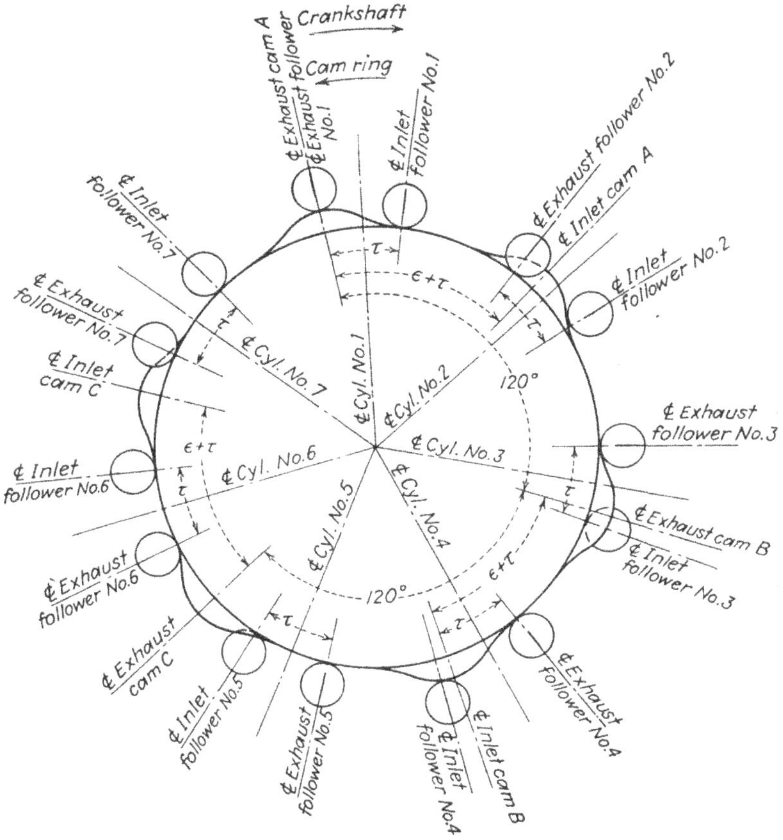

FIG. 10-29.—Layout for locating inlet and exhaust cams on radial-engine cam rings.

cident with the center lines of the push rods, radial-engine followers are usually offset so that their axes form a small angle to the plane containing the corresponding cylinder center line and the crankshaft center line. Thus in Fig. 10-29, the angle $\tau/2$ represents the amount of offset. Obviously, from the geometry of this figure, the angular spacing between inlet and exhaust cams is $\epsilon \pm \tau$, and since all inlet cams are equally spaced, this locates all cams on the cam ring. For the foregoing seven-cylinder engine, if $\tau = 20$ deg.,

$$\epsilon + \tau = 37.5 + 20 = 57.5 \text{ deg.}$$

In in-line and opposed engines, usually all cams for all cylinders are on the same camshaft, and the angular spacing of like cams

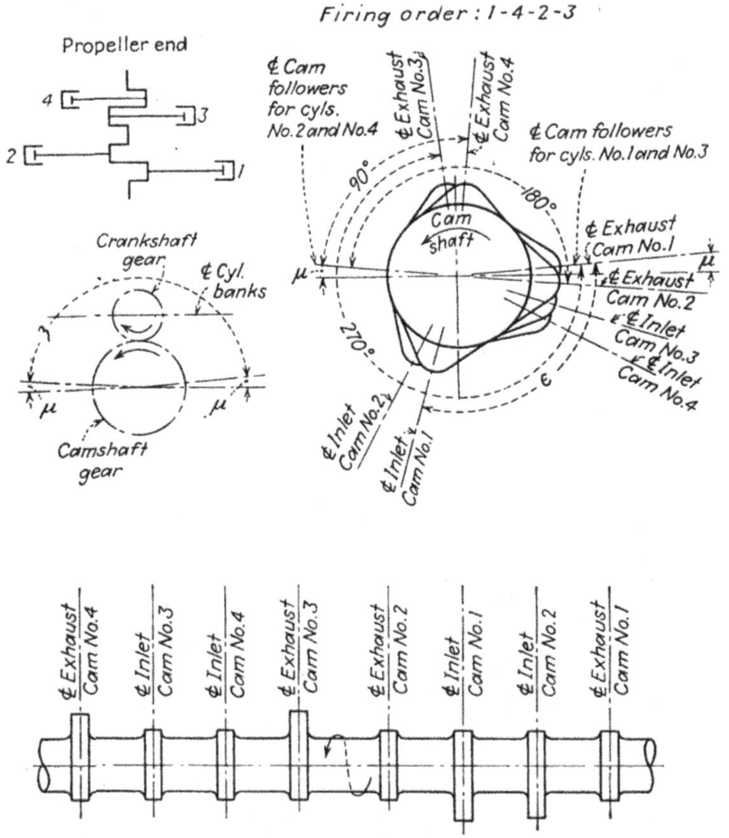

Fig. 10-30.—Method of locating the cams on the camshaft of a four-cylinder opposed engine.

must be determined from a consideration of firing order, arrangement of the crank arms, and, as applies, the angle of the cylinder banks. Thus for a four-cylinder opposed engine having the usual crank-arm arrangement (Fig. 5-13), and a firing order of 1-4-2-3, the cams will be as in Fig. 10-30. Frequently, to provide clearance between the camshaft and crank arms, the camshaft is offset so that the center lines of the cam followers form an angle μ with the plane of the cylinder banks. Then the angle η between the planes of the cam followers is not equal to 180 deg. To locate the cams (Fig. 10-30), center exhaust cam 1 on cam-follower center line 1. Locate inlet cam 1 back against camshaft rotation an amount equal to ϵ [see Eq. (10-41)].

Cylinder 4 fires 180 deg. of crankshaft travel after cylinder 1. Hence exhaust cam 4 will be located $180/2 = 90$ deg. back against rotation from the center line of follower 4. Inlet cam 4 will be located ϵ deg. against rotation from exhaust cam 4. Cylinder 2 fires 360 deg. after cylinder 1. Hence exhaust cam 2 will be located 180 deg. against rotation from the center line of inlet follower 2, and inlet cam 2 will be back ϵ deg. against rotation from exhaust cam 2. Number 3 cylinder fires 540 deg. after No. 1 cylinder. Hence exhaust cam 3 will be located 270 deg. back against rotation from the center line of exhaust follower 3, and, as before, inlet cam 3 will be ϵ deg. back of exhaust cam 3. In Fig. 10-30, $\epsilon = 112.5$ deg., $\mu = 5$ deg., and

$$\eta = 180 - 2 \times 5 = 170 \text{ deg.}$$

Location of cams on in-line engine camshafts differs from the preceding mainly in that $\eta = 0$ deg., *i.e.*, all cam followers are in the same plane and point in the same direction. For more than four cylinders in a line, however, the angle of the engine cranks is not usually 180 deg. V-engines using a separate camshaft for each bank of cylinders may be treated similarly to in-line engines. V-engines having one camshaft may be treated as opposed engines with the added consideration of angle of the banks different from 180 deg.

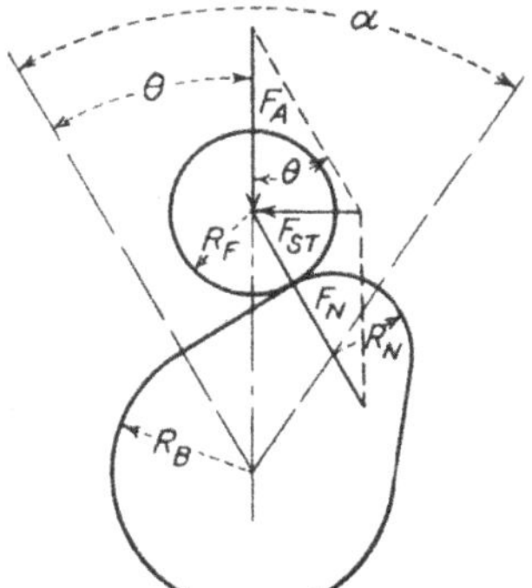

FIG. 10-31.—Forces acting on tangent cam flank.

10-13. Cam Loads.—The load on a cam is a force normal to the common tangent to the cam and follower. This normal force may be broken down into a force along the cam-follower axis and a side-thrust force perpendicular to the cam-follower axis and in the plane of the cam. Thus in Fig. 10-31, F_A is the force along the cam-follower axis, F_N is the force normal to the cam, and

F_{ST} is the resultant side thrust. For the straight flanks of tangent cams, the relations between the forces are

$$F_N = \frac{F_A}{\cos\theta} \tag{10-42}$$

and

$$F_{ST} = F_N \sin\theta \tag{10-43}$$

where θ = angular cam travel from initial opening.

For mushroom cams with flat-faced followers, $F_N = F_A$ and $F_{ST} \approx 0$.

For the flanks of hollow-faced cams (Fig. 10-32), the angle ψ relating F_A, F_{ST}, and F_N is a function of θ. Thus, in Fig. 10-20

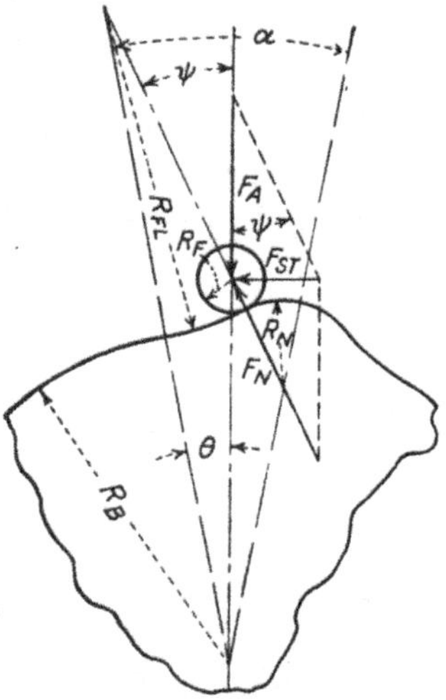

FIG. 10-32.—Forces acting on hollow-faced-cam flank.

$$CK = (R_{FL} + R_B)\sin\theta$$

also

$$CK = (R_{FL} - R_F)\sin\psi$$

hence

$$\psi = \text{arc sin}\left[\left(\frac{R_{FL} + R_B}{R_{FL} - R_F}\right)\sin\theta\right] \tag{10-44}$$

And from Fig. 10-32,

$$F_N = \frac{F_A}{\cos\psi} \tag{10-45}$$

$$F_{ST} = F_N \sin\psi \tag{10-46}$$

The total force along the cam-follower axis F_A is the resultant of the *equivalent* mass (referred to the cam) of the reciprocating part of the valve gear times the acceleration of the follower plus the spring load plus the gas pressure acting on the valve head. Considering these in order, the equivalent mass of the reciprocating parts of the valve gear may be found as follows:

Let M_{EC} = equivalent mass of the reciprocating parts referred to the cam (*i.e.*, M_{EC} is a mass such that it will produce the same force along the cam-follower axis that all the various individual masses together produce).

W_1 = weight of cam follower + one-half weight of push rod considered concentrated at the cam, lb.

W_2 = one-half weight of push rod considered concentrated at the push-rod end of the rocker arm, lb.
W_3 = weight of the rocker arm.
W_4 = weight of valve + spring retainer + one-third of the valve spring considered concentrated at the valve end of the rocker arm, lb.
K = radius of gyration of the rocker arm, in.
d_h = distance at mid lift of the valve between the rocker-arm fulcrum and the axis of the push rod.
d_i = distance at mid lift of the valve between the rocker-arm fulcrum and the axis of the valve stem.
R_{RA} = d_i/d_h ($\approx$ valve lift/cam lift when Λ is small).
Λ = the angle between the center line of the cam follower and the center line of the push rod at mid-lift position.

The arrangement of the various weights, dimensions, etc., is shown diagrammatically in Fig. 10-33.

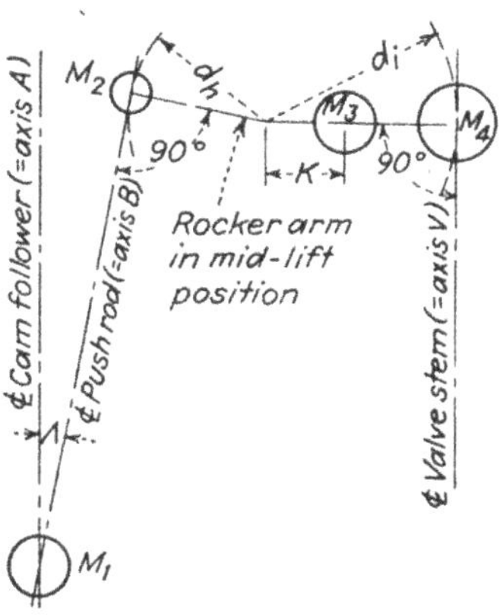

FIG. 10-33.—Diagrammatic arrangement for push-rod and rocker-arm type of overhead valve gear.

For the equivalent mass M_{E3C} referred to the cam follower (axis A), the kinetic energy of rotation of the rocker arm must be considered. Thus the torque at the push-rod end of the rocker arm to accelerate mass M_3 is

$$T_3 = F_{3B} \times d_h = \frac{F_{3A}}{\cos \Lambda} \times d_h = I_3 \times \alpha_a = \frac{W_3}{g} \times K^2 \times \frac{A_A \cos \Lambda}{d_h}$$

where subscripts A and B refer to axis A and B and subscript 3 refers to mass 3.

F = force along an axis.
A = acceleration along an axis.
I_3 = moment of inertia of mass M_3.
α_a = angular acceleration of the rocker arm.

Hence

$$M_{E3C} = \frac{F_{3A}}{A_A} = \frac{W_3 K^2}{g d_h^2} \cos^2 \Lambda \qquad (a)$$

where M_{E3C} = the equivalent of mass M_3 referred to the cam.

For the part of the push rod considered concentrated at the rocker arm, *i.e.*, mass M_2,

$$T_2 = \frac{F_{2A}}{\cos \Lambda} \times d_h = I_2 \times \alpha_a = \frac{W_2}{g} \times d_h^2 \times \frac{A_A \cos \Lambda}{d_h}$$

$$M_{E2C} = \frac{F_{2A}}{A_A} = \frac{W_2 \cos^2 \Lambda}{g} \qquad (b)$$

where M_{E2C} = equivalent of mass M_2 referred to the cam.

For the valve, spring retainer, etc., considered concentrated at the valve end of the rocker arm, *i.e.*, mass M_4,

$$T_4 = \frac{F_{4A}}{\cos \Lambda} \times d_h = I_4 \times \alpha_a = \frac{W_4}{g} \times d_i^2 \times \frac{A_A \cos \Lambda}{d_h}$$

or

$$M_{E4C} = \frac{F_{4A}}{A_A} = \frac{W_4}{g} \times \frac{d_i^2}{d_h^2} \times \cos^2 \Lambda = \frac{W_4 R_{RA}^2 \cos^2 \Lambda}{g} \qquad (c)$$

For the cam follower, tappet, and one-half of the push rod,

$$M_{E1C} = \frac{W_1}{g} \qquad (d)$$

The total equivalent mass at the cam is

$$M_{EC} = M_{E3C} + M_{E2C} + M_{E4C} + M_{E1C}$$

or

$$M_{EC} = \frac{1}{g}\left[\left(\frac{W_3 K^2}{d_h^2} + W_2 + W_4 R_{RA}^2\right) \cos^2 \Lambda + W_1\right] \qquad (10\text{-}47)$$

During the acceleration period X, the inertia force along the cam-follower axis is

$$F_{AX} = A_X \times M_{EC} \qquad (10\text{-}48)$$

where A_X = acceleration as in Eqs. (10-3), (10-16), or (10-27) as applies.
M_{EC} = equivalent mass at the cam as in Eq. (10-47).
F_{AX} = accelerating or inertia force, lb.

During the deceleration period Y, the force along the cam-follower axis is

$$F_{AY} = A_Y \times M_{EC} \qquad (10\text{-}49)$$

where A_Y = deceleration as in Eqs. (10-6), or (10-19), as applies.
M_{EC} = equivalent mass at the cam as in Eq. (10-47).
F_{AY} = decelerating or inertia force, lb.

During the deceleration, the maximum force F_{AYM} necessary to hold the follower on the cam is the resultant of the maximum force F_{VYM} exerted along the valve axis V by the valve spring. From Fig. 10-33,

$$F_{VYM} \times d_i = F_{BYM} \times d_h = \frac{F_{AYM}}{\cos \Lambda} \times d_h + F_F$$

or the force necessary at the valve spring is

$$F_{VYM} = \frac{F_{AYM}}{e_m \times R_{RA} \times \cos \Lambda} = \frac{A_{YM} \times M_{EC}}{e_m \times R_{RA} \times \cos \Lambda} \qquad (10\text{-}50)$$

where F_F = force necessary to overcome friction in the valve linkage.

e_m = mechanical efficiency of the valve linkage and may be taken as 85 to 90 per cent.

subscript M denotes maximum values.

The spring rate is

$$S_R = Q\left(\frac{F_{VYM} - S_c}{L_{V'}}\right) \qquad (10\text{-}51)$$

where S_R = spring rate, lb. per in.

S_c = load on the spring when the valve is closed.

$L_{V'}$ = lift of the valve corresponding to A_{YM}.

Q = 1.1 to 1.5 usually.

The initial load must be (for exhaust valves) sufficient to hold the valve on its seat at maximum intake depression in the cylinder. This intake depression is of the order of 20 in. Hg ($\approx$ 10 lb. per sq. in.). Hence $S_c \geqq 10 \times A_v$ where A_v is the area of the valve head. However, the initial spring load may have to be $S_c = 1$ to $3 \times 10 \times A_v$ in order to satisfy the requirements for a suitable valve spring (see Par. 10-18).

From Fig. 10-33,

$$L_{V'} = L_{B'} \times R_{RA} = L_{A'} \times R_{RA} \times \cos \Lambda$$

and

$$L_{A'} = L_{Y'} = L_Y \text{ for } \phi \text{ corresponding to } A_{YM}$$

where L_Y is as in Eq. (10-4) or (10-17) as applies.

Then Eq. (10-51) becomes

$$S_R = \frac{Q(F_{VYM} - S_c)}{L_{Y'} \times R_{RA} \times \cos \Lambda} \qquad (10\text{-}52)$$

With the spring rate S_R and the initial loading S_e known, the force exerted by the spring along the valve axis V for any value of valve lift L_V becomes

$$F_V = S_R \times L_V + S_e$$

and the corresponding force along the cam-follower axis A due to the spring is

$$\frac{F_{AS}}{e_m \times R_{RA} \times \cos \Lambda} = S_R \times L_A \times R_{RA} \times \cos \Lambda + S_e$$

From which

$$F_{AS} = e_m(S_R \times L_A \times R_{RA} \times \cos \Lambda + S_e)R_{RA} \times \cos \Lambda \quad (10\text{-}53)$$

where F_{AS} = force along axis A due to the spring, lb.

Thus the total force along the inlet cam-follower axis A during acceleration is

$$F_{AITX} = F_{AX} + F_{AS} \quad (10\text{-}54)$$

where F_{AX} is as in Eq. (10-48).
F_{AS} is as in Eq. (10-53).

And the total force along the inlet valve cam-follower axis A during deceleration is

$$F_{AITY} = F_{AY} + F_{AS} \quad (10\text{-}55)$$

where F_{AY} is as in Eq. (10-49).
F_{AS} is as in Eq. (10-53).

To keep the follower in contact with the cam during the deceleration, F_{AITY} must be greater than zero.

For the exhaust valve, in addition to the inertia force and spring force on the cam, there will be a gas-pressure force during the first few degrees of the lift period due to the pressure in the cylinder being greater than in the exhaust manifold. This pressure $P_{EX} \approx P_D - P_A$ [see Eq. (3-2)] and the corresponding force along the cam-follower axis A is

$$F_{AEX} = e_m \times P_{EX} \times A_{EXV} \times R_{RA} \times \cos \Lambda \quad (10\text{-}56)$$

where A_{EXV} = area of the exhaust valve head, sq. in., and the total initial force on the exhaust valve cam is

$$F_{AETX} = F_{AX} + F_{AS} + F_{AEX} \quad (10\text{-}57)$$

where F_{AX} is as in Eq. (10-48), lb.
F_{AS} is as in Eq. (10-53), lb.
F_{AEX} is as in Eq. (10-56), lb.

The gas pressure P_{EX} drops quickly to zero after the exhaust valve has started to open so that $F_{AEX} \approx 0$ for $\theta \geq 5$ to 10 deg. of cam motion.

10-14. Example of Cam-load Calculations.—Determine (*a*) the valve spring rate and (*b*) the cam loads for an overhead-valve engine using push rods and rocker arms. Available data are as follows: maximum cam lift, 0.5 in.; length of rocker arm from fulcrum to push rod, 1.5 in.; length of rocker arm from fulcrum to valve, 2 in.; angle between cam-follower axis and push-rod axis at mid lift, 5 deg.; weight of cam follower, 0.4 lb.; weight of push rod, 0.3 lb.; weight of rocker arm, 0.5 lb.; weight of valve springs, 0.24 lb.; weight of valve, 0.7 lb.; weight of spring retainer, 0.1 lb.; radius of gyration of rocker arm, 0.3 in. Cam same as for Par. 10-5, Example *a*; diameter of valve head, 2 in.

Procedure a.—For Eq. (10-47), $g = 32.2$ ft. per sec.², $W_3 = 0.5$ lb., $K = 0.3$ in., $d_h = 1.5$ in., $d_i = 2$ in., $R_{RA} = \frac{2}{1.5} = 1.33$, $W_2 = 0.15$ lb., $W_4 = 0.7 + 0.1 + (0.24/3) = 0.88$ lb., $\Lambda = 5$ deg.,

$$W_1 = 0.4 + 0.15 = 0.55 \text{ lb.}$$

Then

$$M_{EC} = \frac{1}{32.2}\left[\left(\frac{0.5 \times 0.3^2}{1.5^2} + 0.15 + 0.88 \times 1.33^2\right) \times 0.9962^2 + 0.55\right]$$

$$M_{EC} = 0.0702 \text{ slugs}$$

For Eq. (10-50), $A_{YM} = 1{,}488$ ft. per sec.² (see Fig. 10-13), assume $e_m = 0.9$. Then

$$F_{VYM} = \frac{1{,}488 \times 0.0702}{0.9 \times 1.33 \times 0.9962} = 87.5 \text{ lb.}$$

For Eq. (10-52), $S_c = 2 \times 10 \times 0.785 \times 2^2 \approx 63$ lb., $L_{Y'} = 0.5$ in. ($\phi = 0$ for $A_{YM} = 1{,}488$, from Fig. 10-13).
Then, assuming $Q = 1.3$

$$S_R = 1.3\left(\frac{87.5 - 63}{0.5 \times 1.33 \times 0.9962}\right) \approx 48 \text{ lb. per in.}$$

Procedure b.—For the inlet-valve cam loads, for Eqs. (10-48) and (10-49), values of A_X and A_Y were read from Fig. 10-13 (more accurately from the data for Fig. 10-13) and multiplied by M_{EC} to get the values of F_{AX} and F_{AY} shown in Table 10-3.

For Eq. (10-53), $e_m = 0.9$, $S_R = 48$, $R_{RA} = 1.33$, $\cos \Lambda = 0.9962$, $S_c = 63$, and L_A read from Fig. 10-13 (or more accurately from the data for Fig. 10-13) is given in Table 10-3. Then for $L_A = 0.5$ in. (corresponds to A_{YM}),

$$F_{AS} = 0.9[(48 \times 1.33 \times 0.9962 \times 0.5) + 63]1.33 \times 0.9962 = 109.5 \text{ lb.}$$

From Eq. (10-55),

$$F_{AITY} = -104.4 + 109.5 = 5.1 \text{ lb.}$$

Since F_{AITY} is positive, it indicates that the spring is adequate, but to be certain of contact between the follower and the cam at all values of ϕ, assume $S_R = 50$ lb. per in. and calculate the corresponding values of F_{AITY}. In Table 10-3, it is seen that the follower will not leave the cam at any point during deceleration.

For the exhaust valve at the start of lift, assuming a cylinder pressure of $P_{EX} = 50$ lb. per sq. in. gage, from Eq. (10-56)

$$F_{AEX} = 0.9 \times 50 \times 0.785 \times 2^2 \times 1.33 \times 0.9962 = 187 \text{ lb.}$$

Assuming the pressure equalizes in 10 deg. of cam travel, for $\theta = 5$ deg. $F_{AEX} \approx 93.5$ lb. Values of F_{AETX} [from Eq. (10-57)] are shown in Table 10-3. Values of the force normal to the cam as shown in Table 10-3 were found by means of Eq. (10-42). The results are shown in graphical form in Fig. 10-34.

TABLE 10-3.—DATA FOR EXAMPLE, PAR. 10-14

θ	A_X	F_{AX}	L_A	F_{AS}	F_{AITX}	F_{AETX}	F_N
0	1235	86.9	0	72.5	159.4	346.4	346.4
5	1260	88.5	0.004	73.0	161.5	255.0	256
10	1365	96.0	0.024	74.5	170.5	170.5	173
15	1460	102.8	0.051	76.5	179.3	179.3	186
20	1665	117.0	0.092	79.7	196.7	196.7	210
25	1958	137.8	0.140	83.3	221.1	221.1	244
30	2378	167.0	0.2105	88.6	255.6	255.6	295
30°9.5′	2395	168.1	0.216	89.4	257.6	257.6	298

ϕ	A_Y	F_{AY}	L_A	F_{AS}	F_{AITY}	F_{AETY}
0	−1488	−104.4	0.5	111.2	6.8	6.8
5	−1480	−104.0	0.494	110.8	6.8	6.8
10	−1460	−102.7	0.473	109.5	6.8	6.8
15	−1420	− 99.9	0.442	106.8	6.9	6.9
20	−1368	− 96.1	0.3995	103.5	7.4	7.4
25	−1293	− 91.0	0.349	99.7	8.7	8.7
30	−1211	− 85.3	0.284	94.6	9.3	9.3
30°50.5′	−1112	− 78.4	0.216	89.4	11.0	11.0

10-15. Camshaft Stiffness.—For in-line and V-engines, the camshaft acts as a beam supported by two or more bearings, and to minimize noise and distortion of parts, the deflection should not exceed 0.002 to 0.004 in. The loads causing deflection are the cam reactions from moving the valves, and for one cam between bearings, the maximum deflection would occur when F_N was a maximum. For several cams between bearings, the maximum deflection occurs at the worst vector combination of F_N for the several cams taken together. An accurate analysis is long and tedious; and for the usual arrangement, the designer will generally be on the safe side if he selects a shaft diameter

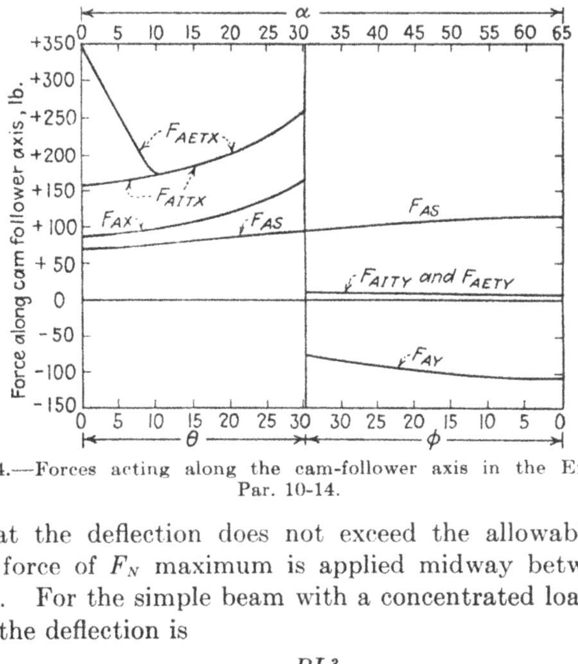

FIG. 10-34.—Forces acting along the cam-follower axis in the Example of Par. 10-14.

such that the deflection does not exceed the allowable value when a force of F_N maximum is applied midway between the bearings. For the simple beam with a concentrated load at the middle, the deflection is

$$Y = \frac{PL^3}{48EI}$$

where Y = deflection, in.
P = load, lb.
L = length between supports, in.
E = 30,000,000 (for steel).
I = moment of inertia of the section.

For hollow circular sections,

$$I = \frac{\pi}{64}(D^4 - d^4)$$

where D = outside diameter, in.
d = internal diameter, in.

Hence

$$(D^4 - d^4) = \frac{0.425PL^3}{EY}$$

usually $d \approx 0.5 \times D$, and on this basis

$$D = \left(\frac{0.454PL^3}{EY}\right)^{0.25} \qquad (10\text{-}58)$$

For the data of Par. 10-14, P = 346 lb., and if we assume Y = 0.003 in., and L = 15 in.

$$D = \left(\frac{0.454 \times 346 \times 15^3}{30{,}000{,}000 \times 0.003}\right)^{0.25} \approx 1.56 \text{ in.}$$

This leaves $2 \times R_B - 1.56 = 2 \times 0.85 - 1.56 = 0.14 \approx 1/8$ in., the originally assumed value for cam-follower clearance (see Par. 10-5).

10-16. Cam and Follower Details.—The width of the cam parallel to the camshaft or cam-ring axis should be such that, at the line of contact between the cam and follower, the material is not stressed beyond the fatigue limit in compression.

For a cylinder on a flat plate, Roark[5] suggests

$$\max S_C = 0.591\left(\frac{PE}{D}\right)^{1/2} \qquad (10\text{-}59)$$

where max S_C = maximum allowable compressive stress, lb. per sq. in. (see Table 8-6).
P = load per linear inch.
E = modulus of elasticity (= 30,000,000 for steel).
D = diameter of the cylinder.

Applying Eq. (10-59) to a tangential cam with the roller on the straight flank, $D = 2R_F$, $P \times W_c = F_N$, and assuming the cam and follower are made of steel,* for *tangential* cams

$$\max S_C = 2{,}295\left(\frac{\max F_N}{W_c \times R_F}\right)^{1/2} \qquad (10\text{-}60)$$

* See footnote, p. 265.

where max F_N = maximum force normal to the cam [see Eq. (10-42)], lb.

W_c = width of the cam parallel to the camshaft or cam ring axis, in.

R_F = radius of the roller follower, in.

Applying Eq. (10-60) to the example of Par. 10-14, $F_N = 346$ lb., $R_F = 1$ in., and if we assume $W_c = 0.25$ in., the maximum stress is

$$\max S_C = 2{,}295 \left(\frac{346}{0.25 \times 1}\right)^{1/2} = 85{,}500 \text{ lb. per sq. in.}$$

From Table 8-6, it is seen that a cam width of ¼ in. would require hardened or heat-treated steel.

Applying Eq. (10-59) to a mushroom cam with flat follower on the flank circle, $P = 2R_{FL}$, $P \times W_c = F_N$, and $E = 30{,}000{,}000$* as before; for *mushroom* cams

$$\max S_C = 2{,}295 \left(\frac{F_N}{W_c \times R_{FL}}\right)^{1/2} \tag{10-61}$$

where F_N and W_c are as in Eq. (10-60).

R_{FL} = radius of the flank circle, in.

For a cylinder in a circular groove, Roark[5] suggests

$$\max S_C = 0.591 \left(PE \frac{D_1 - D_2}{D_1 \times D_2}\right)^{1/2} \tag{10-62}$$

where P and E are as in Eq. (10-59).

D_1 = diameter of the circular groove.

D_2 = diameter of the cylinder.

Applying Eq. (10-62) to hollow-faced cams with a roller follower on the flank circle, $D_1 = 2R_{FL}$, $D_2 = 2R_F$, $P \times W_c = F_N$, and $E = 30{,}000{,}000$.* For *hollow-faced* cams,

$$\max S_C = 2{,}295 \left[\frac{F_N(R_{FL} - R_F)}{W_c \times R_{FL} \times R_F}\right]^{1/2} \tag{10-63}$$

where F_N, W_c, and R_F are as in Eq. (10-60).

R_{FL} = radius of the flank circle, in.

For mushroom cams, the diameter of the flat-faced follower should be sufficient to provide full line contact at all cam angles.

* If the camshaft or cam ring is to be made of cast iron, use $E = 15{,}000{,}000$ and alter Eqs. (10-60), (10-61), and (10-63).

To attain this condition, the diameter of the flat face on the follower will have to be greater than the contact width, W_D (Fig. 10-35). From Fig. 10-17,

$$W_D = 2(R_{FL} - R_B)\sin\theta_M \tag{10-64}$$

The axis of the follower should be in the same plane as the axis of the camshaft, but the follower may be symmetrical with the cam as in Fig. 10-35*A* or offset as in Fig. 10-35*B* to produce rotation of the follower about its own axis and distribute wear. Case *B* is the more common arrangement.

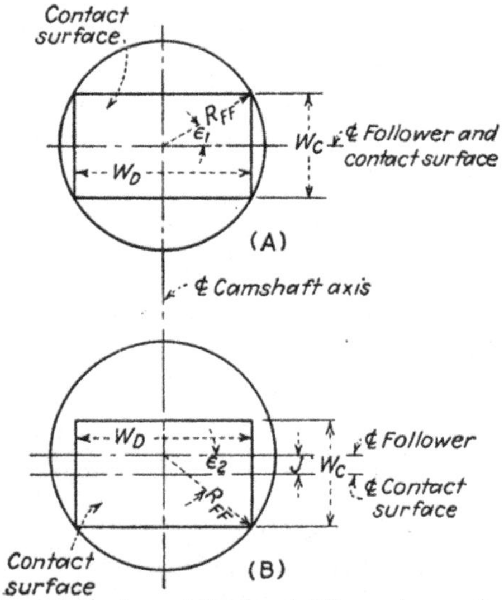

FIG. 10-35.—Dimensions of flat-faced follower for mushroom cams.

For the symmetrical follower, Fig. 10-35*A*,

$$\epsilon_1 = \text{arc tan}\,\frac{W_c}{W_D}$$

and

$$R_{FF} = \frac{W_c}{2\sin\epsilon_1} \tag{10-65}$$

where R_{FF} = radius of the flat follower surface, in.

W_c = width of the cam parallel to the camshaft axis, in. [see Eq. (10-61)].

W_D = contact width, in.

For the offset follower, Fig. 10-35*B*,

$$\epsilon_2 = \text{arc tan} \frac{(W_c/2) + J}{W_D/2} = \text{arc tan} \frac{W_c + 2J}{W_D}$$

and

$$R'_{FF} = \frac{W_D}{2 \cos \epsilon_2} \tag{10-66}$$

where R'_{FF} = radius of the offset flat follower surfaces, in.
W_c = width of the cam parallel to the camshaft axis, in. [see Eq. (10-61)].
W_D = contact width, in.
J = offset of the center line of the follower to the center line of the cam measured parallel to the camshaft axis, in.

10-17. Push Rods and Rocker Arms.—Push rods as used with overhead valve-gear arrangements generally have a slenderness ratio such that they fall into the classification of long columns. Hence the Euler column formula may be used to analyze them, and since they usually use spherical seated bearings, they may also be classed as pin end columns. Thus

$$P = \frac{\pi^2 EI}{L^2} \tag{10-67}$$

where P = load at which the rod buckles, lb.
E = modulus of elasticity, lb. per sq. in. (= 30,000,000 for steel and 10,000,000 for aluminum alloys).
I = moment of inertia of the push rod-section, in.^{4}
L = length of the push rod, in.

For tubular push-rod sections,

$$I = \frac{\pi}{64}(D^4 - d^4)$$

where D = diameter of the push rod, in.
d = diameter of the hole through the push rod, in.

Assuming $d = 0.8 \times D$, and the allowable load $= 0.5P$, Eq. (10-67) reduces to

$$D = 1.625 \left(\frac{PL^2}{E}\right)^{1/4} \tag{10-68}$$

For the example of Par. 10-14, the maximum force along the

push-rod center line is

$$P \approx F_{AETX} \times \cos A \approx 346 \times 0.9962 \approx 345 \text{ lb.}$$

Hence, for a length of 14 in., a hollow steel push rod having a diameter of

$$D = 1.625 \left(\frac{345 \times 14^2}{30{,}000{,}000}\right)^{0.25} = 0.353 \approx \frac{3}{8} \text{ in.}$$

should be satisfactory. By the same procedure, an aluminum-alloy rod would have a diameter of $D \approx \frac{1}{2}$ in. The diameter of the spherical or ball ends of the push rod may be made approximately equal to the diameter of the push rod.

Rocker arms are essentially cantilever beams usually tapered and having a T section. Generally, when they are large enough to meet other requirements, they are not critical in bending, but if there is any doubt, they may be analyzed by simple beam formulas. Frequently, small rollers are fitted to the valve-stem end of the rocker arm, and the size of these rollers may be found by the same formulas used for cam-follower rollers [*i.e.*, the equivalent of Eq. (10-60)]. Clearance adjustment screws with locking nuts are usually built into the push rod end of the rocker arms.

Rocker-arm bearings are either plain or antifriction (*i.e.*, ball or needle bearings). The maximum load is approximately the force in the push rod times the rocker-arm leverage ratio. In addition, an end thrust exists when the push-rod axis is not in the plane of the rocker arm and valve stem.

For the example of Par. 10-14, the maximum rocker-arm bearing reaction is

$$F_{RA} \approx 345 \times \frac{1.5 + 2}{2} \approx 600 \text{ lb.}$$

On the assumption that a ball bearing is to be used, from Table A1-22, $F_{RA} = L = 600$ lb., $Z \approx 0.9$, $K \approx 2$, and if it is assumed that the end thrust will not exceed 10 per cent of the radial load, $F = 0.99$ for nonfilling notch-type bearings. Then, $C = 600 \times 0.99 \times 0.9 \times 2 = 1{,}070$ lb. Since the equivalent r.p.m. of the rocker arm should be low (say 200 r.p.m.), S.A.E. bearing 304 should be adequate (see Table A1-22*I*). However, before final selection is made, the recommendations of the bearing manufacturer should be obtained.

10-18. Valve Springs.—Nearly all modern aircraft engines use cylindrical helical compression springs of round wire, and most engines use two or three concentric springs per valve. Final approval for a proposed design should be made by a spring manufacturer, but the engine designer should make preliminary calculations if for no other reason than to determine necessary dimensions of adjacent parts of the valve gear.

For cylindrical helical compression springs of round wire, let

G = torsional modulus of elasticity ($\approx$ 11,500,000 for steel).
S = stress, lb. per sq. in.
S_A = maximum allowable stress, lb. per sq. in.
f = deflection per turn, in.
n = number of effective turns.
f_n = total deflection, in.
L_{MW} = minimum working length, in.
S_R = spring rate, lb. per in. deflection.
d = diameter of the wire, in.
D_o = outside diameter of the spring, in.
D_i = inside diameter of the spring, in.
L_S = solid length, in.
L_F = free length, in.
F_V = force along the valve axis, lb.
F_{VO} = force along valve axis when valve is open, lb.
F_{VC} = force along the valve axis when valve is closed, lb.
P_S = pitch of spring, in.
F_{VA} = maximum allowable force on the spring, lb.
Z = Wahl factor.
$C = \dfrac{D_o - d}{d}$.

Then[6]

$$L_{MW} = 1.1dn + 2.25d \tag{10-69}$$

$$L_S = (n + 2.25)d \tag{10-70}$$

$$L_F = f_n + L_{MW} \tag{10-71}$$

$$S_R = \frac{F_{VO}}{f_n} \tag{10-72}$$

$$P_S = \frac{L_F - 2.25d}{n} \tag{10-73}$$

$$n_{max} = \frac{L_{MW} - 2.25d}{1.1d} \tag{10-74}$$

$$f_n = \frac{8 \times n \times F_{VO}(D_o - d)^3}{G \times d^4} \tag{10-75}$$

$$F_{VA} = \frac{0.3927 \times S_A \times d^3}{(D_o - d)Z} \tag{10-76}$$

$$Z = \frac{4C - 1}{4C - 4} + \frac{0.615}{C} \tag{10-77}$$

$$S = \frac{(D_o - d) \times F_V \times Z}{0.3927 \times d^3} \tag{10-78}$$

Valve springs are subjected to rapidly varying loads, hence the allowable stress S_A should be based on the torsional endurance limit rather than on the torsional elastic limit. This torsional

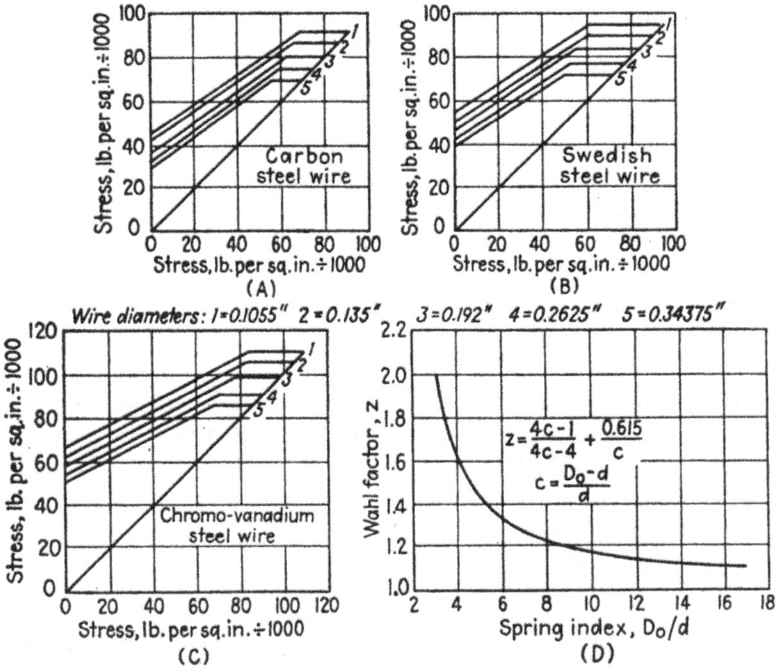

FIG. 10-36.—(*A*), (*B*) and (*C*) Allowable torsional-stress range for the most commonly used valve-spring steels. (*D*) Values of the Wahl stress factor. (*From Griffith, "Standards for Spring Design," Product Engineering Design Work Sheets, Fourth Series.*)

endurance limit depends upon the range of stress, diameter of the wire, and material of which the wire is made. Steel wire is usually drawn to Washburn and Moen gage (Table 10-4), and where possible standard wire diameters should be used. Materials most commonly used are medium- or high-carbon steel, Swedish steel, or chrome-vanadium steel (S.A.E. 6150). Allowable ranges of torsional stress as determined by Griffith[6] are shown in Fig. 10-36*A*, *B*, and *C*. Stresses increase rapidly at low values of the spring index ($= D_o/d$) as is seen in Fig. 10-36*D*, hence the spring index should be relatively large. Griffith[6] recommends a value of $D_o/d = 9$ as optimum, but some departure from this value may be necessary.

TABLE 10-4.—WASHBURN AND MOEN STEEL-WIRE GAGE DIAMETERS

Gage No.	Wire diameter, in.	Gage No.	Wire diameter, in.	Gage No.	Wire diameter, in.	Gage No.	Wire diameter, in.
000	0.3625	3	0.2437	8	0.1620	13	0.0915
00	0.3310	4	0.2253	9	0.1483	14	0.0800
0	0.3065	5	0.2070	10	0.1350	15	0.0720
1	0.2830	6	0.1920	11	0.1205	16	0.0625
2	0.2625	7	0.1770	12	0.1055	17	0.0540

Spring surging or vibration is a frequent cause of valve-gear trouble and may result in noisy operation owing to the follower jumping off the cam, and this in turn may affect the performance. Severe vibration can cause spring breakage. The usual methods of vibration control are (*a*) friction dampers, (*b*) design for high natural frequency of the spring, and (*c*) use of multiple springs having different natural frequencies.

The natural frequency of a valve spring in vibrations per minute is[7]

$$f_v \approx \frac{250 \times d \times G^{1/2}}{(D_o - d)^2 \times n} \tag{10-79}$$

where d = diameter of the wire, in.
D_o = outside diameter of the spring, in.
n = number of effective turns.
G = torsional modulus of elasticity ($\approx$ 11,500,000 for steel).

Values of f_v should be as high as possible, preferably above 15,000, but this is often quite difficult to attain. By using two or more springs per valve which have different natural frequencies, erratic operation and damage from spring breakage will be less likely. Determination of the best possible combination of design factors is usually a long and tedious problem and one which specialists in spring manufacture are best qualified to solve. The following example, however, indicates some of the steps involved.

10-19. Example of Valve-spring Calculations.—Determine (*a*) dimensions of a carbon-steel valve spring for the engine in the example of Par. 10-14, (*b*) dimensions if two concentric springs are to be used.

Procedure a.—From Par. 10-14, $S_R = 50$ lb. per in., maximum lift of the valve $= L_{VO}$ ($= L_{V'}$ for this example) $= L_{A'} \times R_{RA} \times \cos A = 0.5 \times 1.33 \times 0.9962 = 0.663$ in., $F_{VC} = S_c = 63$ lb., $F_{VO} = F_{VC} + S_R \times L_{VO} = 96.2$ lb.

Assume $D_o = 2$ in. ($=$ diameter of valve), and for the optimum value of Z, the spring index $D/d \approx D_o/d = 9$. Hence $d = 2/9 = 0.222$ in. Let $d = 0.2253$ in. which corresponds to Washburn and Moen gage 4 (see Table 10-4). From Fig. 10-36D, for $D_o/d = 2/0.2253 = 8.88$, $Z = 1.19$. From Eq. (10-78), for the valve closed

$$S = \frac{(2 - 0.2253) \times 63 \times 1.19}{0.3927 \times 0.2253^3} = 29{,}550 \text{ lb. per sq. in.}$$

and for the valve open

$$S = \frac{(2 - 0.2253) \times 96.2 \times 1.19}{0.3927 \times 0.2253^3} = 45{,}200 \text{ lb. per sq. in.}$$

From Fig. 10-36A, reading up from $S = 29{,}550$ to the wire size (interpolated between 0.192 and 0.2625), it is seen that the allowable stress is $S_A \approx 57{,}000$ lb. per sq. in. Hence the spring stresses are within the allowable range.

From Eq. (10-72),

$$f_n = \frac{F_{VO}}{S_R} = \frac{96.2}{50} = 1.925 \text{ in.}$$

From Eq. (10-75),

$$n = \frac{1.925 \times 11{,}500{,}000 \times 0.2253^4}{8 \times 96.2 \times (2 - 0.2253)^3} = 7.45 \text{ effective turns}$$

From Eq. (10-79),

$$f_v = \frac{250 \times 0.2253 \times 11{,}500{,}000^{1/2}}{(2 - 0.2253)^2 \times 7.45} = 8{,}120 \text{ vibrations per minute}$$

This value of f_v is lower than is desirable and might cause trouble from noise or even spring breakage.

From Eq. (10-69),

$$L_{MW} = 1.1 \times 0.2253 \times 7.45 + 2.25 \times 0.2253 = 2.347 \text{ in.}$$

From Eq. (10-70),

$$L_S = (7.45 + 2.25) \times 0.2253 = 2.181 \text{ in.}$$

From Eq. (10-71),

$$L_F = 1.925 + 2.347 = 4.272 \text{ in.}$$

From Eq. (10-73),

$$P_S = \frac{4.272 - 2.25 \times 0.2253}{7.45} = 0.505 \text{ in.}$$

Layout of the spring is shown in Fig. 10-37*A*.

Procedure b.—Assuming two concentric springs per valve with the outside diameter $D_o \leq 2$ in. for the outside spring and $D_i \geq 0.8$ in. for the inside spring (to ensure clearance with the valve guide), then for the inside spring $D_o - D_i = 2d$ and $D_o/d \approx 9$. From these relations and assumptions,

$d = 0.1142$, say 0.1055 in. (= Washburn and Moen gage 12, Table 10-4).

$D_o = 2 \times 0.1055 + 0.8 = 1.011$ in.

$D_o/d = 1.011/0.1055 = 9.58$

and from Fig. 10-36*D*, $Z = 1.175$.

The combined force of the two springs (valve open) should be not less than $F_{VO} = 96.2$ lb., and (valve closed) $F_{VC} = 63$ lb. The minimum working length of the two springs can be somewhat different by step cutting the spring retainers, and this may aid in properly proportioning the springs.

Assume an allowable stress $S_A = 60{,}000$ lb. per sq. in. From Eq. (10-78), the corresponding load that the inner spring can carry is

$$F_{VO} = \frac{0.3927 \times 60{,}000 \times 0.1055^3}{(1.011 - 0.1055) \times 1.175} = 26 \text{ lb.}$$

Assume $n = 10$ effective coils. From Eq. (10-75),

$$f_n = \frac{8 \times 10 \times 26 \times (1.011 - 0.1055)^3}{11{,}500{,}000 \times 0.1055^4} = 1.08 \text{ in.}$$

From Eq. (10-72),

$$S_R = \frac{26}{1.08} = 24 \text{ lb. per in.}$$

The load that the inner spring can carry, valve closed, is

$$F_{VC} = F_{VO} - S_R \times L_{VO} = 26 - 24 \times 0.663 = 10 \text{ lb.}$$

The stress, valve closed, is, from Eq. (10-78)

$$S = \frac{(1.011 - 0.1055) \times 10 \times 1.175}{0.3927 \times 0.1055^3} = 23{,}100 \text{ lb. per sq. in.}$$

From Fig. 10-36*A* reading up from 23,100 to the wire size,

$$S_A \approx 60{,}000 \text{ lb. per sq. in.}$$

Therefore the originally assumed value for maximum stress is satisfactory.

From Eq. (10-79),

$$f_v = \frac{250 \times 0.1055 \times 11{,}500{,}000^{1/2}}{(1.011 - 0.1055)^2 \times 10} = 10{,}900$$

This frequency is still less than desirable but better than for case *a*. From Eq. (10-69), the minimum working length is

$$L_{MW} = 1.1 \times 0.1055 \times 10 + 2.55 \times 0.1055 = 1.429 \text{ in.}$$

From Eq. (10-71), the free length is

$$L_F = 1.08 + 1.429 = 2.509 \text{ in.}$$

From Eq. (10-73), the pitch is

$$P_S = \frac{2.509 - 2.25 \times 0.1055}{10} = 0.224 \text{ in.}$$

The outside spring must be capable of carrying a load (valve open) of $F_{VO} = 96.2 - 26 = 70.2$ lb., and (valve closed) $F_{VC} = 63 - 10 = 53$ lb. Hence, the spring rate must be

$$S_R = \frac{70.2 - 53}{0.663} = 26 \text{ lb. per in.}$$

From Eq. (10-72), the total deflection is

$$f_n = \frac{70.2}{26} = 2.7 \text{ in.}$$

The inside diameter of the outside spring must be greater than the outside diameter of the inside spring. Assuming this difference is 0.25 in.; for the outside spring

$$D_i = 1.011 + 0.25 = 1.261 \text{ in.}$$

also

$$D_o - D_i = 2d \qquad \text{and} \qquad D_o/d \approx 9.$$

Hence

$d = 0.18$, say 0.177 in. (= Washburn and Moen gage 7, Table 10-4).

$D_o = 2 \times 0.177 + 1.261 = 1.615$ in.

$$\frac{D_o}{d} = \frac{1.615}{0.177} = 9.11$$

From Fig. 10-36*D*, $Z = 1.18$

From Eq. (10-78), for the valve closed

$$S = \frac{(1.615 - 0.177) \times 53 \times 1.18}{0.3927 \times 0.177^3} = 41{,}500 \text{ lb. per sq. in.}$$

and for the valve open

$$S = \frac{(1.615 - 0.177) \times 53 \times 1.18}{0.3927 \times 0.177^3} = 55{,}000 \text{ lb. per sq. in.}$$

From Fig. 10-36*A*, reading up from $S = 41{,}500$ to the wire size, it is seen that the allowable stress is $S_A \approx 66{,}500$ lb. per sq. in. Therefore the spring stresses are within the allowable range.

Assume $n = 8$ effective coils. Then from Eq. (10-79),

$$f_v = \frac{250 \times 0.177 \times 11{,}500{,}000^{1/2}}{(1.615 - 0.177)^2 \times 8} = 9{,}060 \text{ vibrations per min.}$$

This is lower than desirable but better than case a, and since f_v (outside spring) $\neq f_v$ (inside spring), vibration in either spring will tend to be neutralized by the other.

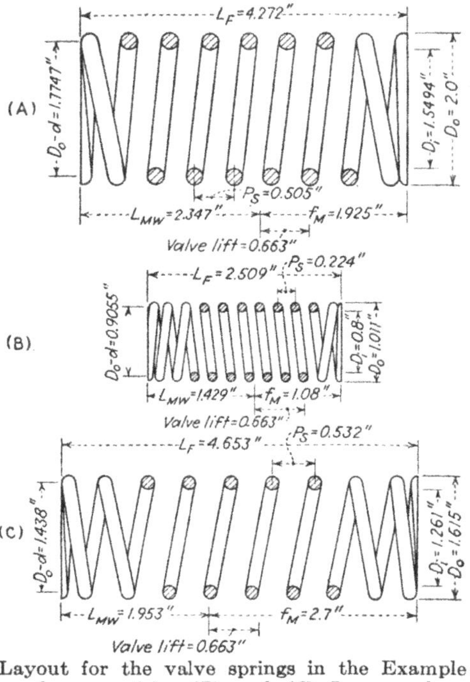

FIG. 10-37.—Layout for the valve springs in the Example of Par. 10-19. (*A*) Single spring for case (*a*). (*B*) and (*C*) Inner and outer springs for case (*b*).

From Eq. (10-69), the minimum working length is

$$L_{MW} = 1.1 \times 0.177 \times 8 + 2.25 \times 0.177 = 1.953 \text{ in.}$$

From Eq. (10-71), the free length of the outside spring is

$$L_F = 2.7 + 1.953 = 4.653 \text{ in.}$$

From Eq. (10-73), the pitch is

$$P_S = \frac{4.653 - 2.25 \times 0.177}{8} = 0.532 \text{ in.}$$

The layout of the two springs for case b is shown in Fig. 10-37B and C.

Since L_{MW} = 1.429 in. for the inside spring and 1.953 in. for the outside spring, the spring retainers will have to be step cut. If the difference is equally divided, each retainer will have an offset of

$$\frac{1.953 - 1.429}{2} = 0.262 \text{ in.}$$

The chief weakness of the springs in this example is the low frequency of vibration, and this may be increased by *reducing* the number of effective turns (Eq. 10-79). Thus, commonly used values of $n = 4$ to 6 would reduce the chances of vibration failure, but to avoid exceeding the allowable stress of the wire, a larger over-all diameter would be necessary. In general, a short, large-diameter spring with very few turns is the best answer.

10-20. Valve-gear Details.—With the dimensions of the principal parts of the valve gear determined, the detail dimension of the valve-spring retainers, valve-stem length, rocker-arm bearing supports, and rocker-arm housing may be determined, and the completion of the drawings of the cylinder (Suggested Design Procedure, items 9, 10, and 11, page 210) can be made.

Lubrication of the rocker arms, etc., formerly was intermittent, but continuous circulation of oil is preferable and is now quite common practice. One method of doing this is to provide a passageway to the cam-follower tappet and through the hollow push rod. A tubular sleeve and rocker-arm cover or rocker box may be used to enclose the gear. Detail arrangements may be best studied by reference to available sectioned drawings of successful engines.

All supports for the valve gear should be as rigid as possible, and such parts as rocker-arm bearing supports should be designed with this requirement in mind. In general, the best arrangement is one that, with the least complexity, fulfills all requirements.

Suggested Design Procedure

Important. Include sample calculations of all items (as applies). Make layouts to a large enough scale to permit accuracy of measurements.

1. Make preliminary pencil sketches approximately to scale showing the desired arrangement of the valve gear to be used. Check to be sure the arrangement represents good practice and that the parts will fit together without interference.

2. Determine all necessary dimensions for the cam drive to be used including gear tooth numbers, and layout the arrangement to large (preferably full) scale.

3. Determine all necessary dimensions for the cams, and make a dimensioned drawing of the cams.

4. Calculate data, and plot lift, velocity, and acceleration curves for the cams to be used. Alter the cam dimensions as necessary to avoid excessively high accelerations.

5. Determine the proper spacings for inlet and exhaust cams, and make a dimensioned drawing showing positions of all cams on the camshaft or cam ring.

6. Determine cam loads, and plot curves of forces acting along the cam-follower axes.

7. Determine all remaining dimensions of the camshaft or cam ring, and alter the dimensioned drawing of the camshaft or cam ring (item 5) as necessary.

8. Determine all necessary dimensions of the cam follower and tappet, and make a dimensioned drawing.

9. Determine all necessary dimensions of the push rods and rocker arms, and make dimensioned drawings.

10. Determine all necessary dimensions for suitable valve springs, and make dimensioned drawings.

11. Complete the details of the valve gear, and make dimensioned drawings of all detail parts.

12. Complete the detailed dimensioned drawing of the cylinder head as started under item 10 of Suggested Design Procedure, page 210.

13. Complete the assembly drawing of the cylinder as started under item 11 of Suggested Design Procedure, page 210.

14. When items 1 to 13 have been completed and put in proper form, submit for checking and approval.

Problems

1. Determine the dimensions, layout the cam, and plot curves of lift, velocity, and acceleration for the tangent cam of Par. 10-5, use the same data but increase the diameter of the roller follower to 1.5 in. Compare the results with the answers in Par. 10-5.

2. Determine dimensions, layout the cam, and plot curves of lift, velocity, and acceleration for a hollow-faced cam to operate the inlet valves on a five-cylinder radial engine. Available data are as follows: inlet-valve opening, 15 deg. before top center; inlet-valve closing, 65 deg. after bottom center; maximum lift of the roller follower, 0.5 in.; diameter of roller follower, 1.5 in.; diameter of the engine crankshaft, 2.5 in.; opposite rotation of crankshaft and cam ring; speed of engine, 2,000 r.p.m.

3. Repeat Problem 2 for a nine-cylinder radial engine, use the same data but increase the diameter of the engine crankshaft to 3.25 in.

4. Determine the necessary data, and layout a cam ring for the inlet and exhaust cams of the five-cylinder engine in Problem 2. Additional data are as follows: exhaust-valve opening, 50 deg. before bottom center; exhaust-valve closing, 10 deg. after top center; cam follower offset to the plane containing the cylinder and crankshaft center lines, 15 deg.

5. Determine the necessary data, and layout a camshaft showing the location of the cams on a six-cylinder in-line engine camshaft. Available data are as follows: inlet-valve opening, 10 deg. before top center; inlet-valve closing, 50 deg. after bottom center; exhaust-valve opening, 55 deg. before bottom center; exhaust-valve closing, 15 deg. after top center; crank-arm arrangement as in Fig. 7-15; firing order 1-5-3-6-2-4; angle of plane of follower center lines to plane of cylinder center lines, 8 deg. Opposite rotation of camshaft and crankshaft.

6. Prove that the flank and nose portions of a constant-acceleration cam are parabolas. Hint, see reference 1.

7. Derive expressions for the relation of parts of a constant acceleration cam. Hint, see reference 1.

8. By using the same valve timing and any other data that applies, determine the necessary data, layout the cam, and plot curves of lift, velocity, and acceleration for a constant-acceleration cam to replace the hollow-faced cams in Problems 2 and 4.

9. Prove that, for the same direction of rotation of cam ring and crankshaft, the number of inlet cams required will be as in Eq. (10-36) and the ratio crankshaft to cam-ring speed will be as in Eq. (10-37).

References

1. Huebotter: "Mechanics of the Gasoline Engine."
2. Heldt: "Automotive Engines."
3. Swan: "Handbook of Aeronautics," Vol. 2, Aero Engines, Design and Practice.
4. Judge: "Automobile and Aircraft Engines."
5. Roark: "Formulas for Stress and Strain."
6. Griffith: "Standards for Spring Design," Product Engineering Design Work Sheets, Fourth Series.
7. Jehle and Spiller: Idiosyncrasies of Valve Mechanisms and Their Causes, *S.A.E. Trans.*, p. 197, Vol. 24, 1929.
8. Frederick: The Single Sleeve Valve Engine, *S.A.E. Trans.*, p. 102, Vol. 22, Part 1, 1927.
9. Fedden: The Single Sleeve as a Valve Mechanism for the Aircraft Engine, *S.A.E. Jour.*, Vol. 43, No. 3, September, 1938.
10. Young: Aircraft Engine Valve Mechanisms, *S.A.E. Jour.*, Vol. 44, No. 3, March, 1939.

CHAPTER 11

THE CRANKCASE, SUPERCHARGERS, AND ACCESSORIES

11-1. Crankcase Materials and Arrangements.—The crankcase serves to hold the various parts of the engine in proper position, to retain part or all of the lubricant, and usually, though not always, to transfer the external forces to the engine mounting. The crankcase should have a high rigidity-weight ratio and high fatigue resistance.

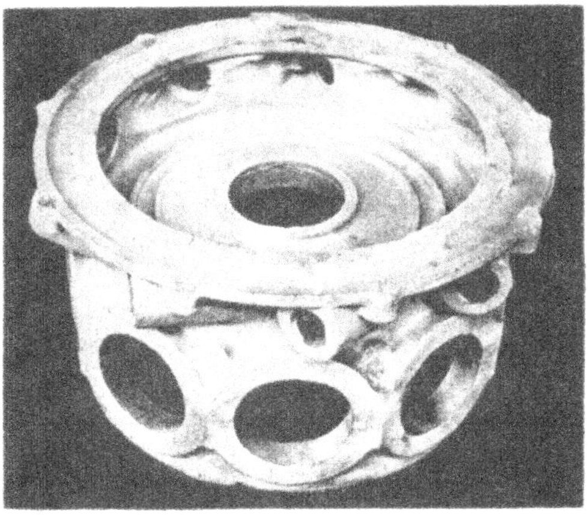

FIG. 11-1.—American Magnesium Corporation, magnesium alloy, AM7-4HT, crankcase for a Lycoming nine-cylinder radial engine.

Rigidity or crankcase stiffness with low weight may be attained by

1. Compact arrangement.
2. Adequate trussing.
3. Arrangement for straight-line transmittal of forces.
4. Use of materials having a high ratio of modulus of elasticity and modulus of rigidity to specific weight.

The crankcase, like the piston, cylinder head, and numerous other stressed parts, is so complex in shape and subjected to such a variety of varying forces as to make the exact calculation of stresses in its parts very difficult if not impossible. Hence the designer will find useful the procedure suggested in connection with the design of connecting-rod ends, *i.e.*, refer the complex structures to similar simple structures with known relations between external forces, dimensions, stresses, deflection, etc. Thus the crankcase may be likened to a cantilever box beam subjected to varying bending and twisting forces and designed for stiffness.

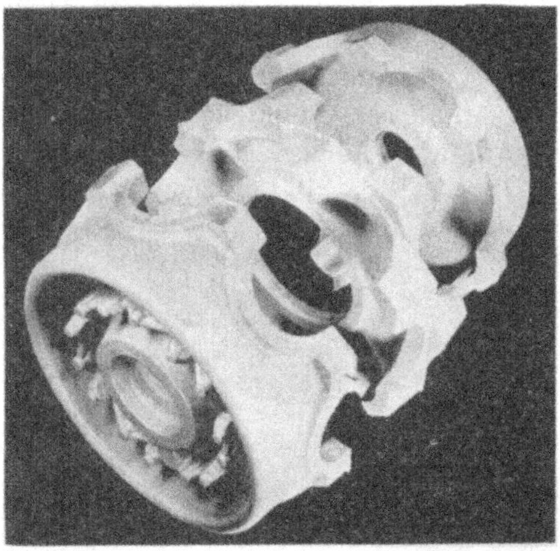

FIG. 11-2.—Aluminum Company of America forged aluminum-alloy crankcase for a Pratt and Whitney 14-cylinder radial engine.

The deflection of simple cantilever beams varies with the cube of the length; hence we would expect the in-line and V-engines to be at a weight disadvantage with a radial engine having the same crankcase stiffness. This disadvantage can, however, be at least partly offset in liquid-cooled designs by using a cooling jacket common to all cylinders in a bank and designing the jacket for stiffness. Deflection also varies directly with the load and inversely with the modulus of elasticity and moment of inertia of the section. This variation indicates the desirability of a larger number of smaller cylinders and large transverse dimensions for the crankcase. Modulus of elasticity is greater for steel than for other possible crankcase materials, but here the lower unit weight of less rigid materials must be considered. Also buckling of very thin sections and fabrication difficulties in forging thin steel sections make it difficult to take full advantage of steel characteristics.

Table 11-1 indicates that on a specific-weight basis, steel has little if any advantage over aluminum forgings, and for small production where forging dies are not justified, aluminum- and magnesium-alloy castings are definitely superior to cast iron.

FIG. 11-3.—Machining operation on a Pratt and Whitney forged aluminum-alloy crankcase.

However, as the engine size goes up, the need for thin sections diminishes, and steel compares more favorably. Also, a fair appraisal should take account of the relative corrosion resistance and strength at operating temperatures as well as relative cost. In the first two, steel has some advantage, and in the last cast iron stands out, particularly with reference to magnesium, whereas aluminum is intermediate. Thus present practice which, for stressed crankcase parts, uses cast iron only for lowest cost engines, cast aluminum and some magnesium for medium-powered, low-production, and forged aluminum and steel for high-powered, high-production engines appears to be basically

FIG. 11-4.—Wright Cyclone steel crankcase showing (left to right) design refinements that have greatly reduced cost and weight.

sound. For lightly stressed parts, light alloy castings are suitable even for high-powered engines.

TABLE 11-1.—COMPARATIVE DATA ON CRANKCASE MATERIALS

Crankcase material	Designation	Modulus of elasticity, E, millions of lb. per sq. in.	Modulus of rigidity, G, millions of lb. per sq. in.	Density, δ, lb. per cu. in.	E/δ	G/δ	Endurance limit, lb. per sq. in.	Endurance limit ÷ 1,000 × δ
Aluminum sand-casting alloy..	195-T6	10.3	3.85	0.100	103	38.5	6,500	65.0
Aluminum sand-casting alloy..	355-T6	10.3	3.85	0.097	106	39.7	8,500	87.5
Aluminum forging alloy......	A51S-T	10.3	3.85	0.097	106	39.7	10,500	108.2
Magnesium sand-casting alloy..	AM260-T4	6.5	.....	0.066	98.5	....	10,000	151.5
Steel............	....	30.0	12.0	0.308	97.5	39.0	30,000	97.5
Cast iron........	....	15.0	6.2	0.261	57.5	23.7	15,000	57.5

11-2. Crankcase Details.—In the Suggested Design Procedure of preceding chapters, it has been recommended that, as the designs were completed, detail parts should be transferred to the assembly drawings. This gradual "assemblying" of the engine aids in checking on errors (no matter how well a part is

FIG. 11-5.—Exploded view of Wright Cyclone engine crankcase sections showing forged aluminum-alloy nose piece, forged main section, cast blower section, and cast rear section.

FIG. 11-6.—Exploded view of Wright cyclone G-200 engine crankcase sections showing details of the steel main section.

designed, it is useless if it will not fit into its proper place), gives the designer a better mental picture of what the final design will look like, and now, for the crankcase, he is able to see what detail shape of the crankcase will be necessary in order to support and hold together the previously designed parts.

As a first step, it is probably best to sketch in the outlines of the crankcase, either directly on the assembly drawings or on superimposed tracing paper. This procedure will aid in fixing

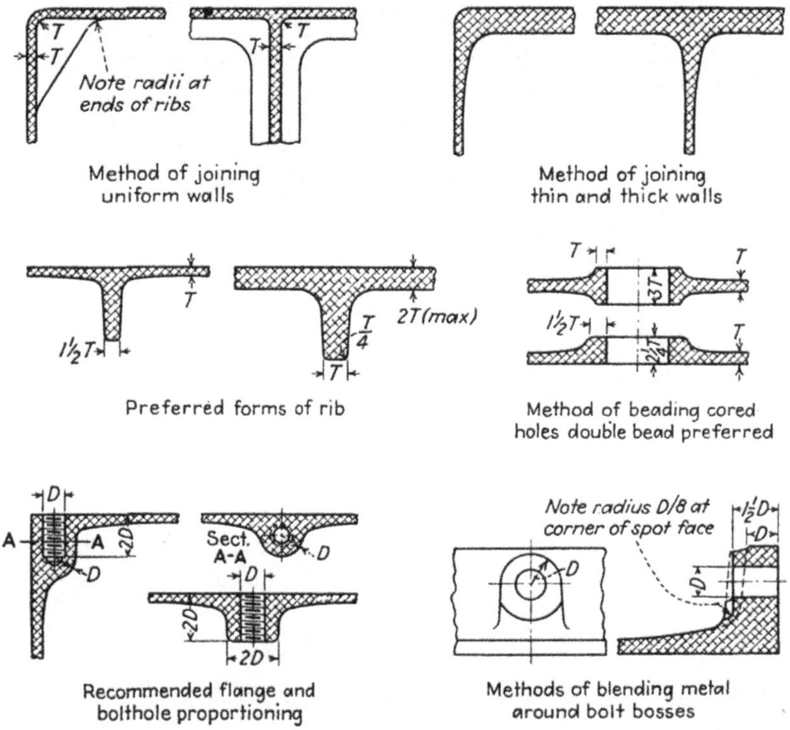

FIG. 11-7.—Aluminum Company of America recommendations on details of casting design.

in mind the necessary shape. Then the general arrangement should be carefully scrutinized for possible weak points while keeping in mind that the stiffness, etc., of complex shapes usually can be estimated by comparing them with simpler beams, columns, etc., of known characteristics. Thus the so-called *stress path* should be as direct as possible, and sharp reentrant corners should be avoided. Thin sections are always possible sources of buckling, and to avoid the excess weight of thicker sections, adequate structural ribbing is a good alternative.

Points at which loads are concentrated are potential sources of local failure, particularly with the nonferrous alloys, and such points should be carefully examined with a view to distributing the load more effectively. For example, too few cylinder hold-down studs may put too great a strain on the crankcase metal immediately surrounding the studs, or the threads may be inaccurate and concentrate most of the load on a small portion of the threaded surface. Such possibilities justify an ample number of accurately ground coarse threads on studs to be

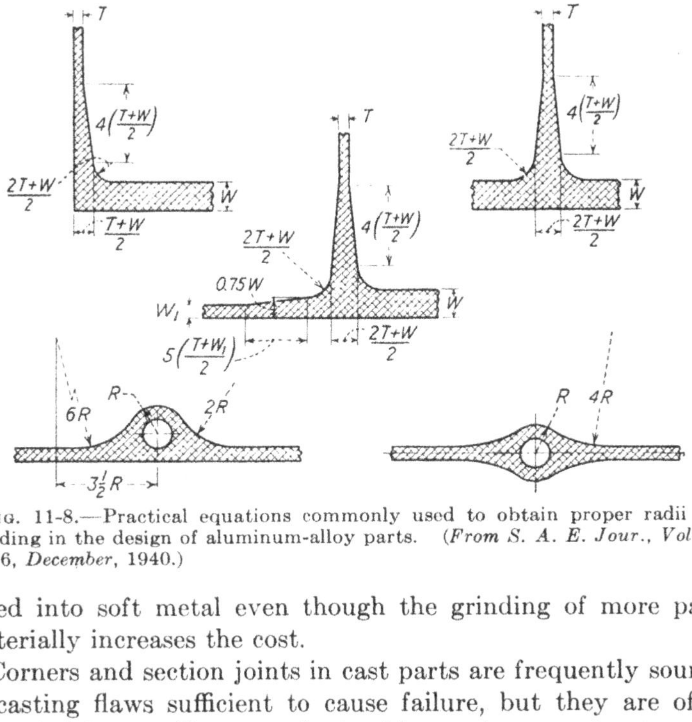

FIG. 11-8.—Practical equations commonly used to obtain proper radii and blending in the design of aluminum-alloy parts. (*From S. A. E. Jour., Vol.* 47, *No.* 6, *December*, 1940.)

fitted into soft metal even though the grinding of more parts materially increases the cost.

Corners and section joints in cast parts are frequently sources of casting flaws sufficient to cause failure, but they are often concealed from ordinary methods of inspection or at least escape detection prior to expensive machining operations. Such troubles can be reduced by utilizing the experience of casting specialists and adhering to their recommendations on casting design. Thus for their alloys, the Aluminum Company recommends the proportioning shown in Figs. 11-7 and 11-8.

11-3. Oil Pumps.—Lubricating oil pumps for aircraft engines are, almost without exception, of the spur-gear type. Oil drawn into the pump (Fig. 11-9) fills the space between the teeth and is carried around to the discharge side where it is squeezed out against the discharge pressure by the meshing of the gear teeth. Pump capacity, for safety, should be appreciably greater than the maximum circulation requirements of the engine, and the excess oil may be by-passed through a pressure relief valve to the inlet side of the pump.

Use of a dry-sump crankcase requires one or more scavenger pumps to transfer the lubricant back to the external supply tank. The scavenger pump, also of the spur-gear type, takes the oil from one or more collection points in the bottom of the crankcase; and to ensure a dry sump at all times, it is usually built with a capacity somewhat greater than that of the pressure pump. For simplicity, the scavenger pump is usually combined

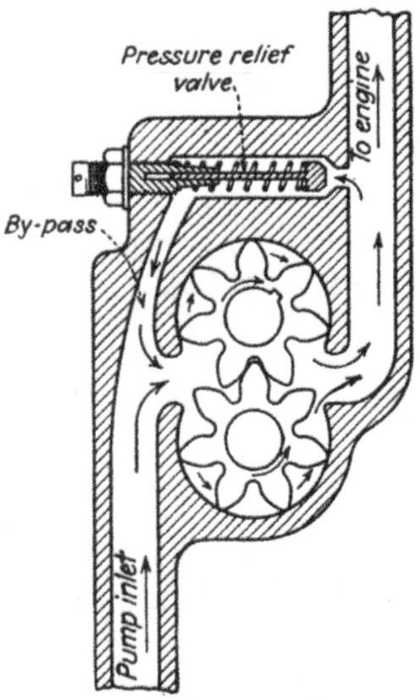

FIG. 11-9.—Schematic diagram of a gear-type oil pump.

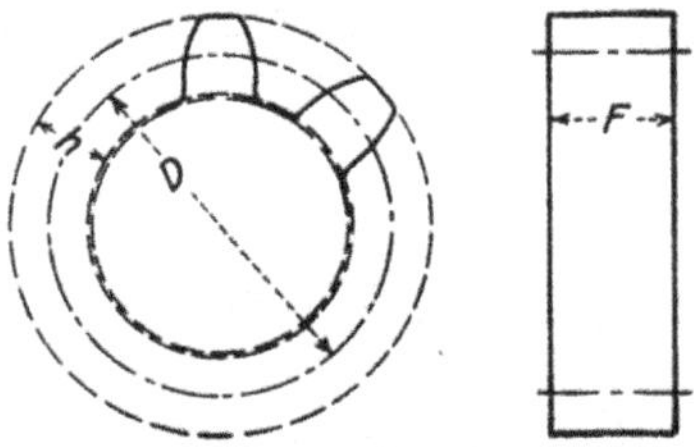

FIG. 11-10.—Diagram showing a method of determining the capacity of a gear pump.

with the pressure pump in a common housing and driven from the same shaft. The scavenger pump preferably should be located near the level of the sump collection point to avoid an excessive suction lift.

To arrive at suitable proportions for the gear pump, the displacement may be determined from the volume of space between the teeth. Thus, in Fig. 11-10, the volume of a cylindrical shell having a length equal to the face width and a thickness equal to the working depth is

$$V = [(D + h)^2 - (D - h)^2]\frac{\pi}{4}F = \pi DhF \qquad (11\text{-}1)$$

where D = pitch diameter of the gear, in.
h = working depth of the tooth, in.
F = face width of the gear, in.

Since the volume of space between the teeth is approximately one-half the volume of this cylindrical shell, the theoretical displacement per gear per revolution is $V_S \approx \frac{V}{2}$, and since for full-depth teeth (Table A3-3) $h = \frac{2}{P_d}$, the displacement of a pump with two gears is

$$Q = \frac{2\pi DFN}{P_d} \text{ cu. in. per min.} \tag{11-2}$$

where P_d = diametral pitch.
N = r.p.m. of the gears.

Data on pump proportions for typical engines are given in Table 11-2.

TABLE 11-2.—OIL-PUMP PROPORTIONS
(All dimensions in inches)

Engine type	Face width	Diametral pitch	Number of teeth	Pump/Crankshaft r.p.m.	Clearances End	Clearances Side	Pump displacement* cu. in. per min. ÷ engine displacement, cu. in.
Pressure Pump							
4 cyl.† opposed	1.125	10	13	1:2	0.005	0.002	7.04
5 cyl. radial	0.75	7	7	5:4	0.002	0.002	2.24
7 cyl. radial	0.875	10	13	5:6	0.004	0.003	2.355
9 cyl. radial	0.5625	6	7	1:1	0.003	0.005	1.135
12 cyl. V	0.75	8	10	1:1	0.003	0.008	1.406
14 cyl. radial	1.25	6	7	1.5:1	0.004	0.003	3.26
Scavenger Pump							
5 cyl. radial	0.75	7	7	5:4	0.002	0.002	2.24
7 cyl. radial	0.875	10	13	5:6	0.003	0.003	2.355
	1.25	10	13	5:6	0.003	0.003	3.36
9 cyl. radial	0.9375	6	7	1:1	0.003	0.005	1.89
12 cyl. V	1.125	8	10	1:1	0.003	0.008	2.11
14 cyl. radial	0.375	6	7	1.5:1	0.004	0.003	0.978
	0.9375	6	7	1.5:1	0.004	0.003	2.45
	0.9375	6	7	1.5:1	0.004	0.003	2.45

* At rated engine speed.
† Wet sump crankcase.

11-4. Blowers and Superchargers.—From the relation

$$\text{b.h.p.} = P_B LAN_p n_c/33{,}000 = P_B DN/792{,}000,$$

it is apparent that horsepower is proportional to size, speed, and b.m.e.p. Size is a function of cylinder displacement and number of cylinders. Cylinder diameter in aircraft engines is limited by cooling requirements to about 6 or 7 in., and stroke, for reasonable stroke-bore ratios, seldom exceeds 6.5 to 7.5 in. Also the stroke is limited by inertia forces.

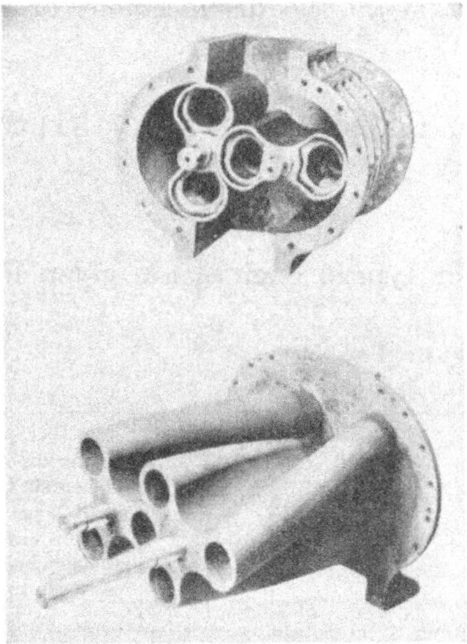

FIG. 11-11.—Roots-type superchargers. Showing two-lobe straight impeller (above) and three-lobe spiral type (below). (*From Roots-Connersville Blower Corporation.*)

Speed in direct-drive engines is ordinarily limited by propeller-efficiency requirements to 2,600 or 2,800 r.p.m., and in large geared engines, speed is usually limited by valve gear or other reciprocating parts to 3,200 to 3,800 r.p.m.

Mean effective pressure mainly is limited by detonation characteristics of the fuel, but with present available octane ratings, fuels can withstand mean pressures well in excess of attainable values in naturally aspirated engines. Hence, with the better fuels, considerable supercharging can be used before detonation limitations are reached.

Superchargers are compact and light-weight fluid pumps capable of handling large volumes of air or air-fuel mixture. They may be used to offset pressure losses in the induction manifold and carburetor and thereby maintain atmospheric pressure at the inlet valve at sea level only or up to some predetermined critical altitude, or they may be used to increase the manifold pressure above atmospheric pressure. Some sources maintain that true supercharging constitutes only the increasing of manifold pressure above atmospheric sea-level values, but there is no very definite line of demarkation, and general usage often calls

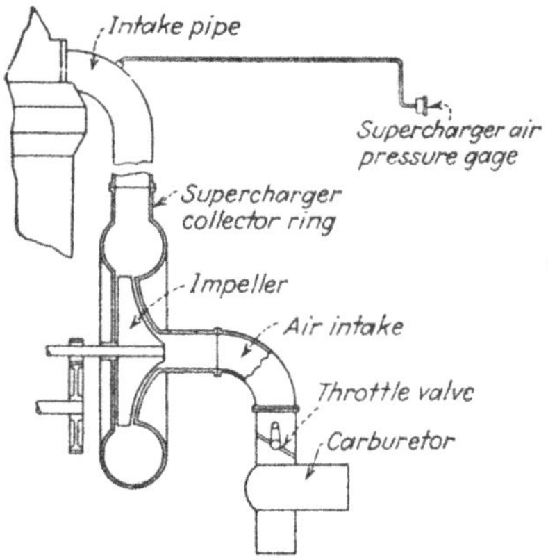

FIG. 11-12.—Single-stage geared centrifugal supercharger. (*From S. A. E. Jour., Vol. 43, No. 5, November, 1938.*)

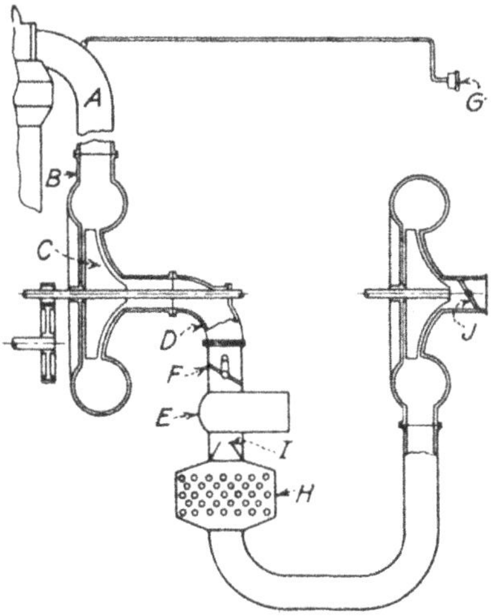

FIG. 11-13.—Two-stage geared centrifugal supercharger. *A*, intake pipe; *B*, super collector ring; *C*, impeller; *D*, air intake; *E*, carburetor; *F*, throttle valve; *G*, super air-pressure gage; *H*, intercooler; *I*, automatic suction valve; *J*, air inlet. (*From S. A. E. Jour., Vol. 43, No. 5, November, 1938.*)

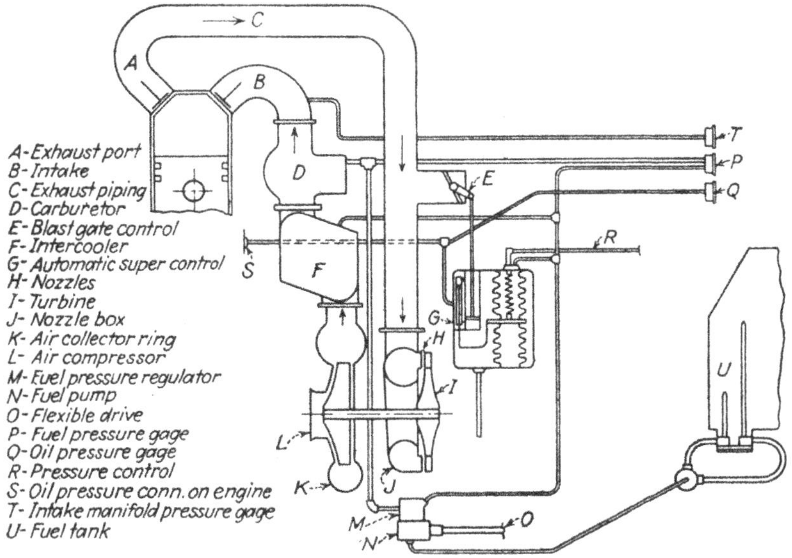

FIG. 11-14.—Exhaust-turbine-drive centrifugal supercharger. (*From S. A. E. Jour., Vol. 43, No. 5, November, 1938.*)

any type of pump for increasing the amount of charge taken into a cylinder a supercharger.

Most applicable types of superchargers are the Roots positive-displacement pump (Fig. 11-11) and the centrifugal or fan type. Roots-type superchargers are gear driven from the engine shaft, but centrifugal types may be either gear driven from the shaft (Figs. 11-12 and 11-13) or exhaust-turbine driven (Fig. 11-14). At present, the great majority of superchargers are of the gear-driven centrifugal type, although there is a definite trend to the exhaust-turbine-driven type for high-altitude operation (20,000 to 35,000 ft.).

11-5. Supercharger Power Requirements.—In the Roots supercharger, air at pressure P_E is trapped in the spaces between the rotors and the housing and carried from the inlet to the exit side. When the exit opening is uncovered by the rotor, air at pressure P_L rushes back and compresses the trapped air. Further movement of the rotors then forces the trapped air at pressure P_L into the exit or discharge pipe.

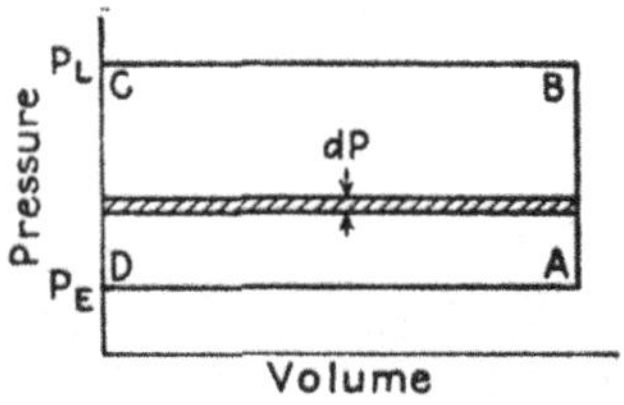

FIG. 11-15.—Theoretical P-V diagram for a Roots supercharger.

A theoretical pressure-volume diagram for a Roots supercharger is shown in Fig. 11-15. In this figure, net work ∝ area $ABCD$ or $W_{net} = \int_{P_E}^{P_L} V\,dP$.

But

$$V = V_E = V_L$$

Hence

$$W_{net} = V_E \int_{P_E}^{P_L} dP = V_E(P_L - P_E) = P_E V_E(R_p - 1)$$

$$= WRT_E(R_p - 1) \tag{11-3}$$

where V_E = volume of entering air, cu. ft. per unit time.
W = weight of entering air, lb. per unit time.
P_E = inlet pressure, lb. per sq. ft. abs.
P_L = discharge pressure, lb. per sq. ft. abs.
$R_p = P_L/P_E$ = the compression ratio.
R = gas constant (= 53.3 for air).
T_E = inlet air temperature, deg. F. abs.

If W is in pounds per minute, the theoretical power required is

$$\text{hp.} = \frac{WRT_E}{33{,}000}(R_p - 1) \qquad (11\text{-}4)$$

For $V_E = V_L$, the temperature varies directly with the pressure, hence the compression temperature is

$$T_L = T_E \times R_p \qquad (11\text{-}5)$$

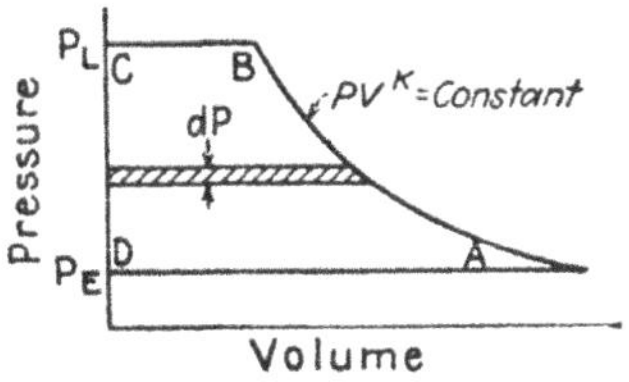

FIG. 11-16.—Theoretical P-V diagram for a centrifugal supercharger.

In the centrifugal supercharger, the air attains a high velocity owing to centrifugal force, and then the kinetic energy acquired contributes to increasing the pressure as the air slows down in the exit space. A theoretical PV diagram for a centrifugal supercharger is shown in Fig. 11-16. In this figure, net work $\propto$ area $ABCD$ or

$$W_{\text{net}} = \int_{P_E}^{P_L} V\,dP$$

But

$$PV^K = P_E V_E^K = P_L V_L^K$$

From which

$$V = \frac{P_E^{1/K} V_E}{P^{1/K}}$$

Hence

$$W_{\text{net}} = P_E^{1/K} V_E \int_{P_E}^{P_L} \frac{dP}{P^{1/K}} = P_E^{1/K} V_E \left[\frac{P_L^{(K-1)/K} - P_E^{(K-1)/K}}{\frac{K-1}{K}} \right]$$

$$W_{\text{net}} = \frac{K}{K-1} P_E V_E [R_p^{(K-1)/K} - 1]$$

$$= \frac{K}{K-1} WRT_E [R_p^{(K-1)/K} - 1]$$

$$W_{\text{net}} = WJC_pT_E[R_p^{(K-1)/K} - 1] \qquad (11\text{-}6)$$

where $K = C_p/C_V$ = adiabatic compression exponent (= 1.4 for air).

V_E = volume of entering air, cu. ft. per unit time.

P_E = inlet pressure, lb. per sq. ft. abs.

P_L = discharge pressure, lb. per sq. ft. abs.

$R_p = P_L/P_E$ = the compression ratio.

W = weight of entering air, lb. per unit time.

$R = J(C_p - C_V)$ = gas constant (= 53.3 for air).
T_E = inlet air temperature, deg. F. abs.
J = 778 ft. lb. per b.t.u.
C_p = specific heat at constant pressure (= 0.24 for air).
C_V = specific heat at constant volume (= 0.17 for air).

If W is in pounds per minute, the power required for theoretical or adiabatic compression in the centrifugal supercharger is

$$\text{Adiabatic hp.} = \frac{\frac{K}{K-1} WRT_E}{33{,}000} [R_p^{\frac{K-1}{K}} - 1] \quad (11\text{-}7a)$$

$$= \frac{WJC_pT_E}{33{,}000} [R_p^{\frac{K-1}{K}} - 1] \quad (11\text{-}7b)$$

and the compression temperature is

$$T_L = T_E \times R_p^{(K-1)/K} \quad (11\text{-}8a)$$

Frequently it is useful to know the adiabatic temperature rise Δt_a which is

$$T_L - T_E = \Delta t_a = T_E[R_p^{(K-1)/K} - 1] \quad (11\text{-}8b)$$

Due to turbulence, friction, etc., the actual fluid horsepower required by the centrifugal supercharger is considerably greater than the adiabatic horsepower. When the actual compression temperature T_L or the temperature rise Δt_a is known, a polytropic exponent n may be found from Eq. (11-8) and then used in Eq. (11-7a) to get the actual fluid horsepower. Then

$$e_a = \frac{\text{adiabatic hp.}}{\text{fluid hp.}} \quad (11\text{-}9)$$

$$e_m = \frac{\text{fluid hp.}}{\text{input hp. to supercharger}} \quad (11\text{-}10)$$

$$e_s = e_a \times e_m \quad (11\text{-}11)$$

where e_a = compression or adiabatic temperature efficiency.
e_m = mechanical efficiency of the supercharger drive.
e_s = over-all adiabatic efficiency.

Values of the polytropic exponent n have been found from practice usually to range between 1.6 and 1.8, and corresponding adiabatic temperature efficiencies usually range between 70 and 80 per cent. Mechanical efficiencies may be taken as 85 to 90 per cent. Hence, an over-all adiabatic efficiency of 65 per cent is a reasonable assumption for design.

Another efficiency expression sometimes used in connection with Roots superchargers is the Roots-type efficiency which may be defined as the ratio of the work to compress adiabatically, as

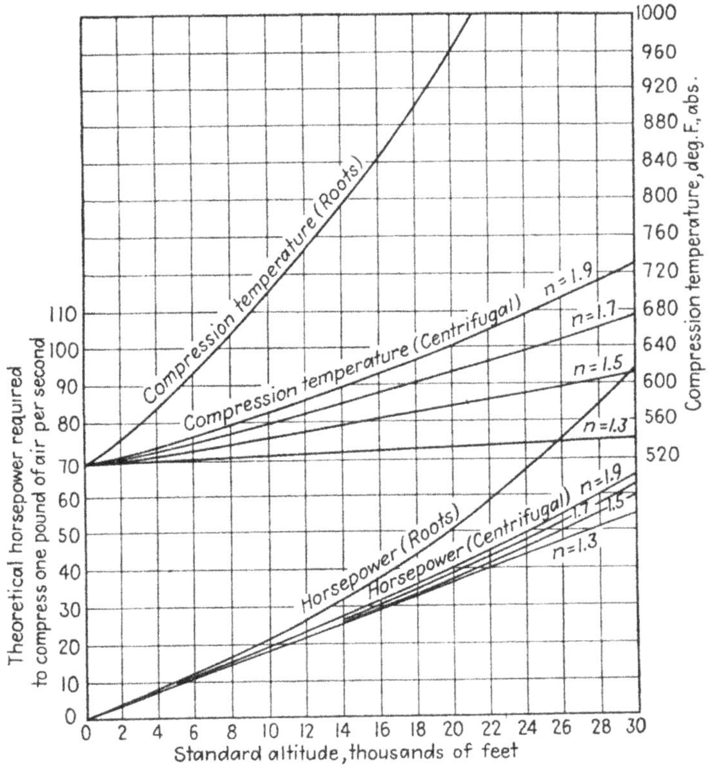

FIG. 11-17.—Compression, temperature, and theoretical horsepower required to compress 1 lb. of air per sec. from standard pressure and temperature at altitude to 29.92 in. Hg. (*Data from NACA Tech. Rept.* 384.)

in the centrifugal supercharger, to the work required at constant volume, as in the Roots supercharger. Thus, from Fig. 11-16,

$$\text{Area } ABCD \propto W_{\text{net}} = \frac{K}{K-1} P_E V_E [R_p{}^{(K-1)/K} - 1]$$

And from Fig. 11-15,

$$\text{Area } ABCD \propto W_{\text{net}} = P_E V_E (R_p - 1)$$

or

$$\text{Roots-type efficiency} = \frac{K}{K-1} \left[\frac{R_p{}^{(K-1)/K} - 1}{R_p - 1} \right] \qquad (11\text{-}12)$$

Inspection of Eq. (11-12) shows that the centrifugal supercharger is less wasteful of power than the Roots and becomes increasingly so as the compression ratio is increased. Inlet pressure P_E decreases with altitude; hence to maintain sea-level power, R_p must be increased with altitude. Thus, the centrifugal supercharger is preferable for engines with high critical altitudes (Fig. 11-17).

Example 1.—A 5.25- by 5.25-in., 2,300-r.p.m., nine-cylinder, 5.5:1-compression-ratio engine rated 500 b.hp. at 5,000 ft. altitude (P_{atm} = 24.9 in. Hg abs., t_{atm} = 41.2°F.) has a fuel rate of 0.53 lb. per b.hp. hr. and an air-fuel ratio of 12.5:1 by weight, an over-all adiabatic efficiency of 65 per cent being assumed, estimate the horsepower required to drive the supercharger.

Solution.—The b.m.e.p. is

$$P_B = \frac{500 \times 33{,}000 \times 12 \times 12}{5.25^3 \times 0.785 \times 2{,}300 \times 9} = 169 \text{ lb. per sq. in.}$$

From Fig. 1-9A, the intake-manifold pressure is approximately 42 in. Hg abs. = P_L, hence

$$R_p = \frac{42}{24.9^*} = 1.688$$

The weight of charge is†

$$W = \frac{500 \times 0.53 \times (12.5 + 1)}{60} = 59.5 \text{ lb. per min.}$$

From Eq. (11-7b),

$$\text{Adiabatic hp.} = \frac{59.5 \times 778 \times 0.24 \times (41.2 + 460)}{33{,}000} [1.688^{(1.4-1)/1.4} - 1]$$
$$= 27.45$$

And for an over-all efficiency of 65 per cent

$$\text{Input hp.} = \frac{27.45}{0.65} = 42.2$$

11-6. Impeller Speed.—To attain desired pressure ratios with compactness and light weight, it is necessary to rotate centrifugal impellers at speeds of the order of 15,000 to 30,000 r.p.m. At such speeds, centrifugal stresses are very high, and to attain the

* For a supercharger with a Venturi-type carburetor in the inlet, slightly greater accuracy will be had if allowance is made for carburetor drop which will reduce P_E by 0.5 to 1.0 in. Hg.

† An alternate procedure for design purposes is to assume an air requirement of 0.11 to 0.12 lb. of air per b.hp. per min.

necessary strength and precision of balance, impeller forms must be used that are structurally strong and can be accurately fabricated without prohibitive cost. These requirements appear to be most feasible of attainment with forged-steel or aluminum-alloy impellers having straight radial blades even though theory indicates that a suitably curved blade may be superior from a fluid-flow standpoint.

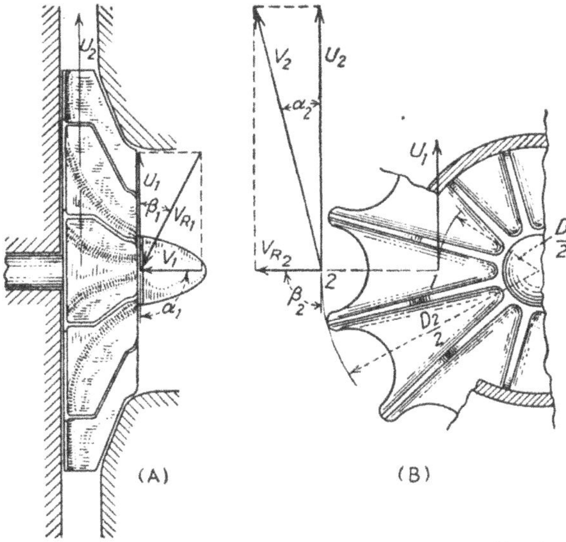

FIG. 11-18.—Vector diagram of air flow into and out of a centrifugal-supercharger impeller.

Referring to Fig. 11-18, let

V_1 = velocity of the air in the inlet pipe to the impeller, f.p.s.

α_1 = angle between the direction of the incoming air and the plane of rotation of the impeller.

U_1 = tangential velocity of the impeller at the point of entry of the air to the impeller, f.p.s.

$U_1 = \pi D_1 N$.

D_1 = diameter of the impeller at the point of entry of the air, ft.

N = impeller speed, r.p.s.

V_{R1} = velocity of the air relative to the impeller at the point of entry of the air to the impeller, f.p.s.

β_1 = angle between V_{R1} and a line normal to the blades at the point of entry of the air to the impeller.

V_{R2} = velocity of the air relative to the impeller at the point of exit from the impeller, f.p.s.

β_2 = angle between V_{R2} and a line normal to the blades at the point of exit of the air from the impeller.

U_2 = tangential velocity of the impeller at the point of exit of the air from the impeller, f.p.s.

$U_2 = \pi D_2 N$.

D_2 = diameter of the impeller at the point of exit of the air, ft.

V_2 = velocity of the air at the point of entry to the diffuser space, f.p.s.

α_2 = angle between V_2 and U_2.

Considering first an ideal frictionless case and using a unit weight of 1 lb. of air as a basis, the impulse force given the air leaving the impeller will be

$$F_2 = \frac{V_2 \cos \alpha_2}{g}$$

The torque to produce this force is

$$Q_2 = \frac{D_2}{2} \times \frac{V_2 \cos \alpha_2}{g}$$

But

$$U_2 = \pi D_2 N$$

and the work required is

$$W_2 = 2\pi N Q_2$$

Hence

$$W_2 = \frac{U_2 V_2 \cos \alpha_2}{g} \tag{11-13}$$

The tangential velocity U_2 is quite large in terms of the relative velocity V_{R2}, so α_2 is small. If the air leaves in a true radial direction relative to the impeller, $\beta_2 = 90$ deg. and $V_2 \cos \alpha_2 \approx U_2$; hence

$$W_2 \approx \frac{U_2^2}{g} \tag{11-14}$$

Actually, the air passing through the impeller tends to pile up against the following blade (Fig. 11-19) so that $\beta_2 < 90$ deg., and this effect combined with a finite value for V_{R2} (*i.e.*, $\alpha_2 > 0$) results, for radial blades, in $V_2 \cos \alpha_2$ always being less than U_2. According to Pye,[6] $V_2 \cos \alpha_2 = 0.85$ to $0.98 \times U_2$; hence

$$W_2 = C\frac{U_2^2}{g} = C\frac{\pi^2 D_2^2 N^2}{g} \tag{11-15}$$

where $C = 0.85$ to 0.98 and varies inversely with the quantity of flow.

Air entering the impeller (Fig. 11-18*A*) meets the blades at the relative velocity V_{R1} and at an angle β_1 to a line normal to the

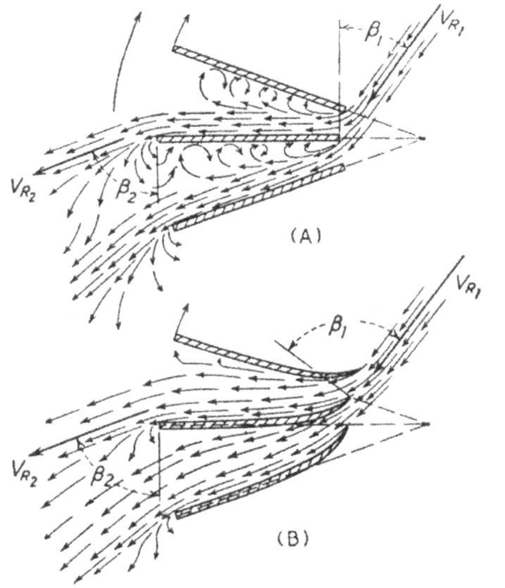

FIG. 11-19.—Air-flow path between the blades of a centrifugal supercharger impeller: (*A*) with flat radial blades; (*B*) with curved entering edges.

blade surface. Hence the component of relative entering velocity normal to the blade is $V_{R1} \cos \beta_1$, and the impact per unit weight of air is $V_{R1}^2 \cos^2 \beta_1/2g$. If this force is considered to act at the center of area of the passageway between the entering edges of the blades, the work to overcome this impact loss will be

$$W_1 = 2\pi NQ \approx \pi D_A N \frac{V_{R1}^2 \cos^2 \beta_1}{2g} \tag{11-16}$$

If flat radial blades are used and the air approaches the impeller from a direction normal to the plane of rotation (*i.e.*, $\alpha_1 = 90$ deg.), the impact component $V_{R1} \cos \beta_1$ will be equal to U_1 and

$$W_1 \approx \pi D_A N \frac{U_1^2}{2g} \approx \frac{\pi^3 D_A^3 N^3}{2g} \qquad (11\text{-}17)$$

Several methods may be used to reduce entering-impact losses. (*a*) By giving the air a helical or whirling motion in the inlet passageway to the impeller, α_1 can be made less than 90 deg. This will reduce V_{R1}, but the pressure loss in the inlet passage incident to producing this whirling will reduce P_1, and for a

FIG. 11-20.—Impeller and diffuser arrangement for a Pratt and Whitney centrifugal supercharger.

given desired manifold pressure, increase the compression ratio R_p, hence the work of compression [Eq. (11-6)], so that the net gain may be small. (*b*) V_1 may be increased with resulting increase in β_1 and decrease in $\cos \beta_1$, but here again an accompanying pressure loss in the inlet pipe will be had. (*c*) Probably the most effective way to reduce the entering-impact loss is to curve the entering edges of the blades (Fig. 11-19*B*) so that they are in line with V_{R1}. This amounts to increasing β_1 to 90 deg., and as $\cos 90$ deg. $= 0$, this theoretically would eliminate the entering-impact loss. Actually, with blades of finite thickness and a practical radius of curvature, some turbulence would remain, but by using thin warped blades,* *i.e.*, by varying the curvature with the distance from the impeller shaft

$$[\beta_1 = f(U_1) = f(D_1)],$$

the net gain would still be considerable. This warping may be accomplished without curvature in the plane of rotation (*i.e.*, without sacrifice in strength) by the arrangement shown in Figs. 11-20 and 11-21.

Since the cross-sectional area of the inlet pipe is fixed, V_1 will vary with the quantity of air flowing, *i.e.*, with the load on the engine. Hence, for a constant impeller speed (U_1 = a constant), β_1 will decrease with V_1, and the change in direction of the air at entry will increase. When β_1 reaches some minimum value depending upon the design, the motion of the air will change from a streamline to a turbulent flow, and a condition analogous to burbling over an airfoil at high angles of attack will result. When an airfoil is increased in angle of attack beyond the stall point, the lift drops off rapidly, and the equivalent effect in centrifugal superchargers is an abrupt decrease in the compression ratio. For an airfoil, as the angle of attack decreases below the stall angle, the lift decreases gradually and the equivalent effect in a centrifugal supercharger is a gradual decrease in compression ratio with increase in the quantity of air flowing.

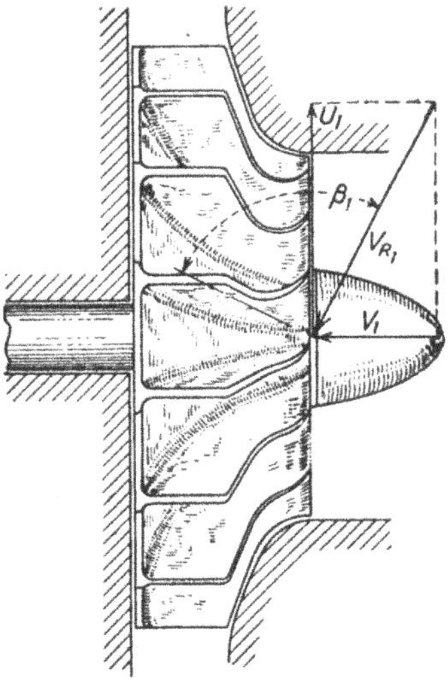

FIG. 11-21.—Impeller with curved entering edges shaped to reduce entering-impact and turbulence losses without sacrifice in strength.

This analogy with airfoils would indicate the desirability of an impeller with blades having an entering edge that could be adjusted in curvature as the quantity of air flowing was varied. An automatic mechanism (comparable to the automatic pitch-varying mechanism in propellers) would be ideal, but complexity and cost might more than offset the gain in performance. Pos-

* This warping is analogous to the warping of propeller blades.

sibly the best compromise is detachable curved entering edges that can be interchanged for different desired performance ratings. Suitably slotted entering edges (like leading-edge wing slots) might also merit investigation.

By assuming that the entering edges of the impeller blades are suitably shaped to minimize entering impact losses and turbulence, a relation among R_p, D_2, and N can be obtained by combining Eqs. (11-6) and (11-15). Thus, per pound of air flowing

$$\frac{C\pi^2 D_2^2 N^2}{g} = \frac{JC_pT_1}{e_a}\left(R_p^{\frac{K-1}{K}} - 1\right)$$

or

$$N = \sqrt{\frac{gJC_pT_1}{C\pi^2 D_2^2 e_a}\left(R_p^{\frac{K-1}{K}} - 1\right)} \tag{11-18}$$

where $T_1 = T_E$.

It should be borne in mind when planning a supercharger design that air-flow characteristics are adversely altered as the moving air approaches the velocity of sound, and since $V_2 \approx U_2$, it is advisable to keep the impeller-tip speed below 1,200 to 1,500 f.p.s.* Thus to attain compression ratios greater than about 2.5 to 3 or to maintain sea-level pressures to critical altitudes greater than about 20,000 to 25,000 ft., two or more stages of compression should be used.

Example 2.—For the engine in Example 1, determine the impeller speed, tip speed, and drive-gear ratio if the impeller diameter is 7½ in.

Solution.—For Eq. (11-18), $g = 32.2$, $J = 778$, $C_p \approx 0.24$,

$$T_1 = 41.2 + 460 = 501.2,$$

$C \approx 0.85$, $D_2 = 7.5/12 = 0.625$, $e_a = 0.7$, $R_p = 1.688$, $K = 1.4$.

$$N = \sqrt{\frac{32.2 \times 778 \times 0.24 \times 501.2}{0.85 \times \pi^2 \times 0.625^2 \times 0.7}(1.688^{0.286} - 1)} = 461 \text{ r.p.s.}$$

$$\text{r.p.m.} = 461 \times 60 = 27,660$$

$$\text{Tip speed} = \pi \times 0.625 \times 461 = 905 \text{ f.p.s.}$$

$$\text{Drive-gear ratio} = \frac{27,660}{2,300} \approx 12:1$$

11-7. Impeller Details.—To avoid restricting the flow into the impeller, the area of passageway into the impeller should be

* The velocity of sound will be greater than in standard air because of the increased temperature of the air in the impeller.

approximately equal to the area of the inlet pipe. The area of the inlet pipe may be found from

$$\frac{Q}{V_0} = A_0 = \pi R_0^2 \tag{11-19}$$

where Q = quantity of charge, c.f.s.
V_0 = mean velocity in inlet pipe, f.p.s.
A_0 = area of inlet pipe.
R_0 = radius of inlet pipe.

The entry area normal to the direction of flow (Fig. 11-22) is the lateral area of the frustrum of a right circular cone in which S is the slant height, R_h is the radius of one base, and $R_h + S\cos\gamma$ is the radius of the other base. Then

$$A_1 = \pi S(2R_h + S\cos\gamma) - Sn_bt = \pi R_0^2$$

or

$$R_0^2 = S^2\cos\gamma + 2R_hS - \frac{Sn_bt}{\pi} \tag{11-20}$$

where γ = angle of approach to the impeller,
n_b = number of blades,
t = thickness of the blades,

also

$$S\cos\gamma = R_1 - R_h$$

Hence

$$R_1 = R_h + S\cos\gamma \tag{11-21}$$

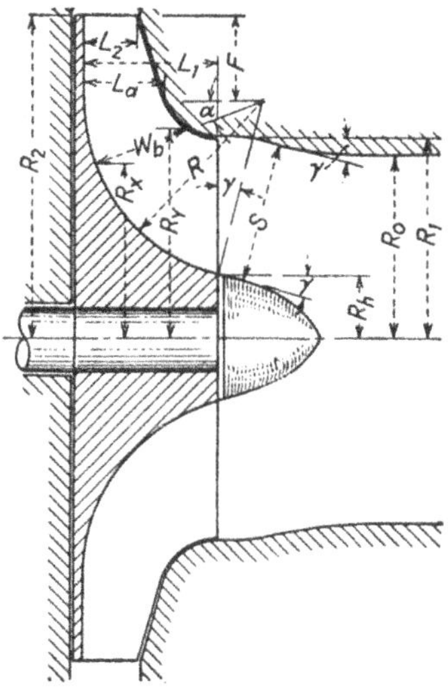

Fig. 11-22.—Diagram for relating impeller dimensions.

Example 3.—For the engine in Examples 1 and 2, determine the inlet diameter of the impeller if the angle of approach is 15 deg., the impeller hub diameter is 1.5 in., there are 14 impeller blades having an entering-edge thickness of 0.0625 in., and the mean velocity in the inlet pipe is 150 f.p.s.

Solution.—At 5,000 ft., the specific volume of entering charge is approximately

$$\text{Specific volume} = \frac{V}{W} = \frac{RT}{P} = \frac{53.3 \times 501.2}{24.9^* \times 0.491 \times 144} = 15.2 \text{ cu. ft. per lb.}$$

Then

$$A_0 = \pi R_0^2 = \frac{Q}{V_0} = \frac{15.2 \times 59.5}{150 \times 60} \approx 0.1 \text{ sq. ft.}$$

* See footnote(*). p. 294.

or

$$R_0 = \sqrt{\frac{0.1 \times 144}{\pi}} = 2.14 \text{ in.}$$

Assume that the diameter of the inlet pipe ($= 2 \times R_0$) is 4.25 in. Then, from Eq. (11-20),

$$4.57 = 0.9659S^2 + 1.5S - \frac{14 \times 0.0625}{\pi} S$$

from which $S = 1.656$ in. Substituting this value in Eq. (11-21)

$$R_1 = 0.75 + 1.656 \times 0.9659 = 2.35 \text{ in.}$$

Hence, the diameter of the impeller at the inlet is

$$D_1 = 2R_1 = 4.7 \text{ in.}$$

To minimize turbulence, the air passing through the impeller should be turned gradually into the plane of rotation. This indicates the desirability of a large radius of curvature R, Fig. 11-22. In this figure,

$$R_2 = F + R \cos \gamma + R_h$$

$$R = \frac{R_2 - F - R_h}{\cos \gamma} \qquad (11\text{-}22)$$

also

$$R = L_1 + R \sin \gamma$$

or

$$L_1 = R(1 - \sin \gamma) \qquad (11\text{-}23)$$

Hub curvature R will be a maximum when $F = 0$, but this may give a hub length L_1 to the impeller that is greater than desirable, *i.e.*, the weight of the impeller may be excessive. Hence a compromise between turbulence [$= f(R)$] and impeller weight may be in order.

Example 4.—For the engine in the preceding three examples, assume $F = 1$ in. and determine the hub length and curvature.

Solution.—For Eq. (11-22), $R_2 = D_2/2 = 3.75$ in., $R_h = 0.75$ in., and $\gamma = 15$ deg. Then

$$R = \frac{3.75 - 1 - 0.75}{0.9659} = 2.075 \text{ in.}$$

and from Eq. (11-23),

$$L_1 = 2.075(1 - 0.2588) = 1.539 \text{ in.}$$

The over-all hub length will be increased by the thickness of the back side of the impeller disk, but some weight may be saved by dishing in the back end of the hub.

Assuming a constant area of passage, the width of the blades W_b normal to the moving air in the impeller may be found as follows:

In the passages where the air is being turned into the plane of the impeller, the cross-sectional area of passages is the lateral area of a frustrum of a right circular cone minus the area occupied by the blades. Thus, in Fig. 11-22,

$$A = \pi W_b(R_x + R_y) - n_b t W_b \tag{11-24}$$
$$R_x = R_2 - F - R \sin \alpha \tag{11-25}$$
$$R_y = R_x + W_b \sin \alpha \tag{11-26}$$

and in the straight radial passage

$$A = 2\pi R_x W_b - n_b t W_b \tag{11-27}$$

Example 5.—For the engine in the preceding four examples, determine the width of the blades for constant area of passage.

Solution.—A ($= A_0$) $= 14.4$ sq. in., $n_b = 14$, $t = 0.1$ in., $R_2 = 3.75$ in., $F = 1$ in., $R = 2.075$ in. For $\alpha = 0$

$$R_x = 3.75 - 1 - 0 = 2.75\ (= R_y)$$
$$14.4 = \pi W_b(2.75 + 2.75) - 14 \times 0.1 \times W_b$$
$$W_b = 0.905 \text{ in. } (= L_a)$$

For $\alpha = 10$ deg., $\sin \alpha = 0.1736$, and

$$R_x = 3.75 - 1 - 2.075 \times 0.1736 = 2.39 \text{ in.}$$
$$R_y = 2.39 + 0.1736 W_b$$
$$14.4 = \pi W_b(2.39 + 2.39 + 0.1736 W_b) - 14 \times 0.1 \times W_b$$
$$W_b = 1.05 \text{ in.}$$
$$R_y = 2.39 + 1.05 \times 0.1736 = 2.572 \text{ in.}$$

Similarly,

For $\alpha = 20$ deg., $R_x = 2.041$ in., $R_y = 2.434$ in., $W_b = 1.15$ in.
For $\alpha = 30$ deg., $R_x = 1.7125$ in., $R_y = 2.345$ in., $W_b = 1.265$ in.
For $\alpha = 40$ deg., $R_x = 1.417$ in., $R_y = 2.316$ in., $W_b = 1.4$ in.
For $\alpha = 50$ deg., $R_x = 1.16$ in., $R_y = 2.32$ in., $W_b = 1.5125$ in.
For $\alpha = 60$ deg., $R_x = 0.952$ in., $R_y = 2.337$ in., $W_b = 1.5975$ in.
For $\alpha = 90 - \gamma = 75$ deg. and $t = 0.0625$ in., $W_b = 1.644 \approx 1.656 = S$,

thus checking the previous relations involving R_1.

In the straight radial portion of the impeller, the blade width varies inversely as the radius [Eq. (11-27)]. Hence the tip width may be calculated directly as

$$L_2 = \frac{14.4}{2\pi \times 3.75 - 14 \times 0.1} = 0.65 \text{ in.}$$

If it is assumed the energy imparted to the air in the impeller is all retained by the air, the work done by the impeller will be divided between increasing the kinetic energy and the enthalpy of the air. Thus

$$W_2 = \left(\frac{V_2^2}{2g} - \frac{V_1^2}{2g}\right) + JC_p(T_2 - T_1)$$

or

$$T_2 - T_1 = \frac{W_2 - [(V_2^2/2g) - (V_1^2/2g)]}{JC_p}$$

But from Fig. 11-18,

$$V_2^2 = V_{R2}^2 + (V_2 \cos \alpha_2)^2 = V_{R2}^2 + C^2U_2^2 \approx V_1^2 + C^2U_2^2$$

And from Eq. (11-15),

$$W_2 = C\frac{U_2^2}{g}$$

Hence, the temperature rise in the impeller is

$$T_2 - T_1 = \frac{2CU_2^2 - C^2U_2^2}{2gJC_p} \qquad (11\text{-}28)$$

and the corresponding temperature ratio is

$$\frac{T_2}{T_1} = \frac{2CU_2^2 - C^2U_2^2}{2gJC_pT_1} + 1 \qquad (11\text{-}29)$$

By applying the temperature-volume relation

$$\frac{T_2}{T_1} = \left(\frac{v_1}{v_2}\right)^{K-1}$$

in Eq. (11-29), the change in the specific volume of the air passing through the impeller can be expressed as

$$\frac{v_1}{v_2} = \left(\frac{2CU_2^2 - C^2U_2^2}{2gJC_pT_1} + 1\right)^{1/(K-1)} \qquad (11\text{-}30)$$

Example 6.—Determine the effect of change in specific volume of the charge on the tip width L_2 of the impeller in Example 5.

Solution.—The specific volume of the air entering the impeller is 15.2 cu. ft. per lb. (Example 3), $C \approx 0.85$, $U_2 = 905$ f.p.s., $T_1 = 501.2$°F. abs. From Eq. (11-30),

$$\frac{v_1}{v_2} = \left(\frac{2 \times 0.85 \times 905^2 - 0.85^2 \times 905^2}{2 \times 32.2 \times 778 \times 0.24 \times 501.2} + 1\right)^{1/(1.4-1)} = 1.1329^{2.5} = 1.366$$

$$v_2 = \frac{15.2}{1.362} = 11.112 \text{ cu. ft. per lb.}$$

The volume per second is

$$Q_2 = \frac{59.5}{60} \times 11.112 = 11.02 \text{ c.f.s.}$$

For $V_{R2} = V_1 = 150$ f.p.s.

$$A_2 = \frac{11.02}{150} \times 144 = 10.6 \text{ sq. in.}$$

From Eq. (11-27), the tip width is

$$L_2 = \frac{10.6}{2\pi \times 3.75 - 14 \times 0.1} = 0.48 \text{ in.}$$

This value for tip width is based on assumed adiabatic flow through the impeller. Actually, impeller losses will place the tip width somewhere between 0.65 and 0.48, say at about 0.55 in., but experimental data are necessary to determine the best value for L_2.

11-8. Diffusers.—Diffusers are designed to convert the kinetic energy of the air leaving the impeller into pressure energy, and they may be made either with or without guide vanes. Generally for in-line and V-engines, the air or mixture is led off from the diffuser housing through one or two pipes, and this makes possible the use of a long spiral diffuser of gradually increasing cross section (Fig. 11-23). In radial engines, usual, though not universal, practice is to provide a diffuser with vanes which guide the air into a collector ring to which individual pipes leading to each cylinder are attached, usually at an angle (Fig. 11-20). The long spiral diffuser permits a more gradual conversion of kinetic energy, and this would indicate less turbulence and perhaps slightly greater efficiency, but the arrangement of cylinders and space limitations make the application difficult in radial engines.

If not restricted radially, air leaving the impeller will form a free vortex. If the axial width of the diffuser is held constant, say at impeller tip width L_2, the area in a radial direction will vary directly as the radius; hence the radial component of velocity V_R will vary inversely as the radius. For constant angular momentum MVR = constant. The tangential component V_T ($= V \cos \alpha$) will also vary inversely with the radius. Thus the resultant velocity will decrease from impeller-tip velocity V_2 inversely as the radius, and the air will follow a logarithmic spiral path, *i.e.*, $\alpha = \alpha_2 =$ constant. With increasing density,

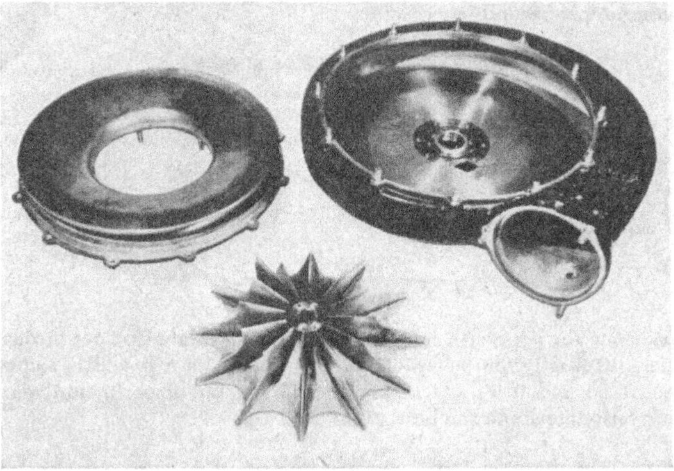

FIG. 11-23.—Impeller and spiral volute type of diffuser for a Mercedes-Benz inverted V-12 engine. (*From S.A.E. Jour., Vol. 49, No. 4, October, 1941.*)

however, the radial component of velocity will vary inversely as the product of the radius and density, *i.e.*,

$$Q = \frac{W}{d} = A_R v_R = 2\pi R L_2 v_R$$

or

$$v_R = \frac{W}{2\pi R L_2 d} = \frac{Wv}{2\pi R L_2} \tag{11-31}$$

where W = air flow, lb. per sec.

v = specific volume, cu. ft. per lb.

The tangential component will not be affected by change in the density since the mass M is unaffected. Hence, the air will deviate from a logarithmic spiral path. However, by using the mean density d_m or mean specific volume v_m in the diffuser, α will remain constant and the mean path of the air can be approximately represented by

$$R = R_2 e^{\theta \tan \alpha_2} \tag{11-32}$$

where $e = 2.718+$.

θ = angle between R_2 and R expressed in radians.

Example 7.—For the supercharger in the preceding examples, determine the absolute velocity and the approximate mean path that the air would follow in the diffuser, if it was not restricted by guide vanes.

Solution.—For the mean specific volume, experimental data on diffuser entering and leaving pressures and temperatures are needed, but if unavailable, an approximation may be made as follows:

From Example 1, the manifold pressure is 42 in. Hg abs., $R_p = 1.688$, $T_1 = 501.2$, the adiabatic horsepower $= 27.45$,

$$W = 59.5 \text{ lb. charge per minute,}$$

and for an assumed adiabatic temperature efficiency of $e_a = 0.7$,

$$\text{Fluid hp.} = \frac{27.45}{0.7} = 39.3$$

Hence an equivalent polytropic exponent based on Eq. (11-7*b*) is,

$$1.688^{(n-1)/1} = \frac{39.3 \times 33{,}000}{59.5 \times 778 \times 0.24 \times 501.2} + 1 = 1.232$$

from which $n = 1.669$ and $\dfrac{n-1}{n} = 0.4$

From Eq. (11-8), the manifold temperature is

$$T_{\text{man.}} = 501.2 \times 1.688^{0.4} = 619°\text{F. abs.}$$

and the specific volume in the manifold is

$$v_{\text{man.}} = \frac{RT}{P} = \frac{53.3 \times 619}{42 \times 0.491 \times 144} = 11.1 \text{ cu. ft. per lb.}$$

For adiabatic conditions, the specific volume leaving the impeller (Example 6) was $v_2 = 11.112$, but for $n = 1.669$, $v_2 = 15.2/1.1329^{1.495} = 12.6$. These two values for v_2 probably bracket the actual value since most sources consider the impeller to be more efficient than the diffuser. Hence it seems reasonable to assume $v_2 \approx 12$, and, for the diffuser,

$$v_m = \frac{12 + 11.1}{2} \approx 11.5 \text{ cu. ft. per lb.}$$

From Eq. (11-31), the radial component of velocity at the impeller tip is

$$v_{R2} = \frac{59.5 \times 11.5 \times 144}{60 \times 2\pi \times 3.75 \times 0.55} = 127 \text{ f.p.s.}$$

The tangential component is

$$V_{T2} = CU_2 \approx 0.85 \times 905 = 769 \text{ f.p.s.}$$

and

$$\alpha_2 = \text{arc tan } {}^{127}\!/\!_{769} = \text{arc tan } 0.1652 = 9°22'$$

$$V_2 = \frac{769}{\cos 9°22'} = 780 \text{ f.p.s.}$$

From Eq. (11-32), for $\theta = 30° = \frac{\pi}{6} = 0.525$ radian

$$R = 3.75 \times 2.718^{0.525 \times 0.1652} = 4.09 \text{ in.}$$

For constant angular momentum, for $\theta = 30°$

$$V_T = V_{T2} \times \frac{R_2}{R} \times 769 \times \frac{3.75}{4.09} = 705 \text{ f.p.s.}$$

and the absolute velocity of the air is

$$V = \frac{V_T}{\cos \alpha_2} = \frac{705}{0.9867} = 715 \text{ f.p.s.}$$

Similarly for

$\theta =$	60	90	120	566	deg.
$R =$	4.45	4.85	5.3...................	19.25	in.
$V =$	665	601	551	150	f.p.s.

These calculations show that to slow the air down to initial velocity $V_0 = V = 150$ f.p.s. in a free vortex, the diffuser would have to be abnormally large in diameter and the diffuser housing would have to be quite long. Wall friction and turbulence tend to reduce both R and θ, and if, in addition, the axial width is increased by passing the air to a spiral volute chamber of uniformly increasing circular or oval cross section, the vaneless diffuser can be used with in-line or V-engine superchargers.

Example 8.—By assuming that the radial-engine supercharger in the previous examples is to be adapted to a V-engine of equivalent size and performance, determine the diameter of the outlet from the diffuser.

Solution.—From Example 7, the specific volume at the diffuser outlet is $v_{\text{man.}} = 11.1$ cu. ft. per lb.; hence, if a mean outlet velocity equal to the inlet velocity ($V = V_0 = 150$ f.p.s.) is assumed, the diameter of the outlet is

$$\frac{\pi}{4} D^2 = \frac{11.1 \times 59.5}{150 \times 60}$$

or

$$D = 0.3055 \text{ ft.} = 3.665 \text{ in.}$$

For vaned diffusers, the passageways through the diffuser may be likened to the exit cone of a wind tunnel wherein energy losses are due to[7] wall friction, angular divergence of the cone, and kinetic energy losses as the air leaves the exit cone. Applied to diffusers, these losses are somewhat conflicting in that excessive length of the passageway will increase skin friction and the bulk of the diffuser housing, a short length with a large apex angle will increase divergence losses, and a short length with a small apex angle will increase exit losses. In addition, with diffusers, any abrupt change of direction of the air entering or leaving will be accompanied by impact losses. Hence, usual practice is to allow the air to form a free vortex from tip radius R_2 outward ½ to 1 in. to the diffuser-vane entry radius R_3 and then set the entering edges of the vanes parallel to the path of the entering air, *i.e.*, tangent to the logarithmic spiral at radius R_3. Since

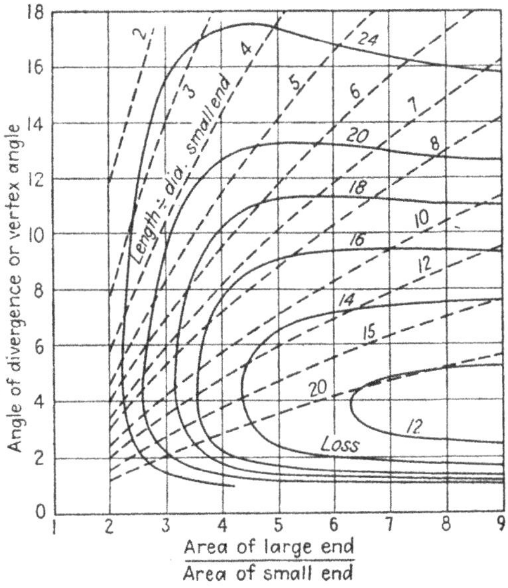

FIG. 11-24.—Percentage losses in exit cones of various forms. (*From NACA Tech. Rept.* 73.)

V_R varies with the quantity of flow, fixed vanes can be set correctly for one condition of operation only.

Diffuser vanes may be straight or curved toward the natural free path and of uniform thickness or they may be wedge shaped (Fig. 11-20). Straight vanes give the greatest deviation from the natural path, whereas curved vanes (usually arcs of circles) allow a more gradual conversion of kinetic energy, *i.e.*, the effective length of the passageway is increased and this would indicate a reduction in the losses (Fig. 11-24). However, with a limited number of vanes of constant thickness, it is difficult to attain a low enough angle of divergence or vertex angle (Fig. 11-24) to avoid excessive turbulence in the passageway. By gradually increasing the thickness of the vanes from radius R_3 to radius R_4, the angle of divergence can be kept low, and this would indicate an increased efficiency, but the increased losses resulting from the higher exit velocity from the diffuser passageway might largely offset the gain. Apparently, experimental methods are necessary to determine the best proportions.

Example 9.—By assuming circular-arc vanes of uniform thickness, lay out diffuser proportions for the radial-engine supercharger in the preceding examples.

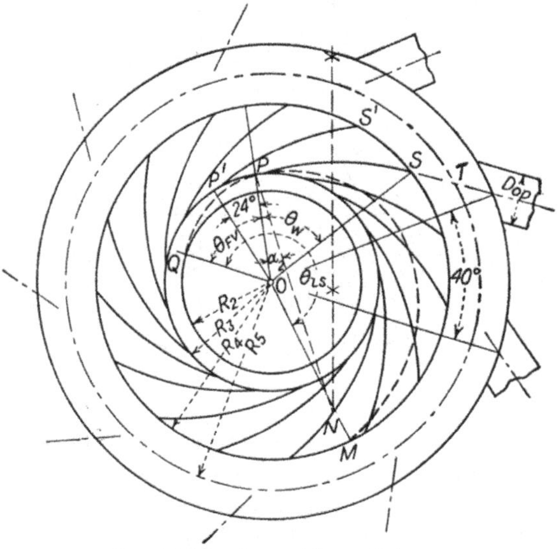

FIG. 11-25.—Layout for a diffuser.

Solution.—Referring to Fig. 11-25, assume a free vortex space of

$$R_3 - R_2 = 0.75 \text{ in.}$$

From Example 7, $\alpha_2 = 9°22'$. From Eq. (11-32), $\theta_{FV} = 63$ deg. and dotted curve *QPM* is the free path of the air without guide vanes. Lay off line *PN* normal to the free path of the air at *P*. Angle $OPN = \alpha_2$. Assume an angle of "wrap" of the vane $\theta_w = 60$ deg. and an outer radius

$$R_4 = 1.6 \times R_3 = 1.6 \times 4.5 = 7.2 \text{ in.}$$

Construct a perpendicular bisector to a line connecting *P* and *S*. The intersection of this bisector with *PN* is at *N* which is the center of arc *PS*. Arc *PS*, the desired vane shape, is parallel to the entering air at *P* and delivers the air to the diffuser collector ring along the line *ST*. Assume 15 vanes. Then angle $POP' = 360/15 = 24$ deg., and vane $P'S'$ is laid off as before. A larger number of vanes will decrease the angle of divergence but increase the skin friction; hence the optimum number appears to be a matter of experiment. Vane thickness may be made about ⅛ in. with sharp entering edges.

Air leaving the diffuser passages is usually allowed to enter a collector ring tangentially (*i.e.*, the ring is slightly offset axially) where it may be assumed to slow down to initial velocity, V_0. The volume of the collector ring may be determined by considering it as a circular torus wherein the parts are related by

$$V_{CT} = 2\pi^2 R_5 R_{CT}^2 \qquad (11\text{-}33)$$

where V_{CT} = volume of the torus ring.
R_5 = radius to center of the torus ring $= R_4 + R_{CT}$.
R_{CT} = radius of the ring.

The diameter of the diffuser collector ring ($= 2R_{CT}$) may be made four to six times the axial width of the diffuser.

Air entering tangentially into the collector ring from the diffuser passageways will move in a helical path approximately along the line *ST*, Fig. 11-25. Hence, to minimize impact losses at entry to the offtake pipes leading from the collector ring to the intake valves, these pipes should join the collector ring at an angle such that their center lines at entry are approximately in a direction corresponding to line *ST*. Turns in the offtake pipes should be of large radius, and the diameter may be made such as to maintain the velocity approximately equal to the initial velocity V_0. To accomplish this last, however, it should be borne in mind that the offtake pipes lead to individual cylinders which take in their charge during a part of the cycle only, *i.e.*, the flow is intermittent.

Example 10.—For the radial engine in the preceding examples, determine a suitable diameter of connecting pipe between the diffuser collector ring and the intake-valve port if the intake valve open time is 240 deg. of crankshaft travel per cycle.

Solution.—The flow for the engine is $\frac{59.5 \times 11.1}{60} = 11$ c.f.s.; hence the equivalent rate of flow to each cylinder is $^{11}/_{9} = 1.221$ c.f.s., but for a four-stroke-cycle engine, this amount of charge flows into the cylinder in

$$^{240}/_{720} = \text{one-third of the time.}$$

Therefore the equivalent continuous flow is $1.221 \times 3 = 3.663$ c.f.s. Then for a mean velocity of 150 f.p.s., the offtake pipe diameter is

$$D_{OP} = 12\sqrt{\frac{4 \times 3.663}{\pi \times 150}} = 2.12 \text{ in.}$$

11-9. Supercharger Drives.—For critical altitudes up to 20,000 ft. or a little more, gear-driven centrifugal superchargers have been most favored. The gearing is usually of the spur type and in two steps, although a variety of possibilities exist.

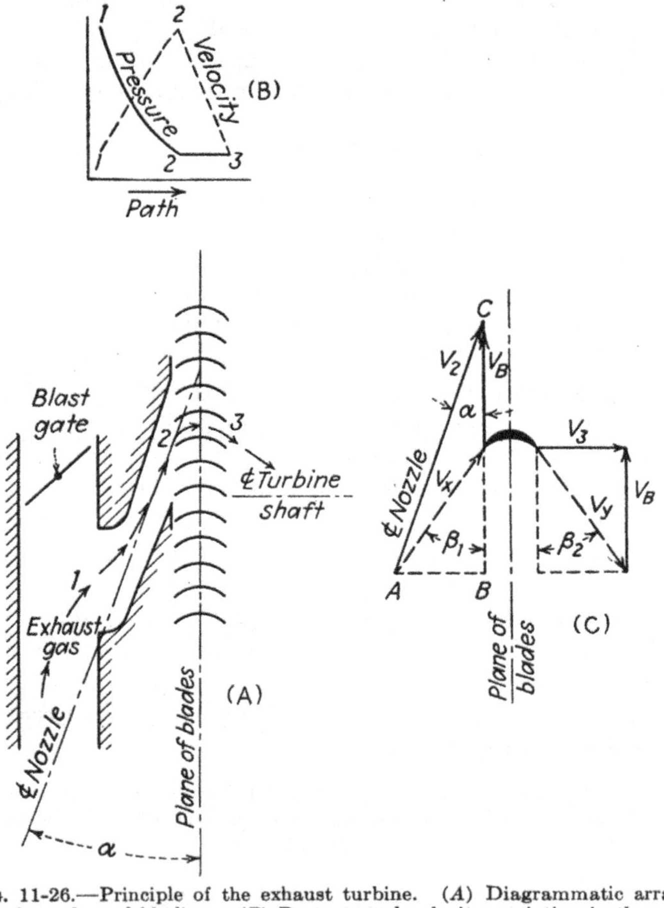

FIG. 11-26.—Principle of the exhaust turbine. (*A*) Diagrammatic arrangement of nozzle and blading. (*B*) Pressure and velocity variation in the nozzle and blading. (*C*) Velocity diagram of gases passing through the turbine.

At the speeds involved, the gears must be quite accurately made and care must be exercised to ensure precise balance and freedom from critical vibration.[9,10]

At any given speed, the problem of transmitting the necessary horsepower to the impeller is not difficult, but under conditions of rapid engine acceleration, supercharger drive-gear stresses may become excessively large, and to avoid failure, slip clutches, springs, fluid couplings, etc., are often used. When a constant-speed propeller is used, however, the problem of protecting the supercharger drive gearing against sudden acceleration loads is greatly reduced. Impeller speed and horsepower requirements have been considered in Pars. 11-5 and 11-6. Detail design of the drive gearing is generally similar to that indicated for reduction gearing in Chap. 8.

For very high altitude operation, exhaust-turbine supercharger drives have many advantages. The exhaust turbine is quite similar to a single-stage impulse steam turbine in so far as the general arrangement is concerned. Thus, in Fig. 11-26, the combustion gases at exhaust pressure P_1 and temperature T_1 enter the nozzle. At the mouth or outlet of the nozzle, the gases have expanded to a lower pressure P_2 and have increased in velocity to V_2. In passing through the turbine wheel, the velocity drops to V_3 and some of the kinetic energy given up is utilized in producing a torque on the turbine wheel.

Referring to Fig. 11-26, for the turbine nozzle, from the basic relation for adiabatic flow of gases and for an assumed negligible entering velocity, *i.e.*, $V_1 = 0$

$$\frac{V_2^2}{2g} = 778[C_p(T_1 - T_2)] \tag{11-34}$$

or

$$V_2 = 223.8\sqrt{C_p(T_1 - T_2)} \tag{11-35}$$

where V_2 = velocity of exhaust gas leaving the nozzle, f.p.s.
C_p = specific heat at constant pressure for the exhaust gases, B.t.u./(lb.)(deg. F.).
T_1 = temperature of gases in the exhaust manifold, deg. F.
T_2 = temperature of exhaust gases leaving the nozzle.

But

$$\frac{T_2}{T_1} = \left(\frac{P_2}{P_1}\right)^{(K-1)/K}$$

Hence

$$V_2 = 223.8\sqrt{C_pT_1\left[1 - \left(\frac{P_2}{P_1}\right)^{(K-1)/K}\right]} \tag{11-36}$$

where P_1 = pressure in the exhaust manifold, lb. per sq. in. abs.
P_2 = pressure of the gases leaving the nozzle, lb. per sq. in. abs.
$K = C_p/C_v$.
C_v = specific heat at constant volume for the exhaust gases, B.t.u./(lb.)(deg. F.).

For maximum absorption of kinetic energy by the turbine blades

$$\frac{V_2 \cos \alpha}{2} = V_B = \pi DN \tag{11-37}$$

where α = the angle between the center line of the nozzle and the plane of the turbine blades.
V_B = mean velocity of the blades, f.p.s.
D = mean diameter of the turbine wheel, ft.
N = r.p.s. of the turbine wheel [= r.p.s. of the impeller, Eq. (11-18)].

The efficiency of the turbine blades may be expressed in terms of the kinetic energy change, thus

$$e_B = \frac{(W_G V_2^2/2g) - (W_G V_3^2/2g)}{W_G V_2^2/2g} = 1 - \left(\frac{V_3}{V_2}\right)^2 = 1 - \sin^2 \alpha \tag{11-38}$$

where W_G = flow of exhaust gas through the turbine, lb. per sec.
V_3 = velocity of the gases leaving the turbine.

The theoretical horsepower imparted to the turbine blades is proportional to the change in kinetic energy, thus

$$\text{hp.} = \frac{W_G(V_2^2 - V_3^2)}{2g \times 550} = \frac{W_G V_2^2(1 - \sin^2 \alpha)}{2g \times 550} = \frac{W_G V_2^2}{1{,}100g} \times e_B \tag{11-39}$$

11-10. Accessories.—Accessories such as carburetors, fuel pumps, magnetos, starters, generators, tachometers, and various kinds of pressure and vacuum pumps are usually purchased from specialty manufacturers and assembled on the engine. Hence the engine designer is at liberty to select rather than design such parts. However, to ensure proper fitting and operation, he should be generally familiar with these accessories, and he must have mounting pad and drive data.

The design of mountings and drives would appear at first thought to be merely a matter of routine layout, but accessories have become so numerous in recent years that the problem of crowding them all into the available space is no longer a simple one. In fact, in many recent large installations, the space between the rear end of the crankcase and the fire wall has been literally jammed with a maze of fittings, pipes, wiring, etc., to the point where installation, inspection, and maintenance is most difficult. A trend toward relieving this condition is observable in some very large airplanes where space permits the use of auxiliary power plants, but in small and medium-sized planes, the main engine is still required to support and drive practically all the accessories. Some parts, which might be classed as primary accessories, are essential to engine operation, and in any type of airplane, these should logically be mounted close by, if not actually on, the engine.

11-11. Carburetors and Fuel Pumps.—A purely rational approach to the design of a suitable size of carburetor for an engine is complicated by a large number of variables such as the intermittent character of the flow, the Venturi characteristics, and the fuel-flow characteristics at various conditions of operation. For a preliminary selection, however, the following two formulas are recommended by the Bendix-Stromberg Carburetor Company.

For one carburetor barrel feeding three or fewer cylinders,

$$V_A = \frac{D_C \times N_R}{133{,}000} + A_{DN} \tag{11-40}$$

and for one carburetor barrel feeding four or more cylinders,

$$V_A = \frac{D_C \times n_C \times N_R}{480{,}000} + A_{DN} \tag{11-41}$$

where V_A = Venturi area, sq. in.
D_C = displacement per cylinder, cu. in.
n_C = number of cylinders.
N_R = engine speed, r.p.m.
A_{DN} = discharge nozzle area, sq. in.

Example.—Select a suitable carburetor and mounting flange for a 4.625- by 4.5-in., nine-cylinder radial engine rated 210 b.hp. at 2,000 r.p.m.

Solution.—From Eq. (11-41), the net Venturi area is

$$V_A - A_{DN} = \frac{4.625^2 \times 0.785 \times 4.5 \times 9 \times 2{,}000}{480{,}000} = 2.82 \text{ sq. in.}$$

From Table A1-24, a model Na-R7A single-barrel carburetor having a barrel diameter of $2^{11}\!/_{16}$ in. should be suitable, but for engines actually to be built, the carburetor manufacturer should be consulted for confirmation of the selection. From Table A1-25, a No. 7, $2\frac{1}{2}$-in. nominal diameter, single-barrel S.A.E. Standard carburetor flange is indicated.

Fuel may be supplied to the carburetor by gravity flow from the fuel tank or by means of a fuel pump. Fuel pumps are usually of the positive-displacement vane type with a by-pass relief valve which may be set to maintain the desired carburetor fuel supply pressure.

Fuel pumps are usually built to fit S.A.E. Standard fuel pump mounting pads (Table A1-27), the square-type pad being favored. Drive shafts designed to transmit 0.1 to 0.5 hp. (depending on the size of pump) will be adequate.

Fuel-pump capacity should be sufficient to supply at least 1 lb. of fuel per b.hp. per hour at maximum power output. Thus for an engine rated 750 b.hp. at 2,700 r.p.m. and using a fuel of 0.7 specific gravity, the fuel-pump capacity should not be less than

$$\frac{750 \times 1.0}{8.33 \times 0.7} = 129 \text{ gal. per hr.}$$

To attain this flow, a Pesco R-400 series pump (Table A1-26) would have to turn at not less than 1,600 r.p.m., *i.e.*, the ratio of pump shaft to crankshaft r.p.m. would have to be at least $1{,}600/2{,}700 \approx 0.6$. However, a somewhat higher ratio, say the normally used value of 0.875, would merely by-pass more excess fuel and allow for greater flexibility, in the control of vapor lock, but Pesco pump shaft speeds in excess of 2,500 r.p.m. are not recommended.

11-12. Magnetos, Starters, and Generators.—Aircraft-engine ignition systems may be either of the magneto or battery-generator type; magneto ignition is by far the most common. Government requirements for engines over 100 hp. dictate the use of two separate ignition systems. In the case of magneto ignition, this means two separate units, or the equivalent, *i.e.*, a double magneto.

Magnetos ordinarily have two, four, or eight poles, *i.e*, they are capable of producing two, four, or eight sparks per revolution. Thus, for a two-pole magneto on a five-cylinder, four-stroke-cycle engine, the magneto shaft should turn at 1.25 times crankshaft speed, and for a four-pole magneto on a seven-cylinder, four-stroke-cycle engine, the magneto shaft should turn at 0.875 times crankshaft speed. Tables A1-28 and A1-29 give magneto selection and mounting-pad data suitable for preliminary design, but for engines actually to be built, the magneto manufacturer should be consulted for confirmation of the selection.

The simplest method of starting an aircraft engine is by swinging the propeller. This method, though somewhat dangerous, is feasible on small, low starting torque engines, but in the larger engines, starters are very desirable and in many cases necessary.

Starters may be classed as hand-turning gear, hand inertia, hand and electric inertia, direct-cranking electric, air, and combustion types. Each type has advantages and disadvantages such as cost, convenience, and continuous availability, and selection should depend on engine size, type of energy available, type and application of the airplane, etc.

Fortunately, however, S.A.E. Standard starter motor mountings are generally used by starter manufacturers so that considerable leeway is available to the engine user. Table A1-30 lists a variety of available starters together with data on S.A.E. mounting-flange number, weight, maximum engine horsepower that the starter will handle, and voltage requirements. Table A1-31 gives data on standard S.A.E. starter motor mountings.

Engine-driven generators are usually built to fit standard S.A.E. mounting pads, and a sizable range of types and capacities can be had for a given mounting. Table A1-32 gives some generator selection and mounting-pad data.

11-13. Tachometers and Miscellaneous Accessories.—To facilitate the checking of operating conditions, practically all aircraft engines are arranged to permit the observation of crankshaft r.p.m. This involves a tachometer-drive connection in which the drive shaft generally turns at one-half crankshaft speed, presumably because early engines were of the in-line or V type, and a convenient point of attachment for the tachometer drive was the end of the camshaft. Table A1-33 gives S.A.E. Standard tachometer-drive data.

In addition to the accessories more or less commonly found on all aircraft engines, additional drives for hydraulic pumps, vacuum pumps, propeller governors, etc., frequently are needed. For the most of these miscellaneous accessories, the S.A.E. has endeavored to standardize mounting pads and flanges. Hence, the designer confronted with the problem of selecting and mounting these accessories usually will find standard mounting data in the S.A.E. "Handbook." For data on these special accessories, the designer is referred to the current literature of the accessory manufacturers.

11-14. Accessory-drive Details.—In view of the number of accessories usually required, some ingenuity on the part of the designer is necessary to obtain a simple and effective arrangement. Power for driving must come either directly or indirectly from the crankshaft, and to keep the arrangement as simple as possible, a main accessory shaft usually is splined to the rear end of the crankshaft. Drive shafts for each accessory are then geared for desired speed ratios to the main accessory shaft. Such drives can be very skillfully arranged or badly cluttered, depending upon the ingenuity of the designer. For the student, a probable best approach is to study the arrangements used in current successful engines even to the extent of sketching such arrangements as most nearly conform to the needs for his engine. This will enable him to visualize the problem more clearly and to draw on previous experience.

Suggested Design Procedure

1. Select materials and sketch in the main crankcase section on the assembly drawings or on superimposed tracing paper. Check the arrangement for weak points such as fabrication difficulties, sharp reentrant corners in highly stressed parts, unnecessarily indirect stress paths, interference of parts, assembly difficulties, etc.
2. Make detail drawings of each part of the crankcase.
3. Design and make detail drawings of the oil pump and connecting parts.
4. If a supercharger is to be built into the engine, sketch a general layout of the proposed arrangement.
5. If a supercharger is to be used, determine speed and power requirements.

6. If a supercharger is to be used, determine detail dimensions for the impeller, and make detail drawings.

7. If a supercharger is to be used, determine detail dimensions for the diffuser and make detail drawings.

8. If a supercharger is to be used, determine detail dimensions of the drive gearing and make detail drawings.

9. List all accessories that are to be included or provided for, and make sketches approximately to scale showing the proposed arrangement for attaching and operating. Study of arrangements used in current successful engines will be very helpful in planning the accessory grouping. "Mock-ups" may even be necessary as aids to visualizing the desired arrangement.

10. With the accessory grouping planned to ensure proper and effective functioning of each unit, determine necessary detail dimensions, and make detail drawings of drives, mounting pads, etc., properly arranged in relation to adjacent parts.

11. Lay out the accessory section of the crankcase, and check for rigidity, ease of fabrication, assembly difficulties, structural weaknesses, etc.

12. Make detail drawings of the accessory section of the crankcase.

13. Design and make detail drawings of all remaining miscellaneous parts.

14. Transfer all remaining detail parts to complete the assembly drawings of the engine.

15. When items 1 to 14 have been completed and put in proper form, submit for checking and approval.

Problems

1. A $5\frac{3}{16}$- by $5\frac{3}{16}$-in., 14-cylinder, 6.75-compression-ratio engine is rated 750 b.hp. at 2,550 r.p.m. at 9,500 ft. altitude (P_{atm} = 20.98 in. Hg abs., t_{atm} = 25.1°F.). The fuel rate is 0.58 lb. per b.hp. hr., and the air:fuel ratio is 12.3:1 by weight. Assuming an over-all adiabatic efficiency of 65 per cent, estimate the horsepower required to drive the supercharger.

2. For the engine in Problem 1, the impeller-crankshaft speed ratio is 11:1. Determine the impeller diameter, tip speed, and r.p.m.

3. For the engine in Problems 1 and 2, determine the inlet diameter of the impeller if the angle of approach is 10 deg., the impeller-hub diameter is 1.5 in., there are 16 impeller blades having an entering edge thickness of 0.05 in., and the mean velocity in the inlet pipe is 150 f.p.s.

4. For the engine in the preceding three problems, assume F = 1.2 in. (see Fig. 11-22), and determine the hub length and curvature.

5. For the engine in the preceding four problems, determine the width of the impeller blades for a constant area of passage.

6. Determine the effect of change in specific volume of the charge on the tip width L_2 of the impeller in Problem 5.

7. For the supercharger in the preceding problems, determine the absolute velocity and the approximate mean path that the air would follow in the diffuser if it was not restricted by guide vanes.

8. Assuming the radial engine supercharger in the preceding problems is to be adapted to a V-engine of equivalent size and performance, determine the diameter of the outlet from the diffuser.

9. Assuming circular-arc vanes of uniform thickness, determine and lay out diffuser proportions for the radial-engine supercharger in the preceding problems.

10. For the radial-engine supercharger in the preceding problems, determine suitable dimensions for the diffuser collector ring.

11. For the radial engine in the preceding problems, determine a suitable diameter of connecting pipe between the diffuser-collector ring and the intake-valve port if the intake-valve open time is 250 deg. of crankshaft travel per cycle.

12. To increase the rated altitude of the engine in Problem 1 to 25,000 ft. (P_{atm} = 11.1 in. Hg abs., t_{atm} = −30.15°F.), it is planned to use an exhaust-turbine supercharger ahead of the gear-driven supercharger. Exhaust manifold temperature = 1340°F.,

exhaust manifold pressure = 30 in. Hg abs.,

turbine-wheel mean diameter = 9 in.,

angle of nozzle to plane of blades = 20 deg.,

specific heats of exhaust gas, C_p = 0.24, C_v = 0.17. Adiabatic temperature efficiency of the air impeller driven by the turbine = 70 per cent.

a. Find the turbine speed in r.p.m.

b. Find the diameter of the air impeller necessary to compress the air from 25,000 ft. to the equivalent of 9,500 ft.

c. Find the fluid horsepower necessary for the air impeller.

d. What proportion of the exhaust gases must pass through the turbine if the over-all turbine efficiency is 50 per cent?

References

1. C. F. Taylor: An Analysis of Critical Stresses in Aircraft Engine Parts, *S.A.E. Jour.*, Vol. 37, No. 5, November, 1935.
2. Heldt: Engine Roughness—Its Cause and Cure, *S.A.E. Jour.*, Vol. 38, No. 2, February, 1936.
3. Harvey: Manufacture, Characteristics, and Uses of Magnesium Castings, *S.A.E. Jour.*, Vol. 42, No. 1, January, 1938.
4. Lansing: The Accessory-drive Problem of Aircraft Engines, *S.A.E. Jour.*, Vol. 42, No. 5, May, 1938.
5. Caminez: Fatigue Strength of Castings Improved by New Screw Thread System, *Aero Digest*, October, 1940.
6. Pye: "The Internal Combustion Engine," Vol. 2, The Aero Engine.
7. *NACA Tech. Rept.* 73.
8. Kerton: "Turbo-blowers and Compressors."
9. Wilson: "Practical Solution of Torsional Vibration Problems."
10. Tuplin: "Torsional Vibration."

TABLES IN APPENDICES

APPENDIX 1

TABLE A1-1.—AMERICAN AIRCRAFT ENGINES
(From *Automotive Ind.*, Vol. 82, No. 5, Mar. 1, 1940)

Engine make and model (*Data continued on pp. 324 and 325*)	Civil Aeronautics License or A. T. C. No.	Cylinder data: Arrangement	Cooling medium	Number of cylinders	Bore and stroke, in.	Total piston displacement, cu. in.	Compression ratio	B.m.e.p. at cruising hp., lb. per sq. in.	Blower ratio	Cylinder material	No. of valves per cylinder: Intake	No. of valves per cylinder: Exhaust	Valve arrangement	Ratings: Maximum (except take-off): Horsepower	Maximum: R.p.m.	Maximum: At sea level or altitude, ft.	Take-off: Horsepower	Take-off: R.p.m.
Akron[1]-Funk, E	201	IV-L	W	4	3⅞×4¼	200.5	6.25	117[a]	No	3	1	1	L	63	2,125	SL		
Allison, C-15	......	V-60	Liq	12	5½×6	1,710.0	6.65	146[a]	8.77	8	2	2	OH	960	2,600	12,000	1,090	3,000
Continental, Series 7, 8, 9-A-50	190	Hor	Air	4	3⅞×3⅝	171.0	5.40	122[a]	No	5	1	1	I	50	1,900	SL	50	1,900
Continental, Series 7, 8, 9-A-65	205	Hor	Air	4	3⅞×3⅝	171.0	6.30	131[a]	No	5	1	1	I	65	2,300	SL	65	2,300
Continental, Series 8, 9-A-75	213	Hor	Air	4	3⅞×3⅝	171.0	6.30	134[a]	No	5	1	1	I	75	2,600	SL	75	2,600
Continental, Series 8, 9-A-80	217	Hor	Air	4	3⅞×3⅝	171.0	7.50	137[a]	No	5	1	1	I	80	2,700	SL	80	2,700
Continental, W670-M1	168	Rad	Air	7	5⅛×4⅝	668.0	6.10	135[a]	No	5	1	1	I	250	2,200	SL	250	2,200
Continental, W670-K	162	Rad	Air	7	5⅛×4⅝	668.0	5.40	123[a]	No	5	1	1	I	225	2,175	SL	225	2,175
Continental, A-40-4	72	Hor	Air	4	3⅛×3¾	115.0	5.14	90	No	4	1	1	L	42	2,700	SL	40	2,575
Franklin,[6] 4AC150	194	Hor	Air	4	3⅝×3⅝	150.0	6.60	94	No	6	1	1	I	50	2,300	SL		
Franklin, 4AC150	194	Hor	Air	4	3⅝×3⅝	150.0	5.50	94	No	6	1	1	I	50	2,300	SL		
Franklin, 4AC171	206	Hor	Air	4	3⅞×3⅝	171.0	6.25	97	No	6	1	1	I	60	2,350	SL		
Franklin, 4AC171	206	Hor	Air	4	3⅞×3⅝	171.0	6.25	97	No	6	1	1	I	60	2,350	SL		
Franklin, 4AC171	206	Hor	Air	4	3⅞×3⅝	171.0	6.25	97	No	6	1	1	I	60	2,350	SL		
Jacobs, L-5	156	Rad	Air	7	5½×5	831.0	6.00	105	No	5	1	1	I	285	2,000	SL	300	2,125
Jacobs, L-5M	156	Rad	Air	7	5½×5	831.0	6.00	105	No	5	1	1	I	285	2,000	SL	300	2,125
Jacobs, L-5MB	156	Rad	Air	7	5½×5	831.0	6.00	105	No	5	1	1	I	285	2,000	SL	300	2,125
Jacobs, L-5MA7	156	Rad	Air	7	5½×5	831.0	6.00	105	No	5	1	1	I	235	2,000	SL	300	2,125
Jacobs, L-6	195	Rad	Air	7	5½×5½	914.0	6.00	100	No	5	1	1	I	300	2,100	3,700	330	2,200
Jacobs, L-6M	195	Rad	Air	7	5½×5½	914.0	6.00	96	No	5	1	1	I	300	2,100	3,700	330	2,200
Jacobs, L-6MB	195	Rad	Air	7	5½×5½	914.0	6.00	96	No	5	1	1	I	300	2,100	3,700	330	2,200
Jacobs, L-6MA	195	Rad	Air	7	5½×5½	914.0	6.00	96	No	5	1	1	I	300	2,100	3,700	330	2,200
Jacobs, L-4	121	Rad	Air	7	5¼×5	757.0	5.38	97	No	5	1	1	I	225	2,000	SL	245	2,200
Jacobs, L-4M	121	Rad	Air	7	5¼×5	757.0	5.38	97	No	5	1	1	I	225	2,000	SL	245	2,200
Jacobs, L-4MB	121	Rad	Air	7	5¼×5	757.0	5.38	97	No	5	1	1	I	225	2,000	SL	245	2,200

Jacobs, L-4MA7	121	Rad	Air	7	5¼×5	757.0	5.38	97	No	5	1	1	I	225	2,000	SL	245	2,200
Kinner, K-5	3	Rad	Air	5	4¼×5¼	372.0	5.00	90	No	5	1	1	I	100	1,810	SL		
Kinner, B-5	51	Rad	Air	5	4⅝×5¼	441.0	5.25	93	No	5	1	1	I	125	1,925	SL		
Kinner, R-5	153	Rad	Air	5	5×5½	540.0	5.50	92	No	5	1	1	I	160	1,850	SL		
Kinner, C-5	62	Rad	Air	5	5⅝×5¾	715.0	5.20	98	No	5	1	1	I	210	1,900	SL		
Kinner, C-7	122	Rad	Air	7	5⅝×6	1,044	5.25	98	No	5	1	1	I	300	1,800	SL		
Kinner, SC-7	175	Rad	Air	7	5⅝×6	1,044	5.50	111	9.70	5	1	1	I	350	1,900	5,000	350	1,900
Lambert,[2] R-266	38	Rad	Air	5	4¼×3¾	265.9	5.55	113[a]	No	8	1	1	I	90	2,375	900	72	
Lenape-Papoose,[3] LM-3	203	Rad	Air	3	4⅛×4	160.4	4.90	111[a]	No	5	1	1	I	52	2,325	SL		
Lenape-Papoose, LM-365	203	Rad	Air	3	4⅛×4	160.4	5.80	137[a]	No	5	1	1	I	65	2,350	SL		
Lenape-Papoose, LM-375	215	Rad	Air	3	4⅛×4	160.4	5.80	151[a]	No	5	1	1	I	75	2,150	SL		
Lenape-Brave, LM-5	212	Rad	Air	5	4⅛×4	267.0	5.00	128[a]	No	5	1	1	I	95	2,200	SL		
Lycoming,[4] O-145-A1	199	Hor	Air	4	3⅝×3½	144.5	5.65	99	No	7	1	1	I	50	2,300	SL	50	2,300
Lycoming, O-145-A2	199	Hor	Air	4	3⅝×3½	144.5	5.65	110	No	7	1	1	I	55	2,300	SL	55	2,300
Lycoming, O-145-A3	199	Hor	Air	4	3⅝×3½	144.5	5.65	110	No	7	1	1	I	55	2,300	SL	55	2,300
Lycoming, O-145-B1	210	Hor	Air	4	3⅝×3½	144.5	6.50	112	No	7	1	1	I	65	2,550	SL	65	2,550
Lycoming, O-145-B2	210	Hor	Air	4	3⅝×3½	144.5	6.50	112	No	7	1	1	I	65	2,550	SL	65	2,550
Lycoming, O-145-B3	210	Hor	Air	4	3⅝×3½	144.5	6.50	112	No	7	1	1	I	65	2,550	SL	65	2,550
Lycoming, O-145-C1	210	Hor	Air	4	3⅝×3½	144.5	6.50	108	No	7	1	1	I	75	3,100	SL	75	3,100
Lycoming, O-145-C2	210	Hor	Air	4	3⅝×3½	144.5	6.50	108	No	7	1	1	I	75	3,100	SL	75	3,100
Lycoming, O-145-C3	210	Hor	Air	4	3⅝×3½	144.5	6.50	108	No	7	1	1	I	75	3,100	SL	75	3,100
Lycoming, GO-145-C1	210	Hor	Air	4	3⅝×3½	144.5	6.50	106	No	7	1	1	I	75	3,200	SL	75	3,200
Lycoming, GO-145-C2	210	Hor	Air	4	3⅝×3½	144.5	6.50	106	No	7	1	1	I	75	3,200	SL	75	3,200
Lycoming, GO-145-C3	210	Hor	Air	4	3⅝×3½	144.5	6.50	106	No	7	1	1	I	75	3,200	SL	75	3,200
Lycoming, R-530-D1	182	Rad	Air	7	4⅝×4½	529.2	6.50	124	1.00	5	1	1	I	210	2,200	SL	220	2,300
Lycoming, R-530-D2	183	Rad	Air	7	4⅝×4½	529.2	5.50	116	1.00	5	1	1	I	200	2,200	SL	210	2,300
Lycoming, R-680-B4C	108	Rad	Air	9	4⅝×4½	680.4	5.50	101	1.00	5	1	1	I	225	2,100	SL	225	2,100
Lycoming, R-680-D5	172	Rad	Air	9	4⅝×4½	680.4	6.50	122	1.00	5	1	1	I	245	2,100	SL	260	2,300
Lycoming, R-680-D6	173	Rad	Air	9	4⅝×4½	680.4	5.50	113	1.00	5	1	1	I	225	2,100	SL	245	2,300
Lycoming, R-680-E1	202	Rad	Air	9	4⅝×4½	680.4	6.20	122	1.00	5	1	1	I	275	2,200	SL	290	2,300
Lycoming, R-680-E2	202	Rad	Air	9	4⅝×4½	680.4	5.50	116	1.00	5	1	1	I	265	2,200	SL	280	2,300
Lycoming, R-680-E3	202	Rad	Air	9	4⅝×4½	680.4	7.00	129	1.00	5	1	1	I	285	2,200	SL	300	2,300
Menasco-Pirate, C4	67	IV-L	Air	4	4¾×5⅛	363.0	5.50	105	No	5	1	1	I	125	2,175	SL	125	2,175
Menasco-Pirate, C4S	134	IV-L	Air	4	4¾×5⅛	363.0	5.50	121	9.60	5	1	1	I	150	2,260	3,000	158	2,250
Menasco-Buccaneer, B6	68	IV-L	Air	6	4½×5⅛	489.0	5.50	108	No	5	1	1	I	160	1,975	SL	160	1,975
Menasco-Buccaneer, B6S	139	IV-L	Air	6	4½×5⅛	489.0	5.50	121	8.75	5	1	1	I	200	2,250	4,500	212	2,250
Menasco-Super Buccaneer, C6S-4	197	IV-L	Air	6	4¾×5⅛	544.0	5.50	132	10.90	5	1	1	I	260	2,300	7,500	290	2,400
Menasco-Unitwin, 2-544	Pend.	IV-L[g]	Air	12	4¾×5⅛	1,088.0	5.50	147	10.90	5	1	1	I	600	2,400	5,400	660	2,480
Pratt & Whitney-Wasp Jr., SB2	....	Rad	Air	9	5 3/16×5 3/16	985.0	6.00	121	10.00	5	1	1	I	400	2,000	5,000	450	2,300
Pratt & Whitney-Wasp Jr., TB2	....	Rad	Air	9	5 3/16×5 3/16	985.0	6.00	121	8.00	5	1	1	I	420	2,200	SL	440	2,300
Pratt & Whitney-Wasp, S1H1	129	Rad	Air	9	5¾×5¾	1,344.0	6.00	118	12.00	5	1	1	I	550	2,200	8,000	600	2,250
Pratt & Whitney-Wasp, S2H1	163	Rad	Air	9	5¾×5¾	1,344.0	6.00	111	12.00	5	1	1	I	500	2,200	10,500	600	2,250
Pratt & Whitney-Wasp, S3H1	143	Rad	Air	9	5¾×5¾	1,344.0	6.00	118	10.00	5	1	1	I	550	2,200	5,000	600	2,250
Pratt & Whitney-Wasp, S1H1-G	129	Rad	Air	9	5¾×5¾	1,344.0	6.00	118	12.00	5	1	1	I	550	2,200	8,000	600	2,250
Pratt & Whitney-Wasp, S2H1-G	141	Rad	Air	9	5¾×5¾	1,344.0	6.00	111	12.00	5	1	1	I	500	2,200	10,500	600	2,250
Pratt & Whitney-Wasp, S3H1-G	142	Rad	Air	9	5¾×5¾	1,344.0	6.00	118	10.00	5	1	1	I	550	2,200	5,000	600	2,250

TABLE A1-1.—AMERICAN AIRCRAFT ENGINES.—(*Continued*)

Engine make and model (*Data continued from pp. 322 and 323*)	Ratings				Weight, lb.		Carburetors		Ignition system			Starting		Installation dimensions, in.				
	Cruising													Over-all				
	Horsepower	R.p.m.	Octane rating of fuel required	Propeller drive	Engine, dry, without hub or starter	Per cruising hp.	Number	Make fitted	Make	Current source	Number	Make	Method	Length	Height or O. D.	Width	Height above engine bed, in.	Diameter mount. ring or distance between bearers
Akron[1]-Funk, E	.....	.....	73	D	255	4.05[a]	1	Zen	Bos	Mag	1	.....	PS	35	25	16		
Allison, C-15	720	2,230	100	G	1,340	1.86	1	Str	Scin	Mag	1	Ecl	EM	98 17/32	41 3/16	29 9/32	23 5/16	18 3/8
Continental, Series 7, 8, 9-A-50	.....	.....	65	D	160	3.20[a]	1	Str	Scin	Mag	1	Ecl[b]	EM	30 13/32	23	31 11/16		
Continental, Series 7, 8, 9-A-65	.....	.....	73	D	160	2.46[a]	1	Str	Scin	Mag	1	Ecl[b]	EM	30 13/32	23	31 11/16		
Continental, Series 8, 9-A-75	.....	.....	73	D	170	2.27[a]	1	Str	Scin	Mag	2	Ecl[b]	EM	30 13/32	23	31 11/16		
Continental, Series 8, 9-A-80	.....	.....	80	D	173	2.16[a]	1	Str	Scin	Mag	2	Ecl[b]	EM	30 13/32	23	31 11/16		
Continental, W670-M1	.....	.....	80	D	460	1.84[a]	1	Mar	Scin	M-B	2	Ecl[b]	EM	34 3/16	42 1/2	.....	.....	20
Continental, W670-K	.....	.....	65	D	460	2.04[a]	1	Str	Scin	M-B	2	Ecl[b]	EM	34 3/16	42 1/2	.....	.....	20
Continental, A-40-4	30	2,300	65	D	145	4.83[a]	1	.....	Scin	Mag	1	.....	.....	14 1/4	21 5/16	25 1/2	.....	8 1/4
Franklin,[6] 4AC150	37 1/2	2,100	70	D	161	4.30	1	MS	SE	Mag	1	.....	PS	27 31/32	18 3/8	34 3/8	11 3/16	10
Franklin, 4AC150	37 1/2	2,100	70	D	172	4.60	1	MS	SE	Mag	2	.....	PS	28	18 3/8	34 3/8	11 3/16	10
Franklin, 4AC171	45	2,150	70	D	161	3.58	1	MS	SE	Mag	1	.....	PS	27 31/32	18 3/8	34 3/8	11 3/16	10
Franklin, 4AC171	45	2,150	70	D	172	3.82	1	MS	SE	Mag	2	.....	PS	28	18 3/8	34 3/8	11 3/16	10
Franklin, 4AC171	45	2,150	70	D	204	4.55	1	MS	AE	BM	2	Au	EM	28 9/16	22 3/16	34 3/8	14 15/16	10
Jacobs, L-5	210	1,900	73	D	510	2.43	1	Str	Scin	Bat	2	Ecl	EM	19	43 1/2	.....	.....	16 1/2
Jacobs, L-5M	210	1,900	73	D	504	2.40	1	Str	Scin	Mag	2	Ecl	EM	19	43 1/2	.....	.....	16 1/2
Jacobs, L-5MB	210	1,900	73	D	525	2.50	1	Str	Scin	[l]	[l]	Ecl	EM	19	43 1/2	.....	.....	16 1/2
Jacobs, L-5MA7	210	1,900	73	D	510	2.43	1	Str	Scin	Mag	2	Ecl	EM	19	43 1/2	.....	.....	16 1/2
Jacobs, L-6	220	1,900	80	D	535	2.43	1	Str	Scin	Bat	2	Ecl	EM	19 7/8	45 3/8	.....	.....	16 1/2
Jacobs, L-6M	210	1,900	80	D	550	2.62	1	Str	Scin	Mag	2	Ecl	EM	19 7/8	45 3/8	.....	.....	16 1/2
Jacobs, L-6MB	210	1,900	80	D	550	2.62	1	Str	Scin	[l]	[l]	Ecl	EM	19 7/8	45 3/8	.....	.....	16 1/2
Jacobs, L-6MA	210	1,900	80	D	550	2.62	1	Str	Scin	Mag	2	Ecl	EM	19 7/8	45 3/8	.....	.....	16 1/2
Jacobs, L-4	175	1,900	73	D	480	2.74	1	Str	Scin	Bat	2	Ecl	EM	19	43 1/2	.....	.....	16 1/2
Jacobs, L-4M	175	1,900	73	D	473	2.70	1	Str	Scin	Mag	2	Ecl	EM	19	43 1/2	.....	.....	16 1/2
Jacobs, L-4MB	175	1,900	73	D	494	2.82	1	Str	Scin	[l]	[l]	Ecl	EM	19	43 1/2	.....	.....	16 1/2

Jacobs, L-4MA7	175	1,900	73	D	476	2.72	1	Str	Scin	Mag	2	....	....	19	43 1/2	......	......	16 1/2
Kinner, K-5	70	1,650	73	D	275	3.93	1	Str	BS	Mag	2	Ecl	Air	32 5/16	45 3/4			
Kinner, B-5	89	1,725	73	D	295	3.31	1	Str	Scin	Mag	2	Ecl	A-DE	32 5/16	45 3/8			
Kinner, R-5	113	1,800	73	D	330	2.92	1	Str	Scin	Mag	2	Ecl	A-DE	32 5/16	45 11/16			
Kinner, C-5	150	1,700	73	D	420	2.80	1	Str	Scin	Mag	2	Ecl	El	34 13/16	50			
Kinner, C-7	215	1,650	73	D	600	2.79	1	Str	Scin	Mag	2	Ecl	DE	44 1/8	51 5/16			
Kinner, SC-7	250	1,700	73	D	650	2.60	1	Str	Scin	Mag	2	Ecl	DE	45 13/32	47 3/4			
Lambert,[2] R-266	63	....	73	D	224	3.56	1	Str	Ben	Mag	2	....	PE	30 3/8	34 1/4	......	11	13 3/4
Lenape-Papoose,[3] LM-3	....	....	73	D	138	2.65[a]	1	Str	ES	Mag	2	Ecl	EM	24	......	36	......	12 1/2
Lenape-Papoose, LM-365	....	....	73	D	140	2.15[a]	1	Str	ES	Mag	2	Ecl	EM	24	......	36	......	12 1/2
Lenape-Papoose, LM-375	....	....	73	D	142	1.89[a]	1	Str	ES	Mag	2	Ecl	EM	24	......	36	......	12 1/2
Lenape-Brave, LM-5	....	....	73	D	192	2.02[a]	1	Str	ES	Mag	2	Ecl	EM	24	......	36	......	12 1/2
Lycoming,[1] O-145-A1	38	2,100	73	D	152	4.00	1	Str	Scin	Mag	1	....	....	23 3/4	19 5/16	29 9/16	9 15/32	10 1/4
Lycoming, O-145-A2	42	2,100	73	D	162	3.86	1	Str	Scin	Mag	2	....	....	24 5/8	19 5/16	29 9/16	9 15/32	10 1/4
Lycoming, O-145-A3	42	2,100	73	D	164	3.91	1	Str	Scin	Mag	2	Ecl	EM	24 5/8	22 1/8	29 9/16	9 15/32	10 1/4
Lycoming, O-145-B1	47	2,300	73	D	155	3.30	1	Str	Scin	Mag	1	....	....	23 3/4	20 1/8	29 9/16	9 15/32	10 1/4
Lycoming, O-145-B2	47	2,300	73	D	165	3.51	1	Str	Scin	Mag	2	....	....	25 5/8	20 1/8	29 9/16	9 15/32	10 1/4
Lycoming, O-145-B3	47	2,300	73	D	167	3.55	1	Str	Scin	Mag	2	Ecl	EM	24 5/8	22 1/8	29 9/16	9 15/32	10 1/4
Lycoming, O-145-C1	55	2,800	73	D	155	2.82	1	Str	Scin	Mag	1	....	....	23 3/4	20 1/8	29 9/16	9 15/32	10 1/4
Lycoming, O-145-C2	55	2,800	73	D	165	3.00	1	Str	Scin	Mag	2	....	....	24 5/8	22 1/8	29 9/16	9 15/32	10 1/4
Lycoming, O-145-C3	55	2,800	73	D	167	3.04	1	Str	Scin	Mag	2	Ecl	EM	24 5/8	22 1/8	29 9/16	9 15/32	10 1/4
Lycoming, GO-145-C1	56	2,900	73	G	181	3.24	1	Str	Scin	Mag	1	....	....	28 5/16	20 1/8	29 9/16	9 35/64	10 1/4
Lycoming, GO-145-C2	56	2,900	73	G	191	3.41	1	Str	Scin	Mag	2	....	....	29 3/16	20 1/8	29 9/16	9 35/64	10 1/4
Lycoming, GO-145-C3	56	2,900	73	G	193	3.45	1	Str	Scin	Mag	2	Ecl	EM	29 3/16	22 1/8	29 9/16	9 35/64	10 1/4
Lycoming, R-530-D1	165[h]	2,000	80	D	430	2.61	1	Str	Scin[i]	Mag	1	Ecl	EM	34 17/64	......	43 1/2	......	19 1/4
Lycoming, R-530-D2	155[h]	2,000	73	D	430	2.77	1	Srt	Scin[i]	Mag	1	Ecl	EM	34 17/64	......	43 1/2	......	19 1/4
Lycoming, R-680-B4C	165[h]	1,900	73	D	492	2.98	1	Str	Scin[i]	Mag	1	Ecl	EM	37 3/64	......	43 1/2	......	19 1/4
Lycoming, R-680-D5	210[h]	2,000	80	D	505	2.40	1	Str	Scin[i]	Mag	1	Ecl	EM	37 3/64	......	43 1/2	......	19 1/4
Lycoming, R-680-D6	195[h]	2,000	73	D	505	2.59	1	Str	Scin[i]	Mag	1	Ecl	EM	37 3/64	......	43 1/2	......	19 1/4
Lycoming, R-680-E1	210[h]	2,000	80	D	514	2.45	1	Str	Scin[i]	Mag	1	Ecl	EM	37 3/64	......	43 1/2	......	19 1/4
Lycoming, R-680-E2	200[h]	2,000	73	D	514	2.57	1	Str	Scin[i]	Mag	1	Ecl	EM	37 3/64	......	43 1/2	......	19 1/4
Lycoming, R-680-E3	222[h]	2,000	87	D	514	2.31	1	Str	Scin[i]	Mag	1	Ecl	EM	37 3/64	......	43 1/2	......	19 1/4
Menasco-Pirate, C4	95	1,975	73	D	284	2.99	1	Str	Ben	Mag	2	Ecl	HC	47 17/32	28 9/16	17 25/32	......	16 5/8
Menasco-Pirate, C4S	112	2,025	73	D	299	2.67	1	Str	Ben	Mag	2	Ecl	HC	47 17/32	28 9/16	14 1/2	......	16 5/8
Menasco-Buccaneer, B6	120	1,800	73	D	398	3.32	1	Str	Bos	Mag	2	....	PS	59 7/16	28 5/16	11 7/8	......	11 7/8
Menasco-Buccaneer, B6S	150	2,025	73	D	423	2.82	1	Str	Bos	Mag	2	....	PS	60 1/8	28 5/16	14 29/32	......	11 7/8
Menasco-Super Buccaneer, C6S-4	195	2,150	87	D	550	2.82	1	Str	Scin	Mag	2	Ecl	EM	67 15/32	30 19/32	16 9/16	......	13 1/2
Menasco-Unitwin, 2-544	450	2,230	87	G	1,350	3.00	2	Str	Scin	Mag	4	Ecl	EM	87 1/8	31 1/2	34 5/8	......	30
Pratt & Whitney-Wasp Jr., SB2	300	2,000	80[c]	D	690	2.30	1	Str	Scin	Mag	2	Opt	Opt	41 19/32	45 3/4	......	......	23 3/8
Pratt & Whitney-Wasp Jr., TB2	300	2,000	80	D	690	2.30	1	Str	Scin	Mag	2	Opt	Opt	41 19/32	45 3/4	......	......	23 3/8
Pratt & Whitney-Wasp, S1H1	400	2,000	87	D	864	2.16	1	Str	Scin	Mag	2	Opt	Opt	42 15/16	51 7/16	......	......	23 3/8
Pratt & Whitney-Wasp, S2H1	375	2,000	80[c]	D	864	2.30	1	Str	Scin	Mag	2	Opt	Opt	42 15/16	51 7/16	......	......	23 3/8
Pratt & Whitney-Wasp, S3H1	400	2,000	80[c]	D	864	2.16	1	Str	Scin	Mag	2	Opt	Opt	42 15/16	51 7/16	......	......	23 3/8
Pratt & Whitney-Wasp, S1H1-G	400	2,000	87	G	930	2.33	1	Str	Scin	Mag	2	Opt	Opt	47 47/64	51 7/16	......	......	23 3/8
Pratt & Whitney-Wasp, S2H1-G	375	2,000	80[c]	G	930	2.48	1	Str	Scin	Mag	2	Opt	Opt	47 47/64	51 7/16	......	......	23 3/8
Pratt & Whitney-Wasp, S3H1-G	400	2,000	80[c]	G	930	2.33	1	Str	Scin	Mag	2	Opt	Opt	47 47/64	51 7/16	......	......	23 3/8

TABLE A1-1.—AMERICAN AIRCRAFT ENGINES.—(*Continued*)

Engine make and model (*Data continued on pp. 328 and 329*)	Civil Aeronautics License or A. T. C. No.	Cylinder data: Arrangement	Cooling medium	Number of cylinders	Bore and stroke, in.	Total piston displacement, cu. in.	Compression ratio	B.m.e.p. at cruising hp., lb. per sq. in.	Blower ratio	Cylinder material	No. of valves per cylinder: Intake	Exhaust	Valve arrangement	Ratings: Maximum (except take-off): Horsepower	R.p.m.	At sea level or altitude, ft.	Take-off: Horsepower	R.p.m.
Pratt & Whitney-Hornet, S1E	Mil	Rad	Air	9	$6\frac{1}{8} \times 6\frac{3}{8}$	1,690.0	6.50	123	12.00	5	1	1	I	750	2,250	7,000	875	2,300
Pratt & Whitney-Hornet, S5E	164	Rad	Air	9	$6\frac{1}{8} \times 6\frac{3}{8}$	1,690.0	6.50	123	12.00	5	1	1	I	700	2,050	6,000	700	2,050
Pratt & Whitney-Hornet, T2E	99	Rad	Air	9	$6\frac{1}{8} \times 6\frac{3}{8}$	1,690.0	6.00	117	10.00	5	1	1	I	650	2,000	SL	650	2,000
Pratt & Whitney-Hornet, S1E3-G	193	Rad	Air	9	$6\frac{1}{8} \times 6\frac{3}{8}$	1,690.0	6.50	123	10.00	5	1	1	I	750	2,250	7,000	875	2,300
Pratt & Whitney-Twin Wasp Jr., SB4-G	179	Rad	Air	14	$5\frac{3}{16} \times 5\frac{3}{16}$	1,535.0	6.75	120	11.00	5	1	1	I	750	2,550	9,500	825	2,625
Pratt & Whitney-Twin Wasp, SC3-G	186	Rad	Air	14	$5\frac{1}{2} \times 5\frac{1}{2}$	1,830.0	6.70	125	7.15	5	1	1	I	900	2,550	12,000	1,050	2,700
Pratt & Whitney-Twin Wasp, S1C3-G	186	Rad	Air	14	$5\frac{1}{2} \times 5\frac{1}{2}$	1,830.0	6.70	130	7.15	5	1	1	I	1,050	2,550	7,500	1,200	2,700
Pratt & Whitney-Twin Wasp, S3C3-G	Mil	Rad	Air	14	$5\frac{1}{2} \times 5\frac{1}{2}$	1,830.0	6.70	137	8.00	5	1	1	I	950	2,700	14,300	1,100	2,700
Pratt & Whitney-Twin Wasp, S3C4-G	Mil	Rad	Air	14	$5\frac{1}{2} \times 5\frac{1}{2}$	1,830.0	6.70	139	7.15	5	1	1	I	1,100	2,550	6,200	1,200	2,700
	(2-speed supercharger engine)					......	....	139	8.47	..	..	..	....	1,000	2,700	14,500		
Pratt & Whitney-Twin Wasp, S4C4-G	Mil	Rad	Air	14	$5\frac{1}{2} \times 5\frac{1}{2}$	1,830.0	6.70	139	7.15	5	1	1	I	1,050	2,550	7,500	1,200	2,700
	(2-speed supercharger engine)					......	....	125	8.47	..	..	..	....	900	2,550	15,400		
Ranger, 6-440C-2	216	IV-L	Air	6	$4\frac{1}{8} \times 5\frac{1}{2}$	441.0	6.00	102	No	5	1	1	OH	175	2,450	SL		
Ranger, 6-410B-3	187	IV-L	Air	6	$4\frac{1}{8} \times 5\frac{1}{8}$	411.0	6.50	107	No	5	1	1	OH	165	2,450	SL		
Ranger, V-770B-4	184	IV-V	Air	12	$4 \times 5\frac{1}{8}$	773.0	6.50	110	No	5	1	1	OH	305	2,300	SL	315	2,400
Ranger, V-820C-1	......	IV-V	Air	12	$4\frac{1}{8} \times 5\frac{1}{8}$	822.0	7.50	116	No	5	1	1	OH	375	2,400	SL		
Ranger, SGV-770B-3	84	IV-V	Air	12	$4 \times 5\frac{1}{8}$	773.0	6.00	123	7.71	5	1	1	OH	420	2,800	SL		
Ranger, SGV-770B-6	207	IV-V	Air	12	$4 \times 5\frac{1}{8}$	773.0	6.00	123	8.84	5	1	1	OH	420	2,850	5,700	500	2,950
Ranger, SGV-770B-5	185	IV-V	Air	12	$4 \times 5\frac{1}{8}$	773.0	6.00	123	8.84	5	1	1	OH	420	2,800	3,000	450	2,900
Rover, L-267	37	IV-L	Air	4	$4\frac{1}{8} \times 5$	267.2	5.10	113[a]	No	3	1	1	I	75	1,975	SL		
Sky,[5] 70	200	IV-L	Air	4	$3\frac{7}{8} \times 4\frac{1}{4}$	200.0	6.25	94	No	1	1	1	OH	60	2,050	SL		
Warner-Super Scarab, 50	104	Rad	Air	7	$4\frac{5}{8} \times 4\frac{1}{4}$	499.0	5.55	112[a]	No	5	1	1	I	145	2,050	SL	145	2,050
Warner-Scarab, 50	2	Rad	Air	7	$4\frac{1}{4} \times 4\frac{1}{4}$	422.0	5.20	115[a]	No	5	1	1	I	125	2,050	SL	125	2,050
Warner-Scarab Jr., 50	54	Rad	Air	5	$4\frac{5}{8} \times 4\frac{1}{4}$	301.0	5.20	117[a]	No	5	1	1	I	90	2,025	SL	90	2,025
Warner-Super Scarab, 165	214	Rad	Air	7	$4\frac{5}{8} \times 4\frac{1}{4}$	500.0	6.40	123[a]	No	5	1	1	I	165	2,100	SL	175	2,250
Wright, Cyclone GR-2600-A2B	176	Rad	Air	14	$6\frac{1}{8} \times 6\frac{5}{16}$	2,603.0	6.30	144	7.00	5	1	1	I	1,350	2,300	5,800	1,600	2,400

Wright, Cyclone GR-2600-A5B	Pend	Rad	Air	14	6⅛×6 5/16	2,603.0	6.30	144	7.14	5	1	1	I	1,350	2,300	5,000	1,600	2,400
	(2-speed supercharger engine)								10.00					1,275	2,300	11,500		
Wright, Cyclone GR-1820-G202A	Pend	Rad	Air	9	6⅛×6⅞	1,823.0	6.70	141	7.00	5	1	1	I	1,000	2,300	4,500	1,200	2,500
Wright, Cyclone GR-1820-G203A	Pend	Rad	Air	9	6⅛×6⅞	1,823.0	6.70	141	8.30	5	1	1	I	950	2,300	9,100	1,100	2,500
Wright, Cyclone GR-1820-G205A	Pend	Rad	Air	9	6⅛×6⅞	1,823.0	6.70	141	7.14	5	1	1	I	1,000	2,300	4,500	1,200	2,500
	(2-speed supercharger engine)								10.00					900	2,300	14,000		
Wright, Cyclone GR-1820-G102A	169	Rad	Air	9	6⅛×6⅞	1,823.0	6.30	143	7.00	5	1	1	I	900	2,300	6,700	1,100	2,350
Wright, Cyclone GR-1820-G103A	180	Rad	Air	9	6⅛×6⅞	1,823.0	6.30	143	8.31	5	1	1	I	860	2,300	11,100	1,000	2,350
Wright, Cyclone GR-1820-G105A	192	Rad	Air	9	6⅛×6⅞	1,823.0	6.30	143	7.14	5	1	1	I	900	2,300	6,700	1,100	2,350
	(2-speed supercharger engine)								10.00					775	2,300	17,300		
Wright, [e]Cyclone R-1820-G2	157	Rad	Air	9	6⅛×6⅞	1,823.0	6.45	137	7.00	5	1	1	I	850	2,100	5,800	1,000	2,200
Wright, [e]Cyclone R-1820-G3	171	Rad	Air	9	6⅛×6⅞	1,823.0	6.45	137	8.31	5	1	1	I	840	2,100	8,700	875	2,200
Wright, [e]Cyclone R-1820-G3B	171	Rad	Air	9	6⅛×6⅞	1,823.0	6.20	137	8.31	5	1	1	I	820	2,100	8,800	900	2,350
Wright, [e]Cyclone R-1820-G5	154	Rad	Air	9	6⅛×6⅞	1,823.0	6.45	137	7.14	5	1	1	I	800	2,100	6,000	1,000	2,200
	(2-speed supercharger engine)								10.00					750	2,100	15,200		
Wright, [f,e]Cyclone R-1820-F52	148	Rad	Air	9	6⅛×6⅞	1,823.0	6.40	126	7.00	5	1	1	I	775	2,100	5,800	890	2,200
Wright, [e]Cyclone R-1820-F53	149	Rad	Air	9	6⅛×6⅞	1,823.0	6.40	126	8.31	5	1	1	I	745	2,100	9,600	785	2,200
Wright, [e]Cyclone R-1820-F55	198	Rad	Air	9	6⅛×6⅞	1,823.0	6.40	126	[n]	5	1	1	I	760	2,100	5,800	875	2,200
Wright, [e]Cyclone R-1820-F56	159	Rad	Air	9	6⅛×6⅞	1,823.0	6.40	126	8.83	5	1	1	I	755	2,100	11,300	785	2,200
Wright, [f,e]Cyclone R-1820-F62	148	Rad	Air	9	6⅛×6⅞	1,823.0	6.40	126	7.00	5	1	1	I	760	2,100	5,800	900	2,350
Wright, Whirlwind R-975-E1	87	Rad	Air	9	5×5½	973.0	6.00	121	7.81	5	1	1	I	365	2,100	SL	365	2,100
Wright, Whirlwind R-975-E3	125	Rad	Air	9	5×5½	973.0	6.30	132	10.15	5	1	1	I	420	2,200	1,400	450	2,250
Wright, Whirlwind R-760-ET	126	Rad	Air	7	5×5½	756.0	6.10	110	No	5	1	1	I	235	2,000	SL	235	2,000
Wright, Whirlwind R-760-E1	94	Rad	Air	7	5×5½	756.0	6.10	118	7.05	5	1	1	I	285	2,100	SL	300	2,250
Wright, Whirlwind R-760-E2	155	Rad	Air	7	5×5½	756.0	6.30	118	9.17	5	1	1	I	320	2,200	SL	300	2,400

Symbols and Abbreviations

(*Continued on p.* 329)

General

[a] Based on maximum horsepower.

[b] Optional.

[c] 87 octane rating for take-off.

[d] Applies to motor with 0.5625 propeller-gear ratio.

[e] Gear-drive engines also available at same ratings.

[f] May be equipped with two-speed supercharger.

[g] Unit consists of two six-cylinder motors placed side by side at 8 deg. angle.

[h] Rate with controllable propeller at 2,000 ft.

[i] Dual type.

[j] May also be obtained for use with 87 octane fuel with slightly lower ratings.

[k] May also be obtained for use with 95 octane fuel with same ratings.

[l] One magneto, one battery.

[m] 75 per cent power allowable.

[n] 7.14 and 10.00:1 blower ratios.

Liq—Liquid.

Mil—Military.

Pend—A. T. C. pending.

W—Water.

Cylinder Arrangement

Hor—Horizontal opposed.
IV-L—Inverted-in-line.
IV-V—Inverted-V type.
Rad—Radial.

Cylinder Material

1—Aluminum with cast-iron liner.
2—Aluminum with steel liner.
3—Cast iron.
4—Cast iron with aluminum head.
5—Steel with aluminum head.
6—Aluminum with nickel iron liner.
7—Cast semisteel with aluminum head.
8—Nickel iron with aluminum head.

TABLE A1-1.—AMERICAN AIRCRAFT ENGINES.—(*Continued*)

Engine make and model (*Data continued from pp. 326 and 327*)	Ratings			Propeller drive	Weight, lb.		Carburetors		Ignition system			Starting		Installation dimensions, in.			Height above engine bed, in.	Diameter mount. ring or distance between bearers
	Cruising		Octane rating of fuel required		Engine, dry, without hub or starter	Per cruising hp.	Number	Make fitted	Make	Current source	Number	Make	Method	Over-all				
	Horsepower	R.p.m.												Length	Height or O. D.	Width		
Pratt & Whitney-Hornet, S1E	525	2,000	87	D	975	1.86	1	Str	Scin	Mag	2	Opt	Opt	44 15/64	54 7/16	......	......	23 3/8
Pratt & Whitney-Hornet, S5E	500	1,900	87	D	975	1.95	1	Str	Scin	Mag	2	Opt	Opt	44 15/64	54 7/16	......	......	23 3/8
Pratt & Whitney-Hornet, T2E	450	1,800	80	D	975	2.16	1	Str	Scin	Mag	2	Opt	Opt	44 15/64	54 7/16	......	......	23 3/8
Pratt & Whitney-Hornet, S1E3-G	525	2,000	87	G	1,087	2.07	1	Str	Scin	Mag	2	Opt	Opt	50 33/64	54 7/16	......	......	23 3/8
Pratt & Whitney-Twin Wasp Jr., SB4-G	525	2,250	87	G	1,133	2.16	1	Str	Scin	Mag	2	Opt	Opt	53 17/64	44 1/8	......	......	23 3/8
Pratt & Whitney-Twin Wasp, SC3-G	650	2,250	87	G	1,450[d]	2.23	1	Str	Scin	Mag	2	Opt	Opt	60 5/16[d]	48	......	......	27
Pratt & Whitney-Twin Wasp, S1C3-G	700	2,325	90	G	1,450[d]	2.07	1	Str	Scin	Mag	2	Opt	Opt	60 5/16[d]	48	......	......	27
Pratt & Whitney-Twin Wasp, S3C3-G	700	2,200	100	G	1,465[d]	2.09	1	Str	Scin	Mag	2	Opt	Opt	61 1/2[d]	48	......	......	27
Pratt & Whitney-Twin Wasp, S3C4-G	700	2,175	100	G	1,490[d]	2.13	1	Str	Scin	Mag	2	Opt	Opt	63 5/16[d]	48	......	......	27
	700	2,175																
Pratt & Whitney-Twin Wasp, S4C4-G	700	2,175	90	G	1,490[d]	2.13	1	Str	Scin	Mag	2	Opt	Opt	63 5/16[d]	48	......	......	27
	650	2,250	...	...	......	2.29												
Ranger, 6-440C-2	130	2,300	65	D	355	2.73	1	Str	Scin	Mag	2	Ecl	EM	51 1/16	31 3/4	20 1/4	12 1/8	11 7/8
Ranger, 6-410B-3	125	2,250	80	D	345	2.76	1	Str	Scin	Mag	2	Ecl	EM	50 3/32	31 9/16	20 13/32	12 1/8	11 7/8
Ranger, V-770B-4	230	2,150	80	D	565	2.46	2	Str	Scin[i]	Mag	1	Ecl	EM	61	27 3/16	28	10 1/2	12 1/4
Ranger, V-820C-1	260	2,150	87	D	565	2.17	2	Str	Scin[i]	Mag	1	Ecl	EM	61	27 3/16	28	10 1/2	13 1/4
Ranger, SGV-770B-3	300	2,500	87	G	640	2.13	1	Str	Scin[i]	Mag	1	Ecl	EM	61 5/16	32 1/4	28	12 1/2	13 1/4
Ranger, SGV-770B-6	300	2,500	87	G	640	2.13	1	Str	Scin[i]	Mag	1	Ecl	EM	61 5/16	32 1/4	28	12 1/2	13 1/4
Ranger, SGV-770B-5	300	2,500	87	G	640	2.13	1	Str	Scin[i]	Mag	1	Ecl	EM	61 5/16	32 1/4	28	12 1/2	13 1/4
Rover, L-267	.....	.....	78	D	225	3.00[a]	1	Str	Scin	Mag	2	.....	.....	42 1/4	31 7/8	16 1/2	4 19/32	14
Sky,[5] 70	45	1,900	73	D	178	3.96	1	Link	ES	Mag	2	.....	PS	40	27	19	5	13 1/2
Warner-Super Scarab, 50	[m]	.....	73	D	305	2.10[a]	1	Str	BS	M-B	2	Ecl	EM	14	36 9/16	......	......	17
Warner-Scarab, 50	[m]	.....	73	D	285	2.28[a]	1	Str	BS	M-B	2	Ecl	EM	14	36 9/16	......	......	17
Warner-Scarab Jr., 50	[m]	.....	73	D	230	2.56[a]	1	Hol	Scin	Mag	1	Ecl	EM	14	36 9/16	......	......	17
Warner-Super Scarab, 165	[m]	.....	73	D	332	2.01[a]	1	Str	Scin	Mag	2	Ecl	EM	30 25/64	37 1/4	......	......	17 1/2
Wright, Cyclone GR-2600-A2B	900	1,900	90[k]	G	1,935	2.15	1	SC	Scin	Mag	2	Opt	Opt	62 1/16	55	......	......	25

Wright, Cyclone GR-2600-A5B	900	1,900	90^{k}	G	1,950	2.17	1	SC	Scin	Mag	2	Opt	Opt	$62\frac{1}{16}$	55			25
Wright, Cyclone GR-1820-G202A	650	2,000	90^{i}	G	1,290	1.98	1	SC	Scin	Mag	2	Opt	Opt	$50\frac{3}{64}$	$55\frac{7}{64}$			$23\frac{3}{8}$
Wright, Cyclone GR-1820-G203A	650	2,000	90^{i}	G	1,305	2.01	1	SC	Scin	Mag	2	Opt	Opt	$50\frac{3}{64}$	$55\frac{7}{64}$			$23\frac{3}{8}$
Wright, Cyclone GR-1820-G205A	650	2,000	90^{i}	G	1,302	2.00	1	SC	Scin	Mag	2	Opt	Opt	$50\frac{3}{64}$	$55\frac{7}{64}$			$23\frac{3}{8}$
Wright, Cyclone GR-1820-G102A	625	1,900	90^{i}	G	1,260	2.02	1	SC	Scin	Mag	2	Opt	Opt	$48\frac{1}{8}$	$55\frac{7}{64}$			$23\frac{3}{8}$
Wright, Cyclone GR-1820-G103A	625	1,900	90^{i}	G	1,260	2.02	1	SC	Scin	Mag	2	Opt	Opt	$48\frac{1}{8}$	$55\frac{7}{64}$			$23\frac{3}{8}$
Wright, Cyclone GR-1820-G105A	625	1,900	90^{i}	G	1,272	2.04	1	SC	Scin	Mag	2	Opt	Opt	$48\frac{1}{8}$	$55\frac{7}{64}$			$23\frac{3}{8}$
Wright, [e]Cyclone R-1820-G2	600	1,900	87	D	1,103	1.84	1	SC	Scin	Mag	2	Opt	Opt	$43\frac{1}{4}$	$54\frac{1}{4}$			$23\frac{3}{8}$
Wright, [e]Cyclone R-1820-G3	600	1,900	87	D	1,103	1.84	1	SC	Scin	Mag	2	Opt	Opt	$43\frac{1}{4}$	$54\frac{1}{4}$			$23\frac{3}{8}$
Wright, [e]Cyclone R-1820-G3B	600	1,900	87	D	1,103	1.84	1	SC	Scin	Mag	2	Opt	Opt	$43\frac{1}{4}$	$54\frac{1}{4}$			$23\frac{3}{8}$
Wright, [e]Cyclone R-1820-G5	600	1,900	87	D	1,115	1.86	1	SC	Scin	Mag	2	Opt	Opt	$43\frac{1}{4}$	$54\frac{1}{4}$			$23\frac{3}{8}$
Wright, [f,e]Cyclone R-1820-F52	550	1,900	87	D	1,000	1.82	1	SC	Scin	Mag	2	Opt	Opt	$43\frac{3}{8}$	$54\frac{3}{32}$			$23\frac{3}{8}$
Wright, [e]Cyclone R-1820-F53	550	1,900	87	D	1,000	1.82	1	SC	Scin	Mag	2	Opt	Opt	$43\frac{3}{8}$	$54\frac{3}{32}$			$23\frac{3}{8}$
Wright, [e]Cyclone R-1820-F55	550	1,900	87	D	1,012	1.84	1	SC	Scin	Mag	2	Opt	Opt	$43\frac{3}{8}$	$54\frac{3}{32}$			$23\frac{3}{8}$
Wright, [e]Cyclone R-1820-F56	550	1,900	87	D	1,000	1.82	1	SC	Scin	Mag	2	Opt	Opt	$43\frac{3}{8}$	$54\frac{3}{32}$			$23\frac{3}{8}$
Wright, [f,e]Cyclone R-1820-F62	550	1,900	87	D	1,000	1.82	1	SC	Scin	Mag	2	Opt	Opt	$43\frac{3}{8}$	$54\frac{3}{32}$			$23\frac{3}{8}$
Wright, Whirlwind R-975-E1	290	1,950	73	D	660	2.28	1	Str	Scin	Mag	2	Opt	Opt	$41\frac{13}{64}$	45			$23\frac{3}{8}$
Wright, Whirlwind R-975-E3	325	2,000	80	D	675	2.08	1	Str	Scin	Mag	2	Opt	Opt	$41\frac{13}{64}$	45			$23\frac{3}{8}$
Wright, Whirlwind R-760-ET	195	1,850	73	D	540	2.77	1	Str	Scin	Mag	2	Opt	Opt	$42\frac{7}{16}$	45			$19\frac{1}{4}$
Wright, Whirlwind R-760-E1	225	2,000	73	D	570	2.53	1	Str	Scin	Mag	2	Opt	Opt	$42\frac{7}{16}$	45			$19\frac{1}{4}$
Wright, Whirlwind R-760-E2	225	2,000	80	D	570	2.53	1	Str	Scin	Mag	2	Opt	Opt	$42\frac{7}{16}$	45			$19\frac{1}{4}$

Symbols and Abbreviations

(Continued from page 327)

Valve Location

I—In head with push rods and rocker arms.
L—Valves at side.
OH—Overhead camshaft.

Rating

SL—Sea level.

Propeller Drive

D—Direct.
G—Geared.

Carburetor Make

Hol—Holley.
Lin—Linkert.
MS—Marvel or Stromberg.
Mar—Marvel.
SC—Stromberg or Chandler-Evans.
Str—Stromberg.
Zen—Zenith.

Ignition System Make

Ben—Bendix.
Bos—Bosch.
BS—Bosch or Scintilla.
ES—Edison-Splitdorf.
Scin—Scintilla.
SE—Scintilla or Eisemann.

Current Sources

B.M.—Battery and magneto.
Bat—Battery.
Mag—Magneto.
M-B—Magneto, battery optional.

Starter Make

Au-Auto-Lite.
Ecl—Eclipse.
Opt—Optional.

Method of Starting

A-DE—Air or direct cranking electric.
DE—Direct cranking electric.
EI—Electric inertia.
EM—Electric motor.
HC—Hand crank.
PE—Propeller swing or electric motor.
PS—Propeller swing.

Engine Manufacturers

[1] Akron Aircraft Inc.
[2] Monocoupe Corp.
[3] Lenape Aircraft Motors, Inc.
[4] Aviation Mfg. Corp.
[5] Sky Motors.
[6] Aircooled Motors Corp.

TABLE A1-2.—FOREIGN AIRPLANE-ENGINE DATA
(From *Automotive Ind.*, Feb. 27, 1937)

Make and model (*Data continued on pp. 332 and 333*)	Cylinder data						
	Arrangement	Cooling	No. of cylinders	Bore, stroke, in.	Piston displacement, cu. in.	Compression ratio	B.m.e.p. at cruising hp.
British							
A.B.C.	Hor	Air	4	4.2 × 4.8	244.0	5.60	130
Armstrong Sid, Tiger IX	Rad	Air	14	5½ × 6	1,996.0	6.20	104
Armstrong Sid, Tiger VI	Rad	Air	14	5½ × 6	1,996.0	6.20	104
Armstrong Sid, Cheetah IX	Rad	Air	7	5¼ × 5¼	834.0	6.35	104
Armstrong Sid, Cheetah IX	Rad	Air	7	5¼ × 5½	834.0	6.35	106
Armstrong Sid, Cheetah VA	Rad	Air	7	5¼ × 5½	834.0	5.20	102
Bristol, Mercury VIII	Rad	Air	9	5¾ × 6½	1,519.0	....	159*
Bristol, Mercury IX	Rad	Air	9	5¾ × 6½	1,519.0	....	159*
Bristol, Pergasus X	Rad	Air	9	5¾ × 7½	1,753.0	....	159*
Bristol, Pegasus Xc	Rad	Air	9	5¾ × 7½	1,753.0	....	137*
Bristol, Pegasus XI	Rad	Air	9	5¾ × 7½	1,753.0	....	159*
Bristol, Pegasus XII	Rad	Air	9	5¾ × 7½	1,753.0	....	159*
Bristol, Pegasus XX	Rad	Air	9	5¾ × 7½	1,753.0	....	161*
Cirrus-Hermes, Major	I	Air	4	4.72 × 5.12	386.1	5.80	102
Cirrus-Hermes, Minor	I	Air	4	3.74 × 5.00	229.6	5.80	95
Napier, Dagger Series II	H6	Air	24	3$\frac{13}{16}$ × 3¾	1,027.5	7.75	125
Napier, Dagger Series III	H6	Air	24	3$\frac{13}{16}$ × 3¾	1,027.5	7.75	139
Napier, Rapier Series V	H4	Air	16	3½ × 3½	538.8	7.00	109
Napier, Rapier Series VI	H4	Air	16	3½ × 3½	538.8	7.00	130
Pobjoy, Niagara IV	Rad	Air	7	3.19 × 3.43	191.4	6.50	122
Rolls-Royce, Kestrel IV, V, VI	V60	Liq	12	5.00 × 5.50	1,296.0	6.00	119
Rolls-Royce, Kestrel VII, VIII, IX	V60	Liq	12	5.00 × 5.50	1,296.0	6.00	134

Rolls-Royce, Kestrel X, XI, XII	V60	Liq	12	5.00 × 5.50	1,296.0	7.00	127
Rolls-Royce, Kestrel XIV, XV, XVI	V60	Liq	12	5.00 × 5.50	1,296.0	6.00	127
Rolls-Royce, Kestrel IV, V, VI (V.P.)	V60	Liq	12	5.00 × 5.50	1,296.0	6.00	119
Rolls-Royce, Kestrel VII, VIII, IX (V.P.)	V60	Liq	12	5.00 × 5.50	1,296.0	6.00	134
Rolls-Royce, Kestrel XIV, XV, XVI (V.P.)	V60	Liq	12	5.00 × 5.50	1,296.0	6.00	127
Czechoslovakia							
Walter, Atom	H	Air	2	3.35 × 3.78	67.1	5.20	113
Walter, Mikron	I	Air	4	3.35 × 3.78	133.0	5.20	117
Walter, Minor	I	Air	4	4.14 × 4.53	244.0	5.30	122
Walter, Junior	I	Air	4	4.53 × 5.51	354.8	5.20	117
Walter, Major-4	I	Air	4	4.65 × 5.51	373.6	5.20	121
Walter, Major-6	I	Air	6	4.65 × 5.51	561.2	5.20	128
Walter, Sagitta I-RC	V60	Air	12	4.65 × 5.51	1,120.6	5.50	118
Walter, Sagitta II-RC	V60	Air	12	4.65 × 5.51	1,120.6	5.50	118
Walter, Gemma I	Rad	Air	9	4.13 × 4.72	570.4	5.30	117
Walter, Scolar	Rad	Air	9	4.13 × 3.94	475.8	5.40	121
Walter, Bora II	Rad	Air	9	4.13 × 4.72	570.4	6.30	133
Walter, Bora II-R	Rad	Air	9	4.13 × 4.72	570.4	6.30	133
Walter, Castor II	Rad	Air	7	5.32 × 6.69	1,038.8	6.00	111
Walter, Pollux II	Rad	Air	9	5.32 × 6.69	1,335.9	6.00	112
Walter, Pollux II-R	Rad	Air	9	5.32 × 6.69	1,335.9	6.00	103
Walter, Super Castor I-RC	Rad	Air	9	5.32 × 5.75	1,146.8	5.50	119
Walter, Super Castor II-RC	Rad	Air	9	5.32 × 5.75	1,146.8	5.50	110
French							
Farman, 7EAr	Rad	Air	7	4.53 × 5.32	600.22	5.20	73
Farman, 7ED	Rad	Air	7	4.53 × 5.32	600.22	5.20	82
Farman, 9EBr	Rad	Air	9	4.53 × 5.32	768.33	5.20	83
Farman, 12WIrs	WI	Liq	12	5.32 × 5.12	1,365.72	6.40	111
Farman, 12WKrs	WI	Liq	12	5.32 × 5.12	1,365.72	6.40	115

TABLE A1-2.—FOREIGN AIRPLANE-ENGINE DATA.—(*Continued*)

Make and model (*Data continued on pp. 334 and 335*)	Cylinder data					Ratings									Weight, lb.	
	Blower ratio	Cylinder material	Valves per cylinder			Maximum except take-off			Take-off		Cruising		Octane rating	Propeller drive	Dry, no hub or starter	Per cruising hp.
			Intake	Exhaust	Arrangement	Hp.	R.p.m.	At sea level or altitude	Hp.	R.p.m.	Hp.	R.p.m.				
British																
A.B.C.	.....	10	1	1	OH	...	.....	SL	...	.....	75	1,875	..	D	219.0	2.92
Armstrong Sid	5.40	7	1	1	I	804	2,450	7,200	880	2,375	560	2,150	87	G	1,220.0	2.18
Armstrong Sid	5.40	7	1	1	I	810	2,450	6,400	840	2,150	560	2,150	87	G	1,180.0	2.11
Armstrong Sid	5.40	7	1	1	I	345	2,425	7,800	368	2,300	230	2,100	87	D	635.0	2.76
Armstrong Sid	6.52	7	1	1	I	350	2,425	7,300	340	2,100	235	2,100	87	D	635.0	2.70
Armstrong Sid	1.00	7	1	1	I	326	2,400	SL	296	2,100	225	2,100	87	D	596.0	2.65
Bristol	.....	..	..	..	...	840	2,750	14,000	730	2,650	...	.....	..	G	980.0	1.17*
Bristol	.....	..	..	..	...	840	2,750	14,000	730	2,650	...	.....	..	G	980.0	1.17*
Bristol	.....	..	..	..	...	915	2,600	6,250	960	2,475	...	.....	..	G	1,005.0	1.09*
Bristol	.....	..	..	..	...	790	2,600	5,500	910	2,475	...	.....	..	G	1,015.0	1.28*
Bristol	.....	..	..	..	...	915	2,600	6,250	960	2,475	...	.....	..	G	1,005.0	1.09*
Bristol	.....	..	..	..	...	915	2,600	6,250	960	2,475	...	.....	..	G	1,005.0	1.09*
Bristol	.....	..	..	..	...	925	2,600	10,000	830	2,475	...	.....	..	G	1,015.0	1.09*
Cirrus-Hermes	.....	7	1	1	I	148	2,450	SL	135	2,100	109	2,200	70	D	325.0	2.98
Cirrus-Hermes	.....	7	1	1	I	90	2,600	SL	75	2,200	63	2,300	70	D	195.0	3.10
Napier	6.47	7	1	1	OH	755	4,000	12,500	710	3,500	565	3,500	87	G	1,280.0	2.27
Napier	5.04	7	1	1	OH	805	4,000	5,000	760	3,500	630	3,500	87	G	1,280.0	2.03
Napier	6.33	7	1	1	OH	340	4,000	13,000	335	3,500	260	3,500	87	G	713.0	2.75
Napier	5.27	7	1	1	OH	395	4,000	5,800	365	3,500	310	3,500	87	G	713.0	2.30
Pobjoy	.....	7	1	1	I	108	3,750	SL	95	3,230	100	3,400	70	G	164.0	1.64
Rolls-Royce	8.83	2	2	2	OH	640	2,900	14,000	700	2,240	485	2,500	87	G	955.0	1.97
Rolls-Royce	6.92	2	2	2	OH	730	2,900	5,250	700	2,375	548	2,500	87	G	955.0	1.74

Rolls-Royce	.....	2	2	2	OH	635	2,900	SL	560	2,375	517	2,500	87	G	900.0	1.74
Rolls-Royce	9.41	2	2	2	OH	745	3,000	14,500	670	2,225	540	2,600	87	G	955.0	1.77
Rolls-Royce	8.83	2	2	2	OH	640	2,900	14,000	760	2,750	485	2,500	87	G	985.0	2.03
Rolls-Royce	6.92	2	2	2	OH	730	2,900	14,000	820	2,750	548	2,500	87	G	985.0	1.80
Rolls-Royce	9.41	2	2	2	OH	745	3,000	14,500	745	2,750	517	2,600	87	G	985.0	1.91
Czechoslovakia																
Walter	.....	7	1	1	I	28	3,000	SL	...	.....	25	2,600	68	D	88.2	3.52
Walter	.....	7	1	1	I	54	2,800	SL	48	2,300	50	2,550	68	D	132.3	2.64
Walter	.....	7	1	1	I	95	2,550	SL	77	2,050	85	2,260	68	D	205.0	2.41
Walter	.....	7	1	1	I	120	2,300	SL	...	.....	105	2,000	68	D	297.6	2.84
Walter	.....	7	1	1	I	130	2,350	SL	120	2,050	120	2,100	73	D	308.6	2.57
Walter	.....	7	1	1	I	205	2,350	SL	190	2,050	190	2,100	73	D	385.8	2.03
Walter	7.55	7	1	1	I	450	2,600	6,562	430	2,490	400	2,400	85	G	815.7	2.05
Walter	10.30	7	1	1	I	470	2,600	12,139	...	.....	400	2,400	85	G	815.7	2.05
Walter	.....	7	1	1	I	165	1,850	SL	...	.....	150	1,785	68	D	359.4	2.39
Walter	1.00	7	1	1	I	180	2,500	SL	...	.....	160	2,200	73	D	341.7	2.13
Walter	1.00	7	1	1	I	225	2,500	SL	...	.....	210	2,200	75	D	363.8	1.73
Walter	1.00	7	1	1	I	245	2,600	SL	...	.....	230	2,400	80	G	379.2	1.65
Walter	1.00	7	1	1	I	340	2,000	SL	300	1,750	260	1,800	75	D	612.9	2.36
Walter	1.00	7	1	1	I	450	2,000	SL	400	1,750	340	1,800	80	D	707.7	2.08
Walter	1.00	7	1	1	I	480	2,250	SL	425	1,950	360	2,070	80	G	758.4	2.10
Walter	7.55	7	1	1	I	450	2,200	SL	...	.....	380	2,200	85	G	782.6	2.06
Walter	10.30	7	1	1	I	450	2,200	SL	...	.....	350	2,200	85	G	782.6	2.24
French																
Farman	1.00	3	1	1	L	150	2,150	SL	190	2,150	150	1,900	..	G	503.0	4.79
Farman	1.00	3	1	1	L	170	2,150	SL	190	2,150	118	1,900	..	D	392.0	3.32
Farman	1.00	3	1	1	L	220	2,150	SL	265	2,150	152	1,900	..	G	531.0	3.49
Farman	2.00	1	2	2	I	550	2,250	16,404	600	2,250	383	2,000	..	G	1,097.0	2.84
Farman	2.60	1	2	2	I	600	2,590	22,966	650	2,500	445	2,250	..	G	1,148.0	2.58

TABLE A1-2.—FOREIGN AIRPLANE-ENGINE DATA.—(*Continued*)

Make and model (*Data continued from pp. 332 and 333*)	Carburetors		Fuel pump supplied	Ignition system			Starting		Installation dimensions over-all, in.			Height above engine bed, in.	Center-to-center engine bearers, in.
	Number	Make		Make	Current source	Number	Make	Method	Length	Height	Width		
British													
A.B.C.	1	Cla	No	BTH	Mag	2	....	PS	25.50	28.00	39.00		
Armstrong Sid.	1	Cla	Yes	BTH	Mag	2	Own	HE	68.30	50.80	33.27		
Armstrong Sid.	1	Cla	Yes	BTH	Mag	2	Own	HE	65.00	50.80	33.80		
Armstrong Sid.	1	Cla	Yes	BTH	Mag	2	Own	HE	46.75	47.70	21.55		
Armstrong Sid.	1	Cla	Yes	BTH	Mag	2	Own	HE	52.90	47.70	22.22		
Armstrong Sid.	1	Cla	Yes	BTH	Mag	2	Own	HE	48.70	47.70	21.85		
Bristol	..	....	...	.....	Mag	2	....	....	.....	51.50			
Bristol	..	....	...	.....	Mag	2	....	....	.....	55.30			
Bristol	..	....	...	.....	Mag	2	....	....	.....	55.30			
Bristol	..	....	...	.....	Mag	2	....	....	.....	55.30			
Bristol	..	....	...	.....	Mag	2	....	....	.....	55.30			
Bristol	..	....	...	.....	Mag	2	....	....	.....	55.30			
Cirrus-Hermes	1	Cla	Yes	BTH	Mag	2	BTH	EM	42.32	29.41	17.05	.....	15.20
Cirrus-Hermes	1	Cla	Yes	BTH	Mag	2	BTH	EM	37.72	25.00	16.50	.....	15.20
Napier	1	Cla	Yes	BTH	Mag	2	Rot	In	80.00	45.12	23.00	22.62	17.12
Napier	1	Cla	Yes	BTH	Mag	2	Rot	In	80.00	45.12	23.00	22.62	17.12
Napier	1	Cla	Yes	BTH	Mag	2	Rot	GD	57.37	37.00	21.62	17.50	18.50
Napier	1	Cla	Yes	BTH	Mag	2	Rot	GD	57.37	37.00	21.62	17.50	18.50
Pobjoy	1	Cla	Yes	Rot	Mag	2	Rot	EM	22.00	27.50	.....	.....	19.75
Rolls-Royce	1	Own	Yes	BW	Mag	2	Own	HA	72.35	35.60	24.40	19.60	17.25
Rolls-Royce	1	Own	Yes	BW	Mag	2	Own	HA	72.35	35.60	24.40	19.60	17.25
Rolls-Royce	2	Own	Yes	BW	Mag	2	Own	HA	63.48	35.60	24.40	19.60	17.25
Rolls-Royce	1	Own	Yes	BW	Mag	2	Own	HA	72.35	36.60	24.40	19.60	17.25

Rolls-Royce	1	Own	Yes	BW	Mag	2	Own	HA	72.35	35.60	24.40	19.60	17.25
Rolls-Royce	1	Own	Yes	BW	Mag	2	Own	HA	72.35	35.60	24.40	19.60	17.25
Rolls-Royce	1	Own	Yes	BW	Mag	2	Own	HA	72.35	35.60	24.40	19.60	17.25
Czechoslovakia													
Walter	2	Ama	Yes	Bos	Mag	2	.....	PS	20.08	16.10	32.09		
Walter	1	Cla	Yes	Scin	Mag	2	Own	HC	31.65	25.00	13.46	.....	9.84
Walter	1	Cla	Yes	Scin	Mag	2	Own	HC	44.09	24.80	17.32	.....	11.89
Walter	1	Cla	Yes	Scin	Mag	2	Own	HC	46.46	30.19	19.49	.....	21.26
Walter	1	Cla	Yes	Scin	Mag	2	Own	HC	46.46	30.19	19.49	.....	21.26
Walter	2	Cla	Yes	Scin	Mag	2	Own	HE	59.84	31.58	19.29	.....	25.20
Walter	1	Str	Yes	Scin	Mag	2	Own	HE	70.36	30.75	28.54	.....	25.20
Walter	1	Str	Yes	Scin	Mag	2	Own	Opt	70.36	30.75	28.54	.....	25.20
Walter	1	Zen	Yes	Scin	Mag	2	Own	CA	32.40	40.95	.....	.....	14.57
Walter	1	Str	Yes	Scin	Mag	2	Own	CA	34.64	38.74	.....	.....	20.28
Walter	1	Str	Yes	Scin	Mag	2	Own	HA	34.64	42.21	.....	.....	20.28
Walter	1	Str	Yes	Scin	Mag	2	Own	HA	40.47	42.21	.....	.....	20.28
Walter	1	Str	Yes	Scin	Mag	2	Own	HA	50.35	48.90	.....	.....	18.90
Walter	1	Str	Yes	Scin	Mag	2	Own	HA	50.35	49.69	.....	.....	19.53
Walter	1	Str	Yes	Scin	Mag	2	Own	HA	58.60	49.69	.....	.....	19.53
Walter	1	Zen	Yes	Scin	Mag	2	Own	HA	55.75	46.93	.....	.....	25.20
Walter	1	Zen	Yes	Scin	Mag	2	Own	HA	55.75	46.93	.....	.....	25.20
French													
Farman	1	Str	...	Scin	....	2	Own	Car	44.29	44.10	41.34		
Farman	1	Str	...	Scin	....	2	Own	Car	33.47	43.31	41.34		
Farman	1	Str	...	Scin	....	2	Own	Car	44.88	43.31	43.31		
Farman	3	Zen	...	Duc	....	2	Own	CA	44.88	33.94	43.70		
Farman	3	Zen	...	Duc	....	2	Own	CA	67.14	42.60	43.70		

TABLE A1-2.—FOREIGN AIRPLANE-ENGINE DATA.—(*Continued*)

Make and model (*Data continued on pp. 338 and 339*)	Cylinder data						
	Arrangement	Cooling	Number of cylinders	Bore, stroke, in.	Piston displacement, cu. in.	Compression ratio	B.m.e.p. at cruising hp.
	French						
Farman, 12 Crs	V60-1	Liq	12	3.94 × 4.72	690.41	7.40	103
Gnome-Rhone, Mistral Major 14No	Rad	Air	14	5.75 × 6.50	2,358.80	6.10	139*
Gnome-Rhone, Mistral Major 18L	Rad	Air	18	5.75 × 7.09	3,308.64	6.10	145*
Gnome-Rhone, Mistral Major 14M	Rad	Air	14	4.80 × 4.57	1,157.78	6.50	148*
Hispano Suiza, 12Y21	V60	Liq	12	5.91 × 6.69	2,196.0	7.00	133*
Hispano Suiza, 12Xirs-1	V60	Liq	12	5.12 × 6.69	1,647.0	5.80	158*
Hispano Suiza, 14AA-04	Rad	Air	14	6.13 × 6.69	2,759.6	6.20	151*
Hispano Suiza, 14AB-02	Rad	Air	14	5.32 × 5.12	1,589.0	6.20	141*
Renault, 4PGI	I	Air	4	4.72 × 5.51	386.1	5.30	103
Renault, 4PEI	I	Air	4	4.72 × 5.51	386.1	5.70	107
Renault, 6Q01	I	Air	6	4.72 × 5.51	579.5	6.40	111
Renault, 6Q03	I	Air	6	4.72 × 5.51	579.5	6.50	116
Renault, 12R01	V60	Air	12	4.72 × 5.51	1,159.0	6.40	111
Renault, 14T	Rad	Air	14	6.06 × 6.93	2,806.0	6.40	122
Salmson, 9AdR-60cv	Rad	Air	9	2.76 × 3.39	181.8	5.60	101*
Salmson, 9AeRS-75cv	Rad	Air	9	2.76 × 3.39	181.8	5.60	126*
Salmson, 9ND-175cv	Rad	Air	9	3.94 × 5.51	603.9	5.30	125*
Salmson, 9Aba-280cv	Rad	Air	9	4.92 × 6.69	1,142.5	5.30	105*
Salmson, 9Na-350cv	Rad	Air	9	5.51 × 6.30	1,350.5	5.30	117*
Salmson, 9NaS-350-630cv	Rad	Air	9	5.51 × 6.30	1,350.5	5.30	112*
Salmson, 18AbS-570-965cv	Rad	Air	18	4.92 × 7.09	2,426.6	5.30	89*
Salmson, 6TE-170cv	I	Air	6	4.53 × 5.04	486.2	5.50	136*
Salmson, 6TES-250cv	I	Air	6	4.53 × 5.04	486.2	5.50	163*

German							
Argus, As10C	V90	Air	8	4.72 × 5.51	772.7	5.90	109
Brandenburg, Sh14A-4	Rad	Air	7	4.25 × 4.72	469.7	6.00	106
Brandenburg, SAM322H2	Rad	Air	9	6.06 × 6.30	1,636.0	6.40	126
Hirth, 60R-2	I	Air	4	4.01 × 4.33	220.0	5.80	106
Hirth, 504A-1	I	Air	4	4.13 × 4.53	244.0	6.00	117
Mercedes-Benz, Diesel OF2	V60	Liq	12	6.50 × 8.27	3,295.2	15.00	100
Junkers, Jumo L5G	I	Liq	6	6.30 × 7.48	1,398.1	5.50	
Junkers, Jumo 205C	I	Liq	6	4.13 × (c)	1,013.8	17.00	100
Junkers, Jumo 205D	I	Liq	6	4.13 × (c)	1,013.8	17.00	105
Italian							
Fiat, A24R	V60	Liq	12	5.51 × 6.89	2,002.0	5.70	115
Fiat, A30RA	V60	Liq	12	5.31 × 5.51	1,713.6	8.00	77
Fiat, A54	Rad	Air	7	4.13 × 4.72	493.5	5.50	84
Fiat, A70	Rad	Air	7	4.53 × 4.53	510.0	5.75	120
Fiat, A74RC	Rad	Air	14	4.53 × 4.53	1,906.3	6.50	145*
Fiat, A80RC	Rad	Air	18	5.51 × 6.50	2,788.9	6.50	135*
Isotta Fraschini, 750R	W	Liq	18	5.12 × 6.69	4,073.1	5.70	70
Isotta Fraschini, 750RC	W	Liq	18	5.12 × 6.69	4,073.1	5.70	70
Isotta Fraschini, XIR	V	Liq	12	5.75 × 6.30	1,991.0	6.70	103
Isotta Fraschini, XIRC40	V	Liq	12	5.75 × 6.30	1,991.0	6.40	105
Isotta Fraschini, Caccia	V	Liq	12	4.92 × 5.12	1,257.2	5.70	119
Isotta Fraschini, Astro 7C-21	Rad	Air	7	6.06 × 6.30	1,272.2	5.95	110

TABLE A1-2.—FOREIGN AIRPLANE-ENGINE DATA.—(*Continued*)

Make and model (*Data continued on pp. 340 and 341*)	Cylinder data					Ratings								Propeller drive	Weight, lb.
	Blower ratio	Cylinder material	Per cylinder valves			Maximum except take-off			Take-off		Cruising		Octane rating		Dry, no hub or starter
			Intake	Exhaust	Arrangement	Hp.	R.p.m.	At sea level or altitude	Hp.	R.p.m.	Hp.	R.p.m.			
French															
Farman	2.60	1	1	1	I	400	3,400	19,680	450	3,400	275	3,050	..	G	650.0
Gnome-Rhone	8.94	7	1	1	I	950	2,300	12,000	900	2,360	...	.....	87	G	1,300.7
Gnome-Rhone	5.65	7	1	1	I	1,300	2,150	12,000	1,400	2,150	...	.....	87	G	1,622.6
Gnome-Rhone	8.24	7	1	1	I	650	3,000	12,000	700	3,000	...	.....	87	G	881.8
Hispano Suiza	10.00	7	1	1	OH	910	.....	11,811	880	2,400	...	.....	85	G	1,036.2
Hispano Suiza	10.00	7	1	1	OH	720	2,200	......	740	2,600	...	.....	85	G	848.8
Hispano Suiza	10.00	7	1	1	I	1,120	2,125	9,350	1,070	2,125	...	.....	85	G	1,311.7
Hispano Suiza	9.38	7	1	1	I	680	2,400	11,483	650	2,400	...	.....	85	D	1,025.1
Renault	.....	7	1	1	I	105	1,800	SL	100	.....	85	1,700	72	D	319.5
Renault	.....	7	1	1	I	150	2,400	SL	140	.....	120	2,300	72	D	319.5
Renault	.....	7	1	1	I	220	2,500	SL	220	.....	190	2,350	72	D	481.0
Renault	7.61	7	1	1	I	220	2,500	6,562	240	.....	200	2,350	85	D	520.0
Renault	11.70	7	1	1	I	450	2,500	11,975	430	.....	380	2,350	85	D	833.0
Renault	10.80	7	1	1	I	1,000	2,000	12,795	940	.....	800	1,850	85	G	1,420.0
Salmson	.....	7	1	1	I	66	2,850	SL	.....	.....	...	.....	75	G	176.4
Salmson	1.30	7	1	1	I	85	2,950	SL	.....	.....	...	.....	80	G	180.8
Salmson	.....	7	1	1	I	205	2,150	SL	.....	.....	...	.....	75	D	328.5

Salmson	.....	7	1	1	I	325	2,150	SL	.....	.....	...	.....	75	D	562.2
Salmson	.....	7	1	1	I	390	1,950	SL	.....	.....	...	.....	75	D	617.3
Salmson	1.60	7	1	1	I	400	2,100	11,483	.....	.....	...	.....	80	D	661.4
Salmson	1.60	7	1	1	I	570	2,100	11,483	.....	.....	...	.....	80	D	1,014.1
Salmson	.....	..	1	1	I	199	2,380	SL	.....	.....	...	.....	75	D	440.9
Salmson	1.30	..	1	1	I	250	2,500	SL	.....	.....	...	.....	80	D	458.6
German															
Argus	.....	7	1	1	I	220	1,940	SL	240	2,000	200	1,880	80	D	469.7
Brandenburg	.....	7	1	1	I	160	2,200	SL	.....	.....	128	2,050	80	D	297.6
Brandenburg	6.25	7	1	2	I	650	2,150	SL	.....	.....	520	2,000	87	G	1,080.3
Hirth	.....	3	1	1	F	80	2,400	SL	72	2,320	66	2,240	74	D	214.0
Hirth	.....	3	1	1	F	100	2,500	SL	90	2,400	80	2,320	77	D	230.0
Mercedes-Benz	.....	8	2	2	...	700	1,750	3,281	800	1,790	720	1,720	..	G	2,061.3
Junkers	.....	8	1	1	I	.....	.....	.....	375	1,700	340	.....	80	D	754.1
Junkers	.....	8	No	No	No	.....	.....	.....	600	2,200	510	2,000	HO	G	1,146.6
Junkers	.....	8	No	No	No	.....	.....	.....	700	2,600	590	2,200	HO	G	1,146.6
Italian															
Fiat	.....	8	2	2	OH	700	2,000	SL	.....	.....	525	1,810	82	G	1,212.5
Fiat	.....	8	2	2	OH	550	2,750	9,842	600	2,600	415	2,500	84	G	1,058.2
Fiat	.....	7	1	1	I	140	2,100	SL	.....	.....	100	1,910	74	D	330.7
Fiat	10.22	7	1	1	I	205	2,200	SL	.....	.....	155	2,000	82	D	368.2
Fiat	8.78	7	1	1	I	840	2,400	12,467	.....	.....	...	.....	87	G	1,245.6
Fiat	9.37	7	1	1	I	1,000	2,100	13,451	1,000	1,995	...	.....	87	G	1,598.4
Isotta Fraschini	.....	8	2	2	I	920	1,900	SL	.....	.....	550	1,530	87	G	1,543.2
Isotta Fraschini	11.46	8	2	2	I	900	1,900	SL	850	1,750	550	1,530	87	G	1,609.4
Isotta Fraschini	.....	8	2	2	I	895	2,250	SL	.....	.....	450	1,740	87	G	1,289.7
Isotta Fraschini	10.71	8	2	2	I	900	2,400	SL	860	2,140	500	1,900	87	G	1,322.8
Isotta Fraschini	.....	8	2	2	I	480	2,550	SL	.....	.....	...	.....	87	D	840.0
Isotta Fraschini	7.65	8	1	1	I	450	2,100	SL	420	2,000	320	1,800	87	D	723.1

TABLE A1-2.—FOREIGN AIRPLANE-ENGINE DATA.—(*Continued*)

Make and model	Weight, lb. per cruising hp.	Carburetors		Fuel pump supplied	Ignition system			Starting		Installation dimensions over-all, in.			Height above engine bed, in.	Center-to-center engine bearers, in.
		Number	Make		Make	Current source	Number	Make	Method	Length	Height	Width		
French														
Farman	2.36	4	Zen	...	Duc	....	2	Own	CA	59.21	29.06	25.75		
Gnome-Rhone	1.37*	1	Str	Yes	RB	Mag	2	Own	EM	.....	50.78			
Gnome-Rhone	1.25*	1	Bro	Yes	RB	Mag	2	Own	EM	.....	55.12			
Gnome-Rhone	1.36*	1	Bro	Yes	RB	Mag	2	Own	EM	.....	37.40			
Hispano Suiza	1.14*	6	Own	Yes	RB	Mag	2	Viet	CA	67.72	36.61	29.92		
Hispano Suiza	1.18*	6	Own	Yes	RB	Mag	2	Viet	CA	61.81	35.92	28.34		
Hispano Suiza	1.17*	1	Own	Yes	RB	Mag	2	Viet	CA	64.96	49.60			
Hispano Suiza	1.51*	1	Own	Yes	RB	Mag	2	Viet	CA	55.51	39.76			
Renault	3.76	1	Zen	Yes	Sev	Mag	2	Viet	CA	50.00	27.56	17.32	50.79	21.26
Renault	2.66	1	Str	Yes	Sev	Mag	2	Viet	CA	50.00	27.56	17.32	50.79	21.26
Renault	2.53	2	Str	Yes	Sev	Mag	2	Viet	CA	60.63	34.25	18.11	11.42	27.56
Renault	2.60	1	Str	Yes	Sev	Mag	2	Viet	CA	63.78	34.25	18.11	11.42	27.56
Renault	2.19	1	Str	Yes	Sev	Mag	2	Viet	CA	72.44	36.22	29.53	14.17	15.04
Renault	1.77	1	Str	Yes	Sev	Mag	2	Viet	CA	.....	50.66	66.54	39.37	38.98
Salmson	2.68*	1	Zen	No	Own	Mag	2	....	HC	32.01	25.91			
Salmson	2.13*	1	Zen	No	Own	Mag	2	....	HC	35.55	25.91			
Salmson	1.60*	1	Zen	Yes	Own	Mag	2	Viet	CA	36.81	37.80			
Salmson	1.73*	1	Zen	Yes	Own	Mag	2	Viet	CA	42.13	45.67			
Salmson	1.58*	1	Zen	Yes	Own	Mag	2	Viet	CA	43.31	46.46			
Salmson	1.64*	1	Zen	Yes	Own	Mag	2	Viet	CA	46.85	46.46			
Salmson	1.78*	1	Zen	Yes	Own	Mag	2	Viet	CA	56.81	48.03			
Salmson	2.21*	2	Str	Yes	Own	Mag	2	Viet	CA	67.64	27.95	15.35	8.66	
Salmson	1.83*	1	Zen	Yes	Own	Mag	2	Viet	CA	72.13	27.95	15.35	8.66	

German														
Argus...........	2.34	2	Sum	Yes	Bos	Mag	2	Own	HA	43.51	28.27	34.64	9.29	18.74
Bradenburg......	2.32	1	Sum	Yes	Bos	Mag	2	Own	CA	38.39	36.85	.....	.....	18.81
Brandenburg.....	2.08	1	Sum	Yes	Bos	Mag	2	Bos	HC	50.59	52.13	.....	.....	22.05
Hirth...........	3.24	1	Sum	Opt	Bos	M & B	e	Own	HC	33.70	27.10	15.40	11.80	6.50
Hirth...........	2.87	1	Pal	Opt	Bos	M & B	e	Own	HC	37.70	28.60	19.90	8.45	11.00
Mercedes-Benz...	2.87	..	...	Yes	....	....	..	DG	....	74.02	42.32	38.58	25.20	29.92
Junkers..........	2.01	1	Sum	Yes	Bos	Mag	2	Own	CA	70.86	47.84	25.59	.....	34.10
Junkers..........	2.24	None		Yes	....	....	..	Own	CA	76.50	52.17	21.54	.....	22.85
Junkers..........	1.94	None		Yes	....	....	..	Own	CA	76.50	52.17	21.54	.....	22.85
Italian														
Fiat.............	2.31	2	Own	Yes	Mar	Mag	2	Own	CA	69.69	41.93	28.93	25.39	
Fiat.............	2.55	3	Own	Yes	Mar	Mag	2	Own	CA	68.94	36.81	25.71	21.81	
Fiat.............	3.31	1	Str	Yes	Mar	Mag	2	Own	CA	21.30	36.81	25.71	21.81	
Fiat.............	2.37	1	Str	Yes	Mar	Mag	2	Mar	CA	20.71	36.61	25.71	21.81	
Fiat.............	1.48*	1	Str	Yes	Mar	Mag	2	Own	CA	41.14	47.05	25.71	21.81	
Fiat.............	1.60*	1	Own	Yes	Mar	Mag	2	Own	CA	45.47	52.56	25.71	21.81	
Isotta Fraschini..	2.81	6	Own	Yes	Mar	Mag	2	Own	CA	85.04	41.92	40.55		
Isotta Fraschini..	2.92	6	Own	Yes	Mar	Mag	2	Own	CA	86.37	47.44	40.55		
Isotta Fraschini..	2.86	4	Own	Yes	Mar	Mag	2	Own	CA	75.16	40.16	32.40		
Isotta Fraschini..	2.64	4	Own	Yes	Mar	Mag	2	Own	CA	83.88	42.84	32.84		
Isotta Fraschini..	1.75*	4	Zen	Yes	Mar	Mag	2	Own	CA	69.92	31.89	29.13		
Isotta Fraschini..	2.26	1	Own	Yes	Mar	Mag	2	Own	CA	19.84	47.25			

For footnotes to table see p. 342.

Symbols and Abbreviations For Table A1-2

General

*—Based on maximum horsepower.

(c)—2 × 6.30 (two cycle).

HO—Heavy oil.

Opt—Optional.

SL—Sea level.

Cylinder Arrangement

Hor—Horizontal.

H4—Four banks of four cylinders each in H formation.

H6—Four banks of six cylinders each in H formation.

I—In line.

Rad—Radial.

V—V.

V60—V type, 60 deg.

V60-I—V type, 60 deg., inverted.

V90—V type, 90 deg.

W—Three banks of cylinders.

WI—Three banks of cylinders —inverted.

Cooling

Liq—Liquid.

Cylinder Material

1—Aluminum with cast-iron liner.

2—Aluminum with steel liner.

3—Cast iron.

7—Steel with aluminum heads.

8—Steel.

10—Cast-iron with steel sleeves.

Valve Arrangement

F—F-head.

I—In head with push rods and rocker arms.

OH—Overhead camshaft.

L—L head—valve at side.

Propeller Drive

D—Direct.

G—Geared.

Carburetor Make

Ama—Amal.

Bro—Bronzania.

Cla—Claudel-Hobson.

Pal—Pallas.

Str—Stromberg.

Zen—Zenith.

Ignition and Starting Systems, Make

Bos—Bosch.

BW—British Thompson Houston or Watford.

BTH—British Thompson Houston.

Duc—Ducellier.

DG—Druckluft & Gluhkerzen.

Mar—Marell.

RB—Robert Bosch.

Rot—Rotax.

Scin—Scintilla.

Starting Method

CA—Compressed air.

Car—Cartridge.
EM—Electric motor.
GD—Gas distributor and hand-turning gear.
HA—Hand crank.
HC—Hand crank from machine.
HE—Hand crank, or electric motor.
In—Inertia.
PS—Propeller swing.

Current Sources

Bat—Battery.
e—One magneto and one battery.
Mag—Magneto.
M & B—Magneto and battery.

TABLE A1-3.—AMERICAN STOCK, MARINE, AND COMMERCIAL VEHICLE ENGINES
(From *Automotive Ind.*, Vol. 82, No. 5, Mar. 1, 1940)

Line number	Make and model	Designed for	Number of cylinders; bore and stroke, in.	Rated hp. (A.M.A.)	Maximum b.hp. at specified r.p.m.	Piston displacement, cu. in.	Compression ratio—to 1
1	Allis-Chalmers, B-15	Tr, Ind	4-3¼×3½	16.9	22-1,800	116.0	4.92
2	Allis-Chalmers, W-25	Tr, Ind	4-4×4	25.6	40-1,800	201.0	5.00
3	Allis-Chalmers, U-40	Tr, Ind	4-4½×5	32.4	51-1,400	318.0	4.74
4	Allis-Chalmers, E-60	Tr, Ind	4-5¼×6½	44.1	70-1,200	563.0	5.20
5	Allis-Chalmers, L-90	Tr, Ind	6-5¼×6½	66.1	117-1,200	844.0	5.20
6	Autocar, 315	T	6-3¾×4¾	33.7	90-2,600	315.0	6.00
7	Autocar, 358	T	6-4×4¾	38.4	100-2,600	358.0	5.80
8	Autocar, 408	T	6-4¹⁄₁₆×5¼	39.6	110-2,400	408.0	5.50
9	Autocar, 447	T	6-4¼×5¼	43.3	116-2,400	447.0	5.50
10	Autocar, 501	T	6-4½×4¼	48.6	127-2,300	501.0	5.50
11	Brennan, Imp. De Luxe	M	4-2×3	7.5	20-3,900	45.0	7.00
12	Brennan, Imp. De Luxe Special	M	4-2¼×3⅛	8.0	25-4,000	50.0	7.40
13	Brennan, 20	Tr, Ind	4-2¼×3⅛	8.0	20-3,900	50.0	7.40
14	Brennan, M	M	4-4×5	25.6	40-2,000	251.0	5.00
15	Brennan, E-4	M	4-4½×5	32.4	50-1,500	318.0	5.00
16	Brennan, CE	T, Tr, Ind	4-4½×5	32.4	50-1,600	318.0	5.00
17	Brennan, 90	M	6-4×5½	38.4	100-2,500	415.0	6.00
18	Brennan, B-70	T, B, Ind	6-4×5½	38.4	85-2,000	415.0	6.00
19	Brennan, 125	M	6-4⅜×5½	45.9	125-2,200	500.0	6.00
20	Brennan, B-100	T, B	6-4⅜×5½	45.9	100-2,000	500.0	6.00
21	Brennan, 150	M	6-4½×6½	48.6	150-2,000	620.3	6.00
22	Brennan, 150	T, B, Tr, Ind	6-4½×6½	48.6	150-2,200	620.3	6.00
23	Bridgeport, F-5	M	1-3¾×4½	.....	6-1,200	49.0	
24	Bridgeport, F-10	M	2-3¾×4½	.....	12-1,200	99.0	
25	Bridgeport, F-20	M	4-2¾×4	.....	25-2,500	95.0	
26	Bridgeport, F-25	M	4-3⁵⁄₁₆×3⅞	.....	50-2,500	134.0	

Data continued on pp. 358 and 372

27	Bridgeport, Pilot	M	$4\text{–}4\frac{1}{2}\times 5$	.....	55–2,000	283.0	
28	Bridgeport, Piloteer	M	$6\text{–}4\frac{3}{8}\times 4\frac{3}{4}$	.....	80–2,000	428.0	
29	Buda, HP-205	T, Tr	$4\text{–}3\frac{13}{16}\times 4\frac{1}{2}$	23.2	51–2,400	205.0	4.76
30	Buda, HP-217	T, Tr	$4\text{–}3\frac{13}{16}\times 4\frac{3}{4}$	23.2	55–2,400	217.0	5.50
31	Buda, 4HM-217-MD	M	$4\text{–}3\frac{13}{16}\times 4\frac{3}{4}$	.....	56–2,400	217.0	5.70
32	Buda, 4HM-217-MHD	M	$4\text{–}3\frac{13}{16}\times 4\frac{3}{4}$	.....	48–1,800	217.0	5.70
33	Buda, 4HM-217-HD	M	$4\text{–}3\frac{13}{16}\times 4\frac{3}{4}$	.....	30–1,200	217.0	5.70
34	Buda, KT-281	Tr	$4\text{–}4\frac{1}{8}\times 5\frac{1}{4}$	27.2	49–1,750	281.0	4.50
35	Buda, YR-425	T	$4\text{–}4\frac{3}{4}\times 6$	36.0	57–1,400	425.3	3.80
36	Buda, BTU	T, B, Tr	$4\text{–}5\times 6\frac{1}{2}$	40.0	61–1,200	510.5	4.65
37	Buda, FR	T, B, Tr	$4\text{–}5\frac{1}{2}\times 6\frac{1}{2}$	48.5	78–1,200	618.0	4.60
38	Buda, JV-4	Tr, Ind	$4\text{–}5\frac{3}{4}\times 7\frac{1}{4}$	52.9	85–1,200	740.0	3.85
39	Buda, JK-4	Tr, Ind	$4\text{–}6\times 7\frac{1}{8}$	57.6	115–1,200	806.0	4.70
40	Buda, JL-877	Tr, Ind	$4\text{–}6\frac{1}{4}\times 7\frac{1}{8}$	62.5	122–1,200	874.0	4.80
41	Buda, HP-260	T, B	$6\text{–}3\frac{1}{2}\times 4\frac{1}{2}$	29.4	68–2,800	260.0	4.75
42	Buda, HP-298	T, B, Tr	$6\text{–}3\frac{3}{4}\times 4\frac{1}{2}$	33.7	77–2,800	298.0	4.75
43	Buda, HP-326	T, B, Tr	$6\text{–}3\frac{13}{16}\times 4\frac{3}{4}$	34.8	78–2,400	326.0	5.40
44	Buda, 6DHM-326-MD	M	$6\text{–}3\frac{13}{16}\times 4\frac{3}{4}$	.....	80–2,400	326.0	5.70
45	Buda, 6DHM-326-MHD	M	$6\text{–}3\frac{13}{16}\times 4\frac{3}{4}$	.....	70–1,800	326.0	5.70
46	Buda, 6DHM-326-HD	M	$6\text{–}3\frac{13}{16}\times 4\frac{3}{4}$	.....	46–1,200	326.0	5.70
47	Buda, K-369	T, B	$6\text{–}4\frac{1}{16}\times 4\frac{3}{4}$	39.6	99–2,800	369.0	4.73
48	Buda, K-393	T, B, Tr	$6\text{–}4\frac{3}{16}\times 4\frac{3}{4}$	42.0	101–2,400	393.0	4.80
49	Buda, K-428	T, B, Tr	$6\text{–}4\frac{3}{8}\times 4\frac{3}{4}$	45.9	107–2,400	428.0	5.33
50	Buda, 6KM-428-MD	M	$6\text{–}4\frac{3}{8}\times 4\frac{3}{4}$	.....	110–2,400	428.0	5.50
51	Buda, 6KM-428-MHD	M	$6\text{–}4\frac{3}{8}\times 4\frac{3}{4}$	.....	96–1,800	428.0	5.50
52	Buda, 6KM-428-HD	M	$6\text{–}4\frac{3}{8}\times 4\frac{3}{4}$	.....	60–1,200	428.0	5.50
53	Buda, L-525	T, B, Tr	$6\text{–}4\frac{1}{2}\times 5\frac{1}{2}$	48.6	110–2,400	525.0	4.75
54	Buda, LO-525	T, B, Tr	$6\text{–}4\frac{1}{2}\times 5\frac{1}{2}$	48.6	135–2,400	525.0	5.00
55	Buda, GF-638	T, B, Tr	$6\text{–}4\frac{3}{4}\times 6$	54.1	134–2,000	638.0	4.75
56	Buda, M-766	T, B, Tr	$6\text{–}5\times 6\frac{1}{2}$	60.0	155–1,800	765.8	5.00
57	Buda, M-712	T, B, Ind	$6\text{–}5\frac{1}{8}\times 5\frac{3}{4}$	63.0	158–2,000	712.0	5.00
58	Buda, M-858	T, B, Ind	$6\text{–}5\frac{5}{8}\times 5\frac{3}{4}$	76.0	190–2,000	858.0	5.00
59	Buda, M-970	T, B, Ind	$6\text{–}5\frac{5}{8}\times 6\frac{1}{2}$	76.0	195–1,800	969.2	5.00
60	Buda, JV-6	Tr, Ind	$6\text{–}5\frac{3}{4}\times 7\frac{1}{4}$	79.3	142–1,200	1,129.5	3.88
61	Buda, JK-6	Tr, Ind	$6\text{–}6\times 7\frac{1}{4}$	86.4	188–1,400	1,230.0	4.70
62	Buda, JL-1335	Tr, Ind	$6\text{–}6\frac{1}{4}\times 7\frac{1}{4}$	93.7	164–1,000	1,334.6	4.40
63	Buda, P-1879	Ind	$6\text{–}6\frac{3}{4}\times 8\frac{3}{4}$	109.0	232–1,000	1,879.0	4.50
64	Buffalo, Navy-BA	M, Ind	$4\text{–}3\frac{1}{2}\times 5$	.....	45–1,800	192.0	5.00
65	Buffalo, RA-4	M, Ind	$4\text{–}5\frac{7}{8}\times 7$	.....	155–1,800	759.0	5.00
66	Buffalo, RAB-4	M, Ind	$4\text{–}6\frac{1}{2}\times 7$	.....	185–1,800	929.0	5.00
67	Buffalo, ATT-4	M, Ind	$4\text{–}8\frac{1}{4}\times 9$	.....	275–1,200	1,925.0	4.50
68	Buffalo, RA-6	M, Ind	$6\text{–}5\frac{7}{8}\times 7$	.....	230–1,800	1,138.0	5.00
69	Buffalo, RAB-6	M, Ind	$6\text{–}6\frac{1}{2}\times 7$	.....	280–1,800	1,393.0	5.00
70	Buffalo, ATT-6	M, Ind	$6\text{–}8\frac{1}{4}\times 9$	.....	425–1,200	2,887.0	4.50
71	Buffalo, RA-8	M, Ind	$8\text{–}5\frac{7}{8}\times 7$	.....	300–1,800	1,518.0	5.00
72	Buffalo, RAB-8	M, Ind	$8\text{–}6\frac{1}{2}\times 7$	.....	375–1,800	1,858.0	5.00
73	Buffalo, ATT-8	M, Ind	$8\text{–}8\frac{1}{4}\times 9$	.....	565–1,200	3,849.0	4.50

Data continued on pp. 359 and 373

TABLE A1-3.—AMERICAN STOCK, MARINE, AND COMMERCIAL VEHICLE ENGINES.—(*Continued*)

Line number	Make and model	Designed for	Number of cylinders; bore and stroke, in.	Rated hp. (A.M.A.)	Maximum b.hp. at specified r.p.m.	Piston displacement, cu. in.	Compression ratio—to 1
74	Capitol, T-12	M	12–5¾×6¼	.....	600–2,000	1,947.0	5.30
75	Chevrolet, 1940	T, B	6–3½×3¾	29.4	....	216.5	6.25
76	Chris-Craft, B[a]	M	4–3¼×4	.....	60–3,200	132.7	7.50
77	Chris-Craft, K[a,b]	M	6–3⅜×4⅛	.....	95–3,200	221.4	7.50
78	Chris-Craft, M[a,b]	M	6–4×4¼	.....	135–3,000	320.4	7.50
79	Chris-Craft, W[a,b]	M	6–4¼×4¾	.....	160–3,000	404.3	6.40
80	Chris-Craft, A-120[b]	M	8–5×5¼	.....	250–2,300	824.7	6.00
81	Chris-Craft, A-120A	M	8–5 1/16×5¼	.....	350–2,850	845.4	7.80
82	Chrysler, ACE-PC[a]	M	6–3⅛×4⅜	.....	78–3,200	201.3	6.70
83	Chrysler, Crown-M2[a]	M	6–3⅜×4½	.....	103–3,200	241.6	6.50
84	Chrysler, Royal-M3[a]	M	8–3¼×4⅞	.....	143–3,200	323.5	6.10
85	Climax, G4C	Tr, Ind	4–4½×5¼	34.2	47–1,200	334.0	4.75
86	Climax, H4B	Tr, Ind	4–5⅛×6¼	42.0	73–1,200	516.0	4.10
87	Climax, N4B	Tr, Ind	4–5¾×6½	52.9	100–1,200	675.0	4.30
88	Climax, R41	Tr, Ind	4–6×7	57.6	112–1,200	792.0	4.20
89	Climax, R61	Tr, Ind	6–6×7	86.4	165–1,200	1,187.4	4.20
90	Climax, R81	Ind	8–6×7	115.2	230–1,200	1,583.0	4.20
91	Continental, Y-4069	C, Tr, Ind	4–2½×3½	10.0	26–3,300	68.7	6.15
92	Continental, Y-4091	C, Tr, Ind	4–2⅞×3½	13.2	34–3,300	90.9	6.00
93	Continental, Y-4112	C, Tr, Ind	4–3 3/16×3½	14.4	41–3,300	111.7	6.00
94	Continental, F-4124	C, T, Tr, Ind	4–3×4⅜	14.4	47–3,300	123.7	6.00
95	Continental, F-4140	C, T, Tr, Ind	4–3 3/16×4⅜	16.2	52–3,200	139.6	6.00
96	Continental, F-4162	C, T, Tr, Ind	4–3 7/16×4⅜	18.9	58–3,300	162.4	5.75
97	Continental, F-6170	C, T, Tr, Ind	6–3×4	21.6	65–3,460	169.6	6.60
98	Continental, F-6199	C, T, Tr, Ind	6–3¼×4	25.4	68–3,300	199.1	6.00
99	Continental, F-6209	C, T, Tr, Ind	6–3 3/16×4⅜	24.3	71–3,100	209.5	5.75

Data continued on pp. 360 and 374

100	Continental, F-6218	C, T, Tr, Ind	6-3 1/4×4 3/8	25.4	73-3,100	217.8	5.95
101	Continental, A-6244	C, T, B, Ind	6-3 7/16×4 3/8	28.4	83-3,200	243.6	5.40
102	Continental, M-6271	T, B, Tr, Ind	6-3 5/8×4 3/8	31.5	85-2,800	270.9	5.75
103	Continental, M-6290	T, B, Ind	6-3 3/4×4 3/8	33.7	88-2,750	289.9	5.70
104	Continental, M-6330	T, B, Ind	6-4×4 3/8	38.4	98-2,700	329.8	5.50
105	Continental, E-600	T, B, Ind	6-3 11/16×4 1/2	32.6	78-2,650	288.3	5.43
106	Continental, E-601	T, B, Ind	6-3 7/8×4 1/2	36.0	86-2,600	318.4	5.48
107	Continental, E-602	T, B, Ind	6-4 1/8×4 1/2	40.8	95-2,500	360.8	5.40
108	Continental, E-603	T, B, Ind	6-4 1/4×4 1/2	43.3	98-2,400	383.0	5.29
109	Continental, 20R	T, B, Ind	6-4 1/8×4 3/4	40.8	106-2,600	380.9	4.76
110	Continental, 21R	T, B, Ind	6-4 3/8×4 3/4	45.9	118-2,500	428.4	4.63
111	Continental, 22R	T, B, Ind	6-4 1/2×5 1/4	48.6	138-2,400	501.0	4.50
112	Dodge, T-105	T	6-3 1/8×4 3/8	23.4	79-3,000	201.3	6.70
113	Dodge, T-94	T	6-3 1/4×4 3/8	25.3	82-3,000	217.7	6.50
114	Dodge, T-98	T	6-3 3/8×4 1/4	27.3	92-3,000	228.1	6.50
115	Dodge, T-100	T	6-3 3/8×4 1/2	27.3	99-3,000	241.5	6.50
116	Dodge, T-104	T	6-3 3/4×5	33.7	100-2,800	331.3	5.20
117	Elco, F-42	M	4-5×6	.....	90-1,600	471.0	5.00
118	Elco, F-62	M	6-5×6	.....	145-1,600	707.0	5.00
119	Ford, 60HP	C, T	8-2.6×3.2	21.6	60-3,500	136.0	6.60
120	Ford, 85HP	C, T, M, Ind	8-3.06×3.75	30.0	85-3,800	221.0	[1]
121	Ford, 95HP	C, T, B	8-3.18×3.75	32.5	95-3,600	239.0	6.15
122	Franklin, 6AH-377	T, B, Tr, Ind	6-4×5	38.4	104-2,500	377.0	4.90
123	Franklin, 6A-377	T, B, Tr, Ind	6-4×5	38.4	104-2,500	377.0	4.90
124	Franklin, 6AH-400	T, B, Tr, Ind	6-4 1/8×5	40.8	110-2,500	400.0	5.00
125	Franklin, 6A-400	T, B, Tr, Ind	6-4 1/8×5	40.8	110-2,500	400.0	5.00
126	Franklin, 4CHO-150	T, Tr	4-3 5/8×3 5/8	.....	43-3,000	150.0	5.50
127	G. M. C., 228	T	6-3 9/16×3 13/16	30.4	78-3,000	228.0	6.15
128	G. M. C., 248	T	6-3 23/32×3 13/16	33.1	89-3,000	248.5	6.15
129	G. M. C., 278	T	6-3 5/8×4 1/2	31.5	100-2,900	278.6	6.00
130	G. M. C., 308	T, B	6-3 13/16×4 1/2	33.9	110-2,800	308.2	6.00
131	G. M. C., 361	T	6-4 1/8×4 1/2	40.8	122-2,800	360.8	6.00
132	G. M. C., 426	T	6-4 1/4×5	43.3	145-2,700	425.6	6.00
133	G. M. C., 451	T	6-4 3/8×5	45.9	149-2,600	450.9	6.00
134	G. M. C., 479	B	6-4 5/8×4 3/4	51.3	141-2,600	478.8	5.50
135	G. M. C., 529	B	6-4 5/8×5 1/4	51.3	159-2,500	529.2	5.50
136	G. M. C., 707	B	6-5×6	60.0	175-2,100	706.8	4.83
137	Gray, Light Four	M	4-2 1/2×3 1/2	.....	16-1,800	69.0	5.50
138	Gray, Sea Scout	M	4-2 7/8×3 1/2	.....	37-3,000	91.0	6.50
139	Gray, Four-22	M	4-............	.....	45-3,000	112.0	6.50
140	Gray, Phantom 4-45	M	4-2 7/8×3 1/2	.....	45-3,600	91.0	7.50
141	Gray, Four-40	M	4-3 3/16×4 3/8	.....	55-3,000	140.0	5.50
142	Gray, Four-52	M	4-3 7/16×4 3/8	.....	57-2,600	162.0	6.00
143	Gray, Phantom 4-62	M	4-3 3/16×4 3/8	.....	62-3,600	140.0	7.20
144	Gray, Phantom 4-75	M	4-3 7/16×4 3/8	.....	75-3,600	162.0	7.00
145	Gray, Six-51	M	6-3 1/4×4	.....	73-3,200	200.0	6.00
146	Gray, Six-71	M	6-3 1/4×4 3/8	.....	84-3,000	218.0	6.50

Data continued on pp. 361 and 375

TABLE A1-3.—AMERICAN STOCK, MARINE, AND COMMERCIAL VEHICLE ENGINES.—(*Continued*)

Line number	Make and model	Designed for	Number of cylinders; bore and stroke, in.	Rated hp. (A.M.A.)	Maximum b.hp. at specified r.p.m.	Piston displacement, cu. in.	Compression ratio—to 1
147	Gray, Phantom 6-90	M	6–3¼×4⅜	.....	90–3,600	218.0	7.00
148	Gray, Phantom 6-103	M	6–3¼×4⅜	.....	103–3,600	218.0	7.00
149	Gray, Six-91	M	6–3 7/16×4⅜	.....	100–3,000	244.0	7.00
150	Gray, Phantom 6-125	M	6–3 7/16×4⅜	.....	125–3,600	244.0	7.00
151	Gray, Fireball 6-140	M	6–3 7/16×4⅜	.....	140–4,000	244.0	7.50
152	Gray, Fireball 6-150	M	6–3 7/16×4⅜	.....	150–4,000	244.0	7.50
153	Gray, Fireball 6-160	M	6–3 7/16×4⅜	.....	160–4,000	244.0	7.50
154	Gray, Fireball 225	M	6–3.30×4⅜	.....	180–5,000	225.0	10.00
155	Gray, Six-101	M	6–3¾×4⅜	.....	110–3,200	290.0	6.50
156	Gray, Six-121	M	6–4×4⅜	.....	124–3,200	330.0	6.50
157	Gray, Super Six	M	6–4×4⅜	.....	145–3,200	330.0	6.50
158	Gray, Six-105	M	6–4¼×4½	.....	101–2,400	383.0	5.50
159	Gray, Eight-160	M	8–3⅝×4½	.....	160–3,200	372.0	6.25
160	Gray, Phantom 8-175	M	8–3⅝×4½	.....	175–3,600	372.0	6.62
161	Hall-Scott, 165	T, B, Ind	4–4¼×5½	28.9	72–2,000	312.0	4.80
162	Hall-Scott, Fisher Jr. 178–179	M	4–4¼×5½	.....	70–1,800	312.0	4.03
163	Hall-Scott, 167	T, B, Ind	4–4¾×5½	36.1	80–2,000	390.0	4.84
164	Hall-Scott, (H)95	T, B	6–4×5	38.4	95–2,400	377.0	5.40
165	Hall-Scott, 147	T, B	6–4×5	38.4	93–2,200	377.0	4.90
166	Hall-Scott, (H)130	T, B	6–4¼×5	43.3	124–2,800	425.6	4.96
167	Hall-Scott, Navigator I, 116–117	M	6–4¼×5½	.....	103–1,800	468.0	4.74
168	Hall-Scott, 160	T, B	6–4¼×5½	43.3	117–2,200	468.0	4.42
169	Hall-Scott, 160-1	T, B	6–4¼×5½	43.3	108–2,200	468.0	4.70
170	Hall-Scott, Navigator II, 163–164	M	6–4¼×5½	.....	107–2,000	468.0	4.43
171	Hall-Scott, (H)135	T, B	6–4½×5	48.6	140–2,800	477.1	5.03
172	Hall-Scott, Explorer 132-33, 157-58	M	6–5×7	.....	190–1,800	824.7	4.85

Data continued on pp. 362 and 376

173	Hall-Scott, 175-1	T, B, FA	6–5×6	60.0	186–2,000	706.8	5.00
174	Hall-Scott, (H)180	T, B	6–5×6	60.0	181–2,100	706.8	4.90
175	Hall-Scott, 176	T, B, FA	6–5¼×6	66.1	191–1,800	779.3	5.02
176	Hall-Scott, (H)190	T, B, RC	6–5¼×6	66.1	200–2,000	779.3	5.00
177	Hall-Scott, 177	T, B, FA	6–5½×6	72.6	203–1,800	855.2	5.00
178	Hall-Scott, Invader 168-169	M	6–5½×7	.....	265–2,100	997.8	4.95
179	Hall-Scott, 230-231	M, Ind	6–5½×7	72.6	245–1,800	997.8	4.90
180	Hall-Scott, Defender 1268-1269	M	12–5½×7	.....	575–2,100	1,996.0	5.80
181	Hall-Scott, Defender 2268-2269	M	12–5¾×7	.....	630–2,100	2,181.0	7.00
182	Hall-Scott, Defender 2286-2287	M	12–5¾×7	.....	605–2,100	2,181.0	7.00
183	Hercules, BXB	Ind	2–2⅞×3	6.6	10–2,300	39.0	6.00
184	Hercules, NXA	M	2–3×4	7.2	13–2,000	56.5	5.50
185	Hercules, NXB	M, Tr, Ind	2–3¼×4	8.9	16–2,000	66.3	5.50
186	Hercules, ZXA	T, Tr, Ind	4–2½×3	10.0	23–3,800	58.8	6.10
187	Hercules, ZXB	T, Tr, Ind	4–2⅝×3	11.0	25–3,800	64.9	6.10
188	Hercules, IX-3	Tr, Ind	4–2½×4	10.0	28–3,200	78.0	5.20
189	Hercules, IX-5	T, Tr, Ind	4–2½×4	10.0	28–3,200	78.0	5.20
190	Hercules, IXA-3	Tr, Ind	4–3×4	14.4	40–3,200	113.0	5.50
191	Hercules, IXA-5	T, Tr, M, Ind	4–3×4	14.4	40–3,200	113.0	5.50
192	Hercules, IXB-3	Tr, Ind	4–3¼×4	16.9	47–3,200	133.0	5.20
193	Hercules, IXB-5	T, Tr, M, Ind	4–3¼×4	16.9	47–3,200	133.0	5.20
194	Hercules, OOA	T, Tr, M, Ind	4–3½×4½	19.6	35–2,000	173.2	4.20
195	Hercules, OOB	T, B, Tr, M, Ind	4–3¾×4½	22.5	38–2,000	198.8	4.20
196	Hercules, OOC	T, B, Tr, M, Ind	4–4×4½	25.6	41–2,000	226.2	4.20
197	Hercules, OX	Tr, Ind	4–4×5	25.6	46–1,800	251.3	4.30
198	Hercules, OXC	Tr, Ind	4–4¼×5	28.9	56–1,800	283.5	4.30
199	Hercules, K	T, Tr, Ind	4–4¼×5¾	28.9	55–1,600	326.3	3.89
200	Hercules, L	Tr, Tr, Ind	4–4½×5¾	32.4	59–1,600	365.8	3.78
201	Hercules, G	T, Tr, Ind	4–4¾×5¾	36.1	63–1,600	407.6	3.89
202	Hercules, E	T, Tr, Ind	4–5×5¾	40.0	74–1,600	451.4	4.00
203	Hercules, TX	Ind	4–5½×7	48.4	88–1,200	665.0	3.84
204	Hercules, TXA	Ind	4–6×7	57.6	98–1,200	792.0	3.84
205	Hercules, TXO	Ind	4–6⅜×7	65.0	112–1,200	894.0	3.84
206	Hercules, QXA-3	T, B, Tr	6–3⅛×4⅛	23.4	29–3,000	190.0	5.50
207	Hercules, QXA-5	T, B, Tr	6–3⅛×4⅛	23.4	59–3,000	190.0	5.50
208	Hercules, QXV-3	T, B, Tr	6–3¼×4⅛	25.3	65–3,500	205.0	5.85
209	Hercules, QXB-5	T, B, Tr, M, Ind	6–3¼×4⅛	25.3	65–3,500	205.0	5.85
210	Hercules, QXC-3	T, B, Tr	6–3⅜×4⅛	27.3	70–3,500	221.0	5.85
211	Hercules, QXC-5	T, B, Tr, M, Ind	6–3⅜×4⅛	27.3	70–3,500	221.0	5.85
212	Hercules, JXA	T, B, Tr, Ind	6–3⅜×4¼	27.3	63–2,800	228.0	5.16
213	Hercules, JXB	T, B, Tr, M, Ind	6–3⅝×4¼	31.5	68–2,800	263.0	5.40
214	Hercules, JXC	T, B, Tr, M, Ind	6–3¾×4¼	33.7	73–2,800	282.0	5.35
215	Hercules, JXD	T, B, Tr, M, Ind	6–4×4¼	38.4	84–2,800	320.0	5.63
216	Hercules, WXC	T, B, Tr, Ind	6–4×4½	38.4	90–2,400	339.0	5.00
217	Hercules, WXC-2	T, B, Tr, Ind	6–4⅛×4½	40.3	95–2,400	360.8	5.00
218	Hercules, WXC-3	T, B, Tr, Ind	6–4¼×4½	43.3	101–2,400	383.0	5.00
219	Hercules, WXLC	T, B, Tr, M, Ind	6–4×4¾	38.4	104–2,800	358.0	5.42

Data continued on pp. 363 and 377

TABLE A1-3.—AMERICAN STOCK, MARINE, AND COMMERCIAL VEHICLE ENGINES.—(*Continued*)

Line number	Make and model	Designed for	Number of cylinders; bore and stroke, in.	Rated hp. (A.M.A.)	Maximum b.hp. at specified r.p.m.	Piston displacement, cu. in.	Compression ratio—to 1
220	Hercules, WXLC-3	T, B, Tr, M, Ind	6–4¼×4¾	43.3	118–2,800	404.0	5.30
221	Hercules, YXC	T, B, Tr, Ind	6–4⅜×4¾	45.9	94–2,200	428.4	4.40
222	Hercules, YXC-2	T, B, Tr, Ind	6–4½×4¾	48.6	98–2,200	453.0	4.77
223	Hercules, YXC-3	T, B, Tr, Ind	6–4⅝×4¾	51.3	104–2,200	478.8	4.40
224	Hercules, RXB	T, B, Tr, Ind	6–4½×5¼	48.6	110–2,200	500.9	4.95
225	Hercules, RXC	T, B, Tr, Ind	6–4⅝×5¼	51.3	114–2,200	529.2	4.95
226	Hercules, RXLC	T, B, Tr, Ind	6–4⅝×5¼	51.3	135–2,200	529.2	5.40
227	Hercules, RXLD	T, B, Tr, Ind	6–4¾×5¼	54.2	142–2,200	558.0	5.40
228	Hercules, HXB	T, B, Tr, Ind	6–5×6	60.0	148–2,000	707.0	4.50
229	Hercules, HXC	T, B, Tr, Ind	6–5¼×6	66.2	164–2,000	779.0	4.50
230	Hercules, HXD	T, B, Tr, Ind	6–5½×6	72.8	180–2,000	855.0	4.50
231	Hercules, HXE	T, B, Tr, Ind	6–5¾×6	79.4	198–2,000	935.0	4.50
232	Hudson, 40-C	C, T	6–3×4⅛	21.6	92–4,000	175.0	7.00
233	Hudson, 48-C	C, T	6–3×5	21.6	98–4,000	212.0	6.50
234	International, U-7	PU	4–3¾×5	22.5	34.5–1,200	220.9	4.80
235	International, U-10	PU	4–4¼×5	28.9	45–1,200	283.7	4.67
236	International, 300	PU	4–4¾×6	36.1	56.5–1,050	425.3	4.74
237	International, U-21	PU	6–3¾×4½	33.7	66–2,000	298.2	5.72
238	International, PA-100	PU	6–5×5½	60.0	110–1,400	648.0	5.30
239	Kermath, 1-10	M	2–4×4	....	10– 800	101.0	
240	Kermath, ZX	M	4–2⅝×3	....	25–3,400	65.0	6.00
241	Kermath, IXL	M	4–3¼×4	....	33–2,200	134.0	5.50
242	Kermath, IXH	M	4–3¼×4	....	50–3,200	134.0	5.50
243	Kermath, 20	M	4–4×4	....	20–1,000	201.0	
244	Kermath, F	M	4–4⅜×5½	....	55–1,500	330.0	4.80
245	Kermath, P-840	M	6–3¼×3¾	....	110–3,600	249.0	7.00

Data continued on pp. 364 and 378

246	Kermath, QXC	M	6–3⅜×4⅛	.....	95–3,600	221.0	6.50
247	Kermath, P-640	M	6–3 7/16×4	.....	95–3,700	223.0	7.20
248	Kermath, JXD	M	6–4×4¼	.....	96–2,600	320.0	5.80
249	Kermath, JXD-HS	M	6–4×4¼	.....	120–3,000	320.0	6.90
250	Kermath, WX	M	6–4¼×4½	.....	115–2,600	383.0	5.80
251	Kermath, WXL	M	6–4¼×4¾	.....	155–3,000	404.0	6.50
252	Kermath, D	M	6–4⅜×5¾	.....	150–2,500	520.0	5.70
253	Kermath, L	M	6–5×5¾	.....	157–2,000	678.0	5.30
254	Kermath, LA	M	6–5×5¾	.....	200–2,400	678.0	5.70
255	Kermath, R	M	6–5×5¾	.....	225–2,400	678.0	5.70
256	Kermath, VF	M	8–3 1/16×3⅜	.....	85–3,800	221.0	6.30
257	Kermath, VM	M	8–3 3/16×3¾	.....	95–3,600	239.0	6.15
258	Kermath, P-8	M	8–3¼×3½	.....	100–3,600	232.0	7.00
259	Kermath, VZ	M	12–2⅞×3¾	.....	120–3,500	292.0	
260	Kermath, V	M	12–5×6	.....	500–2,400	1,414.0	5.70
261	Lathrop, Standard	M	2–5 3/16×6½	.....	16– 600	274.7	
262	Lathrop, Standard	M	2–5½×6½	.....	20– 700	308.8	
263	Lathrop, Standard	M	3–5 3/16×6½	.....	27– 700	412.1	
264	Lathrop, Standard	M	3–5½×6½	.....	34– 800	463.2	
265	Lathrop, LH	M	4–3¼×4	.....	38–2,200	133.0	
266	Lathrop, LH	M	4–4×4½	.....	49–1,900	226.0	
267	Lathrop, Standard	M	4–5 3/16×6½	.....	29– 700	549.5	
268	Lathrop, Standard	M	4–5½×6½	.....	49– 800	617.7	
269	Lathrop, Engineers	M	4–5½×7	.....	64–1,000	665.2	
270	Lathrop, Engineers	M	4–6×7	.....	76–1,000	791.6	
271	Lathrop, LH	M	6–3¾×4¼	.....	62–2,200	282.0	
272	Lathrop, LH-D6	M	6–4×4¼	.....	107–2,500	320.0	5.80
273	Lathrop, Mystic	M	6–4½×5½	.....	96–1,600	524.8	
274	Lathrop, Mystic	M	6–4¾×5½	.....	106–1,600	584.7	
275	Lathrop, Standard	M	6–5½×6½	.....	73– 750	926.5	
276	Lathrop, Standard	M	6–5½×6½	.....	103–1,000	926.5	
277	Lathrop, Mystic	M	6–5½×6½	.....	155–1,500	926.5	
278	Lathrop, Mystic	M	6–5¾×6½	.....	179–1,600	1,012.8	
279	Lehman-Ford, F1	M	4–3⅞×4¼	.....	50–2,500	200.5	4.22
280	Lehman-Ford, D5	M, Ind	8–2.6×3.2	.....	60–3,400	136.0	6.60
281	Lehman-Ford, V5	M, Ind	8–3 1/16×3¾	.....	85–3,400	221.0	6.15
282	Lehman-Mercury, M5	M, Ind	8–3.185×3¾	.....	95–3,400	239.0	6.15
283	Lehman-Zephyr, Z5	M, Ind	12–2⅞×3¾	.....	120–3,400	287.0	6.70
284	Mack, EN-11	T	6–3 3/16×4⅜	24.6	67–3,000	210.0	5.75
285	Mack, FO	T, B, Tr	6–3½×4⅜	29.4	78–3,000	253.0	5.69
286	Mack, FM	T, B, Tr	6–3⅝×4⅜	31.6	83–3,000	271.0	5.65
287	Mack, BG	T, B, Tr	6–3⅝×5	31.6	96–2,800	310.0	5.40
288	Mack, FK	T, B, Tr	6–3¾×4⅜	33.8	94–3,000	290.0	5.68
289	Mack, CU	T, B, Tr	6–3⅞×5	36.0	103–2,600	353.8	5.25
290	Mack, CE	T, Tr	6–4×5½	38.4	108–2,400	415.0	5.00
291	Mack, CF	T, Tr	6–4¼×5½	43.3	118–2,400	468.0	5.00
292	Mack, EO	T, B, Tr	6–4⅜×5¾	45.7	140–2,300	519.0	5.50

Data continued on pp. 365 and 379

TABLE A1-3.—AMERICAN STOCK, MARINE, AND COMMERCIAL VEHICLE ENGINES.—(*Continued*)

Line number	Make and model	Designed for	Number of cylinders; bore and stroke, in.	Rated hp. (A.M.A.)	Maximum b.hp. at specified r.p.m.	Piston displacement, cu. in.	Compression ratio—to 1
293	Mack, CT	T, B, Tr	6–4½×5½	48.6	126–2,400	524.8	4.80
294	Mack, EP	T, B, Tr	6–4¾×5¾	54.1	155–2,200	611.0	5.40
295	Mack, EY	T, B, Tr	6–5×6	60.0	170–2,100	707.0	5.35
296	M-M Twin City,[8] EE	Tr, Ind	4–3⅝×4	20.9	24–1,400	165.1	5.75
297	M-M Twin City, R	Tr, Ind	4–3⅝×4½	20.9	33–1,500	185.8	5.75
298	M-M Twin City, KEC	Tr, Ind	4–4¼×5	28.9	47–1,275	283.7	5.60
299	M-M Twin City, KED	Tr, Ind	4–4¼×5	28.9	47–1,275	283.7	5.60
300	M-M Twin City, GE	Tr, Ind	4–4⅝×6	34.2	59–1,075	403.2	5.25
301	M-M Twin City, TA	M, Ind	4–7¼×9	84.1	119– 650	1,486.0	4.80
302	M-M Twin City, BE	M, Ind	4–7¾×9	96.0	136– 650	1,698.0	4.40
303	M-M Twin City, ME	M, Ind	4–8×9	102.0	145– 650	1,810.0	4.70
304	M-M Twin City, SE	M, Ind	6–7¼×9	120.0	173– 650	2,229.0	4.80
305	M-M Twin City, TE	M, Tr	6–7¾×9	144.0	198– 650	2,547.0	4.40
306	M-M Twin City, NE	M, Ind	6–8×9	153.6	210– 650	2,714.0	4.70
307	Murray & Tregurtha, OC-4	M	4–6½×8	.....	80–1,000	1,062.4	3.33
308	Murray & Tregurtha, M-4	M	4–6½×8	.....	90–1,000	1,062.4	4.20
309	Murray & Tregurtha, K-6	M	6–6¼×7¾	.....	346–1,650	1,426.6	5.25
310	Murray & Tregurtha, OC-6	M	6–6½×8	.....	140–1,100	1,593.6	3.33
311	Murray & Tregurtha, M-6	M	6–6½×8	.....	175–1,100	1,593.6	4.20
312	Murray & Tregurtha, OCX-6	M	6–7¼×8	.....	175–1,100	1,981.4	4.00
313	Plymouth, P9-10	C, T	6–3⅛×4⅜	23.4	84–3,600	201.3	6.70
314	Regal, Y	M	1–3¼×3½	.....	2– 800	29.0	
315	Regal, OA	M	1–4×4½	.....	4– 800	56.0	
316	Regal, HA	M	1–4¾×5½	.....	6– 800	97.0	
317	Regal, EA	M	1–5¼×6½	.....	7– 600	141.0	
318	Regal, DV	M	2–2½×2¾	.....	5–2,000	27.0	

Data continued on pp. 366 and 380

319	Regal, XB	M	2–3¼×4	.....	15–1,900	66.3	
320	Regal, NB	M	2–4×4½	.....	8– 800	112.0	
321	Regal,[7] GB	M	2–4¾×6	.....	16–1,000	213.0	
322	Regal, KB	M	2–5½×7	.....	16– 600	332.0	
323	Regal, LB	M	2–6½×8	.....	20– 600	531.0	
324	Regal, SC	M	4–7½×9	.....	50– 400	1,590.0	
325	Reo, 228	T	6–3⅜×4¼	27.3	78–3,200	228.0	6.20
326	Reo, 245	T, B	6–3½×4¼	29.4	86–3,400	245.0	6.20
327	Reo, 288	T, B	6–3½×5	29.4	87–3,000	288.0	6.20
328	Reo, 310	T, B	6–3⅝×5	31.5	97–2,800	310.0	6.20
329	Reo, 361	T, B	6–4⅛×4½	40.8	100–2,800	361.0	6.00
330	Scripps, 34	M	4–3¼×4	.....	30–.....	134.0	5.20
331	Scripps, 36	M	4–3¼×4	.....	50–3,200	134.0	
332	Scripps, F4	M	4–3¾×5	.....	81–3,000	220.0	6.10
333	Scripps, 94–95	M	6–3⅜×4⅛	.....	70–2,000	221.0	5.85
334	Scripps, 96–97	M	6–3⅜×4⅛	.....	90–3,600	221.0	
335	Scripps, F6	M	6–3¾×5	.....	120–3,000	331.0	6.10
336	Scripps, 104–105	M	6–4×4¼	.....	83–2,000	320.0	5.63
337	Scripps, 106–107	M	6–4×4¼	.....	95–2,600	320.0	
338	Scripps, 150, 1, 2, 3	M	6–4¼×5¼	.....	165–3,000	447.0	6.20
339	Scripps, 154, 5, 6, 7	M	6–4¼×5¼	.....	155–3,000	447.0	6.20
340	Scripps, 160, 1, 2, 3	M	6–4½×5¾	.....	166–2,400	549.0	5.20
341	Scripps, 164, 5	M	6–4½×5¾	.....	125–2,000	549.0	5.20
342	Scripps, 170, 1, 2, 3	M	6–4¾×5¾	.....	186–2,400	611.0	5.20
343	Scripps, 174, 5	M	6–4¾×5¾	.....	146–2,000	611.0	5.20
344	Scripps, 200, 1, 2, 3	M	6–5×5¾	.....	212–2,400	677.0	6.20
345	Scripps, 204, 5, 6, 7	M	6–5×5¾	.....	165–2,000	677.0	6.20
346	Scripps, 208, 9	M	6–5×5¾	.....	220–2,400	677.0	5.85
347	Scripps, 214, 15	M	6–5×5¾	.....	185–2,000	677.0	5.00
348	Scripps, V8	M	8–3 1/16×3¾	.....	85–3,400	221.0	6.16
349	Scripps, V8-Mercury	M	8–3.185×3¾	.....	95–3,600	239.0	6.15
350	Scripps, V12-61, 63, 65, 67, 71, 73, 75, 77	M	12–2¾×3¾	.....	110–3,900	267.3	6.70
351	Scripps, 300, 1, 2, 3	M	12–4¼×5¼	.....	304–2,400	894.0	6.20
352	Scripps, 304, 5, 6, 7	M	12–4¼×5¼	.....	280–2,400	894.0	6.20
353	Seaman, EM-2	M	2–2½×2½	.....	6–1,800	24.5	7.30
354	Seaman, EVM-4	M	4–2½×2½	.....	15–2,600	49.0	7.30
355	Seaman,[8] EC-2	Ind	2–2½×2½	5.0	6–1,800	24.5	7.30
356	Speedway, SW	M	6–4¼×4¾	.....	115–2,500	404.6	5.30
357	Speedway, MP	M	6–5¾×7	.....	190–1,300	1,092.0	4.30
358	Speedway, MC	M	6–5¾×7	.....	260–1,800	1,092.0	5.00
359	Speedway, R	M	6–7×8½	.....	300–1,200	1,963.0	4.20
360	Sterling, Neptune D2-12	M, Ind	2–5½×7	24.2	15– 500	332.6	
361	Sterling, Petrel L-6	M, T, Tr, B, Ind	6–5¼×6	66.1	115–1,200	780.0	4.30
362	Sterling, Petrel L-6	M, T, Tr, B, Ind	6–5¼×6	66.1	145–1,500	780.0	4.68
363	Sterling, Petrel L-6	M, T, Tr, B, Ind	6–5¼×6	66.1	145–1,500	780.0	4.68
364	Sterling, Petrel L-6-6	M, T, Tr, B, Ind	6–5¼×6	66.1	180–1,800	780.0	5.00
365	Sterling, Petrel Reduction L	M, T, Tr, B, Ind	6–5¼×6	66.1	175–1,800	780.0	4.68

Data continued on pp. 367 and 381

TABLE A1-3.—AMERICAN STOCK, MARINE, AND COMMERCIAL VEHICLE ENGINES.—(*Continued*)

Line number	Make and model	Designed for	Number of cylinders; bore and stroke, in.	Rated hp. (A.M.A.)	Maximum b.hp. at specified r.p.m.	Piston displacement, cu. in.	Compression ratio—to 1
366	Sterling, Petrel L-6	M, T, Tr, B, Ind	6-5¼×6	66.1	200-2,000	780.0	5.54
367	Sterling, Petrel L-6	M, T, Tr, B, Ind	6-5¼×6	66.1	225-2,200	780.0	5.50
368	Sterling, Chevron 6	Tr, M, Ind	6-5¼×6¾	72.6	85- 800	962.0	
369	Sterling, Chevron 6	Tr, M, Ind	6-5½×6¾	72.6	130-1,200	962.0	
370	Sterling, Chevron 6	Tr, M, Ind	6-5½×6¾	72.6	150-1,500	962.0	
371	Sterling, Dolphin-Med. GRM-6	Tr, M, Ind	6-5¾×6¾	79.3	165-1,200	1,051.6	3.85
372	Sterling, Dolphin 6-GR-6	Tr, M, Ind	6-5¾×6¾	79.3	225-1,550	1,051.6	4.08
373	Sterling, Dolphin 6-GRS-6	Tr, M, Ind	6-5¾×6¾	79.3	300-2,000	1,051.6	4.70
374	Sterling, Coast Guard M-6	Tr, M, Ind	6-6¼×7¾	93.7	225-1,200	1,426.8	4.13
375	Sterling, Coast Guard 6-M-6	Tr, M, Ind	6-6¼×7¾	93.7	300-1,550	1,426.8	4.13
376	Sterling, Viking 11 T-6	Tr, M, Ind	6-8×9	153.6	190- 600	2,714.3	3.93
377	Sterling, Viking 11 T-6	Tr, M, Ind	6-8×9	153.6	300- 900	2,714.3	4.18
378	Sterling, Viking 11 T-6	Tr, M, Ind	6-8×9	153.6	425-1,200	2,714.3	4.18
379	Sterling, Dolphin-Med. GRM-8	Tr, M, Ind	8-5¾×6¾	105.8	220-1,200	1,402.2	3.85
380	Sterling, Dolphin 8-GR-8	Tr, M, Ind	8-5¾×6¾	105.8	300-1,550	1,402.2	4.08
381	Sterling, Viking 11 8-T-8	Tr, M, Ind	8-8×9	204.8	250- 600	3,619.0	3.93
382	Sterling, Viking 11 8-T-8	Tr, M, Ind	8-8×9	204.8	400- 900	3,619.0	4.18
383	Sterling, Viking 11 8-T-8	Tr, M, Ind	8-8×9	204.8	565-1,200	3,619.0	4.18
384	Thorobred, K	M	1-3¾×4¾		5-1,000	52.5	4.00
385	Thorobred, KK	M	2-3¾×4¾		10-1,100	105.0	4.00
386	Thorobred, Meteor	M	4-2½×3⅛		18-2,800	61.3	5.70
387	Thorobred, DS	M	4-2¾×4		16-1,800	95.0	4.66
388	Thorobred, Arrowhead Jr	M	4-3¼×4		35-2,600	133.0	5.60
389	Thorobred, Arrowhead	M	4-3⅝×4½		42-2,250	186.0	4.60
390	Thorobred, AA	M	4-3¾×4¾		24-1,200	210.0	4.00
391	Thorobred, F	M	4-4 1/16×5		36-1,400	259.0	4.00
392	Thorobred, B	M	4-4½×5		44-1,800	318.0	4.00

Data continued on pp. 368 and 382

393	Thorobred, BB-4	M	4-4½×6	.....	56-1,600	382.0	4.00
394	Thorobred, BC-4	M	4-5×7	.....	56-1,200	550.0	4.00
395	Thorobred, BCS-4	M	4-5¾×7	.....	71-1,100	727.0	4.00
396	Thorobred, BC-Super-4	M	4-6×7	.....	78-1,100	791.0	4.00
397	Thorobred, Hiawatha	M	6-3¾×4¼	.....	82-2,800	282.0	5.70
398	Thorobred, Arrow Super-6	M	6-4¼×4¾	.....	95-2,500	404.0	5.60
399	Thorobred, BB-6	M	6-4½×6	.....	80-1,725	572.5	4.00
400	Thorobred, BBS-6	M	6-5×6	.....	101-1,500	707.0	4.00
401	Thorobred, BC-6	M	6-5×7	.....	90-1,100	825.0	4.00
402	Thorobred, BCS-6	M	6-5¾×7	.....	112-1,100	1,091.0	4.00
403	Thorobred, BC-Super-6	M	6-6×7	.....	124-1,100	1,187.5	4.00
404	Universal, Fisherman-WM	M	1-4⅜×4½	.....	8-1,200	67.6	4.60
405	Universal, Blue Jacket-AFTL	M	2-3×3½	.....	10-2,000	49.5	5.75
406	Universal, Utility Four-BN	M	4-2¾×4	.....	25-2,500	95.0	5.00
407	Universal, Flexifour-FA	M	4-3×3½	.....	40-3,500	99.0	5.00
408	Universal, Superfour-LSG	M	4-3¼×4½	.....	50-3,000	149.3	5.70
409	Universal, American Six-AMS	M	6-3×3½	.....	60-3,500	148.5	6.00
410	Universal, Cruiser Six-HCS	M	6-3½×4½	.....	90-3,000	260.0	5.75
411	Universal, Sea Lion Six-LHS	M	6-3½×4½	.....	110-3,400	260.0	5.75
412	Universal, Cruiser Eight-GCE	M	8-3½×4½	.....	125-3,000	347.0	5.75
413	Universal, Sea Lion Eight-LCE	M	8-3½×4½	.....	141-3,400	347.0	5.75
414	Vimalert, M-12	M	12-5×7	.....	420-1,900	1,650.0	6.25
415	Vimalert, Duplex Unit	M*					
416	Vimalert, V-2500-2	M	12-6⅜×6½	.....	1,200-2,350	2,500.0	6.00
417	Volcano, 12-2200	Ind	1-3⅞×4⅞	.....	12-2,200	57.5	4.75
418	Volcano, 24-2200	Ind	2-3⅞×4⅞	.....	24-2,200	115.0	4.75
419	Volcano, 48.2200	C, T, Tr, Ind	4-3⅞×4⅞	.....	48-2,200	230.0	4.75
420	Waukesha, (H)150	C, Ind	2-3×2¾	7.2	....	39.0	5.60
421	Waukesha, ICK	T, M, Ind	4-2½×3⅛	10.0	18-2,800	61.3	5.70
422	Waukesha, FCS	T, Tr	4-2¾×4	12.1	26-2,600	94.0	4.85
423	Waukesha, FC	T, Tr, Ind	4-3¼×4	16.9	35-2,600	133.5	5.58
424	Waukesha, XAH	T, Tr, Ind	4-3⅝×4½	21.0	37-2,200	186.0	5.10
425	Waukesha, 130-GS	T, Tr, Ind	4-3¾×5	22.5	47-1,900	221.0	6.12
426	Waukesha, 130-GL	T, Tr, Ind	4-4×5	25.6	53-1,900	251.0	5.95
427	Waukesha, VIM	Tr	4-4¼×5¼	29.0	53-1,600	298.0	4.23
428	Waukesha, VIK	Tr, Ind	4-4½×5¼	34.2	58-1,600	334.0	4.30
429	Waukesha, VRZ	T, Tr	4-4⅝×5¼	34.2	....	353.0	5.63
430	Waukesha, CHK	Tr, Ind	4-5⅛×6¼	42.0	79-1,400	516.0	4.20
431	Waukesha, HL	Tr, Ind	4-6×6½	57.6	91-1,200	735.0	3.80
432	Waukesha, WK	Ind	4-6¾×8	73.0	120- 950	1,145.0	4.00
433	Waukesha, WOK	Ind	4-7½×8	90.0	163- 950	1,414.0	4.16
434	Waukesha, 6BM	T, B	6-3⅝×4¼	31.5	77-2,800	263.0	5.70
435	Waukesha, 6BK	T, B, Ind	6-3¾×4¼	33.8	82-2,800	282.0	5.70
436	Waukesha, 6BZ	T, B, Ind	6-4×4¼	38.4	85-2,800	320.0	5.75
437	Waukesha, 6MKR	T, B, Ind	6-4⅛×4¾	40.8	90-2,500	381.0	5.34
438	Waukesha, 6MZR	T, B, Ind	6-4¼×4¾	41.0	95-2,500	404.0	5.38

Data continued on pp. 369 and 383

* Unit is composed of two model M-12 engines placed end to end, with a gear box between, driving one propeller shaft, 800 hp., weight 5,400 lb.

TABLE A1-3.—AMERICAN STOCK, MARINE, AND COMMERCIAL VEHICLE ENGINES.—(*Continued*)

Line number	Make and model	Designed for	Number of cylinders; bore and stroke, in.	Rated hp. (A.M.A.)	Maximum b.hp. at specified r.p.m.	Piston displacement, cu. in.	Compression ratio—to 1
439	Waukesha, 140-GS	T, B, Ind	6–4¼×5½	43.4	122–2,200	468.0	5.78
440	Waukesha, 6SRLR	T, B, Ind	6–4⅜×5⅛	46.0	114–2,250	462.0	5.50
441	Waukesha, 140-GK	T, B, Ind	6–4½×5½	48.6	137–2,200	525.0	5.78
442	Waukesha, 6SRKR	T, B, Ind	6–4⅝×5⅛	51.3	125–2,250	517.0	5.50
443	Waukesha, 145-GS	T, B, Ind	6–4¾×6	54.2	138–1,800	638.0	5.63
444	Waukesha, 6GAL	T, B, Ind	6–5×5½	60.0	136–2,000	648.0	5.00
445	Waukesha, 6RBR	T, B	6–5×5¾	60.0	150–2,000	677.0	5.35
446	Waukesha, 145-GK	T, B, Ind	6–5¼×6	66.2	168–1,800	779.0	5.63
447	Waukesha, 6GSK	T, B, Ind	6–5½×5½	72.5	155–2,200	784.0	4.80
448	Waukesha, 6WAL	T, B, Ind	6–5¾×6½	79.5	172–1,800	1,013.0	5.00
449	Waukesha, 6WAK	T, B, Ind	6–6¼×6½	94.0	202–1,800	1,196.0	5.00
450	Waukesha, 6EL	Ind	6–6½×7	118.0	165–1,125	1,395.0	4.00
451	Waukesha, 6EK	Ind	6–7×7	137.0	190–1,125	1,616.0	4.00
452	Waukesha, 6NK	Ind	6–7×8½	137.0	220–1,050	1,962.0	5.50
453	Waukesha, 6LS	Ind	6–7×8½	137.0	240–1,050	1,962.0	3.80
454	Waukesha, 6LK	Ind	6–7¾×8½	144.0	291–1,050	2,406.0	3.70
455	Waukesha, 6LRO	Ind	6–8½×8½	173.4	343–1,050	2,894.0	3.80
456	White, 250	T, B	6–3 7/16×4½	28.4	76–3,000	250.6	6.13
457	White, 270	T, B	6–3 9/16×4½	30.4	91–2,900	270.0	5.88
458	White, 264	T	6–3⅝×4¼	31.5	78–2,600	264.0	6.11
459	White, 318	T, B	6–3⅞×4½	36.0	110–3,000	318.0	6.30
460	White, 362	T, B	6–3⅞×5⅛	36.0	116–2,900	362.0	5.88
461	White, 460	T, B	6–4 5/16×5¼	44.6	123–2,400	460.0	5.00
462	White, 529	T	6–4⅝×5¼	51.3	134–2,400	529.0	5.00
463	White, (H)681	B	12–4⅛×4¼	81.7	210–3,000	681.0	5.88
464	Willys, 440	C, T	4–3⅛×4⅜	15.6	61–3,600	134.2	6.48

Data continued on pp. 370 and 384

465	Wisconsin, AA	M, Ind	1–2¼×2¾	2.0	1.8–2,600	10.9	4.40
466	Wisconsin, AB	Ind	1–2½×2¾	2.5	3–2,600	13.5	4.40
467	Wisconsin, AD	Ind	1–2¾×3¼	3.0	4.3–2,600	19.3	4.20
468	Wisconsin, AK	Ind	1–2⅞×2¾	3.3	4.2–2,400	17.8	4.59
469	Wisconsin, AE	M, Tr, Ind	1–3×3¼	3.6	4.7–2,400	22.9	4.13
470	Wisconsin, AF	M, Tr, Ind	1–3¼×4	4.2	5.4–2,000	33.1	3.71
471	Wisconsin, AG	M, Tr, Ind	1–3½×4	4.9	6.2–2,000	38.5	3.72
472	Wisconsin, AH	M, Tr, Ind	1–3⅝×4	5.2	8.2–2,200	41.3	4.18
473	Wisconsin, AC-4	M, Tr, Ind	4–2⅝×3¼	11.0	16–2,600	70.4	4.60
474	Wisconsin, AM-4	Tr, Ind	4–3¼×4	16.9	25–2,200	132.0	4.56
475	Wisconsin, AP-4	Tr, Ind	4–3½×4	19.5	32–2,100	154.0	4.60
476	Wisconsin, SU	T, Tr, Ind	4–4×5	25.6	38–1,600	251.0	4.20
477	Wisconsin, X	T, Tr, Ind	4–4½×5	32.4	66–1,900	318.0	4.25
478	Wisconsin, GA-2	T, Tr, Ind	6–3¾×5	33.7	49–1,600	331.0	4.50
479	Wisconsin, L-2	T, Tr, Ind	6–3⅞×5	36.0	62–1,800	354.0	4.27
480	Wisconsin, L-3	T, Tr, Ind	6–4⅛×5	40.7	68–1,800	401.0	4.30
481	Wisconsin, L-4	T, Tr, Ind	6–4¼×5	43.4	71–1,800	426.0	4.26
482	Wisconsin, ZA-1	Tr, Ind	6–4½×5	48.6	78–1,600	477.0	4.50
483	Wisconsin, ZA-2	T, Tr, Ind	6–4⅝×5	.....	82–1,600	504.0	4.88

Data continued on pp. 371 and 385

Symbols and Abbreviations

(Continued on pp. 371 and 385)

[a] Also available in reduction-gear models.
[b] Also available in rated-horsepower rotation.
[c] Pressure also to camshaft thrust bearing.
[d] Weight complete with ignition and carburetor.
[e] Cast-iron pistons also supplied.
[f] Tocco hardened.
[g] Ball or roller bearings used.
[h] Weight per pair.
[1] 6.15 ratio for cars, 5.81 for trucks.
[2] Two used.
[3] Three used.
[4] Four used.
[5] 155 ft.-lb. torque at 2,200 for cars; 150 ft.-lb. at 2,000 for trucks.
[6] Minneapolis-Moline Power Implement.
[7] Also built in 4- and 6-cylinder models.
[8] Also built in 4-cylinder model.

a—Main bearings.
aa—Forked rod, 88 oz.; plain rod, 50 oz.
AI—Alloy iron.
Al—Aluminum alloy.
Ala—Aluminum alloy, anodized.
Als—Aluminum alloy with steel strut.
Alt—Aluminum alloy, tin plated.
AS—Alloy steel.
AUS—Austenitic steel.
b—Connecting-rod bearings.
B—buses.
BG—Bevel gear.
Bo—Used in both intake and exhaust seats.
c—Camshaft bearings.
C—Cars.
CAl—Chrome aluminum.

TABLE A1-3.—AMERICAN STOCK, MARINE, AND COMMERCIAL VEHICLE ENGINES.—(*Continued*)

Line number	Maximum torque at r.p.m., lb.-ft.	Cylinders: No. cast in one piece	Cylinders: Liners, type	Crankcase, upper half integral with cylinders?	Valves: Arrangement	Valves: Exhaust-head material (S.A.E. No.)	Valves: Max. head diameter, in.: Intake	Valves: Max. head diameter, in.: Exhaust	Valves: Min. port diameter, in.: Intake	Valves: Min. port diameter, in.: Exhaust	Valves: Lift, in.: Intake	Valves: Lift, in.: Exhaust	Valves: Stem diameter, in.: Intake	Valves: Stem diameter, in.: Exhaust	Valves: Seats: Angle, deg.	Valves: Seats: Inserts used?	Valves: Seats: Insert material (S.A.E. No.)	Front-end drive, type	Pistons: Material	Pistons: Length, in.	Pistons: Weight with pins, rings, bushings, oz.	Pistons: Piston pin, diameter and length, in.
1	74–1,100	4	W	In	I	Sil	1.43	1.31	1.20	1.03	0.374	0.374	0.341	0.341	45	N	......	HG	CI	3.68	39	0.813×2.87
2	128–1,200	4	W	In	I	Sil	1.68	1.50	1.50	1.32	0.376	0.376	0.372	0.372	45	E	TA	HG	CI	4.43	67	0.989×3.50
3	200–1,000	4	W	In	I	Sil	2.03	1.78	1.75	1.50	0.375	0.375	0.372	0.372	30	E	TA	HG	CI	5.25	92	1.31 ×4.06
4	370– 750	4	W	In	I	Sil	2.21	2.21	2.00	2.00	0.440	0.417	0.497	0.497	45	E	TA	HG	CI	6.78	162	1.50 ×4.87
5	590– 700	6	W	In	I	Sil	2.21	2.21	2.00	2.00	0.440	0.417	0.497	0.497	45	E	TA	HG	CI	6.78	162	1.50 ×4.87
6	220–1,000	6	N	Se	L	Sil	1.75	1.65	1.56	1.43	0.375	0.375	0.437	0.437	45	E	71,360	HG	Al	4.87	36	1.12 ×3.18
7	270–1,000	6	N	Se	L	Sil	1.90	1.78	1.68	1.56	0.375	0.375	0.437	0.437	45	E	71,360	HG	Al	4.87	43	1.12 ×3.43
8	295–1,000	6	N	Se	L	Sil	2.06	1.93	1.87	1.75	0.375	0.375	0.437	0.437	45	E	71,360	HG	Al	5.75	47	1.12 ×3.43
9	333– 900	6	N	Se	L	Sil	2.06	1.93	1.87	1.75	0.375	0.375	0.437	0.437	45	E	71,360	HG	Al	5.75	51	1.12 ×3.68
10	375– 800	6	N	Se	L	Sil	2.06	1.93	1.87	1.75	0.375	0.375	0.437	0.437	45	E	71,360	HG	Al	5.75	57	1.12 ×3.93
11	31–2,500	4	N	Se	I	Sil	1.00	1.00	0.87	0.87	0.250	0.250	0.312	0.312	45	N	......	HG	Al	.....	6	0.62 ×1.87
12	34–3,200	4	N	Se	L	Sil	1.00	1.00	0.87	0.87	0.250	0.250	0.312	0.312	45	N	......	HG	Al	.....	6	0.62 ×2.00
13	34–3,200	4	N	Se	L	Sil	1.00	1.00	0.87	0.87	0.250	0.250	0.312	0.312	45	N	......	HG	Al	.....	6	0.62 ×2.00
14	155–1,000	4	N	Se	L	Sil	.....	.....	.....	.....	.....	.....	.....	.....	..	..	......	HG	SS	4.50	64	1.17 ×3.87
15	203–1,000	4	N	Se	L	Sil	2.00	2.00	1.87	1.87	0.375	0.375	0.375	0.375	45	N	......	HG	SS	5.00	72	1.17 ×4.00
16	203–1,000	4	N	Se	L	Sil	2.00	2.00	1.87	1.87	0.375	0.375	0.375	0.375	45	N	......	HG	SS	5.00	80	1.17 ×4.00
17	278– 900	3	N	Se	I	Sil	2.12	2.12	2.00	2.00	0.375	0.375	0.437	0.437	45	N	......	HG	SS	4.50	64	1.17 ×3.87
18	278– 900	3	N	Se	I	Sil	2.12	2.12	2.00	2.00	0.375	0.375	0.437	0.437	45	N	......	HG	SS	4.50	64	1.17 ×3.87
19	350–1,200	3	N	Se	I	Sil	2.12	2.12	2.00	2.00	0.375	0.375	0.437	0.437	45	N	......	HG	SS	5.50	70	1.25 ×3.87
20	350–1,200	3	N	Se	I	Sil	2.12	2.12	2.00	2.00	0.375	0.375	0.437	0.437	45	N	......	HG	SS	4.50	76	1.25 ×4.00
21	500–1,200	3	N	Se	I	Sil	2.50	2.50	2.12	2.12	0.437	0.437	0.500	0.500	45	N	......	BG	SS	5.00	72	1.37 ×4.00
22	500–1,200	3	N	Se	I	Sil	2.50	2.50	2.12	2.12	0.437	0.437	0.500	0.500	45	N	......	BG	SS	5.00	72	1.37 ×4.00
23	....	1	N	In	I	NS	1.43	1.43	.....	.....	0.312	0.312	.....	.....	45	N	......	SG	CI	4.00	58	0.75 ×3.62
24	....	2	N	In	I	NS	1.43	1.43	.....	.....	0.312	0.312	.....	.....	45	N	......	SG	CI	4.00	58	0.75 ×3.62
25	....	4	N	In	L	CNS	1.12	1.12	.....	.....	0.312	0.312	.....	.....	45	N	......	HG	CI	3.25	41	0.62 ×2.62
26	....	4	N	In	L	CNS	1.37	1.25	.....	.....	0.312	0.312	.....	.....	45	N	......	HG	Al	3.50	50	0.75 ×3.18

27	....	4	N	In	L	NS	1.62	1.62	.....	.....	0.375	0.375	.....	.....	45	N	......	HG	CI	4.00	59	1.37×4.12
28	....	6	N	In	L	Sil	1.75	1.75	.....	.....	0.375	0.375	.....	.....	45	N	......	HG	CI	4.00	61	1.25×4.25
29	132–1,200	4	N	In	L	2,112	1.65	1.53	1.50	1.37	0.344	0.344	0.372	0.372	45	N	......	HG	SS	3.75	42	1.12×3.22
30	148–1,000	4	N	In	L	2,112	1.65	1.53	1.50	1.37	0.344	0.344	0.372	0.372	45	N	......	HG	CI	3.75	42	1.12×3.22
31	148–1,000	4	N	In	L	2,112	1.65	1.53	1.50	1.37	0.344	0.344	0.372	0.372	45	N	......	HG	SS	3.75	42	1.12×3.22
32	148–1,000	4	N	In	L	2,112	1.65	1.53	1.50	1.37	0.344	0.344	0.372	0.372	45	N	......	HG	SS	3.75	42	1.12×3.22
33	148–1,000	4	N	In	L	2,112	1.65	1.53	1.50	1.37	0.344	0.344	0.372	0.372	45	N	......	HG	SS	3.75	42	1.12×3.22
34	173–1,000	4	N	Se	L	Sil	1.87	1.87	1.62	1.62	0.281	0.312	0.372	0.372	45	N	......	HG	CI	5.00	65	1.50×3.31
35	264– 700	4	N	Se	L	2,112	2.37	2.37	2.12	2.12	0.281	0.312	0.434	0.434	45	E	JM	HG	CI	6.12	111	1.43×4.11
36	360– 650	4	N	Se	L	2,112	2.50	2.50	2.25	2.25	0.375	0.375	0.434	0.434	45	E	......	HG	CI	6.75	142	1.37×4.37
37	405– 650	4	N	Se	L	2,112	2.50	2.50	2.25	2.25	0.375	0.375	0.434	0.434	45	E	DC	HG	CI	6.75	144	1.37×4.87
38	472– 750	2	N	Se	L	Sil	2.75	2.78	2.50	2.50	0.375	0.375	0.497	0.497	45	E	JM	HG	CI	6.87	172	2.00×4.87
39	560– 700	2	N	Se	L	2,112	2.93	2.93	2.50	2.50	0.375	0.375	0.497	0.497	30	E	DC	HG	CI	6.87	172	2.00×5.12
40	628– 700	2	N	Se	L	2,112	2.93	2.93	2.50	2.50	0.375	0.375	0.497	0.497	30	E	DC	HG	CI	6.87	199	2.00×5.33
41	165–1,200	6	N	In	L	2,112	1.65	1.53	1.50	1.37	0.344	0.344	0.372	0.372	45	E	DC	HG	CI	3.75	37	1.12×2.97
42	190–1,100	6	N	In	L	2,112	1.65	1.53	1.50	1.37	0.344	0.344	0.372	0.372	45	E	DC	HG	CI	3.75	42	1.12×3.25
43	220–1,000	6	N	In	L	2,112	1.65	1.53	1.50	1.37	0.344	0.344	0.372	0.372	45	E	DC	HG	A1	3.75	42	1.12×3.25
44	220–1,000	6	N	In	L	2,112	1.65	1.53	1.50	1.37	0.344	0.344	0.372	0.372	45	N	......	HG	A1	3.75	42	1.12×3.25
45	220–1,000	6	N	In	L	2,112	1.65	1.53	1.50	1.37	0.344	0.344	0.372	0.372	45	N	......	HG	A1	3.75	42	1.12×3.25
46	220–1,000	6	N	In	L	2,112	1.65	1.53	1.50	1.37	0.344	0.344	0.372	0.372	45	N	......	HG	A1	3.75	42	1.12×3.25
47	234–1,100	6	N	In	L	2,112	1.90	1.78	1.75	1.62	0.400	0.400	0.372	0.372	45	E	DC	HG	CI	4.37	63	1.25×3.47
48	260–1,200	6	N	In	L	2,112	1.90	1.78	1.75	1.62	0.400	0.400	0.372	0.372	45	E	DC	HG	CI	4.37	...	1.25×3.47
49	302–1,000	6	N	In	L	2,112	1.90	1.78	1.75	1.62	0.400	0.400	0.372	0.372	45	E	DC	HG	A1	4.37	...	1.25×3.82
50	280–1,200	6	N	In	L	2,112	1.90	1.78	1.75	1.62	0.400	0.400	0.372	0.372	45	E	DC	HG	CI	4.37	68	1.25×3.82
51	280–1,200	6	N	In	L	2,112	1.90	1.78	1.75	1.62	0.400	0.400	0.372	0.372	45	E	DC	HG	CI	4.37	68	1.25×3.82
52	280–1,200	6	N	In	L	2,112	1.90	1.78	1.75	1.62	0.400	0.400	0.372	0.372	45	E	DC	HG	CI	4.37	68	1.25×3.82
53	340– 800	6	N	In	L	2,112	1.90	1.78	1.75	1.62	0.400	0.400	0.372	0.372	45	E	DC	HG	CI	4.75	88	1.25×3.81
54	384–1,100	6	N	In	I	2,112	1.96	1.68	1.75	1.50	0.468	0.468	0.372	0.372	30	E	JM	HG	A1	4.75	88	1.25×3.81
55	405–1,000	6	N	Se	L	2,112	2.50	2.37	2.25	2.12	0.375	0.375	0.434	0.434	30	E	DC	HG	CI	6.12	94	1.62×4.00
56	500–1,000	6	N	In	I	2,112	2.39	2.14	2.12	1.87	0.540	0.540	0.435	0.435	30	E	DC	HG	A1	6.00	...	1.75×4.15
57	473–1,100	6	W	In	I	CNS	2.64	2.01	2.37	1.75	0.540	0.540	0.435	0.435	30	E	To	HG	CI	6.73	...	1.75×4.35
58	569–1,000	6	W	In	I	CNS	2.64	2.01	2.37	1.75	0.540	0.540	0.435	0.435	30	E	To	HG	CI	6.73	...	1.75×4.88
59	625–1,100	6	W	In	I	CNS	2.64	2.01	2.37	1.75	0.540	0.540	0.435	0.435	30	E	To	HG	CI	6.73	...	1.75×4.88
60	696– 600	2	N	Se	L	Sil	2.75	2.78	2.50	2.50	0.438	0.438	0.497	0.497	45	E	JM	HG	CI	6.87	172	2.00×4.87
61	810– 700	2	N	Se	L	2,112	2.93	2.93	2.50	2.50	0.438	0.438	0.497	0.497	30	E	DC	HG	CI	6.87	196	2.00×5.12
62	915– 600	2	N	Se	L	2,112	2.93	2.93	2.50	2.50	0.438	0.438	0.497	0.497	30	E	DC	HG	CI	6.87	199	2.00×5.33
63	1,305– 750	6	W	In	I	2.112	2.71	2.53	2.50	2.28	0.703	0.703	0.558	0.558	45	E	DC	HG	Ala	9.31	263	2.75×5.53
64	131–1,400	4	N	In	L	Sil	1.93	1.93	1.75	1.75	0.312	0.375	0.375	0.375	45	N	......	HG	CI	3.75	37	1.09×3.31
65	525–1,200	2	N	Se	I	Sil	2.87	2.87	2.62	2.62	0.540	0.540	0.500	0.500	45	E	AS	HG	A1	6.12	132	1.50×5.62
66	635–1,200	2	N	Se	I	Sil	3.12	2.87	2.87	2.62	0.540	0.540	0.500	0.500	45	E	AS	HG	A1	6.50	153	1.69×6.25
67	1,176– 900	2	N	Se	L	Sil	4.12	4.06	3.62	3.44	0.812	0.812	0.683	0.683	45	E	AS	HG	A1	9.00	312	2.25×7.62
68	790–1,200	2	N	Se	I	Sil	2.87	2.87	2.62	2.62	0.540	0.540	0.500	0.500	45	E	AS	HG	A1	6.12	132	1.50×5.62
69	955–1,200	2	N	Se	I	Sil	3.12	2.87	2.87	2.62	0.540	0.540	0.500	0.500	45	E	AS	HG	A1	6.50	153	1.69×6.25
70	1,764– 900	2	N	Se	L	Sil	4.12	4.06	3.62	3.44	0.812	0.812	0.683	0.683	45	E	AS	HG	A1	9.00	312	2.25×7.62
71	1,050–1,200	2	N	Se	I	Sil	2.87	2.87	2.62	2.62	0.540	0.540	0.500	0.500	45	E	AS	HG	A1	6.12	132	1.50×5.62
72	1,275–1,200	2	N	Se	I	Sil	3.12	2.87	2.87	2.62	0.540	0.540	0.500	0.500	45	E	AS	HG	A1	6.50	153	1.69×6.25
73	2,352– 900	2	N	Se	L	Sil	4.12	4.06	3.62	3.44	0.812	0.812	0.683	0.683	45	E	AS	HG	A1	9.00	312	2.25×7.62

TABLE A1-3.—AMERICAN STOCK, MARINE, AND COMMERCIAL VEHICLE ENGINES.—(*Continued*)

Line number	Maximum torque at r.p.m., lb.-ft.	Cylinders		Crankcase, upper half integral with cylinders?	Valves													Front-end drive, type	Pistons			
		No. cast in one piece	Liners, type		Arrangement	Exhaust-head material (S.A.E. No.)	Max. head diameter, in.		Min. port diameter, in.		Lift, in.		Stem diameter, in.		Seats				Material	Length, in.	Weight with pins, rings, bushings, oz.	Piston pin, diameter and length, in.
							Intake	Exhaust	Intake	Exhaust	Intake	Exhaust	Intake	Exhaust	Angle, deg.	Inserts used?	Insert material (S.A.E. No.)					
74	....	3	N	Se	I	Sil	2.06	2.06	....	....	0.500	0.500	0.500	0.500	45	N	......	BG	Ala	5.50	...	1.50 ×5.25
75	....	6	N	In	I	SC	1.64	1.46	1.25	1.20	0.295	0.312	0.341	0.340	30	N	......	HG	CT	3.92	...	0.865×3.15
76	....	4	N	In	L	Sil	1.50	1.37	1.25	1.12	0.312	0.312	0.312	0.312	30	N	......	HG	Al	3.06	....	0.750×2.87
77	....	6	N	In	L	Sil	1.62	1.40	1.46	1.25	0.312	0.312	0.312	0.312	30	N	......	HG	Al	3.50	....	0.875×2.87
78	....	6	N	In	L	Sil	1.84	1.75	1.62	1.50	0.356	0.356	0.375	0.375	30	N	......	HG	Al	4.18	...	1.00 ×3.50
79	....	6	N	In	L	Sil	2.06	1.87	1.81	1.62	0.388	0.388	0.375	0.375	30	N	......	HG	Al	4.00	...	1.12 ×3.68
80	....	2	N	Se	L	Sil	2.53	2.53	2.31	2.31	0.375	0.375	0.437	0.437	45	N	......	HG	Al	5.87	...	1.37 ×4.25
81	....	2	N	Se	L	Sil	2.65	2.65	2.43	2.43	0.531	0.468	0.437	0.437	30	N	......	HG	Al	5.87	...	1.37 ×4.25
82	152–1,700	6	N	In	L	Sil	1.46	1.46	1.25	1.25	0.312	0.312	0.340	0.340	45	E	CNT	HG	Al	3.68	20	0.859×2.62
83	195–1,400	6	N	In	L	Sil	1.65	1.53	1.37	1.31	0.339	0.339	0.340	0.340	45	E	CNT	HG	Al	3.87	25	0.859×2.87
84	260–2,200	8	N	In	L	Sil	1.53	1.35	1.37	1.20	0.376	0.379	0.340	0.340	45	E	CNT	HG	Al	3.87	23	0.859×2.75
85	217– 800	4	W	In	I	Sil	2.00	2.00	1.75	1.75	0.406	0.406	0.375	0.375	45	E	CI	SG	CI	4.87	85	1.31 ×4.06
86	342– 800	4	W	In	I	Sil	2.37	2.37	2.12	2.12	0.437	0.437	0.437	0.437	45	E	CI	SG	CI	5.87	142	1.50 ×4.31
87	463– 650	4	N	In	I	Sil	2.50	2.50	2.25	2.25	0.500	0.500	0.562	0.562	45	N	......	SG	CI	6.75	189	1.48 ×5.25
88	550– 700	2	N	Se	I	Sil	2.50	2.50	2.25	2.25	0.500	0.500	0.562	0.562	45	E	CI	SG	CI	6.94	210	1.48 ×5.37
89	820– 650	2	N	Se	I	Sil	2.50	2.50	2.25	2.25	0.500	0.500	0.562	0.562	45	E	CI	SG	CI	6.94	210	1.48 ×5.37
90	1,140– 650	2	N	Se	I	Sil	2.50	2.50	2.25	2.25	0.500	0.500	0.562	0.562	45	E	CI	SG	CI	6.94	210	1.48 ×5.37
91	49–1,500	4	N	In	L	XCR	1.20	1.01	1.06	0.875	0.291	0.292	0.314	0.313	h	N	......	HG	CT	2.87	...	0.703×2.06
92	66–1,700	4	N	In	L	XCR	1.20	1.01	1.06	0.875	0.291	0.292	0.314	0.313	h	N	......	HG	CT	2.87	...	0.703×2.44
93	84–1,500	4	N	In	L	XCR	1.20	1.01	1.06	0.875	0.291	0.292	0.314	0.313	h	N	......	HG	CT	2.87	...	0.703×2.75
94	94–1,600	4	N	In	L	XCR	1.51	1.32	1.37	1.18	0.281	0.280	0.341	0.339	h	E	WA	HG	CT	3.56	...	0.859×2.50
95	106–1,600	4	N	In	L	XCR	1.51	1.32	1.37	1.18	0.281	0.280	0.341	0.339	h	E	WA	HG	CT	3.56	...	0.859×2.68
96	122–1,600	4	N	In	L	XCR	1.51	1.32	1.37	1.18	0.281	0.280	0.341	0.339	h	E	WA	HG	CT	3.56	...	0.859×2.87
97	124–1,200	6	N	In	L	XCR	1.51	1.32	1.37	1.18	0.284	0.284	0.341	0.339	h	E	WA	Ch	CT	3.75	...	0.859×2.50
98	150–1,200	6	N	In	L	XCR	1.51	1.32	1.37	1.18	0.284	0.284	0.341	0.339	h	E	WA	Ch	CT	3.75	...	0.859×2.66
99	154–1,200	6	N	In	L	XCR	1.51	1.32	1.37	1.18	0.284	0.284	0.341	0.339	h	E	WA	Ch	CT	3.56	...	0.859×2.68

100	162–1,200	6	N	In	L	XCR	1.51	1.32	1.37	1.18	0.284	0.284	0.341	0.339	h	E	WA	Ch	CT	3.56	...	0.859×2.68
101	178–1,200	6	N	In	L	XCR	1.57	1.42	1.43	1.31	0.311	0.311	0.339	0.338	h	E	HS	Ch	Al	3.93	...	0.859×2.87
102	190–1,200	6	N	In	L	XCR	1.76	1.51	1.62	1.37	0.354	0.354	0.404	0.402	h	E	WA	HG	CT	4.75	...	1.10 ×3.06
103	205–1,200	6	N	In	L	XCR	1.76	1.51	1.62	1.37	0.354	0.354	0.404	0.402	h	E	WA	HG	CT	4.75	...	1.10 ×3.18
104	233–1,200	6	N	In	L	XCR	1.76	1.51	1.62	1.37	0.354	0.354	0.404	0.402	h	E	WA	HG	CT	4.75	...	1.10 ×3.43
105	193– 900	6	N	In	L	XCR	2.06	1.87	1.81	1.62	0.361	0.361	0.435	0.432	30	E	HS	HG	CT	5.31	...	1.25 ×3.09
106	214–1,000	6	N	In	L	XCR	2.06	1.87	1.81	1.62	0.361	0.361	0.435	0.432	30	E	HS	HG	CT	5.31	...	1.25 ×3.09
107	253– 850	6	N	In	L	XCR	2.06	1.87	1.81	1.62	0.361	0.361	0.435	0.432	30	E	HS	HG	CT	5.31	...	1.25 ×3.43
108	265–1,000	6	N	In	L	XCR	2.06	1.87	1.81	1.62	0.361	0.361	0.435	0.432	30	E	HS	HG	CT	5.31	...	1.25 ×3.43
109	276–1,200	6	N	In	I	AUS	2.06	1.87	1.81	1.62	0.420	0.420	0.435	0.433	30	E	HS	Ch	Al	5.31	...	1.25 ×3.43
110	308–1,200	6	N	In	I	AUS	2.06	1.87	1.81	1.62	0.420	0.420	0.435	0.433	30	E	HS	Ch	Al	5.31	...	1.25 ×3.68
111	364–1,200	6	N	In	I	AUS	2.06	1.87	1.81	1.62	0.420	0.420	0.435	0.433	30	E	HS	Ch	Al	5.31	...	1.50 ×3.71
112	154–1,200	6	N	In	L	Sil	1.46	1.46	1.31	1.31	0.312	0.312	0.340	0.340	45	E	SA	Ch	Alt	3.68	...	0.859×2.62
113	166–1,200	6	N	In	L	Sil	1.46	1.46	1.31	1.31	0.312	0.312	0.340	0.340	45	E	SA	Ch	Als	3.68	...	0.859×2.75
114	176–1,200	6	N	In	L	Sil	1.65	1.53	1.50	1.37	0.379	0.379	0.340	0.340	45	E	SA	Ch	Al	3.87	20	0.859×2.87
115	188–1,200	6	N	In	L	Tun	1.65	1.53	1.50	1.37	0.379	0.379	0.340	0.340	45	E	SA	Ch	Al	3.87	20	0.859×2.87
116	230– 800	6	N	In	L	Tun	1.93	1.75	1.78	1.59	0.323	0.323	0.371	0.371	45	E	SA	Ch	.....	4.56	28	1.12 ×3.25
117	325–1,050	2	W	Se	F	Sil	2.50	2.50	2.25	2.25	0.303	0.350	0.437	0.437	45	N	......	HG	Als	6.12	77	1.37 ×4.17
118	490–1,050	2	W	Se	F	Sil	2.50	2.50	2.25	2.25	0.303	0.350	0.437	0.437	45	N	......	HG	Als	6.12	77	1.37 ×4.17
119	94–2,500	8	N	In	L	CNS	1.28	1.28	.....	.....	0.251	0.251	0.279	0.279	45	Bo	CNT	HG	CAS	.....	11	0.687×2.36
120	5	8	N	In	L	CNS	1.54	1.54	.....	.....	0.292	0.292	0.311	0.311	45	Bo	CNT	HG	CAS	.....	16	0.750×2.84
121	170–2,100	8	N	In	L	CNS	1.54	1.54	.....	.....	0.292	0.292	0.311	0.311	45	Bo	CNT	HG	CAS	.....	17	0.750×2.84
122	250–1,500	1	..	Se	I	CNS	1.75	1.43	1.62	1.31	0.375	0.375	0.375	0.375	30	..	......	HG	Al	4.37	36	1.25 ×3.12
123	250–1,500	1	..	Se	I	CNS	1.75	1.43	1.62	1.31	0.375	0.375	0.375	0.375	30	..	......	HG	Al	4.37	36	1.25 ×3.12
124	268–1,500	1	..	Se	I	CNS	1.75	1.43	1.62	1.31	0.375	0.375	0.375	0.375	30	..	......	HG	Al	4.37	38	1.25 ×3.12
125	268–1,500	1	..	Se	I	CNS	1.75	1.43	1.62	1.31	0.375	0.375	0.375	0.375	30	..	......	HG	Al	4.37	38	1.25 ×3.12
126	100–1,200	1	..	Se	I	CNS	1.66	1.51	1.50	1.40	0.375	0.375	0.375	0.375	30	..	......	HG	Al	3.21	18	0.859×3.12
127	178–1,000	6	N	In	I	Sil	1.64	1.47	1.25	1.16	0.289	0.307	0.343	0.343	30	..	......	HG	Al	4.20	33	0.990×3.08
128	195–1,100	6	N	In	I	Sil	1.64	1.47	1.25	1.16	0.289	0.307	0.343	0.343	30	E	WR	HG	Al	4.16	34	0.990×3.25
129	223–1,200	6	N	Se	I	CHS	1.81	1.56	1.44	1.37	0.333	0.333	0.375	0.375	45	E	WR	HG	Al	4.39	37	1.00 ×3.18
130	240–1,200	6	N	Se	I	CHS	1.81	1.56	1.44	1.37	0.333	0.333	0.375	0.375	45	E	WR	HG	Al	4.39	40	1.00 ×3.36
131	278– 800	6	N	Se	I	CHS	1.94	1.72	1.50	1.50	0.406	0.406	0.375	0.375	h	E	St	HG	Al	5.39	59	1.25 ×3.59
132	340–1,100	6	N	Se	I	CHS	1.94	1.72	1.50	1.50	0.406	0.406	0.375	0.375	h	E	St	HG	Al	5.14	60	1.25 ×3.71
133	368–1,200	6	N	Se	I	CHS	1.94	1.72	1.50	1.50	0.406	0.406	0.375	0.375	h	E	St	HG	Al	5.14	63	1.25 ×3.84
134	360– 800	6	N	Se	I	CHS	2.12	1.94	1.65	1.62	0.406	0.406	0.375	0.375	30	E	St	HG	Al	5.25	66	1.25 ×4.06
135	405–1,200	6	N	Se	I	CHS	2.12	1.94	1.65	1.62	0.406	0.406	0.437	0.437	30	E	St	HG	Al	5.25	66	1.25 ×4.06
136	540–1,200	6	D	Se	I	CHS	2.44	2.17	1.75	1.75	0.413	0.413	0.437	0.500	30	E	St	HG	Al	5.03	77	1.37 ×4.47
137	....	4	N	In	L	Sil	1.20	1.01	1.06	0.875	0.291	0.292	0.314	0.312	h	N	......	HG	CI	2.87	16	0.703×2.06
138	....	4	N	In	L	Sil	1.20	1.01	1.06	0.875	0.291	0.292	0.341	0.312	h	N	......	HG	CI	2.87	21	0.703×2.43
139	....	4	N	In	L	Sil	1.20	1.01	1.06	0.875	0.291	0.292	0.341	0.312	h	N	......	HG	CI	2.87	...	0.709×2.68
140	....	4	N	In	L	Sil	1.20	1.01	1.06	0.875	0.291	0.292	0.341	0.312	h	N	......	HG	Al	2.87	16	0.703×2.43
141	....	4	N	In	L	Sil	1.51	1.32	1.37	1.18	0.331	0.331	0.341	0.339	h	N	......	HG	Al	3.56	20	0.859×2.68
142	....	4	N	In	L	Sil	1.51	1.32	1.37	1.18	0.331	0.331	0.341	0.341	h	N	......	HG	Al	3.56	24	0.859×2.68
143	....	4	N	In	L	Sil	1.51	1.32	1.37	1.18	0.331	0.331	0.341	0.339	h	N	......	HG	Al	3.56	20	0.859×2.68
144	....	4	N	In	L	Sil	1.51	1.32	1.37	1.18	0.331	0.331	0.341	0.339	h	N	......	HG	Al	3.56	24	0.859×2.68
145	....	6	N	In	L	Sil	1.51	1.32	1.37	1.18	0.284	0.284	0.341	0.339	h	N	......	HG	CI	3.75	29	0.859×2.68
146	....	6	N	In	L	Sil	1.51	1.32	1.37	1.18	0.284	0.284	0.341	0.339	h	N	......	HG	Al	3.56	20	0.859×2.68

TABLE A1-3.—AMERICAN STOCK, MARINE, AND COMMERCIAL VEHICLE ENGINES.—(*Continued*)

		Cylinders			Valves														Pistons			
							Max. head diameter, in.		Min. port diameter, in.		Lift, in.		Stem diameter, in.		Seats							
Line number	Maximum torque at r.p.m., lb.-ft.	No. cast in one piece	Liners, type	Crankcase, upper half integral with cylinders?	Arrangement	Exhaust-head material (S.A.E. No.)	Intake	Exhaust	Intake	Exhaust	Intake	Exhaust	Intake	Exhaust	Angle, deg.	Inserts used?	Insert material (S.A.E. No.)	Front-end drive, type	Material	Length, in.	Weight with pins, rings, bushings, oz.	Piston pin, diameter and length, in.
147	....	6	N	In	L	Sil	1.51	1.32	1.37	1.18	0.284	0.284	0.341	0.339	h	N	......	HG	Al	3.56	20	0.859×2.68
148	....	6	N	In	L	Sil	1.51	1.32	1.37	1.18	0.284	0.284	0.341	0.339	h	N	......	HG	Al	3.56	20	0.859×2.68
149	....	6	N	In	L	Sil	1.57	1.42	1.43	1.31	0.311	0.311	0.339	0.338	h	N	......	Ch	Al	3.97	28	0.859×2.87
150	....	6	N	In	L	Sil	1.57	1.42	1.43	1.31	0.311	0.311	0.399	0.338	h	N	......	HG	Al	3.97	28	0.859×2.87
151	....	6	N	In	L	Sil	1.57	1.42	1.43	1.31	0.311	0.311	0.399	0.338	h	N	......	HG	Al	3.97	28	0.859×2.87
152	....	6	N	In	L	Sil	1.57	1.42	1.43	1.31	0.311	0.311	0.399	0.338	h	N	......	HG	Al	3.97	28	0.859×2.87
153	....	6	N	In	L	Sil	1.57	1.42	1.43	1.31	0.311	0.311	0.339	0.338	h	N	......	HG	Al	3.97	28	0.859×2.87
154	....	6	N	In	L	Sil	1.71	1.58	1.56	1.43	0.360	0.360	0.340	0.371	45	N	......	HG	Al	3.97	25	0.874×2.92
155	....	6	N	In	L	Sil	1.78	1.51	1.62	1.37	0.354	0.354	0.404	0.402	h	N	......	HG	Al	4.75	40	1.10 ×3.18
156	....	6	N	In	L	Sil	1.76	1.51	1.62	1.37	0.354	0.354	0.404	0.402	h	N	......	HG	Al	4.75	45	1.10 ×3.43
157	....	6	N	In	L	Sil	1.76	1.51	1.62	1.37	0.354	0.354	0.404	0.402	h	N	......	HG	Al	4.75	45	1.10 ×3.43
158	....	6	N	In	L	Sil	2.06	1.87	1.81	1.62	0.406	0.406	0.434	0.432	h	E	......	HG	CI	5.31	74	1.25 ×3.43
159	....	8	N	In	L	Sil	1.56	1.42	1.50	1.37	0.372	0.372	0.339	0.338	h	N	......	HG	Al	3.94	30	1.11 ×3.05
160	....	8	N	In	L	Sil	1.56	1.42	1.50	1.37	0.372	0.372	0.339	0.338	h	N	......	HG	Al	3.94	30	1.11 ×3.05
161	212–1,000	4	N	Se	I	Spec	2.13	2.25	1.93	1.89	0.343	0.343	0.435	0.435	45	..	......	HC	Al	5.21	57	1.25 ×3.25
162	213–1,300	4	N	Se	I	Sil	2.13	2.13	1.93	1.93	0.343	0.343	0.435	0.435	45	..	......	HC	Al	5.03	57	1.25 ×3.25
163	265–1,000	4	N	Se	I	Spec	2.25	2.38	2.06	2.03	0.343	0.343	0.435	0.435	45	..	......	HC	Al	5.20	74	1.25 ×3.75
164	255–1,000	6	W	In	I	AESW	2.03	1.91	1.87	1.65	0.421	0.421	0.435	0.434	30	..	......	Ch	Al	5.09	40	1.25 ×3.50
165	247–1,000	6	N	Se	I	Sil	2.13	2.13	1.92	1.92	0.312	0.312	0.435	0.435	45	..	......	HC	Al	4.50	41	1.00 ×3.43
166	310–1,000	6	W	In	I	AESW	2.28	2.16	2.12	1.90	0.421	0.421	0.435	0.435	30	..	......	Ch	Al	5.40	52	1.12 ×3.73
167	333–1,200	6	N	Se	I	Spec	2.25	2.25	1.93	1.93	0.312	0.312	0.435	0.435	45	..	......	HC	Al	4.56	52	1.25 ×3.59
168	324–1,200	6	N	Se	I	Spec	2.25	2.38	2.06	2.06	0.343	0.343	0.435	0.435	45	..	......	HC	Al	5.26	60	1.25 ×3.34
169	310–1,200	6	N	Se	I	Spec	2.13	2.13	1.92	1.92	0.312	0.312	0.435	0.435	45	..	......	HC	Al	5.21	58	1.25 ×3.25
170	319–1,200	6	N	Se	I	AESW	2.25	2.38	2.04	2.04	0.328	0.328	0.435	0.435	45	..	......	HC	Al	5.26	56	1.25 ×3.34
171	330–1,200	6	W	In	I	AESW	2.28	2.16	2.12	1.90	0.421	0.421	0.435	0.434	30	..	......	Ch	Al	5.20	62	1.12 ×4.00
172	593–1,400	6	N	Se	I	AESW	2.73	2.73	2.50	2.50	0.435	0.435	0.406	0.406	30	..	......	Ch	Al	4.92	78	1.25 ×4.00

173	570–1,000	6	N	Se	I	AESW	2.50	2.50	2.31	2.31	0.406	0.343	0.435	0.435	h	..	......	HC	Al	6.31	97	1.37 ×4.18
174	492–1,200	6	W	In	I	AESW	2.62	2.37	2.37	1.99	0.482	0.482	0.497	0.528	30	..	......	Ch	Al	6.17	77	1.37 ×4.18
175	630–1,000	6	N	Se	I	AESW	2.50	2.50	2.31	2.31	0.406	0.343	0.435	0.435	h	..	......	HC	Al	6.18	104	1.37 ×4.44
176	570–1,200	6	W	In	I	AESW	2.87	2.62	2.62	2.24	0.482	0.497	0.497	0.458	30	..	......	Al	Al	6.18	104	1.37 ×4.43
177	690–1,100	6	N	Se	I	AESW	2.50	2.50	2.31	2.31	0.406	0.343	0.435	0.435	h	..	......	HC	Al	5.97	114	1.37 ×4.68
178	754–1,400	6	N	In	I	AESW	2.87	2.87	2.62	2.62	0.482	0.482	0.497	0.528	30	..	......	Ch	Al	6.46	108	1.37 ×4.40
179	745–1,400	6	N	In	I	AESW	2.73	2.73	2.50	2.50	0.406	0.406	0.435	0.435	30	..	......	Ch	Al	6.62	...	1.37 ×4.69
180	....	6	N	Se	I	Sil	2.87	2.75	2.62	2.50	0.482	0.482	0.500	0.531	30	E	St	HC	Al	7.31	...	1.37 ×4.31
181	....	6	N	Se	I	Sil	2.87	2.75	2.62	2.50	0.482	0.482	0.500	0.531	30	E	St	HC	Al	7.31	...	1.37 ×4.31
182	....	6	N	Se	I	Sil	2.87	2.75	2.62	2.50	0.482	0.482	0.500	0.531	30	E	St	HC	Al	7.31	...	1.37 ×4.31
183	28–1,250	2	N	In	L	Sil	1.23	1.10	1.00	0.875	0.200	0.200	0.248	0.248	30	N	......	HG	CI	2.68	17	0.687×2.50
184	39–1,100	2	N	In	L	CNS	1.48	1.35	1.25	1.12	0.250	0.250	0.312	0.312	30	N	......	HG	CI	3.06	28	0.750×2.56
185	46–1,100	2	N	In	L	CNS	1.48	1.35	1.25	1.25	0.250	0.250	0.312	0.312	30	N	......	HG	CI	3.06	29	0.750×2.81
186	40–2,000	4	N	In	L	Sil	1.23	1.10	1.00	1.00	0.200	0.200	0.248	0.248	30	N	......	HG	CI	2.68	17	0.687×2.18
187	....	4	N	In	L	Sil	1.23	1.10	1.00	0.875	0.200	0.200	0.248	0.248	30	N	......	HG	CI	2.68	19	0.687×2.18
188	55–2,000	4	N	In	L	CNS	1.48	1.35	1.25	1.12	0.250	0.250	0.310	0.310	30	N	......	HG	CI	2.43	18	0.750×2.18
189	55–2,000	4	N	In	L	CNS	1.48	1.35	1.25	1.12	0.250	0.250	0.310	0.310	30	N	......	HG	CI	2.43	18	0.750×2.18
190	79–2,000	4	N	In	L	CNS	1.48	1.35	1.25	1.12	0.250	0.250	0.310	0.310	30	N	......	HG	CI	3.06	28	0.750×2.56
191	79–2,000	4	N	In	L	CNS	1.48	1.35	1.25	1.12	0.250	0.250	0.310	0.310	30	N	......	HG	CI	3.06	28	0.750×2.56
192	92–2,000	4	N	In	L	CNS	1.48	1.35	1.25	1.12	0.250	0.250	0.310	0.310	30	N	......	HG	CI	3.06	29	0.750×2.81
193	92–2,000	4	N	In	L	CNS	1.48	1.35	1.25	1.12	0.250	0.250	0.310	0.310	30	N	......	HG	CI	3.06	29	0.750×2.81
194	107–1,200	4	N	In	L	CNS	1.75	1.62	1.50	1.37	0.326	0.326	0.373	0.373	45	N	......	HG	CI	4.31	49	1.00 ×3.12
195	125–1,000	4	N	In	L	CNS	1.75	1.62	1.50	1.37	0.326	0.326	0.373	0.373	45	N	......	HG	CI	4.12	56	1.00 ×3.37
196	143–1,000	4	N	In	L	CNS	1.75	1.62	1.50	1.37	0.326	0.326	0.373	0.373	45	N	......	HG	CI	4.31	56	1.00 ×3.62
197	155–1,000	4	N	In	L	CNS	1.87	1.87	1.62	1.62	0.326	0.326	0.373	0.373	45	E	......	HG	CI	4.87	67	1.37 ×3.50
198	185–1,000	4	N	In	L	CNS	1.87	1.87	1.62	1.62	0.326	0.326	0.373	0.373	45	E	......	HG	CI	4.87	73	1.37 ×3.75
199	202–1,000	4	N	In	L	Sil	2.25	2.25	2.00	2.00	0.326	0.326	0.434	0.434	45	N	......	HG	CI	5.25	82	1.50 ×3.75
200	226–1,000	4	N	In	L	Sil	2.25	2.25	2.00	2.00	0.326	0.326	0.434	0.434	45	N	......	HG	CI	5.25	95	1.50 ×4.00
201	250–1,000	4	N	In	L	Sil	2.25	2.25	2.00	2.00	0.326	0.326	0.434	0.434	45	N	......	HG	CI	5.25	103	1.50 ×4.25
202	288–1,000	4	N	In	L	Sil	2.25	2.25	2.00	2.00	0.326	0.326	0.434	0.434	45	N	......	HG	CI	5.25	106	1.50 ×4.50
203	425– 800	4	N	In	L	Sil	2.90	2.90	2.50	2.50	0.375	0.375	0.497	0.497	45	E	......	HG	CI	7.00	196	1.87 ×4.87
204	485– 800	4	N	In	L	Sil	2.90	2.90	2.50	2.50	0.375	0.375	0.497	0.497	45	E	......	HG	CI	7.00	222	1.87 ×5.37
205	586– 800	4	N	In	L	Sil	2.90	2.90	2.50	2.50	0.375	0.375	0.497	0.497	45	E	......	HG	CI	7.00	240	1.87 ×5.75
206	130–1,000	6	N	In	L	CNS	1.50	1.37	1.31	1.12	0.281	0.281	0.310	0.310	30	N	......	HG	CI	3.50	53	0.875×2.67
207	130–1,000	6	N	In	L	CNS	1.50	1.37	1.31	1.12	0.281	0.281	0.310	0.310	30	N	......	HG	CI	3.50	53	0.875×2.67
208	143–1,000	6	N	In	L	CNS	1.50	1.37	1.31	1.12	0.281	0.281	0.310	0.310	30	N	......	HG	CI	3.50	53	0.875×2.79
209	143–1,000	6	N	In	L	CNS	1.50	1.37	1.31	1.12	0.281	0.281	0.310	0.310	30	N	......	HG	CI	3.50	53	0.875×2.79
210	154–1,000	6	N	In	L	CNS	1.62	1.37	1.43	1.12	0.281	0.281	0.310	0.310	30	N	......	HG	Al	3.50	24	0.875×2.90
211	154–1,000	6	N	In	L	CNS	1.62	1.37	1.43	1.12	0.281	0.281	0.310	0.310	30	N	......	HG	Al	3.50	24	0.875×2.90
212	141–1,000	6	N	In	L	CNS	1.75	1.62	1.50	1.37	0.322	0.322	0.373	0.373	45	N	......	HG	CI	4.37	43	1.00 ×2.90
213	163–1,000	6	N	In	L	CNS	1.75	1.62	1.50	1.50	0.322	0.322	0.373	0.373	45	N	......	HG	CI	4.18	48	1.00 ×3.15
214	175–1,000	6	N	In	L	CNS	1.75	1.62	1.50	1.37	0.322	0.322	0.373	0.373	45	N	......	HG	CI	4.12	56	1.00 ×3.37
215	204–1,000	6	N	In	L	CNS	1.75	1.62	1.50	1.37	0.322	0.322	0.373	0.373	45	N	......	HG	Al	4.18	40	1.00 ×3.51
216	212–1,000	6	N	In	L	Sil	1.75	1.75	1.62	1.50	0.356	0.356	0.373	0.373	45	N	......	HG	CI	4.56	64	1.12 ×3.56
217	233–1,000	6	N	In	L	Sil	1.75	1.75	1.62	1.50	0.356	0.356	0.373	0.373	45	N	......	HG	CI	4.56	65	1.12 ×3.62
218	262–1,000	6	N	In	L	Sil	1.75	1.75	1.62	1.50	0.356	0.356	0.373	0.373	45	N	......	HG	CI	4.56	83	1.12 ×3.68
219	260– 950	6	N	In	L	Sil	1.75	1.75	1.62	1.50	0.356	0.356	0.373	0.373	45	N	......	HG	Al	4.18	47	1.12 ×3.56

TABLE A1-3.—AMERICAN STOCK, MARINE, AND COMMERCIAL VEHICLE ENGINES.—(*Continued*)

Line number	Maximum torque at r.p.m., lb.-ft.	Cylinders: No. cast in one piece	Cylinders: Liners, type	Crankcase, upper half integral with cylinders?	Valves: Arrangement	Valves: Exhaust-head material (S.A.E. No.)	Valves: Max. head diameter, in.: Intake	Valves: Max. head diameter, in.: Exhaust	Valves: Min. port diameter, in.: Intake	Valves: Min. port diameter, in.: Exhaust	Valves: Lift, in.: Intake	Valves: Lift, in.: Exhaust	Valves: Stem diameter, in.: Intake	Valves: Stem diameter, in.: Exhaust	Valves: Seats: Angle, deg.	Valves: Seats: Inserts used?	Valves: Seats: Insert material (S.A.E. No.)	Front-end drive, type	Pistons: Material	Pistons: Length, in.	Pistons: Weight with pins, rings, bushings, oz.	Pistons: Piston pin, diameter and length, in.
220	294– 950	6	N	In	L	Sil	1.75	1.75	1.62	1.50	0.356	0.356	0.373	0.373	45	N	......	HG	Al	4.37	55	1.12 ×3.68
221	281– 800	6	N	In	L	Sil	2.00	2.00	1.75	1.75	0.388	0.388	0.373	0.373	45	N	......	HG	CI	4.87	79	1.25 ×3.93
222	300– 800	6	N	In	L	Sil	2.00	2.00	1.75	1.75	0.388	0.388	0.373	0.373	45	N	......	HG	CI	4.87	85	1.12 ×3.93
223	320– 800	6	N	In	L	Sil	2.00	2.00	1.75	1.75	0.388	0.388	0.373	0.373	45	N	......	HG	CI	4.87	87	1.12 ×4.06
224	330–1,000	6	N	In	L	Sil	2.00	2.00	1.75	1.75	0.388	0.388	0.373	0.373	45	N	......	HG	Al	4.87	60	1.25 ×3.93
225	350–1,000	6	N	In	L	Sil	2.00	2.00	1.75	1.75	0.388	0.388	0.373	0.373	45	N	......	HG	Al	4.87	62	1.25 ×4.06
226	388–1,000	6	N	In	L	Sil	2.00	2.00	1.75	1.75	0.388	0.388	0.373	0.373	45	N	......	HG	Al	4.87	65	1.25 ×4.10
227	407–1,000	6	N	In	L	Sil	2.00	2.00	1.81	1.75	0.388	0.388	0.373	0.373	h	N	......	HG	Al	4.87	69	1.25 ×4.10
228	455– 900	3	N	Se	L	Sil	2.43	2.31	2.12	2.00	0.468	0.468	0.498	0.498	30	E	......	HG	Al	6.50	95	1.50 ×4.43
229	505– 900	3	N	Se	L	Sil	2.43	2.31	2.12	2.00	0.468	0.468	0.498	0.498	30	E	......	HG	Al	6.87	105	1.50 ×4.56
230	555– 900	3	N	Se	L	Sil	2.43	2.31	2.12	2.00	0.468	0.468	0.498	0.498	30	E	......	HG	Al	6.87	117	1.50 ×4.81
231	612– 900	3	N	Se	L	Sil	2.43	2.31	2.12	2.00	0.468	0.468	0.498	0.498	30	E	......	HG	Al	7.25	127	1.50 ×5.06
232	138–1,400	6	N	In	L	Sil	1.37	1.37	1.26	1.23	0.343	0.343	0.341	0.339	45	N	......	HG	Al	3.18	8	0.750×2.44
233	167–1,200	6	N	In	L	Sil	1.37	1.37	1.26	1.23	0.343	0.343	0.341	0.339	45	N	......	HG	Al	3.18	8	0.750×2.44
234	153–1,000	4	W	Se	I	Sil	1.78	1.78	1.56	1.56	0.402	0.402	0.432	0.432	45	..	......	HG	CI	4.71	63	1.29 ×3.37
235	207– 850	4	W	Se	I	Sil	1.90	1.75	1.68	1.48	0.441	0.441	0.432	0.432	45	..	......	HG	CI	6.03	95	1.29 ×3.68
236	300– 750	4	W	Se	I	Sil	2.18	2.18	1.75	1.93	0.381	0.426	0.432	0.432	45	..	......	HG	CI	5.82	122	1.48 ×4.12
237	200–1,200	6	W	Se	I	Sil	1.87	1.75	1.62	1.50	0.343	0.343	0.372	0.372	45	..	......	HG	Al	4.56	34	1.10 ×3.17
238	447– 700	6	W	Se	I	Sil	2.37	2.37	2.12	2.12	0.437	0.437	.....	.....	45	..	......	HG	CI	5.75	130	1.50 ×4.10
239	....	2	N	Se	L	CNS	1.75	1.75	1.50	1.50	.....	.....	0.375	0.375	45	N	......	SG	CI	4.12	73	0.875×3.62
240	40–1,700	4	N	In	L	Sil	1.25	1.12	1.12	0.875	0.250	0.250	0.310	0.310	30	N	......	HG	CI	2.68	19	0.687×2.18
241	97–2,200	4	N	In	L	Sil	1.48	1.35	1.25	1.12	0.250	0.250	0.310	0.310	30	N	......	HG	Al	3.06	20	0.750×2.87
242	97–2,200	4	N	In	L	Sil	1.48	1.35	1.25	1.12	0.250	0.250	0.310	0.310	30	N	......	HG	Al	3.06	20	0.750×2.87
243	....	4	N	Se	L	CNS	1.75	1.75	.....	.....	0.218	0.218	0.375	0.375	45	N	......	SG	CI	4.12	73	0.875×3.17
244	....	4	N	Se	L	CNS	2.25	2.25	2.00	2.00	0.375	0.375	0.375	0.375	45	N	......	HG	Als	5.25	62	1.25 ×3.82
245	....	6	N	In	L	Sil	1.47	1.34	1.31	1.18	0.296	0.296	0.312	0.312	h	N	......	Ch	CNI	3.59	32	0.937×2.87

246	...	6	N	In	L	Sil	1.87	1.87	1.81	1.81	0.281	0.281	0.310	0.310	30	N	...	HG	Al	3.50	73	0.875×2.79
247	...	6	N	In	L	Sil	1.59	1.47	1.37	1.31	0.296	0.296	0.312	0.312	h	N	...	Ch	CI	3.57	...	0.937×3.06
248	225–1,000	6	N	In	L	CNS	1.75	1.62	1.50	1.37	0.322	0.322	0.373	0.373	30	N	...	HG	Al	4.18	40	1.00 ×3.51
249	235–2,000	6	N	In	L	CNS	1.84	1.62	1.62	1.37	0.376	0.376	0.373	0.373	30	N	...	HG	Al	4.18	40	1.00 ×3.51
250	265–1,000	6	N	In	L	Sil	1.87	1.75	1.62	1.50	0.356	0.356	0.373	0.373	45	N	...	HG	Al	4.56	...	1.12 ×3.68
251	...	6	N	In	L	CNS	2.06	1.87	1.81	1.62	0.500	0.500	0.373	0.373	30	N	...	HG	Al	4.00	...	1.12 ×3.62
252	351–1,000	6	N	Se	L	CNS	2.50	2.25	2.28	2.00	0.437	0.375	0.437	0.375	h	E	HS	HG	Als	5.25	62	1.25 ×3.62
253	482–1,000	6	N	Se	L	CNS	2.56	2.37	2.37	2.12	0.375	0.375	0.437	0.437	45	N	...	HG	Als	6.19	82	1.25 ×4.50
254	480–1,000	6	N	Se	L	CNS	2.62	2.37	2.40	2.00	0.437	0.375	0.437	0.437	h	N	...	HG	Als	6.19	82	1.25 ×4.50
255	540–1,700	6	N	Se	I	CNS	1.93	1.93	1.76	1.76	0.375	0.375	0.375	0.375	45	N	...	SB	Als	6.19	82	1.25 ×4.50
256	150–2,000	8	N	In	L	Sil	1.53	1.53	...	...	0.292	0.292	0.310	0.310	45	E	...	HG	Al	...	16	0.750×2.85
257	170–2,100	8	N	In	L	CNS	1.53	1.53	...	...	0.292	0.292	0.311	0.311	45	E	...	HG	Als	...	17	0.750×2.84
258	175–1,600	8	N	In	L	Sil	1.37	1.34	1.25	1.19	0.328	0.328	0.312	0.312	h	N	...	Ch	CI	3.56	32	0.937×2.87
259	...	12	N	In	L	CNS	1.53	1.53	...	...	0.292	0.292	0.311	0.312	45	E	...	HG	S	3.25	11	0.750×2.48
260	1,070–1,400	6	N	Se	I	CNS	1.93	1.93	1.76	1.76	0.375	0.375	0.437	0.437	45	N	...	SB	Als	6.19	82	1.25 ×4.50
261	140– 525	1	N	Se	T	CNS	2.25	2.25	2.00	2.00	0.375	0.375	0.437	0.437	45	..	...	HG	CI	7.00	160	1.37 ×4.62
262	163– 525	1	N	Se	T	CNS	2.25	2.25	2.00	2.00	0.375	0.375	0.437	0.437	45	..	...	HG	CI	7.00	179	1.37 ×5.00
263	206– 600	1	N	Se	T	CNS	2.25	2.25	2.00	2.00	0.375	0.375	0.437	0.437	45	..	...	HG	CI	7.00	160	1.37 ×4.62
264	237– 700	1	N	Se	T	CNS	2.25	2.25	2.00	2.00	0.375	0.375	0.437	0.437	45	..	...	HG	CI	7.00	179	1.37 ×5.00
265	92–2,000	4	N	In	L	CNS	...	...	1.25	1.12	0.312	0.312	0.312	0.312	30	..	...	HG	Al	3.06	...	0.750×2.81
266	152–1,300	4	N	In	L	CNS	...	...	1.50	1.37	0.312	0.312	0.375	0.375	30	..	...	HG	Al	4.12	...	1.00 ×3.50
267	233– 500	1	N	Se	T	CNS	2.25	2.25	2.00	2.00	0.375	0.375	0.437	0.437	45	..	...	HG	CI	7.00	160	1.37 ×4.62
268	342– 600	1	N	Se	T	CNS	2.25	2.25	2.00	2.00	0.375	0.375	0.437	0.437	45	..	...	HG	CI	7.00	179	1.37 ×5.00
269	373– 650	1	N	Se	L	CNS	2.68	2.50	2.31	2.12	0.375	0.375	0.500	0.500	45	..	...	HG	CI	6.25	186	1.50 ×5.00
270	461– 700	1	N	Se	L	CNS	2.68	2.50	2.31	2.12	0.375	0.375	0.500	0.500	45	..	...	HG	CI	6.25	204	1.50 ×5.50
271	173– 550	6	N	In	L	CNS	...	...	1.50	1.37	0.312	0.312	0.375	0.375	30	..	...	HG	Al	4.31	...	1.00 ×3.50
272	226–2,500	6	N	In	L	CNS	1.75	1.62	1.50	1.37	0.356	0.356	0.375	0.375	30	E	Spec	HG	Al	4.18	40	1.00 ×3.50
273	321–1,350	6	N	Se	L	CNS	2.25	2.25	2.00	2.00	0.375	0.375	0.437	0.437	45	..	...	HG	CI	5.75	96	1.37 ×3.87
274	379– 900	6	N	Se	L	CNS	2.25	2.25	2.00	2.00	0.375	0.375	0.437	0.437	45	..	...	HG	CI	5.75	104	1.37 ×4.12
275	515– 675	1	N	Se	T	CNS	2.25	2.25	2.00	2.00	0.375	0.375	0.437	0.437	45	..	...	HG	CI	7.00	179	1.37 ×5.00
276	550– 825	1	N	Se	T	CNS	2.25	2.25	2.00	2.00	0.500	0.500	0.437	0.437	45	..	...	HG	CI	7.00	179	1.37 ×5.00
277	561–1,200	6	N	Se	L	CNS	2.68	2.68	2.31	2.31	0.437	0.437	0.500	0.500	45	..	...	HG	CI	6.21	176	1.50 ×5.00
278	640–1,100	6	N	Se	L	CNS	2.68	2.68	2.31	2.31	0.437	0.437	0.500	0.500	45	..	...	HG	CI	6.21	176	1.50 ×5.12
279	122–1,000	4	N	In	L	CNS	1.37	1.37	...	...	0.287	0.287	0.250	0.250	45	N	...	HG	Al			
280	94–2,500	8	N	In•	L	CNS	1.28	1.28	...	...	0.251	0.251	0.279	0.279	45	Bo	Tun	HG	CAS	...	8	0.687×2.36
281	155–2,200	8	N	In	L	CNS	1.54	1.54	...	...	0.292	0.292	0.311	0.311	45	Bo	Tun	HG	CAS	...	12	0.750×2.85
282	170–2,100	8	N	In	L	CNS	1.54	1.54	...	...	0.292	0.292	0.311	0.311	45	Bo	Tun	HG	CAS	...	13	0.750×2.98
283	186–2,000	12	N	In	L	CNS	1.54	1.54	...	...	0.292	0.292	0.311	0.311	45	Bo	Tun	HG	CAS	...	12	0.750×2.48
284	145–1,100	6	N	In	L	Sil	1.51	1.68	1.37	1.18	0.296	0.296	0.343	0.343	30	E	Al	HG	CT	3.56	...	0.859×2.68
285	166–1,200	6	N	In	L	Sil	1.76	1.51	1.62	1.37	0.354	0.354	0.406	0.406	h	E	HS	HG	Al	4.75	36	1.10 ×2.90
286	188–1,200	6	N	In	L	Sil	1.76	1.51	1.62	1.37	0.354	0.354	0.406	0.406	h	E	HS	HS	Al	4.75	39	1.10 ×3.05
287	210–1,000	6	N	In	L	AUS	1.89	1.76	1.62	1.50	0.365	0.365	0.375	0.375	30	E	Nif	HG	Al	4.31	48	1.00 ×3.25
288	200–1,200	6	N	In	L	Sil	1.76	1.51	1.62	1.37	0.354	0.354	0.406	0.406	h	E	HG	HG	Al	4.75	43	1.10 ×3.18
289	250–1,000	6	N	In	L	AUS	1.89	1.76	1.62	1.50	0.365	0.365	0.375	0.375	30	E	Nif	HG	Al	4.31	40	1.00 ×3.50
290	270–1,000	6	N	Se	L	AUS	2.17	2.01	1.81	1.68	0.375	0.375	0.500	0.500	30	E	Nif	HG	Al	5.10	54	1.12 ×3.62
291	310–1,000	6	N	Se	L	AUS	2.17	2.01	1.81	1.68	0.375	0.375	0.500	0.500	30	E	Nif	HG	Al	5.48	62	1.12 ×3.87
292	368–1,000	6	N	In	I	AUS	2.18	1.89	2.04	1.75	0.500	0.500	0.437	0.437	30	E	Nif	HG	Al	5.75	68	1.43 ×3.78

TABLE A1-3.—AMERICAN STOCK, MARINE, AND COMMERCIAL VEHICLE ENGINES.—(*Continued*)

Line number	Maximum torque at r.p.m., lb.-ft.	Cylinders: No. cast in one piece	Cylinders: Liners, type	Crankcase, upper half integral with cylinders?	Valves: Arrangement	Valves: Exhaust-head material (S.A.E. No.)	Valves: Max. head diameter, in.: Intake	Valves: Max. head diameter, in.: Exhaust	Valves: Min. port diameter, in.: Intake	Valves: Min. port diameter, in.: Exhaust	Valves: Lift, in.: Intake	Valves: Lift, in.: Exhaust	Valves: Stem diameter, in.: Intake	Valves: Stem diameter, in.: Exhaust	Valves: Seats: Angle, deg.	Valves: Seats: Inserts used?	Valves: Seats: Insert material (S.A.E. No.)	Front-end drive, type	Pistons: Material	Pistons: Length, in.	Pistons: Weight with pins, rings, bushings, oz.	Pistons: Piston pin, diameter and length, in.
293	350–1,000	6	N	Se	L	AUS	2.17	2.01	1.81	1.68	0.375	0.375	0.500	0.500	30	E	Nif	HG	Al	5.48	63	1.12 ×4.12
294	443–1,000	6	N	In	I	AUS	2.18	1.89	2.04	1.75	0.500	0.500	0.437	0.437	30	E	Nif	HG	Al	5.75	83	1.43 ×4.15
295	502–1,000	6	N	In	I	AUS	2.18	1.89	2.04	1.75	0.500	0.500	0.437	0.437	30	E	Nif	HG	Al	5.62	95	1.43 ×4.40
296	100–1,100	2	N	Se	HB	Sil	1.46	1.46	1.25	1.25	0.354	0.354	0.343	0.343	45	E	CNM	HG	CI	4.90	56	1.00 ×3.00
297	120–1,150	2	N	Se	HB	Sil	1.46	1.46	1.25	1.25	0.354	0.354	0.343	0.343	45	E	CNM	HG	CI	4.37	54	1.00 ×3.00
298	190–1,000	4	N	Se	I	Sil	1.71	1.59	1.50	1.37	0.488	0.488	0.437	0.437	45	E	CNM	HG	CI	5.00	80	1.25 ×3.87
299	190–1,000	4	N	Se	I	Sil	1.71	1.59	1.50	1.37	0.488	0.488	0.437	0.437	45	E	CNM	HG	CI	5.00	80	1.25 ×3.87
300	295– 950	2	N	Se	I	Sil	1.84	1.71	1.62	1.50	0.488	0.488	0.437	0.437	45	E	CNM	HG	CI	5.50	104	1.25 ×4.25
301	1,050– 300	1	N	Se	I	Sil	3.34	2.84	3.00	3.00	0.687	0.687	0.687	0.687	45	E	CNM	HG	CI	10.00	522	2.18 ×6.68
302	1,150– 300	1	N	Se	I	Sil	3.50	2.84	3.00	3.00	0.687	0.687	0.687	0.687	45	E	CNM	HG	CI	10.25	572	2.18 ×7.18
303	1,250– 300	1	N	Se	I	Sil	3.50	2.84	3.00	3.00	0.687	0.687	0.687	0.687	45	E	CNM	HG	CI	10.25	598	2.18 ×7.18
304	1,450– 400	1	N	Se	I	Sil	3.50	2.84	3.00	3.00	0.687	0.687	0.687	0.687	45	E	CNM	HG	CI	10.00	522	2.18 ×6.68
305	1,660– 450	1	N	Se	I	Sil	3.50	2.84	3.00	3.00	0.687	0.687	0.687	0.687	45	E	CNM	HG	CI	10.25	568	2.18 ×7.18
306	1,780– 450	1	N	Se	I	Sil	3.50	2.84	3.00	3.00	0.687	0.687	0.687	0.687	45	E	CNM	HG	CI	10.25	598	2.18 ×7.18
307	560– 600	2	N	Se	I	CNS	2.46	2.46	2.25	2.25	0.500	0.500	0.437	0.437	30	N	......	HG	Al	.....	...	1.37 ×4.50
308	660– 700	2	N	Se	I	CNS	2.46	2.46	2.25	2.25	0.500	0.500	0.437	0.437	30	N	......	HG	Al	.....	182	1.25 ×6.18
309	1,110–1,525	3	N	Se	I	CNS	2.46	2.46	2.25	2.25	0.375	0.375	0.437	0.437	30	N	......	HG	Al	.....	140	1.37 ×5.62
310	910– 600	2	N	Se	I	CNS	2.46	2.46	2.25	2.25	0.500	0.500	0.437	0.437	30	N	......	HG	Al	.....	...	1.37 ×4.50
311	882– 695	2	N	Se	I	CNS	2.46	2.46	2.25	2.25	0.500	0.500	0.437	0.437	30	N	......	HG	Al	.....	182	1.25 ×6.18
312	1,030– 800	2	N	Se	I	CNS	2.71	2.46	2.37	2.25	0.531	0.531	0.437	0.437	30	N	......	HG	Al	.....	...	1.37 ×4.50
313	154–1,200	6	N	In	L	Sil	1.46	1.46	1.31	1.31	0.312	0.312	0.340	0.340	45	E	SA	Ch	Alt	3.68	19	0.859×2.62
314	....	1	..	Se	L	Cl	....	.....	1.12	1.12	0.312	0.312	0.312	0.312	45	..	......	SG	CI	3.50	...	0.718×3.00
315	....	1	..	Se	L	Cl	....	.....	1.50	1.50	0.312	0.312	0.375	0.375	45	..	......	SG	CI	4.56	...	1.84 ×3.75
316	....	1	..	Se	L	Cl	....	.....	2.00	2.00	0.375	0.375	0.375	0.375	45	..	......	HG	CI	5.50	...	1.09 ×4.50
317	....	1	..	Se	L	Cl	....	.....	2.12	2.12	0.375	0.375	0.375	0.375	45	..	......	SG	CI	6.50	...	1.21 ×5.00
318	....	2	..	In	L	AS	....	.....	.....	.....	.....	.....	.....	.....	..	..	......	HG	Al	.....	...	0.625×

319	...	2	..	In	L	AS	...	...	...	...	...	...	...	...	..	..	...	HG	CI			
320	...	1	..	Se	L	Cl	...	...	1.50	1.50	0.312	0.312	0.375	0.375	45	..	...	SG	CI	4.56	...	1.84 ×3.75
321	...	2	..	Se	L	Cl	...	...	2.18	2.18	0.375	0.375	0.375	0.375	45	..	...	HG	CI	5.50	...	1.09 ×4.50
322	...	.	..	Se	L	Cl	...	...	2.37	2.37	0.437	0.437	0.437	0.437	45	..	...	HG	CI	7.00	...	1.21 ×5.25
323	...	1	..	Se	L	Cl	...	...	2.62	2.62	0.437	0.437	0.500	0.500	45	..	...	SG	CI	8.00	...	1.59 ×6.25
324	...	1	..	Se	L	Cl	...	...	2.62	2.62	0.375	0.375	0.562	0.562	45	..	...	SG	CI	9.87	...	2.00 ×7.25
325	169–1,100	6	..	In	L	Sil	1.71	1.62	...	...	0.312	0.312	0.373	0.373	45	E	...	Ch	Al	4.00	23	0.983×2.90
326	174–1,000	6	..	In	L	Sil	1.71	1.62	...	...	0.312	0.312	0.373	0.373	45	E	...	Ch	Al	4.00	26	0.983×3.02
327	208– 800	6	..	In	L	Sil	1.71	1.62	...	...	0.312	0.312	0.373	0.373	45	E	...	Ch	Al	4.00	26	0.983×3.02
328	226–1,000	6	..	In	L	Sil	1.71	1.62	1.43	1.43	0.312	0.312	0.373	0.373	45	E	...	Ch	Al	4.50	29	0.983×3.13
329	254– 800	6	..	In	L	Tun	2.06	1.87	1.81	1.62	0.359	0.359	0.434	0.432	30	E	...	Ch	Al	5.31	...	1.25 ×3.43
330	92–2,000	4	..	In	L	Sil	1.48	1.35	1.25	1.12	0.250	0.250	0.310	0.310	30	..	...	HG	CI	3.06	29	0.750×2.81
331	...	4	..	In	L	Sil	1.48	1.35	1.25	1.12	0.250	0.250	0.310	0.310	30	..	...	HG	Al	3.06	...	0.750×2.81
332	...	4	..	In	L	Sil	1.93	1.93	...	...	0.406	0.406	0.375	0.375	45	..	...	HG	Al	4.25	33	1.12 ×3.25
333	154–1,000	6	..	In	L	Sil	1.60	1.39	1.43	1.12	0.281	0.281	0.312	0.312	30	..	...	HG	Al	3.50	...	0.875×2.90
334	...	6	..	In	L	Sil	1.60	1.39	1.43	1.12	0.281	0.281	0.312	0.312	30	..	...	HG	Al	3.50	...	0.875×2.90
335	...	6	..	In	L	Sil	1.93	1.93	...	...	0.406	0.406	0.375	0.375	45	..	...	HG	Al	4.25	33	1.12 ×3.25
336	204–1,000	6	..	In	L	Sil	1.75	1.62	1.50	1.37	0.322	0.322	0.373	0.373	45	..	...	HG	Al	4.18	40	1.00 ×3.51
337	...	6	..	In	L	Sil	1.75	1.62	1.50	1.37	0.322	0.322	0.373	0.373	45	..	...	HG	Al	4.18	40	1.00 ×3.51
338	...	2	..	Se	L	Sil	2.25	2.25	...	...	0.375	0.375	0.437	0.437	45	..	...	HG	Al	5.12	49	1.25 ×3.68
339	...	2	..	Se	L	Sil	2.25	2.25	...	...	0.375	0.375	0.437	0.437	45	..	...	HG	Al	5.12	49	1.25 ×3.68
340	...	2	..	Se	L	Sil	2.37	2.28	...	...	0.405	0.405	0.437	0.437	45	..	...	HG	Al	5.12	50	1.37 ×3.84
341	...	2	..	Se	L	Sil	2.37	2.28	...	...	0.405	0.405	0.437	0.437	45	..	...	HG	Al	5.12	50	1.37 ×3.84
342	...	2	..	Se	L	Sil	2.37	2.28	...	...	0.405	0.405	0.437	0.437	45	..	...	HG	Al	5.12	56	1.37 ×4.00
343	...	2	..	Se	L	Sil	2.37	2.28	...	...	0.405	0.405	0.437	0.437	45	..	...	HG	Al	5.12	56	1.37 ×4.00
344	...	2	..	Se	L	Sil	2.56	2.28	...	...	0.405	0.405	0.437	0.437	45	..	...	HG	Al	5.37	62	1.37 ×4.00
345	...	2	..	Se	L	Sil	2.56	2.28	...	...	0.405	0.405	0.437	0.437	45	..	...	HG	Al	5.37	62	1.37 ×4.00
346	...	2	..	Se	L	Sil	2.50	2.37	2.37	2.25	0.406	0.375	0.437	0.437	45	..	...	HG	Al	5.12	56	1.37 ×4.00
347	...	2	..	Se	L	Sil	2.50	2.37	2.37	2.25	0.406	0.375	0.437	0.437	45	..	...	HG	Al	5.12	56	1.37 ×4.00
348	150–2,200	8	N	In	L	Sil	1.53	1.53	...	...	0.296	0.296	0.312	0.312	45	Bo	Tun	HG	Al	...	...	0.750×2.84
349	170–2,100	8	N	In	L	CNS	1.53	1.53	...	...	0.292	0.292	0.311	0.311	45	Bo	Tun	HG	CAS	...	...	0.750×2.84
350	186–2,000	12	N	In	L	CNS	1.53	1.53	...	...	0.292	0.292	0.311	0.311	45	Bo	Tun	HG	CAS	3.27	15	0.750×2.48
351	...	2	..	Se	L	Sil	2.25	2.25	...	...	0.375	0.375	0.437	0.437	45	..	...	HG	Al	5.12	49	1.25 ×3.68
352	...	2	..	Se	L	Sil	2.25	2.25	...	...	0.375	0.375	0.437	0.437	45	..	...	HG	Al	5.12	49	1.25 ×3.68
353	20–1,000	2	..	In	..	Two-cycle motor				...	...	...	...	...	..	..	...	...	Ala	7.25	20	0.850×....
354	20–2,600	2	..	In	..	Two-cycle motor				...	...	...	...	...	..	..	...	...	Ala	7.25	20	0.850×....
355	...	2	..	In	..	Two-cycle motor				...	...	...	...	...	..	..	...	...	Ala	7.25	20	0.850×....
356	265–1,100	6	N	In	L	CNS	...	...	...	...	...	...	...	...	..	..	...	HG	Al			
357	780–1,000	2	N	Se	L	Spec	2.50	2.50	...	...	0.562	0.562	0.531	0.531	45	N	...	HG	CI	6.25	...	1.37 ×5.37
358	835–1,315	2	N	Se	L	Spec	2.50	2.50	...	...	0.562	0.562	0.531	0.531	45	N	...	HG	Al	...	...	1.37 ×5.37
359	1,480– 800	1	N	Se	I	Spec	2.25	2.25	...	...	0.500	0.500	0.562	0.562	45	N	...	HG	CI	7.00	...	1.87 ×5.50
360	158– 400	2	N	Se	L	Sil	2.06	2.06	...	...	0.375	0.375	0.437	0.437	45	..	...	HG	CI	6.00	140	1.25 ×5.12
361	500–1,400	6	N	Se	L	Sil	2.25	2.25	...	...	0.455	0.455	0.437	0.437	45	..	...	HG	Al	5.50	94	1.43 ×4.37
362	500–1,400	6	N	Se	L	Sil	2.25	2.25	...	...	0.455	0.455	0.437	0.437	45	..	...	HG	Al	5.50	94	1.43 ×4.37
363	500–1,400	6	N	Se	L	Sil	2.25	2.25	...	...	0.455	0.455	0.437	0.437	45	..	...	HG	Al	5.50	94	1.43 ×4.37
364	500–1,400	6	N	Se	L	Sil	2.25	2.25	...	...	0.455	0.455	0.437	0.437	45	..	...	HG	Al	5.50	94	1.43 ×4.37
365	500–1,400	6	N	Se	L	Sil	2.25	2.25	...	...	0.455	0.455	0.437	0.437	45	..	...	HG	Al	5.50	94	1.43 ×4.37

TABLE A1-3.—AMERICAN STOCK, MARINE, AND COMMERCIAL VEHICLE ENGINES.—(*Continued*)

		Cylinders			Valves															Pistons			
							Max. head diameter, in.		Min. port diameter, in.		Lift, in.		Stem diameter, in.		Seats							
Line number	Maximum torque at r.p.m., lb.-ft.	No. cast in one piece	Liners, type	Crankcase. upper half integral with cylinders?	Arrangement	Exhaust-head material (S.A.E. No.)	Intake	Exhaust	Intake	Exhaust	Intake	Exhaust	Intake	Exhaust	Angle, deg.	Inserts used?	Insert material (S.A.E. No.)	Front-end drive, type	Material	Length, in.	Weight with pins, rings, bushings, oz.	Piston pin, diameter and length, in.
366	500–1,400	6	N	Se	L	Sil	2.25	2.25	.....	.....	0.455	0.455	0.437	0.437	45	..	......	HG	Al	5.50	94	1.43 ×4.37
367	500–1,400	6	N	Se	L	Sil	2.25	2.25	.....	.....	0.455	0.455	0.437	0.437	45	..	......	HG	Al	5.50	94	1.43 ×4.37
368	590–1,200	2	N	Se	T	Sil	2.25	2.25	.....	.....	0.375	0.375	0.500	0.500	45	..	......	HG	CI	6.00	140	1.25 ×5.12
369	590–1,200	2	N	Se	T	Sil	2.25	1.25	.....	.....	0.375	0.375	0.500	0.500	45	..	......	HG	CI	6.00	140	1.25 ×5.12
370	590–1,200	2	N	Se	T	Sil	2.25	2.25	.....	.....	0.375	0.375	0.500	0.500	45	..	......	HG	CI	6.00	140	1.25 ×5.12
371	785–1,200	2	N	Se	I	Sil	1.87	1.87	.....	.....	0.375	0.375	0.437	0.437	60	..	......	HG	Al	5.81	100	1.25 ×5.12
372	785–1,200	2	N	Se	I	Sil	1.87	1.87	.....	.....	0.375	0.375	0.437	0.437	60	..	......	HG	Al	6.06	100	1.25 ×5.12
373	785–1,200	2	N	Se	I	Sil	1.87	1.87	.....	.....	0.375	0.375	0.437	0.437	60	..	......	HG	Al	6.46	110	1.25 ×5.12
374	986–1,200	6	W	Se	I	Sil	1.87	1.87	.....	.....	0.483	0.483	0.437	0.437	60	..	......	HG	Al	6.06	125	1.50 ×5.37
375	986–1,200	6	W	Se	I	Sil	1.87	1.87	.....	.....	0.483	0.483	0.437	0.437	60	..	......	HG	Al	6.42	129	1.50 ×5.37
376	1,900–1,000	6	W	Se	I	Sil	2.59	2.59	.....	.....	0.556	0.556	0.562	0.562	45	..	......	HG	Al	8.00	280	2.00 ×7.00
377	1,900–1,000	6	W	Se	I	Sil	2.59	2.59	.....	.....	0.556	0.556	0.562	0.562	45	..	......	HG	Al	8.25	290	2.00 ×7.00
378	1,900–1,000	6	W	Se	I	Sil	2.59	2.59	.....	.....	0.556	0.556	0.562	0.562	45	..	......	HG	Al	8.93	290	2.00 ×7.00
379	1,055–1,300	2	N	Se	I	Sil	1.87	1.87	.....	.....	0.375	0.375	0.437	0.437	60	..	......	HG	Al	5.81	100	1.25 ×5.12
380	1,055–1,300	2	N	Se	I	Sil	1.87	1.87	.....	.....	0.375	0.375	0.437	0.437	60	..	......	HG	Al	6.06	100	1.25 ×5.12
381	2,520–1,050	8	W	Se	I	Sil	2.59	2.59	.....	.....	0.556	0.556	0.562	0.562	45	..	......	HG	Al	8.00	280	2.00 ×7.00
382	2,520–1,050	8	W	Se	I	Sil	2.59	2.59	.....	.....	0.556	0.556	0.562	0.562	45	..	......	HG	Al	8.25	290	2.00 ×7.00
383	2,520–1,050	8	W	Se	I	Sil	2.59	2.59	.....	.....	0.556	0.556	0.562	0.562	45	..	......	HG	Al	8.93	290	2.00 ×7.00
384	....	1	N	Se	L	NCI	1.62	1.62	1.43	1.43	0.300	0.300	0.375	0.375	45	N	......	HG	CI	4.12	64	1.10 ×3.25
385	54– 800	2	N	Se	L	NCI	1.62	1.62	1.43	1.43	0.300	0.300	0.375	0.375	45	N	......	HG	CI	4.12	64	1.10 ×3.25
386	40–1,800	4	N	Se	L	CHS	1.12	0.937	1.00	0.812	0.228	0.250	0.312	0.312	45	N	......	HG	Al	2.37	9	0.625×2.12
387	52–1,300	4	N	In	L	CAI	1.46	1.34	1.31	1.18	0.250	0.250	0.312	0.312	45	N	......	HG	CI	3.00	19	0.625×2.40
388	92–1,200	4	N	In	L	CHS	1.34	1.34	1.21	1.21	0.281	0.281	0.312	0.312	45	N	......	HG	CI	3.50	30	0.875×2.62
389	121– 900	4	N	In	L	Sil	1.56	1.56	1.37	1.37	0.281	0.281	0.375	0.375	45	N	......	HG	CI	3.93	45	1.10 ×3.06
390	112– 700	2	N	Se	L	NCI	1.62	1.62	1.43	1.43	0.300	0.300	0.375	0.375	45	N	......	HG	CI	4.12	64	1.10 ×3.25
391	142–1,000	4	N	Se	L	NCI	1.93	1.93	1.75	1.75	0.300	0.300	0.375	0.375	45	N	......	HG	CI	4.75	69	1.10 ×3.56

392	180– 900	4	N	Se	L	Dia	2.09	2.09	1.93	1.93	0.300	0.300	0.375	0.375	45	N	......	HG	CI	4.75	89	1.10 ×3.93
393	224–1,100	4	N	Se	L	Dia	2.34	2.34	2.12	2.12	0.300	0.300	0.434	0.434	45	N	......	HG	CI	5.25	82	1.25 ×3.87
394	292– 900	2	N	Se	L	CAI	2.75	2.75	2.37	2.37	0.375	0.375	0.621	0.621	45	N	......	HG	CIA	6.00	126	1.43 ×4.68
395	408– 900	2	N	Se	L	CAI	2.75	2.75	2.37	2.37	0.375	0.375	0.621	0.621	45	N	......	HG	CIA	6.00	150	1.43 ×5.25
396	465– 700	2	N	Se	L	CAI	2.75	2.75	2.37	2.37	0.375	0.375	0.621	0.621	45	N	......	HG	CIA	6.00	190	1.43 ×5.50
397	185–1,100	6	N	In	L	Sil	1.68	1.43	1.50	1.25	0.375	0.375	0.375	0.375	45	N	......	HG	Al	4.37	37	1.00 ×3.00
398	286– 800	6	N	In	L	Sil	1.93	1.43	1.75	1.25	0.375	0.375	0.375	0.375	45	E	......	HG	CI	4.37	78	1.00 ×4.00
399	379– 900	6	N	Se	L	Dia	2.34	2.34	2.12	2.12	0.300	0.300	0.434	0.434	45	N	......	HG	CIA	5.25	82	1.25 ×3.87
400	420– 900	6	N	Se	L	Dia	2.34	2.34	2.12	2.12	0.300	0.300	0.434	0.434	45	N	......	HG	CIA	5.25	106	1.25 ×4.31
401	451–1,000	2	N	Se	L	CAI	2.75	2.75	2.37	2.37	0.375	0.375	0.621	0.621	45	N	......	HG	CIA	6.00	126	1.43 ×4.68
402	596– 850	2	N	Se	L	CAI	2.75	2.75	2.37	2.37	0.375	0.375	0.621	0.621	45	N	......	HG	CIA	6.00	150	1.43 ×5.25
403	630– 875	2	N	Se	L	CAI	2.75	2.75	2.37	2.37	0.375	0.375	0.621	0.621	45	N	......	HG	CIA	6.00	190	1.43 ×5.50
404	....	1	N	In	L	CNS	1.87	1.87	.....	.....	0.250	0.250	0.375	0.375	45	N	......	HG	CI	4.65	...	1.00 ×3.75
405	....	2	N	In	L	CNS	1.50	1.50	.....	.....	0.250	0.250	0.375	0.375	45	N	......	HG	Al	3.25	...	0.750×2.56
406	....	4	N	In	L	CNS	1.12	1.12	.....	.....	0.234	0.234	0.312	0.312	45	N	......	HG	CI	3.25	...	0.625×2.44
407	....	4	N	In	L	Sil	1.50	1.50	.....	.....	0.312	0.312	0.375	0.375	45	N	......	HG	Al	3.25	15	0.750×2.56
408	....	4	N	Se	L	CNS	1.37	1.37	.....	.....	0.312	0.312	0.375	0.375	45	N	......	HG	Al	3.50	20	0.875×2.75
409	....	6	N	In	L	Sil	1.50	1.50	.....	.....	0.312	0.312	0.375	0.375	45	N	......	HG	Al	3.25	15	0.750×2.56
410	....	6	N	Se	L	Sil	1.50	1.50	.....	.....	0.312	0.312	0.375	0.375	45	N	......	HG	Al	3.87	28	0.875×3.00
411	....	6	N	Se	L	Sil	1.50	1.50	.....	.....	0.312	0.312	0.375	0.375	45	N	......	HG	Al	3.87	28	0.875×3.00
412	....	8	N	Se	L	Sil	1.50	1.50	.....	.....	0.328	0.328	0.375	0.375	45	N	......	HG	Al	3.87	28	0.875×3.00
413	....	8	N	Se	L	Sil	1.50	1.50	.....	.....	0.328	0.328	0.375	0.375	45	N	......	HG	Al	3.87	28	0.875×3.00
414	1,286–1,400	1	N	Se	I	Sil	2.73	2.73	2.25	2.25	0.437	0.375	0.435	0.435	30	E	CNS	...	Ala	4.65	76	1.25 ×4.18
415																						
416	....	6	W	Se	I	CNS	2.20	2.21	1.95	1.96	0.500	0.500	0.403	0.559	45	Bo	CNS	...	Al	4.59	88	1.62 ×5.90
417	36–1,400	1	N	Se	L	Sil	1.53	1.53	1.37	1.37	0.312	0.312	0.312	0.312	45	N	......	HG	Al	3.84	28	1.00 ×3.18
418	72–1,400	1	N	Se	L	Sil	1.53	1.53	1.37	1.37	0.312	0.312	0.312	0.312	45	N	......	HG	Al	3.84	28	1.00 ×3.18
419	144–1,400	1	N	Se	L	Sil	1.53	1.53	1.37	1.37	0.312	0.312	0.312	0.312	45	N	......	HG	Al	3.84	28	1.00 ×3.18
420	....	2	N	In	L	Sil	1.37	1.15	1.25	1.00	0.300	0.300	0.312	0.312	45	N	......	HG	Al	2.62	12	0.625×2.62
421	40–1,800	4	N	In	L	Sil	1.12	0.937	1.00	0.937	0.228	0.250	0.312	0.312	45	N	......	HG	CI	2.34	9	0.625×2.12
422	67–1,100	4	N	In	L	Sil	1.34	1.34	1.18	1.18	0.281	0.281	0.312	0.312	45	N	......	HG	CI	3.25	26	0.875×2.25
423	92–1,200	4	N	In	L	Sil	1.34	1.34	1.18	1.18	0.281	0.281	0.312	0.312	45	N	......	HG	CI	3.50	30	0.875×2.75
424	121– 900	4	N	In	L	Sil	1.56	1.56	1.37	1.37	0.281	0.281	0.375	0.375	45	E	AI	HG	CI	3.93	45	1.12 ×3.03
425	162– 900	4	W	In	I	Sil	1.84	1.40	1.62	1.25	0.445	0.453	0.434	0.434	h	E	AI	HG	CI	5.12	58	1.12 ×3.06
426	184– 900	4	W	In	I	Sil	1.84	1.40	1.62	1.25	0.445	0.453	0.434	0.434	h	E	AI	HG	CI	5.12	...	1.12 ×3.31
427	200–1,000	4	W	In	I	Sil	2.00	1.75	1.75	1.50	0.400	0.400	0.375	0.375	45	E	AI	HG	CI	4.87	76	1.31 ×3.87
428	224–1,000	4	W	In	I	Sil	2.00	2.00	1.75	1.75	0.400	0.400	0.375	0.375	45	E	AI	HG	CI	4.87	85	1.31 ×4.06
429	....	4	W	In	I	Sil	....	.....	1.75	1.50	.....	.....	.....	.....	..	E	AI	HG	CI	5.96	96	1.31× 4.06
430	375– 800	4	W	In	I	Sil	2.37	2.37	2.12	2.12	0.437	0.437	0.437	0.437	45	E	AI	HG	CI	5.87	132	1.50 ×4.48
431	419– 700	2	N	Se	L	Sil	2.40	2.15	2.12	1.87	0.500	0.500	0.437	0.437	45	N	......	HG	CI	6.53	178	1.37 ×5.25
432	762– 600	2	N	Se	L	Sil	3.00	3.00	2.75	2.75	0.500	0.500	0.562	0.562	30	N	......	HG	CI	7.50	370	1.56 ×6.00
433	950– 700	4	N	Se	I	Sil	3.25	2.75	3.00	2.50	0.756	0.756	0.562	0.562	30	E	AI	HG	CI	8.68	447	2.00 ×6.75
434	176–1,100	6	N	In	L	Sil	1.68	1.43	1.50	1.25	0.375	0.375	0.375	0.375	45	E	AI	HG	Al	4.43	34	1.00 ×3.25
435	185–1,100	6	N	In	L	Sil	1.68	1.43	1.50	1.25	0.375	0.375	0.375	0.375	45	E	AI	HG	Al	4.37	37	1.00 ×3.50
436	210–1,200	6	N	In	L	Sil	1.68	1.43	1.50	1.25	0.375	0.375	0.375	0.375	45	E	AI	HG	Al	4.37	46	1.00 ×3.85
437	270– 800	6	N	In	L	Sil	1.93	1.43	1.75	1.25	0.375	0.375	0.375	0.375	45	E	AI	HG	Al	4.37	45	1.00 ×3.85
438	286– 800	6	N	In	L	Sil	1.93	1.43	1.75	1.25	0.375	0.375	0.375	0.375	45	E	AI	HG	Al	4.37	48	1.00 ×4.00

TABLE A1-3.—AMERICAN STOCK, MARINE, AND COMMERCIAL VEHICLE ENGINES.—(*Continued*)

Line number	Maximum torque at r.p.m., lb.-ft.	Cylinders: No. cast in one piece	Cylinders: Liners, type	Crankcase, upper half integral with cylinders?	Valves: Arrangement	Valves: Exhaust-head material (S.A.E. No.)	Valves: Max. head diameter, in.: Intake	Valves: Max. head diameter, in.: Exhaust	Valves: Min. port diameter, in.: Intake	Valves: Min. port diameter, in.: Exhaust	Valves: Lift, in.: Intake	Valves: Lift, in.: Exhaust	Valves: Stem diameter, in.: Intake	Valves: Stem diameter, in.: Exhaust	Valves: Seats: Angle, deg.	Valves: Seats: Inserts used?	Valves: Seats: Insert material (S.A.E. No.)	Front-end drive, type	Pistons: Material	Pistons: Length, in.	Pistons: Weight with pins, rings, bushings, oz.	Pistons: Piston pin, diameter and length, in.
439	360–1,000	6	W	In	I	Sil	2.12	1.56	1.87	1.37	0.531	0.469	0.434	0.434	h	E	AI	HG	CI	6.50	...	1.37 ×3.62
440	307– 600	6	N	Se	L	Sil	1.90	1.65	1.56	1.37	0.386	0.375	0.375	0.375	30	E	AI	HG	Al	4.62	89	1.37 ×4.00
441	406–1,000	6	W	In	I	Sil	2.12	1.56	1.87	1.37	0.531	0.469	0.434	0.434	h	E	AI	HG	CI	6.50	113	1.37 ×3.87
442	368– 600	6	N	Se	L	Sil	1.90	1.65	1.56	1.37	0.386	0.375	0.375	0.375	30	E	AI	HG	Al	5.25	70	1.37 ×4.25
443	506– 800	6	W	In	I	Sil	2.12	1.62	1.87	1.37	0.594	0.531	0.495	0.495	h	E	AI	HG	CI	7.25	...	1.62 ×4.37
444	468– 700	6	D	In	I	Sil	2.21	1.75	2.00	1.50	0.500	0.500	0.437	0.437	45	E	AI	HG	CI	6.43	190	1.87 ×4.25
445	490– 800	6	N	Se	L	Sil	2.40	2.40	2.12	2.12	0.400	0.400	0.437	0.437	45	E	AI	HG	Al	6.00	89	1.37 ×4.60
446	620– 800	6	W	In	I	Sil	2.12	1.62	1.87	1.37	0.594	0.531	0.495	0.495	h	E	AI	HG	CI	7.25	184	1.62 ×4.87
447	567– 700	6	D	In	I	Sil	2.21	1.75	2.00	1.50	0.500	0.500	0.437	0.437	45	E	AI	HG	CI	6.43	196	1.87 ×4.75
448	685– 600	6	W	In	I	Sil	2.65	2.22	2.43	2.00	0.656	0.656	0.500	0.500	h	Bo	AI	HG	CI	8.37	262	1.87 ×5.00
449	810– 600	6	W	In	I	Sil	2.65	2.22	2.37	2.00	0.656	0.656	0.500	0.500	h	Bo	AI	HG	CI	8.37	294	1.87 ×5.50
450	1,000– 550	6	W	In	I	Sil	2.75	2.50	2.50	2.25	0.718	0.718	0.562	0.562	30	E	AI	HG	CI	9.18	294	2.00 ×5.50
451	1,120– 550	6	W	In	I	Sil	2.75	2.50	2.50	2.25	0.718	0.718	0.562	0.562	30	E	AI	HG	CI	9.18	304	2.00 ×6.00
452	1,330– 600	6	W	In	I	Sil	2.75	2.50	2.50	2.25	0.718	0.718	0.562	0.562	30	E	AI	HG	CI	9.18	304	2.00 ×6.00
453	1,310– 720	3	N	Se	L	Sil	3.00	2.75	2.75	2.50	0.740	0.825	0.562	0.562	30	E	AI	HG	CI	8.62	420	2.25 ×6.25
454	1,610– 700	3	N	Se	I	Sil	3.00	2.75	2.75	2.50	0.740	0.825	0.562	0.562	30	E	AI	HG	CI	8.62	489	2.25 ×7.00
455	1,860– 700	3	W	Se	I	Sil	3.50	3.00	3.25	0.275	0.750	0.840	0.562	0.562	30	E	AI	HG	CI	10.37	776	2.25 ×7.75
456	175–1,200	6	N	In	L	AUS	1.69	1.62	1.50	1.37	0.396	0.396	0.375	0.375	45	E	CNM	HG	Al	4.71	33	1.00 ×3.03
457	200–1,200	6	N	In	L	AUS	1.69	1.62	1.50	1.37	0.396	0.396	0.406	0.406	45	E	CNM	HG	CI	4.71	48	1.12 ×3.03
458	187–1,000	6	N	In	L	Sil	1.75	1.62	1.50	1.37	0.356	0.356	0.375	0.375	45	N	......	HG	Al	4.18	39	1.00 ×3.15
459	245–1,300	6	N	In	L	AUS	1.66	1.69	1.50	1.37	0.396	0.396	0.375	0.375	45	E	CNM	HG	Al	4.40	37	1.00 ×3.46
460	280–1,200	6	N	In	L	AUS	1.66	1.69	1.43	1.50	0.381	0.381	0.406	0.406	45	E	CNM	HG	Al	4.40	36	1.00 ×3.46
461	320–1,000	6	N	Se	I	CNT	2.13	2.03	1.87	1.75	0.381	0.381	0.437	0.437	45	E	CNM	Ch	Al	5.46	66	1.18 ×3.87
462	390–1,000	6	N	Se	I	CNT	2.13	2.03	1.87	1.81	0.381	0.381	0.437	0.500	45	E	CNM	Ch	Al	5.34	66	1.25 ×3.93
463	500–1,200	12	D	In	L	AUS	1.85	1.75	1.62	1.62	0.381	0.381	0.401	0.401	h	E	CNM	HG	Al	4.40	44	1.18 ×3.62
464	109–2,200	4	N	In	L	2,112	1.53	1.46	1.34	1.28	0.359	0.359	0.373	0.373	45	N	......	Ch	Al	3.75	18	0.812×2.78

465	4.2–2,000	1	N	In	L	AUS	1.12	1.12	0.875	0.875	0.187	0.187	0.310	0.310	45	E	Mo	HG	Al	2.50	6	0.625×1.81
466	6.8–1,800	1	N	In	L	AUS	1.12	1.12	0.875	0.875	0.187	0.187	0.310	0.310	45	E	Mo	HG	Al	2.50	8	0.625×2.06
467	10–1,600	1	N	Se	L	Sil	1.31	1.31	0.687	0.875	0.275	0.275	0.310	0.310	45	Bo	Mo	HG	Al	2.75	12	0.750×2.29
468	9.5–1,700	1	N	In	L	AUS	1.12	1.12	0.937	0.937	0.187	0.187	0.310	0.310	45	E	Mo	HG	Al	2.96	10	0.625×2.37
469	11.8–1,600	1	N	Se	L	Sil	1.31	1.31	0.687	0.875	0.275	0.275	0.310	0.310	45	Bo	Mo	HG	Al	2.82	14	0.750×2.56
470	16.9–1,200	1	N	Se	L	Sil	1.56	1.56	0.812	1.25	0.275	0.275	0.310	0.310	45	Bo	Mo	HG	Al	3.75	24	0.937×2.75
471	19.6–1,200	1	N	Se	L	Sil	1.56	1.56	0.812	1.25	0.275	0.275	0.310	0.310	45	Bo	Mo	HG	Al	3.56	24	0.937×2.75
472	23.7–1,400	1	N	Se	L	Sil	1.56	1.56	0.812	1.25	0.275	0.275	0.310	0.310	45	Bo	Mo	HG	Al	3.68	26	0.937×3.00
473	39.5–1,600	4	N	Se	I	Sil	1.12	1.12	0.937	0.937	0.232	0.232	0.310	0.310	45	Bo	Mo	HG	Al	3.00	11	0.875×2.17
474	79–1,300	4	N	Se	I	Sil	1.50	1.37	1.12	1.12	0.276	0.256	0.310	0.310	45	Bo	Mo	HG	Al	2.75	24	0.937×2.75
475	94–1,100	4	N	Se	I	Sil	1.50	1.37	1.12	1.12	0.276	0.256	0.310	0.310	45	Bo	Mo	HG	Al	2.75	27	0.937×2.75
476	160–1,000	4	N	In	I	AUS	1.50	1.50	1.68	1.68	0.438	0.379	0.375	0.375	45	Bo	Mo	HG	CI	4.25	50	1.06 ×3.47
477	224–1,000	4	N	In	I	Sil	2.00	2.00	1.81	1.81	0.384	0.393	0.437	0.437	45	N	......	HG	CI	4.75	118	1.18 ×3.93
478	211– 700	6	N	In	I	Sil	1.71	1.71	1.50	1.50	0.379	0.379	0.375	0.375	45	N	......	HG	CI	3.90	53	1.06 ×3.09
479	236– 700	6	N	In	I	Sil	2.00	2.00	1.75	1.75	0.379	0.379	0.434	0.434	45	N	......	HG	CI	4.87	66	1.25 ×3.14
480	260– 700	6	N	In	I	AUS	2.00	2.00	1.62	1.62	0.379	0.379	0.434	0.434	45	N	......	HG	CI	4.71	71	1.25 ×3.40
481	280– 700	6	N	In	I	AUS	2.00	2.00	1.62	1.62	0.379	0.379	0.434	0.434	45	N	......	HG	CI	4.62	81	1.25 ×3.40
482	322– 800	6	N	Se	I	Sil	2.25	2.25	2.06	2.06	0.450	0.450	0.437	0.437	45	N	......	HG	CI	4.75	56	1.18 ×3.93
483	340– 750	6	N	In	I	AUS	2.25	2.25	1.56	1.56	0.450	0.450	0.434	0.434	45	N	......	HG	CI	4.68	82	1.18 ×3.93

Symbols and Abbreviations

(Continued from page 357)

Car—Carter carburetor.
CAS—Cast alloy steel.
CG—Chandler-Groves carburetor.
Ch—Chain.
CHS—Chrome nickel silicon steel.
CI—Cast iron.
CIA—Cast iron, anodized.
CNI—Chrome nickel iron.
CNM—Chromenickel molybdenum.
CNS—Chrome nickel steel.
CNT—Chrome nickel steel with tungsten.
CS—Carbon steel.
CT—Cast iron, tin plated.
CV—Chrome vanadium.
d—Wrist pins.
D—Dry liners.
DC—Durachrome casting.
DFS—Drop-forged steel.
Dia—Diachrome.
Dur—Duralumin.
e—Timing gears or chain.
E—Used on exhaust-valve seats.
Ext—Extruded steel.
f—Accessoried drive.
F—In head and side (F head).
FA—Fire apparatus.
g—Rocker arms and shafts.
h—Intake 30 deg., exhaust 45 deg.
H—Horizontal motor.
HB—Horizontal in block (valves).
HC—Helical gear and chain.
HG—Helical gear
Hol—Holley carburetor.
HS—High-speed steel.
I—In head (valves).
In—Integral.
Ind—Industrial.
JM—Jadson 1-S material.
L—Valves at side (L-head).

(Continued on p. 385)

TABLE A1-3.—AMERICAN STOCK, MARINE, AND COMMERCIAL VEHICLE ENGINES.—(*Continued*)

Line number	Number of rings per piston	Connecting rods			Crankshaft						Oil pressure to—	Spark plug, thread size	Carburetor		Engine weight without carburetor or ignition, lb.	Over-all dimensions, in.		
							Crankpin	Main bearings										
									Diameter and length, in.									
		Material	Center-to-center length, in.	Weight with bushing and cap, oz.	Material	Counterbalance used?	Diameter and length, in.	Number	Front	Rear			Make	Size		Width	Height	Length
1	3	1,040	6½	30	1,045	N	1.93×1.22	3	2.25×1.62	2.25×1.50	ac	14 mm.	Zen	⅞	360	16 5/16	31⅝	27
2	4	1,040	7½	42	1,045	N	2.37×1.54	3	2.43×1.62	2.50×1.75	abce	14 mm.	Zen	1	500	24	31⅜	33⅝
3	4	1,040	9½	92	1,045	N	2.37×2.37	3	2.50×2.31	2.50×2.75	abcde	⅞–18	Zen	1¼	985	26	37½	43¾
4	4	1,040	13	182	1,045	N	2.75×3.24	3	3.00×3.50	3.00×4.75	abcde	⅞–18	Zen	1½	1,835	27	44⅞	55⅝
5	4	1,040	13	182	1,045	N	2.75×3.24	4	3.00×3.50	3.00×4.75	abcde	⅞–18	Zen	1½	2,875	28	50	62½
6	4	2,340	10¼	73	1,050	N	2.25×1.44	7	3.00×1.87	3.00×2.62	abcdef	⅞–18	Str	1½	1,070	23 1/16	34⅝	43 15/16
7	4	2,340	10¼	73	1,050	N	2.25×1.44	7	3.00×1.87	3.00×2.62	abcdef	⅞–18	Str	1½	1,080	23 1/16	34⅝	43 15/16
8	4	2,340	10¼	83	1,050	N	2.50×1.58	7	3.25×1.87	3.25×2.87	abcdef	18 mm.	Str	1¾	1,270	26⅛	38⅞	47 7/16
9	4	2,340	10¾	83	1,050	N	2.50×1.58	7	3.25×1.87	3.25×2.87	abcdef	18 mm.	Str	1¾	1,270	26⅛	38⅞	47 7/16
10	4	2,340	10¼	88	1,050	N	2.50×1.58	7	3.25×1.87	3.25×2.87	abcdef	18 mm.	Str	1¾	1,270	26⅛	38⅞	47 7/16
11	3	1,045	5¾	14	1,045	Y	1.31×1.25	2	2.50×1.50	2.50×1.50	abce	14 mm.	Til	⅝	160[d]	12¾	17⅝	29
12	3	1,045	5¾	14	1,045	Y	1.31×1.25	3	2.50×1.50	2.50×1.50	abce	14 mm.	Zen	⅝	165[d]	12¾	17⅝	29
13	3	1,045	5¾	14	1,045	Y	1.31×1.25	3	2.50×1.50	2.50×1.50	abce	14 mm.	Zen	⅝	128	12¾	17⅝	18¾
14	4	1,045	11	64	1,045	..	2.50×2.00	3	2.12×4.25	2.12×2.25	abcer	......	Str	1¼	650	12 7/16	19½	53¾
15	4	1,045	11	64	1,045	N	2.50×2.50	3	2.50×4.25	2.50×3.50	abcder	......	Str	1¼	950	16	18	53
16	4	1,045	11	64	1,045	N	2.50×2.50	3	2.50×4.25	2.50×3.50	abce	......	Str	1¼	600	21	29¾	37⅛
17	4	1,045	11	64	CNS	Y	2.50×2.00	3	2.75×4.50	2.75×3.00	abceg	......	Str	1½	800	19¾	24¾	65
18	4	CNS	11	65	CNS	Y	2.50×2.00	3	2.75×4.50	2.75×3.00	abce	......	Str	1½	800	25¾	33¼	49
19	5	CNS	11	65	CNS	Y	2.50×2.00	3	2.75×4.50	5.75×4.50	abceg	......	Str	1½	900	19¾	24¾	65
20	4	CNS	11	65	CNS	Y	2.50×2.00	3	2.75×4.50	2.75×3.00	abce	......	Str	1½	875	25¾	33¼	49
21	5	CNS	12	80	CNS	Y	2.62×2.67	7	2.62×5.00	2.62×3.50	abce	......	Str	1¾	1,450[d]	20	30	74
22	5	CNS	12	80	CNS	Y	2.62×2.67	7	2.62×5.00	2.62×3.50	abce	......	Str	1¾	1,000	22	40	54
23	3	CS	9	60	CS	Y	1.37×2.00	2	1.37×2.50	1.37×2.50	splash	⅞–18	Zen	1	155	14	22½	21
24	3	CS	9	60	CNS	Y	1.37×2.00	2	1.50×3.00	1.50×3.00	splash	⅞–18	Zen	1¼	320	14	24	32
25	3	CS	9	47	CS	N	1.50×1.75	3	2.75×1.50	2.75×1.50	abcd	⅞–18	Zen	1¼	397	17	22	35
26	3	AS	9	50	CNS	N	1.75×1.50	3	1.87×2.00	1.87×1.50	abcd	⅞–18	Zen	1¼	460	20	22½	36

27	3	AS	9 1/2	60	CNS	N	2.00×2.25	3	2.00×2.75	2.00×1.12	abcde	7/8–18	Zen	1 1/4	950	20	28	53
28	3	CS	9 5/8	60	CS	N	2.50×1.75	7	3.00×2.00	3.00×3.00	abcd	7/8–18	Zen	1 1/2	1,650	18	31	62
29	4	CS	9 1/2	42	CS	N	2.12×1.62	5	3.00×1.50	3.00×2.12	abcde	18 mm.	Zen	1 1/4	590	26	29 31/32	31 31/32
30	4	CS	9 1/2	42	CS	N	2.12×1.62	5	3.00×1.50	3.00×2.12	abcde	18 mm.	Zen	1 1/4	590	25 3/4	31 1/16	38 9/32
31	4	CS	9 1/2	42	CS	N	2.12×1.62	5	3.00×1.50	3.00×2.12	abcder	18 mm.	Str	1 1/4	770	23 1/4	31 1/16	43 3/8
32	4	CS	9 1/2	42	CS	N	2.12×1.62	5	3.00×1.50	3.00×2.12	abcder	18 mm.	Str	1 1/4	770	23 1/4	31 1/16	43 3/8
33	4	CS	9 1/2	42	CS	N	2.12×1.62	5	3.00×1.50	3.00×2.12	abcder	18 mm.	Str	1 1/4	770	23 1/4	31 1/16	43 3/8
34	4	CS	11 1/4	89	CS	N	2.00×2.25	3	1.87×2.87	2.12×3.44	abcde	7/8–18	Zen	1 1/4	875	25 3/4	33 1/4	40 15/16
35	4	AS	13 1/4	106	CS	N	2.50×2.87	3	2.50×3.00	2.50×4.50	abcde	7/8–18	Zen	1 1/2	1,087	25 3/4	36 15/16	47 1/2
36	4	AS	14 3/8	163	CS	N	2.50×3.12	3	2.25×4.12	2.62×4.69	abcde	7/8–18	Zen	1 1/2	1,409	28 3/4	40 7/16	52 9/16
37	4	AS	14 3/8	163	CS	N	2.50×3.12	3	2.25×4.12	2.62×4.69	abcde	7/8–18	Zen	1 1/2	1,430	28 3/4	40 7/16	52 9/16
38	4	AS	14 5/8	252	CS	N	3.00×3.31	3	3.00×4.75	3.00×4.75	abcde	7/8–18	Zen	1 3/4	1,925	30	47 11/32	58 1/2
39	4	AS	14 5/8	252	CS	N	3.00×3.31	3	3.00×4.75	3.00×4.75	abcde	7/8–18	Zen	1 3/4	1,925	30	44 11/32	58 1/2
40	4	AS	14 5/8	227	CS	N	3.00×3.31	3	3.00×4.75	3.00×4.75	abcde	7/8–18	Zen	1 3/4	1,925	30	40 11/32	58 1/2
41	4	CS	9 1/2	42	CS	N	2.12×1.62	7	3.00×1.50	3.00×2.12	abcde	18 mm.	Zen	1 1/2	825	25 3/4	33 1/16	39 25/32
42	4	CS	9 1/2	42	CS	N	2.12×1.62	7	3.00×1.50	3.00×2.12	abcde	18 mm.	Zen	1 1/2	825	25 3/4	33 1/16	39 25/32
43	4	CS	9 1/2	42	CS	N	2.12×1.62	7	3.00×1.50	3.00×2.12	abcde	18 mm.	Zen	1 1/2	825	25 3/4	33 1/16	39 25/32
44	4	CS	9 1/2	42	CS	N	2.12×1.62	7	3.00×1.50	3.00×2.12	abcder	18 mm.	Str	1 1/2	....	22 5/8	28 11/16	54 1/2
45	4	CS	9 1/2	42	CS	N	2.12×1.62	7	3.00×1.50	3.00×2.12	abcder	18 mm.	Str	1 1/2	....	22 5/8	28 11/16	54 1/2
46	4	CS	9 1/2	42	CS	N	2.12×1.62	7	3.00×1.50	3.00×2.12	abcder	18 mm.	Str	1 1/2	....	22 5/8	28 11/16	54 1/2
47	4	CS	9 1/2	58	CS	N	2.37×1.75	7	3.00×1.75	3.00×2.50	abcde	18 mm.	Zen	1 3/4	900	25 3/4	30 17/32	47 9/32
48	4	CS	9 1/2	58	CS	N	2.37×1.75	7	3.00×1.75	3.00×2.50	abcde	18 mm.	Zen	1 3/4	905	25 3/4	30 17/32	47 9/32
49	4	CS	9 1/2	58	CS	N	2.37×1.75	7	3.00×1.75	3.00×2.50	abcde	18 mm.	Zen	1 3/4	905	25 3/4	30 17/32	47 9/32
50	4	CS	9 1/2	58	CS	N	2.37×1.75	7	3.00×1.75	3.00×2.50	abcder	18 mm.	Str	1 3/4	1,320	24 13/16	29 1/2	60 11/16
51	4	CS	9 1/2	58	CS	N	2.37×1.75	7	3.00×1.75	3.00×2.50	abcder	18 mm.	Str	1 3/4	1,320	24 13/16	29 1/2	60 11/16
52	4	CS	9 1/2	58	CS	N	2.37×1.75	7	3.00×1.75	3.00×2.50	abcder	18 mm.	Str	1 3/4	1,320	24 13/16	29 1/2	60 11/16
53	4	CS	11	66	CS	N	2.37×1.75	7	3.00×1.75	3.00×2.50	abcde	18 mm.	Zen	1 3/4	950	25 3/4	33 7/32	47 9/32
54	5	AS	11	78	CS	N	2.37×1.75	7	3.00×1.75	3.00×2.50	abcdeg	14 mm.	Zen	1 3/4	1,195	27 9/16	38 27/32	49 3/8
55	4	AS	13 1/4	139	CS	N	3.00×2.25	4	3.00×2.25	3.00×3.69	abcde	18 mm.	Zen	1 3/4	1,525	28 3/4	43 27/32	53 13/16
56	5	AS	12 1/2	164	CS	N	3.24×2.12	7	3.50×3.00	3.50×3.00	abcdeg	18 mm.	Zen	1 3/4	2,150	31 17/32	49 9/16	57 1/2
57	5	6,140	12 1/2	...	CS[f]	N	3.24×2.12	7	3.50×2.43	3.50×2.43	abcde	14 mm.	.....	2	2,025	28 1/8	46 1/8	58 17/32
58	5	6,140	12 1/2	...	CS[f]	N	3.24×2.12	7	3.50×2.43	3.50×2.43	abcde	14 mm.	.....	2	2,025	28 1/8	45 7/8	58 17/32
59	5	6,140	12 1/2	...	CS[f]	N	3.24×2.12	7	3.50×2.43	3.50×2.43	abcde	14 mm.	Zen	2	....	28 1/8	46 1/8	58 17/32
60	4	AS	15 1/4	239	CS	Y	3.50×3.31	4	3.50×4.75	3.50×4.75	abcde	7/8–18	Zen	2	3,700	28 5/8	43	74 9/16
61	4	AS	15 1/4	239	CS	Y	3.50×3.31	4	3.50×4.75	3.50×4.75	abcde	7/8–18	Zen	2	3,700	28 5/8	43	74 9/16
62	4	AS	15 1/4	239	CS	Y	3.50×3.31	4	3.50×4.75	3.50×4.75	abcde	7/8–18	Zen	2	3,700	28 5/8	43	74 9/16
63	5	CS	17 3/4	430	CS	Y	3.75×3.50	7	3.75×2.31	3.75×3.48	abcdeg	18 mm.	Zen[2]	2	....	48	68 5/16	86 5/32
64	3	1,035	10 1/4	50	1,045	N	1.87×2.12	3	2.25×3.00	2.19×2.87	Splash	7/8–18	Zen	1	720	22	28	53
65	5	3,135	14	147	3,140	Y	3.00×2.37	5	3.75×4.12	3.75×4.50	abc	7/8–18	Zen	2 1/2	2,500	24	46	56
66	5	3,135	14	147	3,140	Y	3.00×2.37	5	3.75×4.12	3.75×4.50	abc	7/8–18	Zen	2 1/2	2,700	24	46	56
67	5	3,135	21	186	3,140	Y	4.25×4.00	5	5.00×7.00	5.00×7.00	abc	14 mm.	Zen	2 1/2	7,000	47	54	132
68	5	3,135	14	147	3,140	Y	3.00×2.37	7	3.75×4.12	3.75×4.50	abc	7/8–18	Zen	2	3,400	24	46	94
69	5	3,135	14	147	3,140	Y	3.00×2.37	7	3.75×4.12	3.75×4.50	abc	7/8–18	Zen	2	3,500	24	46	94
70	5	3,135	21	186	3,140	Y	4.25×4.00	7	5.00×7.00	5.00×7.00	abc	14 mm.	Zen	2 1/2	11,600	47	54	160
71	5	3,135	14	147	3,140	Y	3.00×2.37	9	3.75×4.12	3.75×4.50	abc	7/8–18	Zen	2 1/2	4,000	24	46	111
72	5	3,135	14	147	3,140	Y	3.00×2.37	9	3.75×4.12	3.75×4.50	abc	7/8–18	Zen	2 1/2	4,200	24	46	111
73	5	3,135	21	186	3,140	Y	4.25×4.00	9	5.00×7.00	5.00×7.00	abc	14 mm.	Zen	2 1/2	15,000	47	54	188

TABLE A1-3.—AMERICAN STOCK, MARINE, AND COMMERCIAL VEHICLE ENGINES.—(*Continued*)

Line number	Number of rings per piston	Connecting rods			Crankshaft						Oil pressure to—	Spark plug, thread size	Carburetor		Engine weight without carburetor or ignition, lb.	Over-all dimensions, in.		
							Crankpin	Main bearings										
									Diameter and length, in.									
		Material	Center-to-center length, in.	Weight with bushing and cap, oz.	Material	Counterbalance used?	Diameter and length, in.	Number	Front	Rear			Make	Size		Width	Height	Length
74	4	AS	10 15/16	...	CNM	N	3.00×2.50	7	3.25×....	3.25×....	abcer	18 mm.	Str	....	2,400[d]	39 1/8	40 5/8	97 1/8
75	3	CS	6 13/16	28	CS	Y	2.31×1.50	4	2.68×1.18	2.78×1.62	abcg	14 mm.	Car	1 1/4	563[d]	20 5/16	29 1/2	38 15/16
76	3	....	6 9/16	...	.....	N	1.75×1.12	3	2.00×1.62	2.00×1.62	ab	14 mm.	Zen	1 1/4	456[d]	24 1/8	23 1/16	31 3/4
77	4	....	7	...	.....	N	2.00×1.25	7	2.50×1.93	2.50×1.37	ab[c]	14 mm.	Zen	1 3/8	626[d]	24 1/8	25 3/4	40
78	..	....	8	...	CS	N	2.00×1.50	7	2.50×2.12	2.50×1.37	ab	14 mm.	Zen	1 3/4	850[d]	24 1/8	27 7/8	46 7/16
79	4	CNS	8 7/8	...	CS	Y	2.25×1.50	7	2.62×2.75	2.62×1.75	ab	14 mm.	Zen	2	1,240[d]	26 7/8	29 3/8	53 11/32
80	4	CS	11 1/4	...	CS	Y	3.00×2.06	3	3.00×3.75	3.00×3.56	abc	18 mm.	Hol	2	1,550[d]	34	41	62 1/2
81	4	CS	11 1/4	...	CS	Y	3.00×2.06	3	3.00×3.75	3.00×3.56	abc	18 mm.	Hol[2]	2	1,590[d]	34	41	62 1/2
82	4	T-1,335	7 15/16	31	1,040	Y	1.93×1.25	4	2.25×1.23	2.25×1.87	abce	18 mm.	Zen	1 1/4	665[d]	22 1/2	25 1/4	43
83	4	T-1,335	7 7/8	34	1,040	Y	2.12×1.37	4	2.50×1.31	2.50×1.87	abce	18 mm.	Zen	1 1/2	785[d]	22 1/2	26 5/8	45
84	4	T-1,335	9	36	1,040	Y	2.18×1.37	5	2.70×1.65	2.70×2.09	abce	18 mm.	Zen	1 1/2	990[d]	22 1/2	28 3/16	54 9/16
85	5	CS	10 1/2	167	CS	N	2.37×2.12	3	2.37×2.12	2.37×2.75	abcde	7/8–18	Zen	1 1/4	925	25 3/4	37	41 1/4
86	5	1,045	11 3/4	139	1,045	N	2.75×2.50	3	3.00×2.87	3.00×3.62	abcde	7/8–18	Zen	1 1/2	1,350	28 3/4	43	47 3/4
87	5	1,035	14	194	1,045	N	3.00×3.00	3	3.25×3.50	3.25×4.75	abcde	7/8–18	Zen	1 3/4	1,800	30 1/2	46	53 1/4
88	5	3,135	16	244	3,140	Y	3.00×3.00	3	3.25×3.68	3.25×4.50	abcde	7/8–18	Zen	2	2,300	30 15/16	49 1/2	57 9/16
89	5	3,135	16	244	4,140	Y	3.00×3.50	4	3.25×3.81	3.25×4.50	abcde	7/8–18	Zen[2]	1 3/4	3,200	29 1/8	51 3/8	70 15/16
90	5	3,135	16	244	3,140	Y	3.37×3.18	5	4.00×3.62	4.00×4.50	abcde	7/8–18	Zen[2]	2	4,500	35 5/16	56 13/16	97 7/8
91	3	CS	5 3/4	...	CS	N	1.50×1.18	3	1.75×1.37	1.75×1.78	abcet	18 mm.	.....	7/8	298	26	22 15/16	25 1/2
92	3	CS	5 3/4	...	CS	N	1.50×1.18	3	1.75×1.37	1.75×1.78	abcet	18 mm.	.....	7/8	305	26	22 15/16	25 1/2
93	3	CS	5 3/4	...	CS	N	1.50×1.18	3	1.75×1.28	1.75×1.65	abcet	18 mm.	.....	1	315	26	22 15/16	25 1/2
94	4	CS	7	...	CS	N	1.93×1.31	3	2.25×1.18	2.25×1.89	abcet	18 mm.	.....	1 1/4	400	26	26 11/32	29 7/16
95	4	CS	7	...	CS	N	1.93×1.31	3	2.25×1.18	2.25×1.89	abcet	18 mm.	.....	1 1/4	405	26	26 11/32	29 7/16
96	4	CS	7	...	CS	N	1.93×1.31	3	2.25×1.18	2.25×1.89	abcet	18 mm.	.....	1 1/4	410	26	26 11/32	29 7/16
97	4	CS	7	...	CS	Y	1.93×1.31	4	2.25×1.21	2.25×1.81	abcet	18 mm.	.....	1 1/4	481	26	27 1/4	36 11/32
98	4	CS	7	...	CS	Y	1.93×1.31	4	2.25×1.21	2.25×1.81	abcet	18 mm.	.....	1 1/4	491	26	27 1/4	36 11/32
99	4	CS	7	...	CS	Y	1.93×1.31	4	2.25×1.21	2.25×1.81	abcet	18 mm.	.....	1 1/4	506	26	27 1/4	36 11/32

100	4	CS	7	...	CS	Y	1.93×1.31	4	2.25×1.21	2.25×1.81	abcet	13 mm.	.....	1 1/4	512	26	27 1/4	36 11/32
101	4	CS	8 3/8	...	CS	Y	2.12×1.37	4	2.37×1.43	2.37×2.06	abce	18 mm.	.....	1 1/2	548			
102	4	CS	8 3/8	...	CS	Y	2.25×1.56	7	2.62×1.56	2.62×2.18	abcet	18 mm.	.....	1 1/2	750	25 3/4	29 1/2	42
103	4	CS	8 3/8	...	CS	Y	2.25×1.56	7	2.62×1.56	2.62×2.18	abcet	18 mm.	.....	1 1/2	760	25 3/4	29 1/2	42
104	4	CS	8 3/8	...	CS	Y	2.25×1.56	7	2.62×1.56	2.62×2.18	abcet	18 mm.	......	1 1/2	770	25 3/4	29 1/2	42
105	4	CS	9	...	CS	Y	2.37×1.81	7	2.62×1.65	2.62×2.65	abcef	18 mm.	.....	1 1/2	925	25 3/4	32 7/32	44 11/16
106	4	CS	9	...	CS	Y	2.37×1.81	7	2.62×1.65	2.62×2.65	abcef	18 mm.	.....	1 1/2	932	25 3/4	32 7/32	44 11/16
107	5	CS	9	...	CS	Y	2.37×1.81	7	2.62×1.65	2.62×2.65	abcef	18 mm.	.....	1 1/2	938	25 3/4	32 7/32	44 11/16
108	4	CS	9	...	CS	Y	2.37×1.81	7	2.62×1.65	2.62×2.65	abcef	18 mm.	.....	1 3/4	951	25 3/4	32 7/32	44 11/16
109	4	CS	9 1/2	...	4,140	N	2.50×1.81	7	2.75×1.75	2.75×2.81	abcefg	18 mm.	.....	1 3/4	1,298	25 3/4	36 35/64	46 31/32
110	4	CS	9 1/2	...	4,140	N	2.50×1.81	7	2.75×1.75	2.75×2.81	abcefg	18 mm.	.....	1 3/4	1,318	25 3/4	36 25/32	46 31/32
111	4	CS	10 1/2	...	CV	Y	2.75×1.81	7	2.75×1.75	2.75×2.81	abcefg	18 mm.	......	2	1,430	25 3/4	39 5/8	46 31/32
112	4	MS	7 15/16	31	1,040	Y	1.93×1.00	4	2.25×1.23	2.25×1.87	abce	14 mm.	Car	1 1/2	500	23 3/4	31 27/32	37 17/32
113	4	MS	7 15/16	31	1,040	Y	2.06×1.00	4	2.50×1.23	2.50×1.87	abce	14 mm.	Str	1 1/2	535	23 3/4	31 27/32	37 17/32
114	4	MS	8	...	1,040	Y	2.12×1.21	4	2.50×1.31	2.50×1.87	abce	14 mm.	Car	1 1/2	580	23 3/4	31 7/32	39 13/16
115	4	MS	7 7/8	35	1,040	Y	2.12×1.21	4	2.50×1.31	2.50×1.87	abce	14 mm.	Car	1 1/2	600	23 3/4	31 7/32	39 13/16
116	4	MS	10 7/16	56	1,040	Y	2.31×1.43	7	3.00×1.87	3.00×3.04	abce	14 mm.	Str	1 1/2	1,050	24 3/4	31 1/2	49 13/32
117	4	1,040	12 3/4	132	CNS	Y	2.37×3.00	3	2.62×3.31	2.62×4.00	abcdeg	7/8–18	Zen	1 1/2	1,450	27	37 11/16	70 15/16
118	4	1,040	12 3/4	132	CNS	Y	2.37×3.00	4	2.62×3.31	2.62×4.00	abcdeg	7/8–18	Zen[2]	1 1/2	1,900	27	37 11/16	87 5/16
119	3	MS	6 1/8	10	CAS	Y	1.70×1.41	3	2.10×1.50	2.10×2.00	abce	14 mm.	Own	0.78	310			
120	3	MS	7	17	CAS	Y	2.00×1.75	3	2.50×1.37	2.50×1.97	abce	14 mm.	Own	0.94	485			
121	3	MS	7	18	CAS	Y	2.14×1.75	3	2.50×1.37	2.50×1.97	abce	14 mm.	Own	0.94	495			
122	5	AS	9 1/2	54	CS	..	2.37×1.75	7	2.70×2.25	2.70×2.87	abcdefg	18 mm.	Zen	1 1/2	1,247	40 1/8	27 3/4	44
123	5	AS	9 1/2	54	CS	..	2.37×1.75	7	2.70×2.25	2.70×2.87	abcdeg	18 mm.	Zen	1 1/2	1,087	24 1/2	37 1/16	44
124	5	AS	9 1/2	54	CS	..	2.37×1.75	7	2.70×2.25	2.70×2.87	abcdefg	18 mm.	Zen	1 1/2	1,247	40 1/8	27 3/4	44
125	5	AS	9 1/2	54	CS	..	2.37×1.75	7	2.70×2.25	2.70×2.87	abcdeg	18 mm.	Zen	1 1/2	1,087	24 1/2	37 1/16	44
126	5	AS	7	74	CS	Y	1.93×1.00	3	2.25×1.25	2.25×1.50	ace	14 mm.	Op	1–1 1/4	315	34	18 1/2	18
127	4	1,040A	7	33	1,050	Y	2.31×1.23	4	2.69×1.19	2.78×1.47	abcdeg	14 mm.	Zen	1 3/8	....	21 13/16	26 5/16	40 3/4
128	4	1,040A	7	33	1,050	Y	2.31×1.23	4	2.69×1.19	2.78×1.47	abcdeg	14 mm.	Zen	1 3/8	....	21 13/16	26 5/16	40 3/4
129	4	1,040A	9 3/4	51	1,050	Y	2.37×1.34	7	2.75×2.09	2.75×2.09	abcdeg	14 mm.	Zen	1 3/8	....	25 1/4	31 29/64	45 1/32
130	4	1,040A	9 3/4	51	1,050	Y	2.37×1.34	7	2.75×2.09	2.75×2.09	abcdeg	14 mm.	Zen	1 3/8	....	25 1/4	31 29/64	45 1/32
131	4	1,040A	10 3/4	73	1,050	Y	2.62×1.47	7	3.00×2.22	3.00×2.22	abcdeg	14 mm.	Zen	1 1/2	....	22 1/2	35 3/16	47 11/16
132	4	1,040A	10 3/4	73	1,050	Y	2.62×1.47	7	3.00×2.22	3.00×2.22	abcdeg	14 mm.	Zen	1 1/2	....	22 1/2	35 3/16	47 11/16
133	4	1,040A	10 3/4	73	1,050	Y	2.62×1.47	7	3.00×2.22	3.00×2.22	abcdeg	14 mm.	Zen	1 3/4	....	22 1/2	35 3/16	47 11/16
134	4	1,040A	9 1/2	81	1,050	Y	2.62×1.72	7	3.50×2.50	3.50×2.50	abcdeg	14 mm.	Zen	2				
135	4	1,040A	10 1/2	82	1,050	Y	2.62×1.72	7	3.50×2.50	3.50×2.50	abcdeg	14 mm.	Zen	2				
136	5	4,140A	12 1/8	77	4,140	Y	2.75×2.06	7	3.00×1.72	3.00×2.50	abdceg	18 mm.	Str	2				
137	3	CS	5 3/4	17	1,045	N	1.50×1.18	3	1.75×1.78	1.75×1.37	abe	14 mm.	Zen	3/4	330[d]	16 1/4	17 1/2	30
138	3	CS	5 3/4	17	1,045	N	1.50×1.18	3	1.75×1.78	1.75×1.37	abce	14 mm.	Zen	1	375[d]	15	18 5/8	30 5/8
139	3	CS	5 3/4	17	1,045	N	1.50×1.18	3	1.75×1.66	1.75×1.28	abce	14 mm.	Zen	1	410[d]	15	18 5/8	31 5/8
140	3	CS	5 3/4	17	1,045	N	1.50×1.18	3	1.75×1.78	1.75×1.37	abce	14 mm.	Zen	1 3/8	340[d]	17 3/8	18 1/2	30
141	4	CS	7	32	1,045	N	1.93×1.31	3	2.25×1.89	2.25×1.18	abce	18 mm.	Zen	1 1/4	510[d]	18 7/16	23 1/2	34 1/8
142	4	CS	7	32	1,045	N	1.93×1.31	3	2.25×1.89	2.25×1.18	abce	18 mm.	Zen	1 1/4	540[d]	18 7/16	23 1/2	35 1/8
143	4	CS	7	32	1,045	N	1.93×1.31	3	2.25×1.89	2.25×1.18	abce	18 mm.	Zen	1 1/4	435[d]	20 1/4	21 7/8	33
144	4	CS	7	32	1,045	N	1.93×1.31	3	2.25×1.89	2.25×1.18	abce	18 mm.	Zen	1 1/4	460[d]	20 1/4	21 7/8	33
145	4	CS	7	32	1,045	Y	1.93×1.31	4	2.25×1.81	2.25×1.21	abce	18 mm.	Zen	1 1/4	650[d]	18 1/2	23	41
146	4	CS	7	32	1,045	Y	1.93×1.31	4	2.25×1.81	2.25×1.21	abce	18 mm.	Zen	1 1/2	665[d]	18 1/2	23	42

TABLE A1-3.—AMERICAN STOCK, MARINE, AND COMMERCIAL VEHICLE ENGINES.—(*Continued*)

Line number	Number of rings per piston	Connecting rods			Crankshaft						Oil pressure to—	Spark plug, thread size	Carburetor		Engine weight without carburetor or ignition, lb.	Over-all dimensions, in.		
							Crankpin	Main bearings										
									Diameter and length, in.									
		Material	Center-to-center length, in.	Weight with bushing and cap, oz.	Material	Counterbalance used?	Diameter and length, in.	Number	Front	Rear			Make	Size		Width	Height	Length
147	4	CS	7	32	1,045	Y	1.93×1.31	4	2.25×1.81	2.25×1.21	abce	18 mm.	Zen	1½	625[d]	23	22	39⅞
148	4	CS	7	32	1,045	Y	1.93×1.31	4	2.25×1.81	2.25×1.21	abce	18 mm.	Zen	1¼	630[d]	23	22	39⅞
149	4	CS	8⅜	34	1,045	Y	2.12×1.37	4	2.37×2.06	2.37×1.43	abce	18 mm.	Zen	1½	799[d]	19⅞	23⅞	49½
150	4	CS	8⅜	34	1,045	Y	2.12×1.37	4	2.37×2.06	2.37×1.43	abce	18 mm.	Str	1½	730[d]	22½	30	43½
151	4	CS	8⅜	34	1,045	Y	2.12×1.37	4	2.37×2.06	2.37×1.43	abce	18 mm.	Str	1½	730[d]	22½	30	42¾
152	4	CS	8⅜	34	1,045	Y	2.12×1.37	4	2.37×2.06	2.37×1.43	abce	18 mm.	Str	1½	730[d]	22½	30	42¾
153	4	CS	8⅜	34	1,045	Y	2.12×1.37	4	2.37×2.06	2.37×1.43	abce	18 mm.	Zen	.....	730[d]	30¾	32½	42¾
154	4	CS	8⅜	32	1,045	Y	2.12×1.37	4	2.39×1.91	2.39×1.50	abce	18 mm.	Zen	1½	595	28⅜	23	38⅞
155	4	CS	8⅜	47	1,045	Y	2.25×1.56	7	2.62×2.15	2.62×1.56	abce	18 mm.	Str	1½	985[d]	23 15/16	26	49⅝
156	4	CS	8⅜	47	1,045	Y	2.25×1.56	7	2.62×2.15	2.62×1.56	abce	18 mm.	Str	1¾	1,015[d]	23 15/16	26	49⅝
157	4	CS	8⅜	47	1,045	Y	2.25×1.56	7	2.62×2.15	2.62×1.56	abce	18 mm.	Str	1½	990[d]	23 15/16	26	49
158	4	CS	9	60	1,045	Y	2.37×1.81	7	2.62×2.65	2.62×1.65	abce	18 mm.	Str	1½	1,250[d]	22	29⅛	55⅝
159	4	CS	9	49	1,045	Y	2.25×1.50	5	2.62×2.53	2.62×1.65	abce	18 mm.	Str	1½	1,275[d]	25¾	28	59½
160	4	CS	9	49	1,045	Y	2.25×1.50	5	2.62×2.53	2.62×1.65	abce	18 mm.	Str	1½	1,250[d]	25¾	28	54⅞
161	6	AS	11	105	CNS	N	2.25×2.00	3	2.75×2.31	2.75×3.31	abcefg	⅞–18	Zen	1½	1,135[d]	27 7/16	40½	46 11/16
162	6	AS	11	103	CNS	N	2.25×2.00	3	2.75×2.31	2.75×2.31	abcefg	18 mm.	Zen	1¾	1,905[d]	24⅝	41⅜	76 7/16
163	6	AS	11	100	CNS	N	2.37×2.00	3	3.00×2.31	3.00×2.31	abcefg	⅞–18	Zen	1½	1,160[d]	27⅛	40¾	49⅜
164	4	AS	11	82	4,140	N	2.62×1.75	7	3.00×1.50	3.00×2.25	abcefg	18 mm.	Zen	1½	1,173[d]	54¾	18 1/16	52⅞
165	4	AS	11	75	4,140	N	2.25×1.45	4	2.75×2.68	2.75×2.68	abcefg	18 mm.	Zen	1½	1,180[d]	25¾	38 15/16	49 1/16
166	4	AS	11	83	4,140	Y	2.62×1.75	7	3.00×1.50	3.00×2.25	abcefg	18 mm.	Zen	1¾	1,252[d]	54¾	18 7/16	52⅞
167	6	AS	11	104	4,140	Y	2.25×2.00	4	2.75×2.18	2.75×3.18	abcefg	18 mm.	Zen	1¾	2,120[d]	23⅛	42	99 15/16
168	5	AS	11	101	4,140	Y	2.37×2.00	4	2.75×2.18	2.75×3.18	abcefg	18 mm.	Zen	1¾	1,407[d]	29 13/16	46⅛	57 7/16
169	5	AS	11	102	4,140	Y	2.37×2.00	4	2.75×2.18	2.75×3.18	abcefg	18 mm.	Zen	1¾	1,310[d]	31 7/32	49 9/16	55 7/16
170	5	AS	11	101	4,140	Y	2.37×2.00	4	2.75×2.18	2.75×3.18	abcefg	18 mm.	Zen	1¾	2,392[d]	28	42½	90
171	4	AS	11	83	3,140	Y	2.62×1.75	7	3.00×1.50	3.00×2.25	abcefg	18 mm.	Zen	1¾	1,265[d]	54¾	18 7/16	52⅞
172	5	AS	11	97	4,140	Y	2.37×2.00	7	2.75×2.43	2.75×4.56	abcefg	18 mm.	Zen	2	2,820[d]	31½	42	90 3/16

173	5	AS	11	113	4,140	Y	2.50×2.00	7	3.25×1.43	3.25×2.43	abcefg	18 mm.	Zen	2	1,830[d]	32 13/16	44 15/16	59 7/8
174	4	AS	11	103	4,140	Y	2.75×2.00	7	3.25×2.18	3.25×2.56	abcefg	18 mm.	Zen	2	1,960[d]	59 9/16	20 11/16	67 19/32
175	5	AS	11	113	4,140	Y	2.50×2.00	7	3.25×1.43	3.25×2.43	abcefg	18 mm.	Zen	2	1,830[d]	32 13/16	44 15/16	59 7/8
176	4	AS	11	103	4,140	Y	2.75×2.00	7	3.25×2.18	3.25×2.56	abcefg	18 mm.	Zen	1 3/4	....	53 1/2	22 1/4	67 19/32
177	5	AS	11	113	4,140	Y	2.50×2.00	7	3.25×1.43	3.25×2.43	abcefg	18 mm.	Zen	2	1,900[d]	13 13/16	44 15/16	59 7/8
178	5	AS	11	114	4,140	Y	2.37×2.00	7	2.75×2.43	2.75×4.56	abcefg	18 mm.	Zen	2	2,015[d]	25 1/32	39 5/8	74 3/16
179	5	AS	11	97	4,140	Y	2.37×2.00	7	2.75×2.43	2.75×4.56	abcefg	18 mm.	Zen	2	....	31 1/2	42	90 3/16
180	6	3,140	12	...	4,140	Y	3.00×2.43	7	3.25×2.09	3.25×3.09	abce	18 mm.	Zen[3]	2 1/2	3,650[d]	45	42 3/4	88
181	6	3,140	12	...	4,140	Y	3.00×2.43	7	3.25×2.09	3.25×2.09	abce	18 mm.	Zen[3]	2 1/2	3,650[d]	45	42 3/4	88
182	6	3,140	12	...	4,140	Y	3.00×2.43	7	3.25×2.09	3.25×2.09	abce	18 mm.	Zen[3]	2 1/2	4,560[d]	45	50 1/4	98 1/4
183	3	3,140	5 1/8	15	1,045	Y	1.50×1.00	2	2.00×1.25	2.00×1.37	abe	14 mm.	Op	5/8	131	14 7/8	16 1/4	18 13/16
184	3	3,140	6 9/16	21	1,045	Y	1.75×1.12	2	2.00×1.56	2.00×1.62	abe	7/8-18	Op	Op	270	15 1/4	18 1/2	18 5/8
185	3	3,140	6 9/16	21	1,045	Y	1.75×1.12	2	2.00×1.56	2.00×1.62	abe	7/8-18	Op	Op	270	15 1/4	18 1/2	18 5/8
186	3	3,140	5 1/8	15	1,045	N	1.50×1.00	3	2.00×1.31	2.00×1.37	abe	14 mm.	Op	Op	179	14 3/4	16 9/16	21 5/16
187	3	3,140	5 1/8	15	1,045	N	1.50×1.00	3	2.00×1.31	2.00×1.37	abe	14 mm.	Op	Op	179	14 3/4	16 9/16	21 5/16
188	3	3,140	6 9/16	21	1,045	N	1.75×1.12	3	2.00×1.56	2.00×1.62	abe	18 mm.	Op	Op	255	16 5/8	18 3/4	24 3/16
189	3	3,140	6 9/16	21	1,045	N	1.75×1.12	3	2.00×1.56	2.00×1.62	abe	18 mm.	Op	Op	250	16 5/8	18 3/4	24 3/16
190	3	3,140	6 9/16	21	1,045	N	1.75×1.12	3	2.00×1.56	2.00×1.62	abe	7/8-18	Op	Op	285	16 5/8	18 3/4	24 3/16
191	3	3,140	6 9/16	21	1,045	N	1.75×1.12	3	2.00×1.56	2.00×1.62	abe	7/8-18	Op	Op	283	16 5/8	18 3/4	24 3/16
192	3	3,140	6 9/16	21	1,045	N	1.75×1.12	3	2.00×1.56	2.00×1.62	abe	7/8-18	Op	Op	293	16 5/8	18 3/4	24 3/16
193	3	3,140	6 9/16	21	1,045	N	1.75×1.12	3	2.00×1.56	2.00×1.62	abe	7/8-18	Op	Op	291	16 5/8	18 3/4	24 3/16
194	4	1,035	8	37	1,045	N	2.00×1.50	3	2.00×2.18	2.00×2.62	abe	7/8-18	Op	...	460	17 1/2	23 1/4	29 7/8
195	4	1,035	8	37	1,045	N	2.00×1.50	3	2.00×2.18	2.00×2.62	abe	7/8-18	Op	Op	460	17 1/2	23 1/4	29 7/8
196	4	1,035	8	37	1,045	N	2.00×1.50	3	2.00×2.18	2.00×2.62	abe	7/8-18	Op	Op	460	17 1/2	23 1/4	29 7/8
197	4	1,035	9 1/2	58	1,045	N	2.00×2.25	3	2.00×3.18	2.00×3.31	abe	7/8-18	Op	1 1/4	655	20 17/32	28 7/32	36 5/32
198	4	1,035	9 1/2	58	1,045	N	2.00×2.25	3	2.00×3.18	2.00×3.31	abe	7/8-18	Op	1 1/4	655	20 17/32	28 7/32	36 5/32
199	4	1,035	10 7/8	83	1,045	N	2.50×2.62	3	3.00×3.37	3.00×3.50	abe	7/8-18	Op	1 1/4	875	21 31/32	30 7/8	41 1/16
200	4	1,035	10 7/8	83	1,045	N	2.50×2.62	3	3.00×3.37	3.00×3.50	abe	7/8-18	Op	1 1/2	880	21 31/32	30 7/8	41 1/16
201	4	1,035	10 7/8	83	1,045	N	2.50×2.62	3	3.00×3.37	3.00×3.50	abe	7/8-18	Op	1 1/2	885	21 31/32	30 7/8	41 1/16
202.	4	1,035	10 7/8	83	1,045	N	2.50×2.62	3	3.00×3.37	3.00×3.50	abe	7/8-18	Op	1 1/2	890	21 31/32	30 7/8	41 1/16
203	5	1,035	13 1/4	178	1,045	N	3.00×3.00	3	3.75×4.37	3.75×4.50	abe	7/8-18	Op	1 3/4	1,800	26 15/16	38	52 1/2
204	5	1,035	13 1/4	178	1,045	N	3.00×3.00	3	3.75×4.37	3.75×4.50	abe	7/8-18	Op	1 3/4	1,815	26 15/16	38	52 1/2
205	5	1,035	13 1/4	178	1,045	N	3.00×3.00	3	3.75×4.37	3.75×4.50	abe	7/8-18	Op	2	1,850	26 15/16	38	52 1/2
206	4	1,035	7	26	CS[f]	N	2.00×1.25	7	2.50×1.31	2.50×1.93	abe	7/8-18	Op	1 1/4	480	17 7/16	20 7/16	33 7/16
207	4	1,035	7	26	CS[f]	N	2.00×1.25	7	2.50×1.31	2.50×1.93	abe	7/8-18	Op	1 1/4	480	17 7/16	20 7/16	33 7/16
208	4	1,035	7	26	CS[f]	N	2.00×1.25	7	2.50×1.31	2.50×1.93	abe	7/8-18	Op	1 1/4	480	17 7/16	20 7/16	33 7/16
209	4	1,035	7	26	CS[f]	N	2.00×1.25	7	2.50×1.31	2.50×1.93	abe	7/8-18	Op	1 1/4	480	17 7/16	20 7/16	33 7/16
210	4	1,035	7	26	CS[f]	N	2.00×1.25	7	2.50×1.31	2.50×1.93	abe	7/8-18	Op	1 1/4	481	17 7/16	20 7/16	30 7/16
211	4	1,035	7	26	CS[f]	N	2.00×1.25	7	2.50×1.31	2.50×1.93	abe	7/8-18	Op	1 1/4	483	17 7/16	20 7/16	33 7/16
212	4	1,035	8	37	CS[f]	N	2.00×1.50	7	2.50×1.31	2.50×2.12	abe	7/8-18	Op	1 1/4	550	17 5/8	23 1/2	37 3/8
213	4	1,035	8	37	CS[f]	N	2.00×1.50	7	2.50×1.31	2.50×2.12	abe	7/8-18	Op	1 1/4	560	17 5/8	23 1/2	37 3/8
214	4	1,035	8	37	CS[f]	N	2.00×1.50	7	2.50×1.31	2.50×2.12	abe	7/8-18	Op	1 1/4	565	17 5/8	23 1/2	37 5/8
215	4	1,035	8	37	CS[f]	N	2.00×1.50	7	2.50×1.31	2.50×2.12	abe	7/8-18	Op	1 1/4	570	17 5/8	23 1/2	37 3/8
216	4	3,140	9 1/8	51	1,045	N	2.25×1.50	7	2.62×1.75	2.62×2.75	abe	7/8-18	Op	1 1/2	805	21	27	41 15/16
217	4	3,140	9 1/8	51	1,045	N	2.25×1.50	7	2.62×1.75	2.62×2.75	abe	7/8-18	Op	1 1/2	810	21	27	41 15/16
218	4	3,140	9 1/8	51	1,045	N	2.25×1.50	7	2.62×1.75	2.62×2.75	abe	7/8-18	Op	1 1/2	820	21	27	41 15/16
219	4	3,140	8 7/8	50	CS[f]	Y	2.25×1.65	7	2.62×1.43	2.62×2.21	abe	7/8-18	Op	Op	811	21	27	41 15/16

TABLE A1-3.—AMERICAN STOCK, MARINE, AND COMMERCIAL VEHICLE ENGINES.—(*Continued*)

Line number	Number of rings per piston	Connecting rods			Crankshaft						Oil pressure to—	Spark plug, thread size	Carburetor		Engine weight without carburetor or ignition, lb.	Over-all dimensions, in.		
							Crankpin	Main bearings										
									Diameter and length, in.									
		Material	Center-to-center length, in.	Weight with bushing and cap, oz.	Material	Counterbalance used?	Diameter and length, in.	Number	Front	Rear			Make	Size		Width	Height	Length
220	4	3,140	8 7/8	50	CS[f]	Y	2.25×1.65	7	2.62×1.43	2.62×2.21	abe	7/8–18	Op	Op	825	21	27	41 15/16
221	4	1,035	9 5/8	64	1,045	N	2.50×1.75	7	3.00×2.00	3.00×3.00	abe	7/8–18	Op	1 3/4	975	21 11/32	31 13/16	45 9/16
222	4	1,035	9 5/8	64	1,045	N	2.50×1.75	7	3.00×2.00	3.00×3.00	abe	7/8–18	Op	1 3/4	975	21 11/32	31 13/16	45 9/16
223	4	1,035	9 5/8	64	1,045	N	2.50×1.75	7	3.00×2.00	3.00×3.00	abe	7/8–18	Op	1 3/4	975	21 11/32	31 13/16	45 9/16
224	5	3,140	9 3/8	81	1,045	N	2.62×2.00	7	3.00×1.93	3.00×2.93	abe	7/8–18	Op	1 3/4	1,000	21 11/32	31 13/16	45 9/16
225	5	3,140	9 3/8	81	1,045	N	2.62×2.00	7	3.00×1.93	3.00×2.93	abe	7/8–18	Op	1 3/4	1,010	21 11/32	31 13/16	45 9/16
226	5	CNM	9 3/8	99	CS[f]	N	3.00×2.00	7	3.50×1.93	3.50×2.93	abe	7/8–18	Op	1 3/4	1,195	22 5/16	30	44 3/4
227	5	CNM	9 3/8	99	CS[f]	N	3.00×2.00	7	3.50×1.93	3.50×2.93	abe	7/8–18	Op	1 3/4	1,195	22 5/16	30	44 3/4
228	4	3,140	12	143	1,045	N	3.00×2.25	7	3.50×2.37	3.50×3.50	abce	7/8–18	Op	2	1,810	24 1/2	40 1/2	54 1/2
229	4	3,140	12	143	1,045	N	3.00×2.25	7	3.50×2.37	3.50×3.50	abce	7/8–18	Op	2	1,810	24 1/2	40 1/2	54 1/2
230	4	3,140	12	143	1,045	N	3.00×2.25	7	3.50×2.37	3.50×3.50	abce	7/8–18	Op	2	1,830	24 1/2	40 1/2	54 1/2
231	4	3,140	12	143	1,045	N	3.00×2.25	7	3.50×2.37	3.50×3.50	abce	7/8–18	Op	2	1,830	24 1/2	40 1/2	54 1/2
232	4	DFS	8 5/8	29	DFS	Y	1.94×1.37	3	2.34×1.62	2.41×2.37	Splash	14 mm.	Car	1 1/4	482	25	37 1/4	29 1/2
233	4	DFS	8 11/64	29	DFS	Y	1.94×1.32	3	2.34×1.62	2.41×2.37	Splash	14 mm.	Car	1 1/4	504	25	37 1/4	29 1/2
234	4	AS	10	73	CNS	N	2.25×2.23	2[g]	SAE-313	SAE-315	Splash	7/8–18	Zen	1 1/4	1,435[d]	22 1/2	39	40 29/32
235	4	AS	11 1/4	102	CNS	N	2.62×2.73	2[g]	SAE-314	SAE-316	Splash	7/8–18	Zen	1 1/4	1,735[d]	25 3/4	42 13/16	42
236	4	AS	13	152	CNS	N	3.12×2.73	2[g]	SAE-315	SAE-318	Splash	7/8–18	Zen	1 1/4	2,160[d]	27 1/4	44 1/2	45 15/16
237	4	AS	9 1/4	49	CS[f]	N	2.25×1.62	7	2.70×1.53	2.70×2.54	abcde	7/8–18	Zen	1 3/8	1,950[d]	28	35 3/8	49 9/16
238	4	AS	11 3/4	137	CNS	N	2.75×2.25	7	3.25×2.56	3.25×3.50	abcde	7/8–18	Zen	1 1/2	3,120[d]	35	50 3/16	59 1/8
239	3	DFS	9	...	DFS	N	1.37×2.25	2	1.37×1.50	1.37×3.00	ae	7/8–18	May	7/8	465[d]	18 1/4	28 3/4	40 5/16
240	3	CS	5 1/8	15	CS	N	1.50×1.00	3	2.00×1.37	2.00×1.31	abcr	14 mm.	Str	3/4	260	18 11/16	17 5/8	28 5/16
241	4	3,140	6 9/16	20	CS	N	1.75×1.12	3	2.00×1.62	2.00×1.58	abcr	7/8–18	Str	1 1/4	370	23 1/8	21 15/16	34 3/16
242	4	3,140	6 9/16	20	CS	N	1.75×1.12	3	2.00×1.62	2.00×1.58	abcr	7/8–18	Str	1 1/4	405	23 1/8	21 15/16	34 3/16
243	3	DFS	9	38	CS	N	1.37×2.25	3	1.37×3.50	1.37×3.00	ae	7/8–18	May	7/8	590[d]	21	27 1/4	51 11/16
244	4	DFS	11	81	DFS	N	2.00×2.25	5	2.00×3.87	2.00×2.62	abcde	7/8–18	Str	1 1/2	830	22 1/2	31 5/8	52 1/4
245	3	DFS	7 9/16	32	DFS	Y	2.00×1.06	5	2.50×1.87	2.37×1.25	abcder	14 mm.	Str	1 1/2	750[d]	24	23 3/16	52 1/2

246	4	CS	7	26	CS	N	2.00×1.25	7	2.50×1.93	2.50×1.87	aber	⅞-18	Str	1½	620[d]	24	24⅛	42 5/16
247	3	DFS	7 9/16	37	DFS	Y	2.12×1.28	4	2.62×1.56	2.50×1.25	abcder	14 mm.	Str	1½	....	24	23 3/16	47⅞
248	4	CS	8	38	CS	N	2.00×1.50	7	2.50×2.12	2.50×1.37	aber	⅞-18	Str	1½	865[d]	21 13/16	27¾	52⅜
249	4	CS	8	38	CS	N	2.00×1.50	7	2.50×2.12	2.50×1.37	aber	⅞-18	Str	1¾	865[d]	21 13/16	27¾	52⅜
250	4	CS	9⅛	51	CS	N	2.25×1.50	7	2.62×1.75	2.62×2.75	aber	14 mm.	Str	1¾	1,125[d]	24¼	29⅝	56 3/16
251	4	CS	9	51	CS	Y	2.25×1.12	7	2.68×2.75	2.68×1.75	aber	14 mm.	Str	2	1,125[d]	24¼	29⅝	56 3/16
252	4	DFS	11	80	CS	N	2.25×2.23	7	2.50×3.91	2.50×2.56	abcder	14 mm.	Str	2	1,390[d]	26¾	33¼	69⅞
253	4	DFS	11	80	CS	N	2.25×2.23	7	2.50×3.91	2.50×2.62	abcder	18 mm.	Str	2	1,570[d]	29 13/16	34 1/16	69⅞
254	4	DFS	11	80	CS	N	2.25×2.23	7	2.50×3.91	2.50×2.62	abcder	14 mm.	Str[2]	1¾	1,620[d]	26⅛	34 1/16	69⅞
255	4	DFS	11	80	DFS	N	2.25×2.23	7	2.50×3.91	2.50×2.62	abcder	18 mm.	Str[2]	1¾	1,500[d]	30½	39¾	69¾
256	3	AS	7	16	CAS	Y	2.00×1.94	3	2.00×1.81	2.00×2.25	ab	⅞-18	Str	1	730	29 1/16	31 7/16	43 15/16
257	3	AS	7	17	CAS	Y	2.14×1.75	3	2.50×1.72	2.50×2.25	abce	14 mm.	Str	0.97	730	29 1/16	31 7/16	43 15/16
258	3	DFS	7 9/16	32	DFS	Y	2.00×1.56	5	2.37×2.25	2.24×1.31	abcder	14 mm.	Str	1½	750[d]	24	23 3/16	52½
259	3	DFS	......	22	CAS	Y	2.12×1.57	4	2.40×1.83	2.40×2.54	abce	14 mm.	Str	1	800	24 11/16	32	50⅝
260	4	CNS	11	184	CNS	Y	2.75×2.25	7	3.00×3.84	3.00×2.25	abcder	18 mm.	Str[4]	2	2,700[d]	42½	43 9/16	75 15/16
261	4	AS	12¼	96	CNS	N	1.87×2.75	3	2.25×5.00	2.25×4.00	abe	⅞-18	Zen	1¼	975	23½	22	55½
262	4	AS	12¼	96	CNS	N	1.87×2.75	3	2.25×5.00	2.25×4.00	abe	⅞-18	Zen	1¼	1,000	23½	22	55½
263	4	AS	12¼	96	CNS	N	1.87×2.75	4	2.25×5.00	2.25×4.00	abe	⅞-18	Zen	1½	1,400	25	22¼	60½
264	4	AS	12¼	96	CNS	N	1.87×2.75	4	2.25×5.00	2.25×4.00	abe	⅞-18	Zen	1½	1,450	25	22¼	60½
265	3	AS	6 9/16	...	CNS	N	1.75×1.25	3	2.00×1.56	2.00×1.62	abe	⅞-18	Zen	1	440	17½	14⅜	40½
266	3	AS	8	...	CNS	N	2.00×1.50	3	2.00×2.18	2.00×2.62	abe	⅞-18	Zen	1¼	690	21⅞	16	51¼
267	4	AS	12¼	96	CNS	N	1.87×2.75	5	2.25×5.00	2.25×4.00	abe	⅞-18	Zen	1½	1,700	25¾	22¼	68½
268	4	AS	12¼	96	CNS	N	1.87×2.75	5	2.25×5.00	2.25×4.00	abe	⅞-18	Zen	1½	1,750	25¾	22¼	68½
269	4	AS	13¼	172	CNS	N	2.75×2.75	5	3.00×3.50	3.00×3.50	abce	⅞-18	Zen	.....	2,100	29¼	24	74¼
270	4	AS	13¼	172	CNS	N	2.75×2.75	5	3.00×3.50	3.00×3.50	abce	⅞-18	Zen	.....	2,290	29¼	24	75
271	3	AS	8	...	CNS	N	2.00×1.50	7	2.50×1.31	2.50×2.12	abe	⅞-18	Zen	1¼	830	21⅛	17	51
272	4	DFS	8	36	Spec	N	2.00×1.50	7	2.50×1.31	2.50×1.12	abe	⅞-18	Zen	1½	820	21⅛	25	51
273	4	Dur	12¼	68	CNS	N	2.25×2.37	7	2.75×3.25	2.75×2.12	abce	⅞-18	Hol	.....	1,700	24½	27¾	68
274	4	Dur	12¼	68	CNS	N	2.25×2.37	7	2.75×3.25	2.75×2.12	abce	⅞-18	Hol	.....	1,700	24½	27¾	68
275	4	AS	12¼	96	CNS	N	2.12×2.75	7	2.62×5.00	2.62×4.00	abe	⅞-18	Zen[2]	1½	2,500	26½	22¾	90½
276	4	AS	12¼	96	CNS	N	2.12×2.75	7	2.62×5.00	2.62×4.00	abe	⅞-18	Zen[2]	1½	2,500	26½	22¾	88½
277	4	AS	13¼	164	CNS	N	2.75×2.75	7	3.00×3.37	3.00×3.25	abce	⅞-18	Hol	.....	2,435	24½	29⅜	78½
278	4	AS	13¼	164	CNS	N	2.75×2.75	7	3.00×3.37	3.00×3.25	abce	⅞-18	Hol	.....	2,460	24½	29⅜	78½
279	3	....	....	...	DFS	Y	1.50×1.00	3	1.62×....	1.62×....	Splash	⅞-18	Zen	1	580	22	25¾	44¾
280	3	....	6⅛	10	CAS	Y	1.60×1.54	3	2.00×1.66	2.00×2.00	abc	14 mm.	CG	.....	500	21¼	29	44
281	3	....	7	17	CAS	Y	2.00×1.75	3	2.50×1.72	2.50×2.25	abce	14 mm.	CG	.....	700	22¾	30½	46½
282	3	....	7	17	CAS	Y	2.14×1.75	3	2.50×1.72	2.50×2.25	abc	14 mm.	CG	.....	700	23¾	30½	46½
283	3	....	7 13/32	22	CAS	Y	2.12×1.75	3	2.40×1.84	2.40×2.25	abc	14 mm.	CG	.....	860	22¾	30½	54
284	4	CS	7	...	CS	Y	1.93×1.31	4	2.25×1.18	2.25×1.81	abceft	18 mm.	Str	1¼	501	26	26	39 9/32
285	5	CS	8⅜	44	CS	Y	2.25×1.56	7	2.62×1.53	2.62×2.15	abcet	18 mm.	Str	1¼	749	18¼	29 3/16	40
286	5	CS	8⅜	44	CS	Y	2.25×1.56	7	2.62×1.53	2.62×2.15	abcet	18 mm.	Str	1¼	738	18¼	29 3/16	40
287	4	CNM	11¼	54	CS	Y	2.37×1.62	7	2.62×1.68	2.62×2.78	abcef	18 mm.	Str	1½	959	26⅞	34⅞	46 17/32
288	5	CS	8⅜	44	CS	Y	2.25×1.56	7	2.62×1.53	2.62×2.15	abcet	18 mm.	Str	1½	759	18¼	29 3/16	40
289	5	CNM	11¼	54	CS	Y	2.37×1.62	7	2.62×1.68	2.62×2.78	abcef	18 mm.	Str	1½	907	26⅞	34⅞	46 17/32
290	4	CNM	12½	83	CS	Y	2.50×1.81	7	3.00×2.25	3.00×3.12	abcef	18 mm.	Str	1¾	1,199	27¼	41	50 13/16
291	5	CNM	12½	83	CS	Y	2.50×1.81	7	3.00×2.25	3.00×3.12	abcef	18 mm.	Str	1¾	1,209	27¼	41	50 13/16
292	5	CNM	11¼	100	CS	Y	3.00×2.09	7	3.50×1.68	3.50×2.37	abcef	14 mm.	Str	1¾	1,687	30½	47½	53 3/32

TABLE A1-3.—AMERICAN STOCK, MARINE, AND COMMERCIAL VEHICLE ENGINES.—(*Continued*)

Line number	Number of rings per piston	Connecting rods			Crankshaft						Oil pressure to—	Spark plug, thread size	Carburetor		Engine weight without carburetor or ignition, lb.	Over-all dimensions, in.		
							Crankpin	Main bearings										
									Diameter and length, in.									
		Material	Center-to-center length, in.	Weight with bushing and cap, oz.	Material	Counterbalance used?	Diameter and length, in.	Number	Front	Rear			Make	Size		Width	Height	Length
293	5	CS	12½	83	CS	Y	2.50×1.81	7	3.00×2.25	3.00×3.12	abcef	18 mm.	Str	1¾	1,214	27¼	41	50 13/16
294	5	CNM	11¼	100	CS	Y	3.00×2.09	7	3.50×1.68	3.50×2.37	abcef	14 mm.	Str	2	1,700	30½	47¼	53 3/32
295	5	CNM	11¼	100	CS	Y	3.00×2.09	7	3.50×1.68	3.50×2.37	abcef	14 mm.	Str	2	1,710	30½	47¼	53 3/32
296	4	1,040	9	54	1,045	N	2.62×1.28	2	SAE-212	3.00×2.18	abet	14 mm.	Sch	1	640	18½	36½	34
297	4	1,040	9	54	1,045	N	2.62×1.28	2[a]	SAE-313	SAE-314	bet	14 mm.	Sch	1	650	18½	36½	34
298	4	1,040	10	97	1,045	N	2.37×2.50	3	2.50×2.50	2.62×3.50	abceg	⅞-18	Sch	1	1,125	24¼	41½	41½
299	4	1,040	10	97	1,045	N	2.37×2.50	3	2.50×2.50	2.62×3.50	abceg	⅞-18	Sch	1	1,125	24¼	44	41½
300	4	1,040	12	116	1,045	N	2.37×2.87	3	2.25×2.96	2.75×4.00	abceg	⅞-18	Sch	1¼	1,110	24½	57½	45
301	4	1,045	20½	656	3,140	N	3.50×4.37	5	3.50×5.21	3.50×6.37	abdeg	⅞-18	Zen	2	4,600	38½	68¼	72
302	4	1,045	20½	656	3,140	N	3.50×4.37	5	3.50×5.21	3.50×6.37	abdeg	⅞-18	Zen	2	......	38½	68¼	72
303	4	1,045	20½	656	3,140	N	3.50×4.37	5	3.50×5.21	3.50×6.37	abdeg	⅞-18	Zen	2	4,700	38½	68¼	72
304	4	1,045	20½	624	1,045	N	3.81×4.25	7	4.00×5.21	4.00×6.37	abdeg	⅞-18	Zen[2]	1¾	6,500	37½	74	97
305	4	1,045	20½	624	1,045	N	3.81×4.25	7	4.00×5.21	4.00×6.37	abdeg	⅞-18	Zen[2]	1¾	6,600	37½	74	97
306	4	1,045	20½	624	1,045	N	3.81×4.25	7	4.00×5.21	4.00×6.37	abdeg	⅞-18	Zen[2]	1¾	6,650	37½	74	97
307	6	3,135	15½	180	CNS	N	2.56×3.16	5	2.56×3.48	2.56×4.25	abcde	18 mm.	Sho[2]	2	2,950	32⅝	54⅞	78⅞
308	5	3,135	17 5/16	212	CNS	N	2.56×3.16	5	2.56×3.48	2.56×4.25	abcde	18 mm.	Hol	1½	2,450	32⅝	49½	78⅞
309	5	3,135	15½	184	CNS	N	2.56×3.15	7	2.56×1.50	2.56×4.25	abcde	18 mm.	Str	2	2,400	27⅞	51⅞	96¼
310	6	3,135	15½	180	CNS	N	2.56×3.16	7	2.56×3.48	2.56×4.25	abcde	18 mm.	Sho[3]	2	3,800	32⅝	54⅞	99⅞
311	5	3,135	17 5/16	212	CNS	N	2.56×3.16	7	2.56×3.48	2.56×4.25	abcde	18 mm.	Zen[2]	1½	3,050	32⅝	49½	98⅞
312	6	3,135	15½	180	CNS	N	2.56×3.16	7	2.56×3.48	2.56×4.25	abcde	18 mm.	Sho[3]	2½	4,050	35	52¼	98⅞
313	4	T-1,335	7 15/16	31	1,040	Y	1.93×1.00	4	2.25×1.23	2.25×1.87	abce	14 mm.	Car	1½	476	19⅞	29 11/16	32 5/32
314	3	AS	......	...	CS	Y	1.37×1.50	2	1.37×2.62	1.37×2.25	Splash	⅞-18	Zen	¾	130	11½	14½	13½
315	4	CS	......	...	CS	Y	1.50×2.00	2[a]	....	....	Splash	⅞-18	Zen	1	245[d]	16	17⅝	21⅝
316	4	......	......	...	CS	Y	1.75×2.50	2	1.62×4.25	1.62×4.12	Splash	⅞-18	Zen	.....	400[d]	17	22½	24½
317	4	CS	......	...	CS	Y	1.87×2.81	2	1.87×5.25	1.87×5.00	Splash	⅞-18	.....	.....	610	20	24¼	44⅜
318	4	Al	......	...	CS	..	....	2[a]	1.25×1.25	1.25×1.25	Splash	14 mm.	Zen	.....	145	13	17½	18¼

319	3	......	......	...	......	..	....	2	2.00×....	2.00×....	ab	......	Zen	.....	375[d]	20	19¾	28½
320	4	......	......	...	CS	Y	1.50×2.00	3	1.37×4.00	1.37×4.00	Splash	⅞–18	Zen	1	540[d]	16	17⅝	42
321	4	CS	......	...	CS	Y	2.25×2.25	3	2.25×5.00	2.25×5.00	Splash	⅞–18	Str	1¼	900	20	24	47¾
322	4	CS	......	...	CS	Y	2.37×2.81	3	2.37×4.00	2.37×4.00	Splash	⅞–18	Str	1¼	1,200	21¾	25½	52 9/16
323	4	CS	......	...	CS	Y	2.37×2.37	3	2.37×5.00	2.37×5.00	Splash	⅞–18	Str	1½	1,850	23½	30½	63
324	5	CS	......	...	CS	N	3.00×3.25	5	2.75×6.00	2.18×5.50	ML	⅞–18	....	....	4,600	28¼	96½	37
325	4	1,040	10½	...	1,045	N	2.18×1.28	7	2.62×1.94	2.62×2.47	abcde	14 mm.	Zen	1¼				
326	4	1,040	10½	...	1,045	N	2.18×1.28	7	2.62×1.94	2.62×2.47	abcde	14 mm.	Zen	1¼				
327	4	1,040	10½	...	1,045	Y	2.18×1.28	7	2.62×1.94	2.62×2.47	abcde	14 mm.	Zen	1½				
328	4	1,040	10½	...	1,045	Y	2.18×1.28	7	2.62×1.94	2.62×2.47	abcd	14 mm.	Zen	1½				
329	4	1,040	9	...	1,045	N	2.37×1.59	7	2.62×1.71	2.62×2.50	abc	14 mm.	Zen	1½				
330	3	3,140	6 9/16	21	1,045	N	1.75×1.12	3	2.00×1.56	2.00×1.62	abr	⅞–18	Zen	1¼	410[d]	15½	21¾	32⅜
331	3	AS	6 9/16	21	1,045	N	1.75×1.12	3	2.00×1.56	2.00×1.62	abr	⅞–18	Zen	1¼	410[d]	15½	21¾	32⅜
332	4	Dur	10½	41	NS	N	2.18×1.87	3	2.25×2.62	2.25×2.62	abc	18 mm.	Str	1¼	660[d]	23	28 13/16	45¼
333	4	CS	7	26	CS[f]	N	2.00×1.25	7	2.50×1.31	2.50×1.93	abr	14 mm.	Zen	1½	675[d]	21	22½	41⅛
334	4	CS	7	26	CS[f]	N	2.00×1.25	7	2.50×1.31	2.50×1.93	abr	14 mm.	Zen	1½	675[d]	21	22½	41⅛
335	4	Dur	10½	41	NS	Y	2.18×1.87	4	2.25×2.62	2.25×2.62	abc	18 mm.	Zen	1½	900[d]	19¾	29 1/16	56½
336	4	1,035	8	37	1,045	N	2.00×1.50	7	2.50×1.31	2.50×2.12	abr	⅞–18	Zen	1¾	860[d]	21	31⅞	49¾
337	4	1,035	8	37	1,045	N	2.00×1.50	7	2.50×1.31	2.50×2.12	abr	⅞–18	Zen	1¾	860[d]	21	31⅞	49¾
338	4	AS	10⅞	70	NS	Y	2.75×2.25	4	3.25×2.25	3.25×2.25	abc	18 mm.	Hol	2	1,050[d]	27¾	32 5/16	56¾
339	4	AS	10⅞	70	NS	Y	2.75×2.25	4	3.25×2.25	3.25×2.25	abc	18 mm.	Str	2	1,050[d]	24¾	29⅜	56¾
340	4	AS	11¼	84	NS	Y	2.87×2.00	4	3.00×3.00	3.00×3.62	ab	18 mm.	Hol	1½	1,295[d]	26¼	39 9/16	66
341	4	AS	11¼	84	NS	Y	2.87×2.00	4	3.00×3.00	3.00×3.62	ab	18 mm.	Sch	1½	1,295[d]	33½	33 7/16	66
342	4	AS	11¼	84	NS	Y	2.87×2.00	4	3.00×3.00	3.00×3.62	ab	18 mm.	Hol	2	1,295[d]	26¼	36 9/16	66
343	4	AS	11¼	84	NS	Y	2.87×2.00	4	3.00×3.00	3.00×3.62	ab	18 mm.	Sch	1½	1,295[d]	33½	33 7/16	66
344	4	AS	11¼	84	NS	Y	2.87×2.00	4	3.00×3.00	3.00×3.62	ab	18 mm.	Hol	2	1,295[d]	26¼	36 9/16	66
345	4	AS	11¼	84	NS	Y	2.87×2.00	4	3.00×3.00	3.00×3.62	ab	18 mm.	Sch	1½	1,295[d]	33½	33 7/16	66
346	4	AS	11¼	85	CS	Y	2.87×2.25	4	3.00×3.00	3.00×3.62	abr	18 mm.	Zen[2]	2	1,175[d]	25	23⅜	66
347	4	AS	11¼	85	CS	Y	2.87×2.25	4	3.00×3.00	3.00×3.62	abr	18 mm.	Zen[2]	2	1,175[d]	25	23⅜	66
348	3	AS	7	17	CS	Y	2.00×1.75	3	2.50×1.37	2.00×1.96	abc	18 mm.	Str[2]	1	660[d]	23½	31⅛	42⅝
349	3	DFS	7	18	....	Y	2.14×1.75	3	2.50×1.37	2.00×1.96	abc	14 mm.	Hol	1	700[d]	23½	31⅜	43 5/16
350	3	DFS	7 13/32	22	CAS	Y	2.12×1.57	4	2.40×1.83	2.40×2.25	abc	14 mm.	Str	1	......	25⅜	29⅞	49 13/16
351	4	AS	10⅞	164[h]	NS	Y	2.75×2.25	4	3.25×2.25	3.25×2.25	abc	18 mm.	Hol	2	1,700[d]	35	33¾	58½
352	4	AS	10⅞	164[h]	NS	Y	2.75×2.25	4	3.25×2.25	3.25×2.25	abc	18 mm.	Hol	2	1,700[d]	33¼	30 11/16	58½
353	3	AS	7⅜	...	AS	Y	1.50×1.87	2[g]	....	....	b	14 mm.	Op	1	100	12	22	18
354	3	AS	7⅜	...	AS	Y	1.50×1.87	2[g]	....	....	b	14 mm.	Op	1	180	20	20	22
355	3	AS	7⅜	...	AS	Y	1.50×1.87	2[g]	....	....	b	14 mm.	Op	1	......	12	22	18
356	..	....	......	...	....	Y	....	..	....	....	abde	14 mm.	Str	1¼	1,150	23¾	17¼	54 9/16
357	4	AS	15½	...	Spec	N	2.62×3.00	7	2.62×3.00	2.62×3.00	abde	⅞–18	Str[2]	2	2,100[d]	28	30	88 1/16
358	6	AS	15½	...	Spec	Y	2.62×3.00	7	2.62×3.00	2.62×3.00	abde	⅞–18	Str[2]	2	2,100[d]	28	30	88 1/16
359	4	AS	19	...	Spec	N	3.50×4.12	7	3.50×4.12	3.50×4.12	abde	⅞–18	Str[2]	2	4,000[d]	34	42	117 7/16
360	4	CS	12½	96	CNS	Y	2.25×2.62	3	2.25×4.62	2.25×4.75	Splash	½	Sch	1½	1,100	24	36½	53⅞
361	4	CS	12½	113	CNS	Y	2.50×2.12	7	3.00×1.75	3.00×2.87	abcdef	⅞–18	Zen[2]	1½	1,400	27½	33½	71¾
362	4	CS	12½	113	CNS	Y	2.50×2.12	7	3.00×1.75	3.00×2.87	abcdef	⅞–18	Zen[2]	1½	1,400	27½	33½	71¾
363	4	CS	12½	113	CNS	Y	2.50×2.12	7	3.00×1.75	3.00×2.87	abcdef	⅞–18	Zen[2]	1½	1,850	27½	33½	71¾
364	4	CS	12½	113	CNS	Y	2.50×2.12	7	3.00×1.75	3.00×2.87	abcdef	⅞–18	Zen[2]	1¾	1,400	27½	33½	71¾
365	4	CS	12½	113	CNS	Y	2.50×2.12	7	3.00×1.75	3.00×2.87	abcdef	⅞–18	Zen[2]	1½	2,000	27 7/16	23	81 1/16

TABLE A1-3.—AMERICAN STOCK, MARINE, AND COMMERCIAL VEHICLE ENGINES.—(*Continued*)

Line number	Number of rings per piston	Connecting rods			Crankshaft						Oil pressure to—	Spark plug, thread size	Carburetor		Engine weight without carburetor or ignition, lb.	Over-all dimensions, in.		
							Crankpin	Main bearings										
									Diameter and length, in.									
		Material	Center-to-center length, in.	Weight with bushing and cap, oz.	Material	Counterbalance used?	Diameter and length, in.	Number	Front	Rear			Make	Size		Width	Height	Length
366	4	CS	12 1/2	113	CNS	Y	2.50×2.12	7	3.00×1.75	3.00×2.87	abcdef	7/8-18	Zen[2]	1 3/4	2,000	27 1/2	33 1/2	71 3/4
367	4	CS	12 1/2	113	CNS	Y	2.50×2.12	7	3.00×1.75	3.00×2.87	abcdef	7/8-18	Zen[2]	1 1/2	2,000	27 1/2	33 1/2	71 3/4
368	5	CS	14	130	CNS	Y	2.50×3.00	4	2.50×4.43	2.50×4.43	abcdef	7/8-18	Zen[2]	1 3/4	2,000	30 1/2	36 1/4	87 1/2
369	5	CS	14	130	CNS	Y	2.50×3.00	4	2.50×4.43	2.50×4.43	abcdef	7/8-18	Zen[2]	1 3/4	2,000	30 1/2	36 1/4	87 1/2
370	5	CS	14	130	CNS	Y	2.50×3.00	4	2.50×4.43	2.50×4.43	abcdef	7/8-18	Zen[2]	1 3/4	2,000	30 1/2	36 1/4	87 1/2
371	3	CS	14	130	CNS	Y	2.50×3.00	4	2.50×4.43	2.50×4.43	abcdef	7/8-18	Zen[2]	1 3/4	2,250	30 1/2	45 1/8	87 1/2
372	3	CS	14	130	CNS	Y	2.50×3.00	4	2.50×4.43	2.50×4.43	abcdef	7/8-18	Zen[2]	1 3/4	2,000	30 1/2	45 1/8	87 1/2
373	3	CS	14	130	CNS	Y	2.50×3.00	4	2.50×4.43	2.50×4.43	abcdef	7/8-18	Zen[2]	2 1/2	2,175	30 1/2	45 1/8	87 1/2
374	4	CS	14 1/2	224	CNS	Y	3.00×3.12	7	3.25×2.62	3.25×3.68	abcdef	7/8-18	SZ[2]	2 1/2	4,000	31	49 1/4	102
375	4	CS	14 1/2	224	CNS	Y	3.00×3.12	7	3.25×2.62	3.25×3.68	abcdef	7/8-18	SZ[2]	2 1/2	3,400	31	49 1/4	102
376	4	CS	18	416	CNS	Y	4.00×3.12	7	4.00×3.37	4.00×5.50	abcdef	7/8-18	Zen[3]	2 1/2	7,100	40 1/8	72 3/4	121 3/8
377	4	CS	18	416	CNS	Y	4.00×3.12	7	4.00×3.37	4.00×5.50	abcdef	7/8-18	Zen[3]	2 1/2	7,100	40 1/8	72 3/4	121 3/8
378	4	CS	18	416	CNS	Y	4.00×3.12	7	4.00×3.37	4.00×5.50	abcdef	7/8-18	Zen[3]	2 1/2	7,100	40 1/8	72 3/4	121 3/8
379	3	CS	14	130	CNS	Y	2.50×3.00	5	2.50×4.43	2.50×4.43	abcdef	7/8-18	Zen[2]	2 1/2	3,400	30 1/2	45 1/8	104 9/16
380	3	CS	14	130	CNS	Y	2.50×3.00	5	2.50×4.43	2.50×4.43	abcdef	7/8-18	Zen[2]	2 1/2	3,400	30 1/2	45 1/8	104 9/16
381	4	CS	18	416	CNS	Y	4.00×3.12	9	4.00×3.37	4.00×5.50	abcdef	7/8-18	Zen[4]	2 1/2	9,000	40 1/8	72 3/4	142 3/8
382	4	CS	18	416	CNS	Y	4.00×3.12	9	4.00×3.37	4.00×5.50	abcdef	7/8-18	Zen[4]	2 1/2	9,000	40 1/8	72 3/4	142 3/8
383	4	CS	18	416	CNS	Y	4.00×3.12	9	4.00×3.37	4.00×5.50	abcdef	7/8-18	Zen[4]	2 1/2	9,000	40 1/8	72 3/4	142 3/8
384	3	1,045	8 1/2	43	1,045	Y	1.50×2.12	2	1.50×3.00	1.50×3.00	Splash	7/8-18	Str	3/4	210	19 1/4	22 9/16	21 3/8
385	3	1,045	8 1/2	43	1,045	N	1.50×2.12	2	1.50×3.00	1.50×3.00	Splash	7/8-18	Str	1	415	19 1/2	22 13/16	36 1/4
386	3	1,045	6	14	1,045	N	1.43×1.25	2[g]	ND 1,207	ND 1,207	ce	14 mm.	Str	3/4	240	18 1/4	19 1/8	27 7/16
387	3	1,045	8	27	1,045	N	1.75×1.50	2	1.76×2.81	1.73×2.87	ab	7/8-18	Str	1	330	15 7/8	21 1/8	38 5/8
388	3	1,045	7 1/4	29	1,045	N	1.75×1.25	3	2.12×1.43	2.12×1.18	abcde	18 mm.	Str	1	490	21 3/8	24 3/16	35
389	4	1,045	8 3/4	46	1,045	N	2.00×1.50	3	2.00×2.50	2.00×1.87	abcde	7/8-18	Str	1	610	19 1/8	26 1/4	41
390	3	1,045	8 1/2	43	1,045	N	1.50×2.12	3	1.50×3.00	1.50×3.00	abe	7/8-18	Str	1	620	19 1/2	22 13/16	46 5/8
391	4	1,045	10 3/8	66	1,045	N	2.00×2.25	3	2.00×4.18	2.00×3.50	abe	7/8-18	Str	1 1/4	830	20 1/4	26 15/16	54 3/8

392	4	1,045	10 3/8	66	1,045	N	2.00×2.25	3	2.00×4.18	2.00×3.50	abe	7/8–18	Str	1 1/2	830	20 1/4	26 15/16	54 3/8
393	4	1,045	11 3/8	87	1,045	N	2.56×2.25	5	2.56×4.25	2.56×4.25	abce	7/8–18	Str	1 1/2	1,175	22 3/4	27 1/16	59 5/16
394	4	1,045	13 1/4	168	1,045	N	2.56×3.00	5	2.62×4.50	2.62×4.50	abcde	7/8–18	Str	2	1,720	25 7/8	35 1/2	74 1/16
395	4	1,045	13 1/4	168	1,045	N	2.56×3.00	5	2.62×4.50	2.62×4.50	abcde	7/8–18	Str	2	1,730	25 7/8	35 1/2	74 1/16
396	4	1,045	13 1/4	168	1,045	N	2.56×3.00	5	2.62×4.50	2.62×4.50	abcde	7/8–18	Str	2	1,740	25 7/8	35 1/2	74 1/16
397	4	1,045	8	40	1,045	N	2.00×1.50	7	2.62×2.00	2.62×1.25	abcde	18 mm.	Str	1 1/4	1,000	22 1/2	28 1/4	53 13/16
398	4	1,045	8 3/4	48	1,045	N	2.25×1.50	7	2.62×2.75	2.62×1.62	abcde	14 mm.	Str	1 1/4	1,185	20 1/4	27 7/8	57 15/16
399	4	1,045	11 3/8	87	1,045	N	2.56×2.25	7	2.56×4.25	2.56×4.25	abce	7/8–18	Str	1 1/2	1,475	33 3/8	30 9/16	72 9/16
400	4	1,045	11 3/8	87	1,045	N	2.56×2.25	7	2.56×4.25	2.56×4.25	abce	7/8–18	Str	2	1,565	24 1/8	30 9/16	72 3/4
401	4	1,045	13 1/4	168	1,045	N	2.56×3.00	7	2.62×4.50	2.62×4.50	abcde	7/8–18	Str	2	2,330	27 1/8	35 1/2	91 3/8
402	4	1,045	13 1/4	168	1,045	N	2.56×3.00	7	2.62×4.50	2.62×4.50	abcde	7/8–18	Str	2	2,360	27 1/8	35 1/2	91 3/8
403	4	1,045	13 1/4	168	1,045	N	2.56×3.00	7	2.62×4.50	2.62×4.50	abcde	7/8–18	Str	2	2,380	27 1/8	35 1/2	91 3/8
404	3	Dur	8 1/2	...	CS	Y	2.00×2.00	2	2.00×2.00	2.00×2.00	Splash	7/8–18	Str	5/8	200	15 3/16	24 3/8	17 9/16
405	4	CS	7 1/4	...	CS	Y	1.75×1.37	2	1.75×1.87	1.75×2.00	abc	18 mm.	Str	5/8	300	20 7/16	22 5/8	28 1/8
406	3	CS	7 1/2	...	CS	N	1.50×1.75	2	1.50×2.75	1.50×2.75	abc	18 mm.	Zen	1	347	17 5/16	21 7/16	34 13/16
407	4	Dur	7 1/4	18	CS	Y	1.75×1.37	3	1.75×2.50	1.75×2.50	abcde	18 mm.	Zen	1	400	20 1/8	22 11/16	34 13/16
408	4	Dur	8 1/2	30	CS	N	2.00×1.75	3	2.00×2.50	2.00×2.50	abc	18 mm.	Zen	1	520	18 1/2	25 1/4	39 5/8
409	4	Dur	7 1/4	18	CS	Y	1.75×1.37	4	1.75×2.50	1.75×2.50	abcde	18 mm.	Zen	1 1/4	515	24	20 3/4	43 7/8
410	4	Dur	8 1/2	28	CS	N	2.00×1.87	7	2.00×2.56	2.00×2.56	abc	18 mm.	Str	1 1/4	775	20 1/16	26 3/4	49 15/16
411	4	Dur	8 1/2	28	CS	N	2.00×1.87	7	2.00×2.56	2.00×2.56	abc	18 mm.	Str	1 1/4	810	22 3/16	26 3/4	49 15/16
412	4	CS	8 1/2	28	CS	N	2.00×1.87	9	2.00×2.56	2.00×2.56	abc	18 mm.	Str	1 3/4	1,050	21 9/16	26 3/16	62 5/8
413	4	CS	8 1/2	28	CS	N	2.00×1.87	9	2.00×2.56	2.00×2.56	abc	18 mm.	Str²	1 1/2	1,100	22 3/4	26 3/16	62 5/8
414	5	AS	12	aa	AS	N	2.37×2.50	7	2.62×4.62	2.62×2.00	abc	18 mm.	Zen	2	1,920	32 1/4	40 3/4	90 1/4
415																		
416	5	4,340	12	228	4,140	Y	3.50×2.87	7	4.00×3.00	4.00×3.00	abce	18 mm.	Str	2 3/4	3,490	49	50	101
417	4	AS	7 1/2	29	CAS	Y	1.87×1.62	2v	BT22,168	BT22,325	Splash	14 mm.	Zen	1	160			
418	4	AS	7 1/2	29	CAS	Y	1.87×1.62	3v	BT22,168	BT22,325	Splash	14 mm.	Zen	1 1/4	250			
419	4	AS	7 1/2	29	CAS	Y	1.87×1.62	3v	BT22,168	BT22,325	Splash	14 mm.	Zen	1 1/4	340			
420	4	1,045	4 5/8	12	1,045	Y	1.50×1.00	2	1.50×1.00	2.25×2.31	abce	14 mm.	Op	5/8	125	20 3/4	14 1/2	15 3/8
421	3	1,045	6	14	1,045	N	1.56×1.25	2v	ND1,207	ND1,207	abcde	14 mm.	Op	5/8	143	14 15/16	20 1/2	21 13/16
422	3	1,045	7 1/4	29	1,045	N	1.75×1.06	3	2.12×1.18	2.12×1.43	abcde	18 mm.	Op	1	280	19	26 3/4	37 3/8
423	3	1,045	7 1/4	29	1,045	N	1.75×1.06	3	2.12×1.18	2.12×1.43	abcde	18 mm.	Op	1	290	19	26 3/4	37 3/8
424	4	1,045	8 3/4	46	1,045	N	2.00×1.50	3	2.00×1.87	2.00×2.50	abcde	7/8–18	Op	1	385	17 7/8	27	32 1/4
425	4	1,045	8 3/4	57	Pro	N	2.25×1.75	3	2.62×1.75	2.62×2.25	abcdeg	18 mm.	Op	1 1/4	670	21 1/2	38 3/4	33 9/16
426	4	1,045	8 3/4	57	Pro	N	2.25×1.75	3	2.62×1.75	2.62×2.25	abcdeg	18 mm.	Op	1 1/4	680	21 1/2	38 3/4	33 9/16
427	4	1,045	10 1/2	86	1,045	N	2.37×2.12	3	2.37×2.12	2.37×2.75	abcdeg	7/8–18	Op	1 1/4	870	21 1/4	35	39
428	4	1,045	10 1/2	86	1,045	N	2.37×2.12	3	2.37×2.12	2.37×2.75	abcdeg	7/8–18	Op	1 1/4	925	21 1/4	35	39
429	4	1,045	10 1/2	...	1,045	N	2.37×2.12	3	2.37×2.31	2.37×2.75	abcdeg	18 mm.	Op	1 1/4	920	......	......	39 9/16
430	4	1,045	11 3/4	141	1,045	N	2.75×2.50	3	3.00×2.87	3.00×3.56	abcdeg	7/8–18	Op	1 1/2	1,600	24	42	46 1/8
431	4	1,045	13 1/4	146	1,045	N	2.75×2.50	3	3.00×3.00	3.00×3.56	abcde	7/8–18	Op	1 3/4	1,600	24 3/8	45 5/16	62
432	4	1,045	18	278	1,045	N	3.25×2.75	5	3.75×3.75	3.75×5.50	abcde	7/8–18	Op	2	2,750	34	51 3/4	59 7/16
433	4	1,045	18	278	1,045	N	3.25×2.75	5	3.75×3.75	3.75×5.50	abcdeg	7/8–18	Op	2 1/2	3,560	34 3/4	63 1/2	59 7/16
434	4	1,045	8	40	1,045	N	2.00×1.50	7	2.62×1.25	2.62×2.00	abcde	18 mm.	Op	1 1/4	685	26	31	39 1/2
435	4	1,045	8	40	1,045	N	2.00×1.50	7	2.62×1.25	2.62×2.00	abcde	18 mm.	Op	1 1/2	690	28	31	39 1/2
436	4	1,045	8	40	1,045	N	2.00×1.50	7	2.62×1.25	2.62×2.00	abcde	18 mm.	Op	1 1/2	706	26	31	39 1/2
437	4	1,045	8 3/4	45	1,045	N	2.25×1.50	7	2.62×1.62	2.62×2.75	abcde	18 mm.	Op	1 1/2	890	20 1/2	31	43 11/16
438	4	1,045	8 3/4	48	1,045	N	2.25×1.50	7	2.62×1.62	2.62×2.75	abcde	18 mm.	Op	1 1/2	920	20 1/2	31	43 11/16

TABLE A1-3.—AMERICAN STOCK, MARINE, AND COMMERCIAL VEHICLE ENGINES.—(*Continued*)

Line number	Number of rings per piston	Connecting rods: Material	Connecting rods: Center-to-center length, in.	Connecting rods: Weight with bushing and cap, oz.	Crankshaft: Material	Crankshaft: Counterbalance used?	Crankshaft: Crankpin: Diameter and length, in.	Crankshaft: Main bearings: Number	Crankshaft: Main bearings: Diameter and length, in.: Front	Crankshaft: Main bearings: Diameter and length, in.: Rear	Oil pressure to—	Spark plug, thread size	Carburetor: Make	Carburetor: Size	Engine weight without carburetor or ignition, lb.	Over-all dimensions, in.: Width	Over-all dimensions, in.: Height	Over-all dimensions, in.: Length
439	4	1,045	10 1/4	85	Pro	N	2.62×2.00	7	3.25×1.75	3.25×3.00	abcdeg	18 mm.	Op	1 3/4	....	21 1/16	43 3/16	50 3/8
440	4	1,045	10 1/4	83	1,045	N	2.75×1.75	7	3.00×1.87	3.00×3.00	abcde	7/8-18	Op	1 3/4	1,185	26	35 7/16	46 9/16
441	4	1,045	10 1/4	85	Pro	N	2.62×2.00	7	3.25×1.75	3.25×3.00	abcdeg	18 mm.	Op	1 3/4	1,340	21 1/16	43 3/16	50 3/8
442	4	1,045	10 1/4	83	1,045	N	2.75×1.75	7	3.00×1.87	3.00×3.00	abcde	7/8-18	Op	1 3/4	1,225	26	35 7/16	46 9/16
443	4	1,045	11 3/4	133	Pro	N	3.00×2.25	7	3.50×2.00	3.50×3.50	abcdeg	18 mm.	Op	1 3/4	1,983			
444	4	1,045	11 3/4	130	1,045	N	3.00×2.00	7	3.62×2.62	3.62×3.50	abcdeg	18 mm.	Op	1 3/4	2,230	26 7/16	49 1/16	58 9/16
445	4	1,045	13 1/4	145	1,045	N	2.75×2.50	7	3.50×2.50	3.50×3.50	abcde	7/8-18	Op	2	1,575	26	41 15/16	54 7/16
446	4	1,045	11 3/4	133	Pro	N	3.00×2.25	7	3.50×2.00	3.50×3.50	abcdeg	18 mm.	Op	1 3/4				
447	4	1,045	11 3/4	130	1,045	N	3.00×2.00	7	3.62×2.62	3.62×3.50	abcdeg	18 mm.	Op	1 3/4	2,250	26 7/16	49 1/16	58 9/16
448	4	1,045	13 1/4	195	Pro	N	3.37×2.37	7	4.00×2.56	4.00×3.56	abcdeg	18 mm.	Op	2 1/2	3,200	28	51 1/2	65 5/8
449	4	1,045	13 1/4	195	Pro	N	3.37×2.37	7	4.00×2.56	4.00×3.56	abcdeg	18 mm.	Op	2 1/2	3,220	28	51 1/2	65 5/8
450	4	1,045	15 3/8	314	1,045	N	3.25×2.75	7	3.75×3.75	3.75×5.50	abcdeg	18 mm.	Op	2	4,950	32 3/8	56 1/4	76 5/8
451	4	1,045	15 3/8	314	1,045	N	3.25×2.75	7	3.75×3.75	3.75×5.50	abcdeg	18 mm.	Op	2 1/2	4,975	32 3/8	56 1/4	76 5/8
452	4	1,045	15 3/8	314	1,045	N	3.25×2.75	7	3.75×3.75	3.75×5.50	abcdeg	18 mm.	Op	2 1/2	4,975	32 3/8	57 5/16	76 5/8
453	4	1,045	18 3/8	474	1,045	N	4.00×3.75	7	4.25×4.81	4.25×5.50	abcde	7/8-18	Op	2 1/2	7,500	46 1/2	59	92 7/16
454	4	1,045	18 3/8	474	1,045	N	4.00×3.75	7	4.25×4.81	4.25×5.50	abcde	7/8-18	Op	2 1/2	7,500	46 1/2	59	92 7/16
455	5	1,045	18 3/8	474	1,045	N	4.00×3.75	7	4.25×4.81	4.25×5.50	abcdeg	7/8-18	Op[2]	2 1/2	7,800	46 1/2	63 3/4	92 7/16
456	4	CS	9 15/32	38	CNS	Y	2.18×1.34	7	2.87×1.84	2.87×2.12	abcde	14 mm.	Zen	1 1/4	850	27	36	45
457	4	CS	9 5/32	46	CNS	Y	2.18×1.34	7	2.87×1.84	2.87×2.12	abcde	14 mm.	Zen	1 3/8	852	29	35	43
458	5	1,035	8	38	1,045	N	2.00×1.50	7	2.50×1.31	2.50×2.12	abe	14 mm.	Zen	1 1/4	560	18	24	40 1/2
459	4	CS	9 25/32	50	CNS	Y	2.18×1.34	7	2.87×1.84	2.87×2.12	abcde	14 mm.	Str	1 1/4	850	29	35	43
460	5	CS	9 15/32	49	CNS	Y	2.18×1.34	7	3.00×1.84	3.00×1.93	abcde	14 mm.	Str	1 1/4	850	29	35	43
461	5	CS	10 1/2	69	CNS	Y	2.37×1.75	7	2.75×2.50	2.75×2.87	abcdeg	18 mm.	Zen	1 3/4	1,279	34	42 1/2	55 1/2
462	5	CS	10 1/2	59	CNS	Y	2.37×1.75	7	2.75×2.56	2.75×2.93	abcdeg	18 mm.	Zen	1 3/4	1,291	34	42 1/2	55 1/2
463	5	CS	8 1/32	51	CS	Y	2.43×3.00	7	2.87×2.18	2.87×2.50	abcd	14 mm.	Zen	1 3/8	2,200	38 1/2	21 1/4	69 1/4
464	3	1,035	9 3/16	34	1,040	Y	1.94×1.30	3	2.33×1.92	2.33×1.75	abce	14 mm.	Car	1 1/8	364	19	26 3/8	26 1/4

465	3	Al	6	6	1,045	Y	1.00×1.00	2[g]	......	......	......	18 mm.	Str	5/8	68	17 1/4	16 3/4	14 1/2
466	3	Al	6	6	1,045	Y	1.00×1.00	2[g]	......	......	......	18 mm.	Str	5/8	68	17 1/4	16 3/4	14 1/2
467	3	1,035	8	20	1,045	Y	1.12×1.25	2[g]	......	......	......	18 mm.	Str	3/4	125	16 3/4	20 5/16	17 1/2
468	4	Al	6	6	1,045	Y	1.00×1.00	2[g]	......	......	......	18 mm.	Str	5/8	70	17 3/4	16 3/4	15
469	3	1,035	8	20	1,045	Y	1.12×1.25	2[g]	......	......	......	18 mm.	Str	3/4	130	16 3/4	20 5/16	17 1/2
470	5	1,035	9 1/2	33	1,045	Y	1.37×1.37	2[g]	......	......	......	18 mm.	Str	3/4	......	18 7/8	24 3/8	19 1/8
471	3	1,035	9 1/2	33	1,045	Y	1.37×1.37	2[g]	......	......	......	18 mm.	Str	3/4	175	18 7/8	24 3/8	19 1/8
472	3	1,035	9 1/2	33	1,045	Y	1.37×1.37	2[g]	......	......	......	18 mm.	Zen	1	180	18 7/8	24 3/8	19 1/8
473	4	1,035	8 3/8	21	1,045	Y	1.75×1.12	2[g]	......	......	g	18 mm.	Str	3/4	230	17	26 1/4	29
474	5	1,035	8 3/8	29	1,045	N	1.75×1.25	3	g	g	ag	18 mm.	Str	1	340	20	29	36 3/8
475	4	1,035	8 3/8	29	1,045	N	1.75×1.25	3	g	g	ag	18 mm.	Str	1	345	20	29	36 3/8
476	3	1,035	10 1/2	64	1,045	N	2.00×2.00	3	1.93×2.50	2.06×3.00	abcdeg	7/8-18	Str	1 1/4	615	25 3/4	34 3/4	35 1/4
477	5	CS	10 1/2	119	CS	N	2.75×2.50	3	2.75×3.00	2.75×3.00	abcdeg	7/8-18	Str	1 1/2	850	25 3/4	36 3/8	47 1/16
478	3	CS	10 1/2	68	CS	N	2.50×1.75	4	2.50×2.50	2.50×3.00	abdeg	7/8-18	Str	1 1/2	975	25 3/4	36 1/4	45 15/16
479	3	CS	10 1/2	75	CS	N	2.62×1.75	4	2.75×2.25	2.75×2.75	abdeg	7/8-18	Str	1 1/2	1,075	25 3/4	37 3/16	53 5/8
480	4	1,035	10 1/2	75	1,045	N	2.62×1.75	4	2.75×2.25	2.75×2.75	abdeg	7/8-18	Str	1 1/2	1,095	25 3/4	37 5/16	53 5/8
481	4	1,035	10 1/2	75	1,045	N	2.62×1.75	4	2.75×2.25	2.75×2.75	abdeg	7/8-18	Str	1 1/2	1,110	25 3/4	37 5/16	53 5/8
482	4	CS	10 1/2	119	CS	N	2.75×2.50	4	2.75×3.00	2.75×3.00	abcdeg	7/8-18	Str	1 3/4	1,260	25 3/4	37 1/8	60 5/16
483	4	1,035	10 1/2	87	1,045	N	2.75×2.50	4	2.75×3.00	2.75×3.00	abcdeg	7/8-18	Str	1 3/4	1,270	25 3/4	37 1/8	60 5/16

Symbols and Abbreviations

(Continued from p. 371)

M—Marine (engine type).
May—Mayer carburetor.
ML—McCord lubricator system.
Mo—Molybdenum.
MS—Manganese steel.
N—No or none.
NCI—Nickel cast iron.
Nif—Niferrite.
NS—Nickel steel.
Op—Optional.
Pro—Proferal.
PU—Power units.
r—Reverse gear.
RC—Rail cars.
S—Steel.
SA—Special alloy.
SB—Spiral bevel gear.
SBG—Spur and bevel gear.
SC-XB—Silicon chromium.
Sch—Schebler carburetor.
Se—Separate.
SG—Spur gear.
Sho—Shore carburetor.
Sil—Silchrome steel.
Spec—Special.
SS—Semisteel.
St—Stellite steel.
SZ—Schebler and Zenith carburetor.
t—Tappets and valve mechanism.
T—Valves opposite (T-head).
T—Trucks.
TA—Tungsten alloy.
Til—Tillotson carburetor
To—Toledo TP-12.
Tr—Tractors.
Tun—Tungsten steel.
W—Wet liners.
WA—Wausau alloy.
WR—Wilcox-Rich-EA5.
Y—Yes.
Zen—Zenith carburetor.

TABLE A1-4.—ROTATING AND RECIPROCATING WEIGHTS IN AIRCRAFT ENGINES

(From Angle, "Engine Dynamics and Crankshaft Design," and *S.A.E. Jour.*, Vol. 29, Nos. 4 and 5, October, November, 1931)

Engine	Type	Bore, in.	Stroke, in.	Piston area, sq. in.	Cylinder displacement, cu. in.	Reciprocating weight per cylinder, lb.	Centrifugal weight per cylinder lb.	Reciprocating weight per sq. in. piston area	Total reciprocating and centrifugal weight per cu. in. displacement
Liberty-6	6-cyl. vertical	5.00	7.00	19.63	137.4	5.80	3.70	0.295	0.069
Rausie-E-6	6-cyl. vertical	5.00	6.00	19.63	117.8	5.99	3.73	0.305	0.074
Hall-Scott, L-6	6-cyl. vertical	5.00	7.00	19.63	137.4	5.95	3.13	0.303	0.066
Isotta-Fraschini	6-cyl. vertical	5.51	7.08	23.86	169.0	7.20	3.80	0.302	0.065
Benz-200	6-cyl. vertical	5.51	7.48	23.86	178.4	6.40	4.40	0.268	0.061
B.hp.-200	6-cyl. vertical	5.71	7.48	25.62	191.3	6.50	3.80	0.254	0.054
Mercedes-200	6-cyl. vertical	5.51	6.30	23.86	150.3	9.40	3.30	0.395	0.085
Aeromarine, U-8	8-cyl. V	4.50	6.50	14.18	92.2	3.60	2.60	0.266	0.067
Wright-E	8-cyl. V	4.72	5.11	17.53	89.9	4.60	2.30	0.262	0.077
Wright-H	8-cyl. V	5.51	5.90	23.82	140.8	7.20	3.10	0.302	0.073
Packard-744	8-cyl. V	4.75	5.25	17.72	93.0	5.00	2.35	0.282	0.079
Curtiss-K12	12-cyl. V	4.50	6.00	15.92	95.4	3.35	1.70	0.210	0.053
Liberty-12	12-cyl. V	5.00	7.00	19.63	137.4	6.20	3.15	0.315	0.068
Rolls-Royce "Eagle"	12-cyl. V	4.50	6.50	15.90	103.4	3.75	1.92	0.236	0.055
Packard-1237	12-cyl. V	5.00	5.25	19.63	103.1	5.83	2.56	0.297	0.081
Packard-2025	12-cyl. V	5.75	6.50	26.00	168.8	8.06	4.01	0.310	0.072
Duesenberg-H	16-cyl. V	6.00	7.50	28.27	212.1	6.29	3.47	0.219	0.046
Napier-"Lion"	12-cyl. W	5.50	5.125	23.76	121.7	5.50	2.13	0.321	0.063
Eng. Div. W-1-A	18-cyl. W	5.50	6.50	23.76	154.4	7.40	3.74	0.312	0.072
Lawrence-L	3-cyl. radial	4.25	5.25	14.20	74.5	3.05	1.50	0.215	0.061
A.B.C. "Wasp"	7-cyl. radial	4.53	5.91	16.12	95.2	2.30	1.41	0.143	0.039
Lawrence R-1	9-cyl. radial	4.25	5.25	14.18	74.5	3.05	2.83	0.215	0.079
ABC "Dragon Fly"	9-cyl. radial	5.50	6.50	23.76	154.4	4.20	2.59	0.177	0.044
Curtiss V-1570	12-cyl. V	5.125	6.25	20.6	129	4.11*	2.68	0.200	0.053
(Conqueror)	...	...	...	...	...	3.9†	2.68	0.189	0.051
Wright R 1750	9-cyl. radial	6.00	6.875	28.25	169.8	7.45*	2.80	0.264	0.061
(Cyclone)	...	...	...	...	...	6.74†	2.80	0.237	0.057

* Master cylinder.

† Articulated cylinder.

TABLE A1-5.—CRANKPIN DATA

(From Angle, "Engine Dynamics and Crankshaft Design," and *S.A.E. Jour.*, Vol. 29, Nos. 4 and 5, October, November, 1931)

Engine	Type	Bore and stroke, in.	Rated r.p.m.	B.hp. per crank-pin	Cylinders per crank pin	Crankpin: Effective length, L, in.	Crankpin: Diameter, D, in.	Crankpin: L/D	Projected bearing area, sq. in.	Bearing pressures, lb. per sq. in.: Max.	Bearing pressures, lb. per sq. in.: Mean	Rubbing velocity, f.p.s.	Rubbing or PV factor	V/P	Oil pressure, lb. per sq. in.	Centrifugal weight/crank-pin, lb.	Reciprocating weight/cyl., lb.
Liberty 6	6-cyl. in-line	5 × 7	1,800	40	1	2.187	2.375	0.922	5.2	1,057	485	18.65	9,050	0.0385	40	3.7	5.8
Chevrolet Master (1938)	6-cyl. in-line	3.5 × 3.75	3,200	14.19	1	1.5	2.3125	0.65	3.465	.....	.....	32.3	.....	.....	13.5		
Nash Ambassador (1938)	6-cyl. in-line	3.375 × 4.375	3,400	17.5	1	1.425	2.0	0.712	2.85	.....	.....	29.7	.....	.....	30		
Hispano 300	8-cyl. V	5.51 × 5.9	1,800	81	2	2.444	2.126	1.15	5.19	1,545	840	16.70	14,020	0.0199	70	6.2	7.2
Wright H-2	8-cyl. V	5.51 × 5.9	2,000	101	2	2.444	2.126	1.15	5.19	1,835	944	18.55	17,520	0.0197	70	6.2	6.06
Cadillac V-8 (1938)	8-cyl. V	3.5 × 4.5	3,400	33.8	2	2.3125	2.46	0.94	5.69	.....	.....	36.5	.....	.....	25		
Curtiss D-12	12-cyl. V	4.5 × 6	2,000	71	2	1.805	2.5	0.723	4.52	1,270	717	21.80	15,640	0.0304	100	5.1	3.52
Liberty 12	12-cyl. V	5 × 7	1,700	70	2	2.187	2.375	0.922	5.2	1,035	740	17.60	13,200	0.0234	40	6.3	6.2
Packard 2500	12-cyl. V	6.375 × 6.5	2,000	133	2	2.5	3.25	0.77	8.12	1,346	836	28.37	23,717	0.0338	.....	8.5	7.6
Curtiss V-1570	12-cyl. V	5.125 × 6.25	2,400	105	2	1.39	2.5	0.556	3.28	2,154	1,585	26.2	41,500	0.0165	.....	5.36	4.11
Lawrence L	3-cyl. radial	4.25 × 5.25	1,600	56	3	2.5	1.875	1.33	4.7	906	464	13.07	6,060	0.0284	10	4.5	3.05
Lucifer	3-cyl. radial	5.75 × 6.25	1,600	108	3	2.12	3.04	0.7	6.42	1,270	727	14.83	10,800	0.0204	.....	7.34	3.89
Kinner K-5	5-cyl. radial	4.25 × 5.25	2,000	110	5	2.5	1.875	1.33	4.69	1,915	982	16.36	16,070	0.01665	.....	11.05	3.06
Lawrence J	9-cyl. radial	4.5 × 5.5	1,800	220	9	3.45	2.0	1.73	6.9	1,188	971	15.7	15,250	0.0162	30	17.19	4.16
Wright R-2	9-cyl. radial	5.625 × 6.5	1,650	400	9	3.433	2.625	1.31	9.02	1,164	934	18.87	17,640	0.0202	50	20.5	5.77
P. & W. Wasp S.C.	9-cyl. radial	5.75 × 5.75	2,100	450	9	3.37	2.62	1.29	8.85	1,720	1,488	24.05	35,800	0.0162	.....	19.8	6.28
P. & W. Hornet B	9-cyl. radial	6.25 × 6.75	1,950	567	9	3.5	2.87	1.22	10.06	2,072	1,835	24.7	45,400	0.0135	.....	28.18	7.93
Wright R-1750	9-cyl. radial	6 × 6.875	1,900	525	9	3.562	3.25	1.1	11.58	1,505	1,312	26.95	35,400	0.0205	.....	25.22	6.82

TABLE A1-6.—CRANKSHAFT MAIN BEARING DATA FOR IN-LINE AND V-ENGINES
(From Angle, "Engine Dynamics and Crankshaft Design," and *S.A.E. Jour.*, Vol. 29, Nos. 4 and 5, October, November, 1931)

Engine	Type	Bore and stroke, in.	Hp. and r.p.m.	Crankshaft bearing	Diameter, in.	Effective length, in.	L/D	Projected area, sq. in.	Rubbing velocity, f.p.s.	Bearing pressure, lb. per sq. in. Max.	Bearing pressure, lb. per sq. in. Mean	Rubbing factor	V/P
Liberty 6	6-cyl. vertical	5 × 7	240 at 1,800	Center	2.625	1.625	0.619	4.27	20.6	1,219	957	19,750	0.02155
				Intermediate	2.625	1.625	0.619	4.27	20.6	984	548	11,300	0.0376
				End	2.625	1.625	0.619	4.27	20.6	691	527	10,860	0.0391
Chevrolet Master (1938).	6-cyl. vertical	3.5 × 3.75	85 at 3,200	No. 1	2.6875	1.1875	0.441	3.185	37.6				
				No. 2	2.719	1.1875	0.436	3.22	38.0				
				No. 3	2.75	1.4375	0.522	3.95	38.4				
				No. 4	2.781	1.625	0.584	4.52	38.9				
Nash Ambassador Six (1938).	6-cyl. vertical	3.375 × 4.375	105 at 3,400	No. 1	2.485	1.25	0.502	3.11	36.9				
				No. 2	2.485	0.9375	0.377	2.33	36.9				
				No. 3	2.485	0.9375	0.377	2.33	36.9				
				No. 4	2.485	1.875	0.754	4.66	36.9				
				No. 5	2.485	0.9375	0.377	2.33	36.9				
				No. 6	2.485	0.9375	0.377	2.33	36.9				
				No. 7	2.485	1.625	0.654	4.04	36.9				
Hispano 300	8-cyl. V	5.51 × 5.9	325 at 1,800	Center	2.52	1.893	0.75	4.765	19.8	1,365	989	19,590	0.0200
				Intermediate	2.52	1.893	0.75	4.765	19.8	1,269	512	10,140	0.0386
				End	2.52	4.255	1.69	10.715	19.8	350	259	5,130	0.0765
Wright H-2	8-cyl. V	5.51 × 5.9	406 at 2,000	Center	2.52	1.893	0.75	4.765	22.0	1,480	1,072	23,585	0.0204
				Intermediate	2.52	1.893	0.75	4.765	22.0	1,450	581	12,782	0.0378
				End	2.52	4.255	1.69	10.715	22.0	361	291	6,402	0.0755
Cadillac V-8 (1938)	8-cyl. V	3.5 × 4.5	135 at 3,400	No. 1	2.5	1.0625	0.425	2.66	37.05				
				No. 2	2.5	1.1562	0.461	2.39	37.05				
				No. 3	2.5	1.969	0.786	4.91	37.05				
Curtiss D-12	12-cyl. V	4.5 × 6	425 at 2,000	Center	3.0	1.781	0.595	5.34	26.18	964	734	19,200	0.0357
				Intermediate	3.0	1.531	0.51	4.563	26.18	1,040	561	14,680	0.0465
				End	3.0	1.531	0.51	4.563	26.18	608	178	12,500	0.0547
Liberty 12	12-cyl. V	5 × 7	420 at 1,700	Center	2.625	1.625	0.619	4.27	19.45	1,580	1,165	22,650	0.0167
				Intermediate	2.625	1.625	0.619	4.27	19.45	1,150	620	14,000	0.0270
				End	2.625	1.625	0.619	4.27	19.45	845	623	12,100	0.0312
Packard 2500	12-cyl. V	6.375 × 6.5	800 at 2,000	Propeller end	3.5	2.25	0.643	7.875	30.55	728	555	16,950	0.0550
				Intermediate	3.5	1.5	0.429	5.25	30.55	1,785	982	30,000	0.0311
				Center	3.5	2.5	0.715	8.75	30.55	1,206	908	27,750	0.0336
				Rear end	3.5	1.437	0.41	5.031	30.55	1,139	868	26,517	0.0352
Curtiss V-1570	12-cyl. V	5.125 × 6.25	....	Center	3.5	1.594	0.455	5.39	36.7	1,754	1,377	50,500	0.0266
				Intermediate	3.5	1.344	0.384	4.52	36.7	1,504	948	34,700	0.0387
				End	3.5	1.344	0.384	4.52	36.7	1,112	863	31,600	0.0425

TABLE A1-7.—RELATIVE CRANKPIN-BEARING LOADS WITH VARIOUS ARRANGEMENTS OF 6- BY 6.875-IN. CYLINDERS
(From *S.A.E. Jour.*, Vol. 29, No. 5, November, 1931)

Cylinders per crankpin	Arrangement	Displacement, cu. in.	Weight per crankpin		Bearing load, lb.		Force per cu. in. of piston displacement, lb.		Required bearing area based on mean load, %
			Rotating, lb.	Reciprocating, lb.	Max.	Mean	Max.	Mean	
1		195	6.00	6.82	7,880	3,740	40.4	19.2	100
2	60- or 45-deg. V	390	8.00	13.64	8,140	5,730	20.9	14.7	153
3	40-deg. W	585	10.00	20.46	10,100	6,790	17.3	11.6	181
3	60-deg. W	585	10.00	20.46	13,050	8,020	22.3	13.7	214
3	80-deg. W	585	10.00	20.46	9,960	7,630	17.0	13.0	204
3	Radial	585	10.00	20.46	10,230	6,860	17.5	11.7	183
5	Radial	975	14.00	34.10	11,880	8,920	12.2	9.2	238
7	Radial	1365	18.00	47.74	13,740	11,520	10.1	8.5	308
9	Radial	1755	22.00	61.38	16,320	14,080	9.3	8.0	376

TABLE A1-8.—FIELD OF USEFULNESS FOR VARIOUS BEARING METALS
(From *S.A.E. Jour.*, Vol. 45, No. 6, December, 1939)

Description of bearing metal	Max. permissible unit pressure, lb. per sq. in.	Min. permissible Zn/P_{max}	Maximum $P_{max}V$	Oil reservoir temperature, deg. F.	Minimum crankshaft hardness	Affected by corrosion
Tin-base babbitt: Copper........... 3.50% Antimony........ 7.50% Tin.............. 89.00 Lead (max.)...... 0.25%	1,000	20	35,000	235	Not important	No
		Standard quality bearings				
Tin-base babbitt: Same composition as above	1,500	15	42,500	235	Not important	No
		Alpha process quality bearings				
High-lead babbitt: Tin............. 5-7% Antimony....... 9-11% Lead............ 82-86% Copper (max.).... 0.25%	1,800	10	40,000	225	Not important	No
Cadmium-silver: Silver............ 0.75% Copper........... 0.50% Cadmium......... 98.75%	Over 1,800 and up to 3,850	3.75	90,000 and upwards	260	250 Brinell	Not likely if temperature is maintained as specified and proper lubricating oil is used
Copper-lead: Copper............ 60% Lead.............. 40%	Over 1,800	3.75	90,000 and upwards	260	300 Brinell	

TABLE A1-9.—PRINCIPAL DIMENSIONS (IN INCHES) OF SIX DIFFERENT AIRCRAFT-ENGINE CRANKSHAFTS
(From *S.A.E. Jour.*, Vol. 28, No. 4, April, 1931)

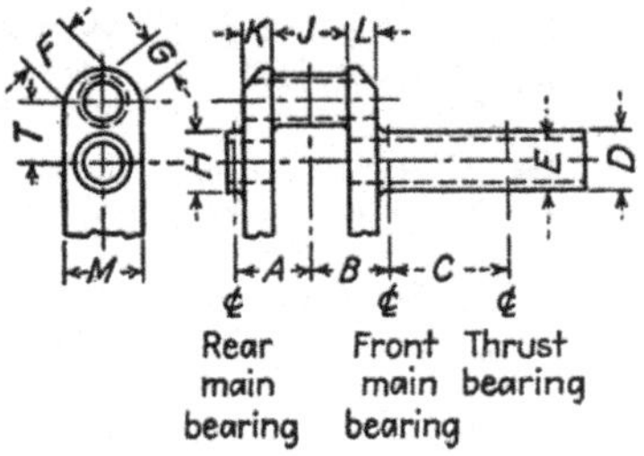

Dimension	Crankshaft number					
	1	2	3	4	5	6
A	3.3125	4.125	3.5	3.75	4.125	4.25
B	3.3125	4.125	3.5	3.75	4.125	4.25
C	6	3	5.5	5.75	7.75	5.75
D	2.5625	3	2.875	2.875	3.25	3.875
E	1.375	1.9375	1.625	1.625	1.75	3.125
F	2	2.25	2.5	2.5	2.75	3.25
G	1	1.4375	0.75	0.75	1	2.5625
H	2.75	3.25	3.125	3.125	3.125	3.5625
J	3.75	3.75	3.375	3.75	4.25	3.9375
K	1.3125	1.4375	1.1875	1.3125	1.375	1.6875
L	1.3125	1.125	1.25	1.375	1.25	1.3125
M 1	2.625	3.5	3.75	3.75	3.875	4.75
M 2	2.375	3.125				4.125
T	2.75	2.75	2.4375	2.875	3.1875	3.4375

1. Front.
2. Rear.

TABLE A1-10.—RADIAL-ENGINE BEARING REACTIONS FOR MAXIMUM RADIAL GAS LOAD ON CRANKPIN (1) AT RATED POWER AND (2) AT RATED POWER COMBINED WITH INERTIA LOAD AT RATED SPEED (Average Values Only)
(From *S.A.E. Jour.*, Vol. 28, No. 4, April, 1931)

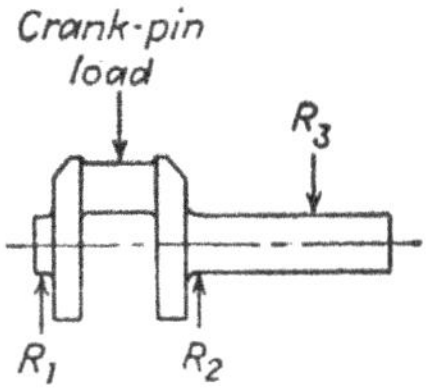

Crank-shaft number	Crank-pin load, lb.	R_1		R_2		R_3	
		lb.	%	lb.	%	lb.	%
1	8,440	3,380	40	6,000	71	930	11
2	10,400	3,740	36	9,100	87.5	2,440	23.5
3	11,200	4,260	38	8,680	77.5	1,740	15.5
4	13,800	5,240	38	10,800	78	2,200	16
5	15,650	6,260	40	11,100	71	1,720	11
6	15,000	5,920	39.5	11,600	77	2,480	16.5
Average (1)		.....	38.6	.....	77	.....	15.6
Variation, % of mean value		.....	+3.6 −6.7	.....	+13.6 − 7.8	.....	+51 −29
Average (2)		.....	40.6	.....	72.6	.....	12.9

TABLE A1-11.—CHARACTERISTICS AND DIMENSIONS OF CURTISS V-1570 CONQUEROR ENGINE

(From *S.A.E. Jour.*, Vol. 29, No. 4, October, 1931)

Number of cylinders	12
Arrangement of cylinders	Two banks at 60-deg. V
Method of numbering cylinders	6, 5, 4, 3, 2, 1—Right
Propeller end	6, 5, 4, 3, 2, 1—Left
Firing order, crankshaft rotation clockwise facing rear of engine	1L, 6R, 5L, 2R, 3L, 4R, 6L, 1R, 2L, 5R, 4L, 3R
Bore, in	5.125
Stroke of master-rod cylinder ($2R$), in	6.250
Stroke of articulated-rod cylinder, in	6.430
Piston area, sq. in	20.63
Total piston displacement, cu. in	1,569.5
Brake horsepower	630
Speed, r.p.m	2,400
Compression-ratio, average	5.80:1
Mechanical efficiency, %	89.4
Brake mean effective pressure, lb. per sq. in	132.4
Indicated mean effective pressure, lb. per sq. in	148.1
Master connecting-rod length, center to center (L), in	10.0
Master connecting rod to crank ratio (L/R), in	3.20
Articulated-rod length, in	7.594
Link-pin radius (R_1), in	2.406
Angle between link-pin radius and master-rod center line (α_1), deg. min	66–30
Master rods are assembled in the left cylinder bank	
Valve timing:	
Inlet valve opens, deg. before top dead center	5
Inlet valve closes, deg. after bottom dead center	55
Exhaust valve opens, deg. before bottom dead center	60
Exhaust valve closes, deg. before top dead center	0
Valve-tappet clearance, intake and exhaust, in	0.014–0.016
Magneto timing:	
Left magneto advance, deg	33
Right magneto advance, deg	38
Reciprocating and rotating weights:	
Reciprocating weight per cylinder of master rod, lb	4.11
Piston, complete with rings and pin, lb	2.97
Upper end of master connecting rod, lb	1.14
Reciprocating weight per cylinder of articulated rod, lb	3.90
Upper end of articulated connecting rod, lb	0.93
Rotating weight of crankpin, lb	5.36
Lower end of master connecting rod, lb	4.53
Lower end of articulated connecting rod, lb	0.83
Crankpin bearing:	
Diameter, in	2.500
Length, total, in	1.500

TABLE A1-11.—CHARACTERISTICS AND DIMENSIONS OF CURTISS V-1570 CONQUEROR ENGINE.—(*Continued*)

Length, effective, in	1.391
Effective projected bearing area, minus oil groove, sq. in	3.28
Crankshaft end and intermediate bearings:	
Diameter, in	3.500
Length, total, in	1.500
Length, effective, in	1.344
Effective projected bearing area, sq. in	4.52
Crankshaft center bearing:	
Diameter, in	3.500
Length, total, in	1.750
Length, effective, in	1.594
Effective projected bearing area, sq. in	5.39
Crankshaft:	
Diameter of journal, in	3.500
Bore through journal, in	2.750
Diameter of crankpin, in	2.500
Length of crankpin, in	1.920
Bore through crankpin, in	1.250
Crankpin fillets, in	0.250
Journal fillets, in	0.187
Width of crank cheek at top of journal, in	3.790
Thickness of crank cheek, in	0.987
Distance between end and intermediate crankpin centers, in	5.750
Distance between center crankpin centers, in	6.000

TABLE A1-12.—ENGINE CHARACTERISTICS AND DIMENSIONS OF WRIGHT R-1750 CYCLONE ENGINE

(From *S.A.E. Jour.*, Vol. 29, No. 4, October, 1931)

Number of cylinders	9
Arrangement of cylinders	Radial
Numbering of cylinders	1–9 consecutively, clockwise facing rear of engine, No. 1 vertical and on top
Firing order	1, 3, 5, 7, 9, 2, 4, 6, 8
Crankshaft rotation	Clockwise facing rear of engine
Bore, in	6.000
Stroke of master-rod cylinder ($2R$), in	6.875
Piston area (Ap), sq. in	28.27
Total piston displacement, cu. in	1,750
Brake horsepower	525
Speed, r.p.m	1,900
Compression ratio	5.0:1
Mechanical efficiency, assumed, %	90
Brake mean effective pressure, lb. per sq. in	125.0
Indicated mean effective pressure, lb. per sq. in	139.0
Master connecting-rod length, center to center (L), in	13.750
Master connecting-rod to crank ratio (L/R)	4.000
Articulated or link-rod length, in	11.046
Master rod is assembled in cylinder 7	

Table A1-12.—Engine Characteristics and Dimensions of Wright R-1750 Cyclone Engine.—(*Continued*)

Valve timing:	
Inlet valve opens, deg. before top dead center	25
Inlet valve closes, deg. after bottom dead center	60
Exhaust valve opens, deg. before bottom dead center	80
Exhaust valve closes, deg. after top dead center	25
Valve tappet clearance, in	0.050
Spark advance, deg. before top dead center	30
Supercharger:	
Type	Geared centrifugal
Impeller speed	8 times crankshaft
Reciprocating and rotating weights:	
Reciprocating weight per cylinder of master rod, lb	7.45
Piston, complete with rings and pin, lb	5.34
Upper end of master connecting rod, lb	2.11
Reciprocating weight per cylinder of link rod, lb	6.74
Upper end of link connecting rod, lb	1.40
Rotating weight at crankpin (W_c), lb	25.22
Lower end of master connecting rod, lb	15.62
Lower end of link connecting rod, lb	1.20
Crankpin-bearing dimensions:	
Diameter, in	3.250
Length, total, in	3.906
Length, effective, in	3.562
Effective projected bearing area, sq. in	11.58
Front main bearing:	
Construction	Steel sheet lined with high-load bronze
Diameter, in	4.375
Effective length, in	1.687
Effective projected bearing area, sq. in	7.38
Rear main bearing:	
Type	Commercial Hoffman R-190-LL or SKF light series roller bearing No. 8216-C
Inner diameter, in	3.5433
Mm	90
Outside diameter, in	62,992
Mm	160
Width, in	1.181
Mm	30
Front thrust bearing:	
Type	Commercial Standard S.A.E. light series ball bearing No. 218
Inner diameter, in	3.3433
Mm	90
Outside diameter, in	6.2992
Mm	160
Width, in	1.181
Mm	30

TABLE A1-13.—RELATION OF CRUISING TO TAKE-OFF HORSEPOWER AND R.P.M. FOR ELEVEN AMERICAN AIRCRAFT ENGINES

Cruising hp.	Take-off hp.	Cruising r.p.m.	Take-off r.p.m.	$\frac{\text{Cruising hp.}}{\text{Take-off hp.}}$ %	$\frac{\text{Cruising r.p.m.}}{\text{Take-off r.p.m.}}$ %
30	40	2,300	2,575	75	89.5
175	225	2,000	2,175	78	92
190	250	2,000	2,200	76	91
160	200	1,750	2,000	80	87.5
200	252	1,900	2,050	79.5	92.8
75	100	1,650	1,810	75	91
94	125	1,725	1,925	75	89.5
120	160	1,775	1,975	75	90
120	160	1,650	1,850	75	89
160	210	1,700	1,900	76	89.5
110	160	2,050	2,260	69	90.5

TABLE A1-14.—AIRCRAFT-ENGINE PISTON DATA

(See page 396 for values)

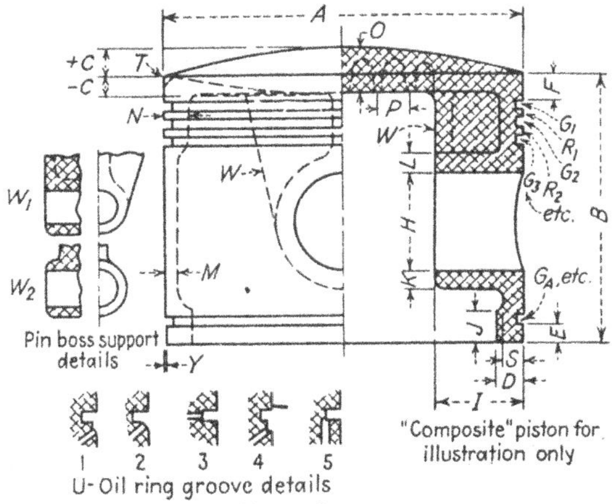

Nomenclature

A—Diameter.
B—Length (not including crown).
C—Crown or cup.
D—Thickness of bottom flange.
E—Distance from bottom of skirt to bottom of lowest groove.
F—Width of top land.
G_A (etc.)—Width and depth of lower ring grooves (G_1—upper).
G_L—Groove containing holes for lubricant.
H—Diameter of wrist-pin hole.
I—Length of wrist-pin hole.
J—Height of bottom flange.
K—Thickness of boss, bottom.
L—Thickness of boss, top.
M—Thickness of skirt above bottom flange.
N—Thickness of wall at upper ring grooves.
O—Thickness of head (to bottom of ribs on ribbed pistons).
P—Spacing of ribs.
Q—Depth of ribs.
R—Width of lands between grooves.
S—Wall thickness under boss.
T—Beveled 60 deg. to crown (about ¼ in. wide).
U—Type of lubrication holes.
V—Full floating piston pin.
W—Web support for pin bosses.
X—Number of rings per groove.

TABLE A1-14.—AIRCRAFT-ENGINE PISTON DATA.—(*Continued*)

(See page 395 for nomenclature)

Item	Piston numbers*							
	1	2	3	4	5	6	7	8
Material	Al	S.A.E. 321	Al	Y-alloy	Lynite	Lynite	S.A.E. 321	Al
Weight	3 lb. 5.5 oz.	4 lb. 12.4 oz.	Alloy 4 lb. 14.4 oz.	4 lb. 2.4 oz.	3 lb. 8.75 oz.	…	1 lb. 13.5 oz.	2.25 lb.
Volume, cu. in.	33.6	49.35	48.28	41.35	33.68	…	19.04	
Density, lb. per cu. in.	0.102	0.97	0.102	0.1	0.105	0.105	0.097	
A	5.59	6.1	6.1	5.95	4.97	4.98	4.5	4.6
B	3.92	4.0	3.92	3.95	5.1	4.3	3.2	3.47
C	−0.8	0	+0.64	0	+0.4	0	0	0
D	0.1	0.4	0.35	0.45	0.15	0.4	0.4	0.125
E	0.88	0.19	0.22	0.1	…	0.3	0.15	
F	0.35	0.47	0.22	0.25	0.4	0.3	0.2	0.35
G_1	0.13 × 0.22	0.13 × 0.22	0.2 × 0.2	0.2 × 0.2	0.25 × 0.19	0.25 × 0.19	0.125 × 0.16	0.15 × 0.20
G_2	0.13 × 0.22	0.13 × 0.22	0.2 × 0.2	0.2 × 0.2	0.25 × 0.19	0.25 × 0.19	0.125 × 0.16	0.15 × 0.20
G_3	0.2 × 0.22	0.13 × 0.22	…	0.2 × 0.2	0.25 × 0.19	0.25 × 0.19	0.125 × 0.16	0.15 × 0.20
C_4	0.2 × 0.22	…	…	…	…	…	…	0.15 × 0.20
C_L	G_4, G_A	G_3, G_A	G_2, G_A	G_3	G_3	None	G_A	G_4
G_A	0.2 × 0.22	0.13 × 0.22	0.125 × 0.2	0.2 × 0.2	…	0.25 × 0.2	0.125 × 0.16	
H	1.25	1.5	1.5	1.5	1.25	1.25	1.2	1.10
I	1.5	1.7	1.9	1.85	1.05	1.13	1.0	1.40
J	…	0.5	0.35	0.4	…	0.65	0.35	0.18
K	0.35	0.4	0.25	0.20	0.25	0.25	0.15	0.1875
L	…	…	…	…	0.25	0.25	…	0.1875
M	Ribbed	0.15	0.1	0.12	Tapered	0.25	0.11	0.125
N	0.5	0.5	0.4	0.4	0.7–0.4	0.8–0.6	0.5	0.50
O (P, H, R)†	0.25(H)	1.0(R)	1.0(H)	1.15(H)	0.4(P)	0.45(1 R)	0.85(1R)	0.32(R)
P	0.8 × 0.86	0.4	0.6 × 0.7	0.6 × 0.7	…	…	0.85	0.375
Q	0.375–0.5	0.4–0.5	0.7	0.85	…	…	0.6	0.30
R_1	0.1	0.15	0.17	0.125	0.125	0.125	0.08	0.125
R_2	0.10	0.15	…	0.125	0.125	0.125	0.10	0.125
R_3	0.10	…	…	…	…	…	…	0.125
S	0.4	0.4	0.35	0.15	0.2	0.25	0.125	0.125
T	…	Yes	…	…	0.125 Rad	…	…	0.125
U	3	2,3	1,5	3	3,4	None	3	3
V	Yes	Yes	Yes	Yes	No	No	Yes	Yes
W	W_1	W_1	W_1	W_1	…	W_2	W_1	W_1
X	1	1	2(G_A1)	2	1	1	1	1
Y	…	0.02	0.015	0.03	…	…	0.02	

* 1. Skirt cut away 0.6 in. at bosses. Well ribbed between bosses. 2 and 3. Valve indentations in crown. 4. Crown domed. 5. Skirt inset at pin bosses to center of boss. Pins are driven in and retainers fitted.

Weights do not include rings or pin.

All dimensions are in inches.

† P, H, R refers to type of crown (underside), P—plain, H—honeycombed, R—ribbed.

TABLE A1-15.—AIRCRAFT-ENGINE PISTON-PIN AND PISTON-RING DATA

Item	Piston-pin numbers*			
	2	4	5	7
A	6.1	5.9		
B	5.5	5.5	4.25	4.1
C	0.18	0.12		
D	1.5	1.5	1.25	1.2
E	1.1	1.23	0.82	0.95
F	1.12	1.25	0.82	1.0
G	1.1	1.1	....	0.7
H	0.9	0.95		
J	0.87	1.0		

* Piston-pin numbers correspond to piston numbers in Table A1-14.
All dimensions are in inches.

Item	Piston-ring numbers*									
	2	3	3	4	4	5	7	7	9	10
A	6.25	6.25	6.3	6.15	6.15	5.12	4.62	4.7	5.18	4.125
B	0.125	0.10	0.10	0.10	0.10	0.25	0.125	0.12	0.25	0.29
C	0.18	0.20	0.20	0.18	0.18	0.18	0.15	0.15	0.15	0.13
D	0.60	0.70	0.75	0.56	0.65	0.45	0.45	0.68	0.65	0.36
Section	*b*	*b*	*e*	*b*	*e*	*d*	*b*	*c*	*a*	*a*
Function†	*c*	*c*	*s*	*c*	*s*	*c*	*c*	*s*	*c*	*c*
Weight, oz	1.7	1.4	1.4	1.3	1.2	2.6	1.0	0.8	2.4	2.0

* Piston-ring numbers correspond to piston numbers in Table A1-14.
† *c* = compression ring. *s* = scraper or oil ring.
All dimensions are in inches.
Dimensions *A* and *D* are for ring free.

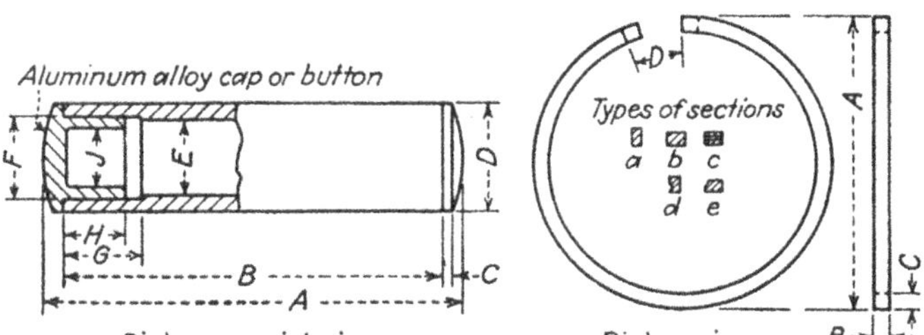

Piston or wrist pins   Piston ring

TABLE A1-16A.—PISTON-RING AND GROOVE WIDTHS
(From S.A.E. "Handbook")

Nom. ring width	All ring widths, in.			Ring-groove widths, in.							
				Oil-ring grooves				Top compression ring grooves only			
	All diam. max.	Under 6 in. diam.	6–8 in. diam.	2–4$\frac{11}{16}$ in. diam.		4$\frac{3}{4}$–8 in. diam.		2–4$\frac{11}{16}$ in. diam.		4$\frac{3}{4}$–8 in. diam.	
		Min.	Min.	Max.	Min.	Max.	Min.	Max.	Min.	Max.	Min.
$\frac{3}{32}$	0.0935	0.0930	0.0925	0.0955	0.0945	0.0960	0.0950	0.0960	0.0950	0.0965	0.0955
$\frac{1}{8}$	0.1240	0.1235	0.1230	0.1260	0.1250	0.1265	0.1255	0.1265	0.1255	0.1270	0.1260
$\frac{5}{32}$	0.1550	0.1545	0.1540	0.1570	0.1560	0.1575	0.1565	0.1575	0.1565	0.1580	0.1570
$\frac{3}{16}$	0.1865	0.1855	0.1855	0.1885	0.1875	0.1890	0.1880	0.1890	0.1880	0.1895	0.1885
$\frac{1}{4}$	0.2490	0.2485	0.2480	0.2510	0.2500	0.2515	0.2505	0.2515	0.2505	0.2520	0.2510

The piston-ring grooves provide for a minimum side clearance of 0.001 in. for cylinders under 4$\frac{3}{4}$ in. diameter and 0.0015 in. for cylinders of 4$\frac{3}{4}$ to 8 in. diameter. The greater clearance for the top compression ring is recommended only in order to give improved ring performance.

TABLE A1-16B.—RING WIDTHS FOR CYLINDER DIAMETERS

Cylinder diameter, in.	Ring width, in.				
	Compression rings	Oil rings			
2 –4$\frac{7}{16}$	$\frac{1}{8}$	$\frac{1}{8}$	$\frac{5}{32}$	$\frac{3}{16}$	
4$\frac{1}{2}$–5$\frac{7}{16}$	$\frac{5}{32}$	. .	$\frac{5}{32}$	$\frac{3}{16}$	$\frac{1}{4}$
5$\frac{1}{2}$–6$\frac{7}{16}$	$\frac{3}{16}$	. .	. . .	$\frac{3}{16}$	$\frac{1}{4}$
6$\frac{1}{2}$–8	$\frac{1}{4}$	. .	. . .	. . .	$\frac{1}{4}$

NOTE: Oil-ring widths or combinations of widths shall be selected from the widths specified between the diameters listed.

The accompanying specifications for piston rings and grooves have in general been used for some time and have been adopted as a standard primarily for the types of internal-combustion engines commonly used in automobiles, motorboats, etc. For pistons used in aircraft engines and engines not of the conventional automobile type, it may be necessary to deviate from the rings and grooves recommended in order to secure most satisfactory performance, but such modifications should not be made by changing the piston-ring width or radial thickness.

TABLE A1-16C.—PISTON-RING RADIAL WALL THICKNESS AND GROOVE DIAMETERS

Cylinder diameter, in.	Ring radial wall thickness, in., max.	Ring groove bottom diameter, in., max.			
		Cast-iron pistons		Aluminum pistons	
		Compression rings	Oil rings	Compression rings	Oil rings
$3\frac{1}{2}$	0.150	3.166	3.126	3.159	3.119
$3\frac{9}{16}$	0.150	3.228	3.188	3.221	3.181
$3\frac{5}{8}$	0.155	3.280	3.240	3.273	3.233
$3\frac{11}{16}$	0.155	3.342	3.302	3.335	3.295
$3\frac{3}{4}$	0.160	3.395	3.355	3.387	3.347
$3\frac{13}{16}$	0.160	3.457	3.417	3.449	3.409
$3\frac{7}{8}$	0.165	3.509	3.469	3.501	3.461
$3\frac{15}{16}$	0.165	3.571	3.531	3.563	3.523
4	0.165	3.634	3.594	3.626	3.586
$4\frac{1}{16}$	0.165	3.696	3.656	3.688	3.648
$4\frac{1}{8}$	0.165	3.758	3.718	3.750	3.710
$4\frac{3}{16}$	0.165	3.820	3.780	3.812	3.772
$4\frac{1}{4}$	0.170	3.873	3.833	3.864	3.824
$4\frac{5}{16}$	0.170	3.935	3.895	3.926	3.886
$4\frac{3}{8}$	0.175	3.987	3.947	3.978	3.938
$4\frac{7}{16}$	0.175	4.049	4.009	4.040	4.000
$4\frac{1}{2}$	0.180	4.102	4.062	4.093	4.053
$4\frac{9}{16}$	0.180	4.164	4.124	4.155	4.115
$4\frac{5}{8}$	0.180	4.226	4.186	4.217	4.177
$4\frac{11}{16}$	0.180	4.288	4.248	4.279	4.239
$4\frac{3}{4}$	0.185	4.341	4.301	4.331	4.291
$4\frac{13}{16}$	0.185	4.403	4.363	4.393	4.353
$4\frac{7}{8}$	0.190	4.455	4.415	4.445	4.405
$4\frac{15}{16}$	0.190	4.517	4.477	4.507	4.467
5	0.195	4.570	4.530	4.560	4.520
$5\frac{1}{16}$	0.195	4.632	4.592	4.622	4.582
$5\frac{1}{8}$	0.195	4.694	4.654	4.684	4.644
$5\frac{3}{16}$	0.195	4.756	4.716	4.746	4.706
$5\frac{1}{4}$	0.200	4.809	4.769	4.798	4.758
$5\frac{5}{16}$	0.200	4.871	4.831	4.860	4.820
$5\frac{3}{8}$	0.205	4.923	4.883	4.912	4.872
$5\frac{7}{16}$	0.205	4.985	4.945	4.974	4.934
$5\frac{1}{2}$	0.210	5.038	4.998	5.027	4.987
$5\frac{9}{16}$	0.210	5.100	5.060	5.089	5.049
$5\frac{5}{8}$	0.215	5.152	5.112	5.141	5.101
$5\frac{11}{16}$	0.215	5.214	5.174	5.203	5.163
$5\frac{3}{4}$	0.220	5.267	5.227	5.255	5.215
$5\frac{13}{16}$	0.220	5.329	5.289	5.317	5.277
$5\frac{7}{8}$	0.225	5.381	5.341	5.369	5.329
$5\frac{15}{16}$	0.225	5.443	5.403	5.431	5.391
6	0.230	5.496	5.456	5.484	5.444
$6\frac{1}{16}$	0.230	5.558	5.518	5.546	5.506
$6\frac{1}{8}$	0.235	5.610	5.570	5.598	5.558
$6\frac{3}{16}$	0.235	5.672	5.632	5.660	5.620
$6\frac{1}{4}$	0.240	5.725	5.685	5.712	5.672
$6\frac{5}{16}$	0.240	5.787	5.747	5.774	5.734
$6\frac{3}{8}$	0.245	5.839	5.799	5.826	5.786
$6\frac{7}{16}$	0.245	5.901	5.861	5.888	5.848

TABLE A1-16D.—RING JOINTS AND DRAIN HOLES

(The following data were adopted as recommended practice only.)

Ring-joint Clearance.—Rings having maximum radial wall thickness as recommended should have a free joint opening of approximately $D/6.75$ to permit assembling them without overstressing individual castings having a mean tensile strength of about 28,000 lb. per sq. in., as piston-ring joint clearance must be determined from the minimum cylinder diameter. It is recommended that joint clearance for rings be as follows:

Ring-joint Clearances

Cylinder Diameter, In.	Joint Clearance, In.
2–4 15/16	0.007–0.017
5–8	0.010–0.020

Number of Oil-ring Groove Drain Holes

Cylinder Diameter, In.	Number of Holes
2 –2 1/2	8
2 9/16 –2 15/16	10
3 –3 7/16	12
3 1/2 –3 15/16	14
4 –4 1/2	12
4 9/16 –5 7/16	14
5 1/2 –5 15/16	16
6 –6 5/8	14
6 11/16–7 15/16	16
8	17

Size of Oil-ring Groove Drain Hole

Nominal Ring Width, In.	Drill Hole Diameter, In.
1/8	3/32
5/32	1/8
3/16	5/32
1/4	7/32

TABLE A1-17.—S.A.E. STANDARD DIMENSIONS FOR CONNECTING-ROD BOLTS
(From S.A.E. "Handbook")

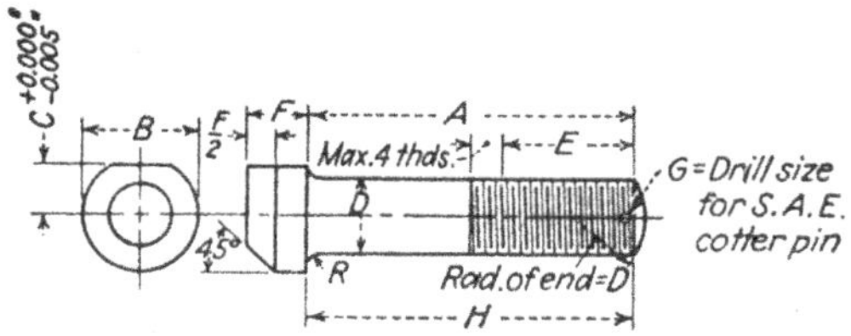

Diameter	A	B	C	D	Threads per inch NF-3	E*	F	G No.	H length minus	R
5/16	Lengths vary by even 1/8 in., preferably by 1/4 in.	3/16	3/16	0.3125 0.3105	24	1/2	3/16	48	3/32	
3/8		5/8	7/32	0.3750 0.3730	24	5/8	7/32	36	9/64	0.01–1/32
7/16		11/16	1/4	0.4375 0.4355	20	11/16	1/4	36	9/64	
1/2		3/4	9/32	0.5000 0.4975	20	3/4	9/32	36	9/64	
9/16		7/8	5/16	0.5625 0.5600	18	7/8	5/16	28	3/16	1/64–3/64
5/8		1	3/8	0.6250 0.6220	18	1	3/8	28	3/16	
3/4		1 1/8	7/16	0.7500 0.7470	16	1 1/8	7/16	28	3/16	
7/8		1 3/8	1/2	0.8750 0.8715	14	1 1/4	1/2	28	3/16	1/32–1/16
1		1 1/2	9/16	1.0000 0.9965	14	1 1/2	5/8	28	3/16	

* Minimum length of usable threads.
Recommended practice for material is S.A.E. steel 2330 or 3130.
Head-treatment should give a Brinell test of 223 to 285.
From the report of the Engine Division, adopted by the Society, August, 1920.
Revised by Gasoline Engine Division, January, 1941.

TABLE A1-18.—AIRCRAFT-ENGINE LINK-ROD DATA

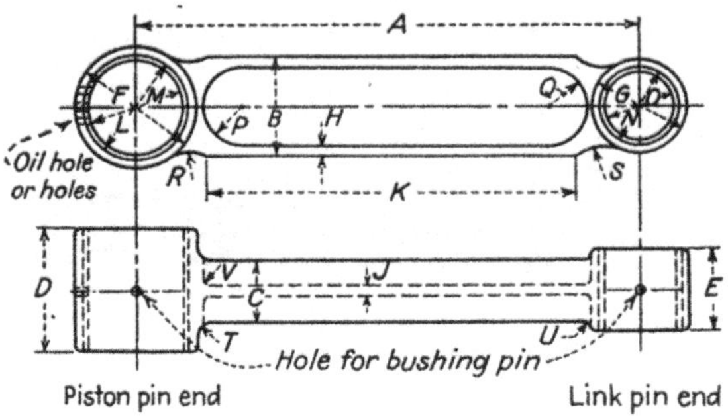

Item	Link-rod number			
	1	2	3	4
A	11.05	10.20	9.00	
B	1.35	1.55	0.85	
C	1.05	0.875	1.00	
D	1.98	2.00	1.55	
E	1.48	1.35	1.55	
F	1.90	1.70	1.50	
G	1.50	1.45		
H	0.14	0.12	0.10	
J	0.125	0.125	0.125	
K	9.30	7.55		
L	1.65	1.53	1.25	
M	1.50	1.37	1.10	
N	1.23	1.23	1.15	
O	1.07	1.07	0.90	
P	0.531	0.675	0.32	
Q	0.531	0.675	0.32	
R	Fillet	1.25	1.00	
S	0.625	1.50	1.00	
T	0.1875	0.125	0.25	
U	0.25	0.125	0.25	
V	0.125	0.125	0.50	
Weight, including bushing, lb	2.75	2.1875		
Weight piston-pin end, lb	1.438	1.1875		
Weight link-pin end, lb	1.312	1.00		
Number oil holes	1	2		

All dimensions in inches.

TABLE A1-19.—PROPELLER HUBS AND SHAFT ENDS, AIRCRAFT, SPLINE TYPE

S.A.E. Recommended Practice

(From S.A.E. "Handbook")

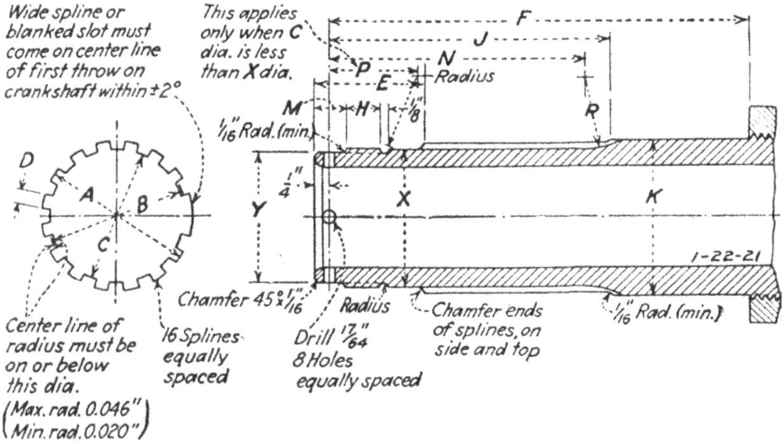

SHAFT ENDS

Propeller Shaft Ends

S.A.E. shaft No.	A +0.000 −0.002	B max.	C min.	D ±0.0008	E	F ±0.015 (extended)	F ±0.015	Thread: Size and threads	Thread: Pitch diam. +0.000 −0.003
10	1.992	1.781	1.689	0.1940	2 5/16	.....	.....	1 11/16—12	1.631
20	2.367	2.156	2.064	0.2310	2 5/16	7.875	6.875	2 1/16 —12	2.006
30	2.617	2.406	2.314	0.2570	2 5/16	8.243	.....	2 5/16 —12	2.256
40	3.117	2.875	2.783	0.3040	2 11/16	7.906	.....	2 13/16—12	2.756
50	3.804	3.554	3.462	0.3750	2 15/16	8.562	.....	3 7/16 —12	3.381

S.A.E. shaft No.	H	J ±0.015	K +0.000 −0.002	M	N ±0.030	R	X +0.000 −0.002	Y	P
10	1 3/16	.....	2.000	...	...	...	1.687	1 9/16	
20	1 3/16	5.781	2.375	9/16	5 1/4	1 1/2	2.062	1 15/16	
30	1 3/16	6.156	2.625	9/16	5 5/8	1 1/2	2.312	2 3/16	
40	1 3/16	5.781	3.125	9/16	5 1/4	1 1/2	2.812	2 11/16	2 1/8
50	1 1/16	6.500	3.812	1/2	6	1 1/2	3.500	3 5/16	2 3/8

Diameters *A*, *K*, and *X* shall be concentric with each other within 0.0003 in. total indicator reading before splining operation.

American Standard 12 pitch threads.

TABLE A1-20.—SHAFT END, TAPER TYPE
(From S.A.E. "Handbook")

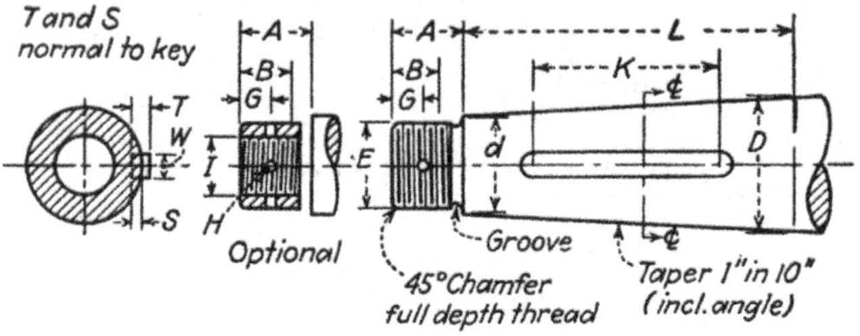

Taper Shaft End

S.A.E. shaft No.	Taper			Key				Locking holes		
	L	*D*	*d*	*K*	*W* +0.0000 −0.0005	*T* +0.000 −0.007	*S* +0.010 −0.000	*G*	*H*	Number
00	3	1.250	0.950	1⅝	0.2495	0.250	0.154	3/16	5/32	1
0	3⅝	1.875	1.512	2¼	0.3750	0.278	0.154	¼	7/32	4
1	5 5/32	2.050	1.535	3	0.3750	0.278	0.154	15/32	13/64	5
2	7	2.362	1.662	5 7/16	0.4730	0.237	0.143			

The taper (included angle) should vary from absolute uniformity by being 0.000 to 0.001 in. larger at large end.

Taper Shaft End Threads

S.A.E. shaft No.	Internal thread (optional)			External thread		
	B	*I*	Pitch diam., min.	*A**	*E*	Pitch diam., min.
00	None	None	None	1	¾" —16	0.7094
0	⅝	⅞" —18	0.8390	1 3/16	1⅜" —18	1.3390
1	⅞	15/16"—24	0.9104	1 1/16	1½" —18	1.4640
2	1 1/16	1" —14	0.9536	1¼	1 9/16"—12	1.5084

Thread Form, American (National) Standard.

* Number 00 shaft end is designed for use of standard S.A.E. ¾-in. castle nut. All other sizes require special nuts.

TABLE A1-21.—SPLINES FOR SOFT BROACHED HOLES IN FITTINGS

S.A.E. STANDARD

(From S.A.E. "Handbook")

The accompanying splines have become established as basic standard for a great many applications in the automotive, machine-tool, and other industries, since they were adopted originally by the Society in 1914.

The dimensions, given in inches, apply only to soft broached holes. The shaft dimensions depend upon the shape and material of the parts, their heat-treatment, methods of machining, etc., to give the required fit. The method and amount of "breaking" sharp corners and edges also depend upon the conditions and requirements of each application.

The tolerances allowed are for good construction and may be readily maintained by usual broaching methods. The tolerances selected for the large and small diameters will depend upon whether the fit between the mating parts, as finally made, is on the large or the small diameter. The other diameter, being designed for clearance, may have a wider manufacturing tolerance. If the final fit between the parts is only on the sides of the spline, wider tolerances may be permitted on both the large and small diameters.

The formula for theoretical torque capacity (pressure on sides of spline) in inch-pounds per inch of bearing length and at 1,000 lb. pressure per square inch, is given in footnotes following the table for each type of spline.

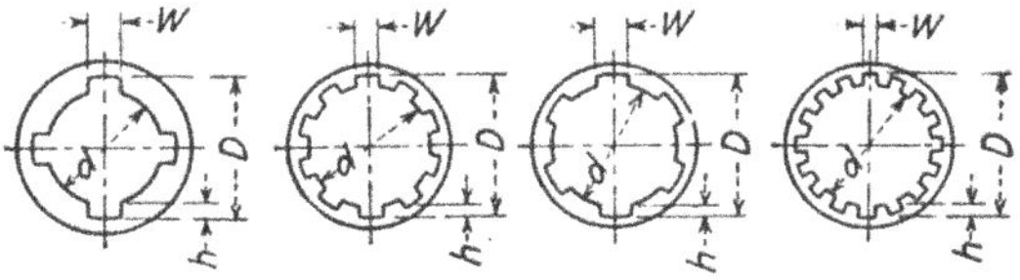

Formulas for W, h, and d, in terms of Large Diameter, D

No. of splines	W, for all fits	A permanent fit		B to slide not under load		C to slide under load	
		h	d	h	d	h	d
4	$0.241D$*	$0.075D$	$0.850D$	$0.125D$	$0.730D$		
6	$0.250D$	$0.050D$	$0.900D$	$0.075D$	$0.850D$	$0.100D$	$0.800D$
10	$0.156D$	$0.045D$	$0.910D$	$0.070D$	$0.860D$	$0.095D$	$0.810D$
16	$0.098D$	$0.045D$	$0.910D$	$0.070D$	$0.860D$	$0.095D$	$0.810D$

* Four splines, for fits A and B only.

Radii on corners of splines not to exceed 0.015 in.

Splines small not be more than 0.006 in. per ft. out of parallel with respect to the axis of the shaft.

No allowance is made for radii on corners or for clearness. Dimensions are intended to apply to only the soft broached hole. Allowance must be made for machining.

TABLE A1-22.—BALL-BEARING SELECTION[1,2]

All standard ball bearings are made to internationally agreed upon dimensions for bore, outer diameter, and width and in three series, *i.e.*, light, medium, and heavy. The American edition of these standards is published in the S.A.E. "Handbook." Detailed dimensions may vary with different manufacturers, but over-all S.A.E. standard dimensions are adhered to. Hence, the following data may be used to select the size of bearing needed. Final approval of the selection should be obtained from the manufacturer of the bearing.

In the following subdivisions of this table, the first digit in the bearing number refers to the type of New Departure bearings, the second identifies the series, and the third and fourth, taken together, are the bearing bore number, which is such that, multiplied by 5, it gives the bore diameter in millimeters, except for the small bores 0, 1, 2, and 3. In the S.A.E. "Handbook," the serial and bore number is used as the bearing identifying number for single-row radial bearings. For example, in bearing 1309, the number 1 indicates that the bearing is a New Departure single-row radial filling-notch type; 3 indicates that the bearing is of medium series; 09 indicates that the bore number is 9, and as $9 \times 5 = 45$, it also indicates that the bore diameter is 45 mm.; 309 is the S.A.E. number for this bearing.

The following methods of selection are adapted from recommended practice in the New Departure "Handbook," 12th edition. It is assumed that the loads and speeds are known.

Let L = calculated radial load on bearing, lb.
T = calculated thrust load on bearing, lb.
n = r.p.m. of shaft through bearing (= r.p.m. of inner ring of bearing).
F = radial equivalent conversion factor (Table A1-22A).
Z = life modifier (Table A1-22B).
M = the speed correction for rotating outer ring (M = 1.46 for the light, 1.61 for the medium, and 1.74 for the heavy series bearings in this table).
K = shock-load correction factor (Table A1-22C).
C = radial or equivalent radial capacity, lb.

For bearings under radial load,

$$C = L \times Z \times K$$

For bearings under thrust and radial load,

$$C = L \times F \times Z \times K$$

For bearings under pure thrust,

$$C = T \times F \times Z \times K$$

For a rotating inner ring (the usual case), locate the value of C found by the preceding methods in the proper speed (n) column of Table A1-22E or

[1] New Departure "Handbook."
[2] S.A.E. "Handbook."

A1-21G as applies (for filling-notch bearings); or Table A1-22I or' A1-22K as applied (for nonfilling-notch bearings). Then read across to the left-hand column of the table for the corresponding New Departure (or S.A.E.) bearing number. Enter this bearing number in Table A1-22D, A1-22F, A1-22H, or A1-22J as applies, and read the over-all bearing dimensions as indicated. An alternate last step is to enter the last three digits (S.A.E. bearing number) in the S.A.E. "Handbook" bearing tables and read the bearing dimensions therein.

If the outer ring of the bearing is rotating, multiply the speed n by the speed correction factor M and use the product to locate C in the load tables.

TABLE A1-22A.—COMBINED LOAD FACTORS F, FOR CONVERSION TO RADIAL EQUIVALENT

(From New Departure "Handbook")

T/L	Single-row filling-notch N.D. type 1000	Single-row nonfilling-notch N.D. type 3000
0.05	0.99	0.99
0.10	1.00	0.99
0.15	1.02	0.99
0.20	1.04	1.00
0.25	1.06	1.00
0.30	1.10	1.01
0.35	1.14	1.02
0.40	1.19	1.04
0.45	1.24	1.06
0.50	1.30	1.09
0.60	....	1.14
0.70	....	1.21
0.80	....	1.28
0.90	....	1.35
1.00	....	1.44
1.25	....	1.66
1.50	....	1.90
1.75	....	2.17
2.00	....	2.45
3.00	....	3.62
4.00	....	4.65
5.00	....	5.63
7.50	....	8.07
10.00	....	10.57
Pure thrust	....	1.00

TABLE A1-22B.—RADIAL LOAD LIFE MODIFIERS, Z; FOR GIVING DESIRED BEARING LIFE

(Based on data from New Departure "Handbook")

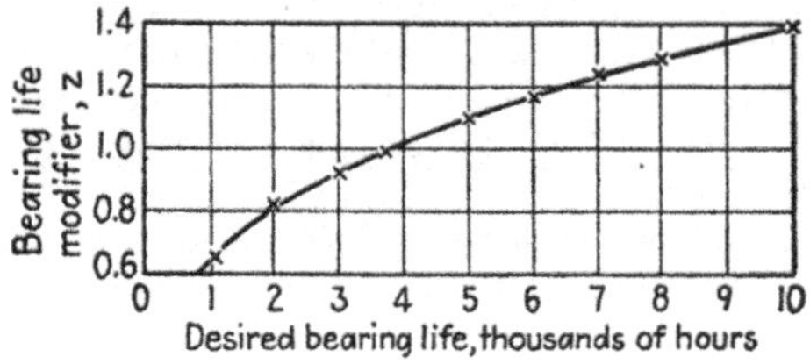

TABLE A1-22C.—SHOCK-LOAD FACTORS

(From Norman, Ault, and Zarobsky, "Fundamentals of Machine Design")

Type of Service	K for Ball Bearings
Uniform steady load	1.00
Light shock load	1.50
Moderate shock load	2.00
Heavy shock load	2.50
Extreme and indeterminate shock load	3.00

TABLE A1-22D.—SINGLE-ROW RADIAL BEARINGS TYPE 1000
(Filling-notch type)
(From New Departure "Handbook")
Principal Dimensions

Provide maximum single-row capacity for radial loads. May be used for combined loads when chosen in accordance with factors F (Table A1-22A).

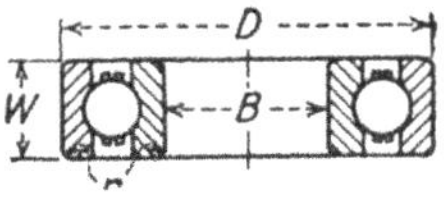

N.D. bearing No.	Bore B		Diameter D		Width W		Balls		Radius r*
	Mm.	In.	Mm.	In.	Mm.	In.	Diam.	No.	
1304	20	0.7874	52	2.0472	15	0.5906	13/32	9	0.04
1404	. .	. . . . . .	72	2.8346	19	0.7480	9/16	8	
1305	25	0.9843	62	2.4409	17	0.6693	13/32	11	0.04
1405	. .	. . . . . .	80	3.1496	21	0.8268	5/8	8	0.06
1206	. .	. . . . . .	62	2.4409	16	0.6299	3/32	. .	0.04
1306	30	1.1811	72	2.8346	19	0.7480	15/32	11	0.04
1406	. .	. . . . . .	90	3.5433	22	0.9055	11/16	9	0.06
1207	. .	. . . . . .	72	2.8346	17	0.6693	7/16	12	0.04
1307	35	1.3780	80	3.1496	21	0.8268	17/32	11	0.06
1407	. .	. . . . . .	100	3.9370	25	0.9043	3/4	9	0.06
1208	. .	. . . . . .	80	3.1496	18	0.7087	15/32	13	0.04
1308	40	1.5748	90	3.5433	23	0.9055	19/32	11	0.06
1408	. .	. . . . . .	110	4.3307	27	1.0630	13/16	9	0.08
1209	. .	. . . . . .	85	3.3465	19	0.7480	15/32	14	0.04
1309	45	1.7717	100	3.9370	25	0.9843	21/32	12	0.06
1409	. .	. . . . . .	120	4.7244	29	1.1417	7/8	10	0.08
1210	. .	. . . . . .	90	3.5433	20	0.7874	15/32	15	0.04
1310	50	1.9685	110	4.3307	27	1.0630	23/32	12	0.08
1410	. .	. . . . . .	130	5.1181	31	1.2205	15/16	10	0.08
1211	. .	. . . . . .	100	3.9370	21	0.8268	17/32	15	0.06
1311	55	2.1654	120	4.7244	29	1.1417	25/32	12	0.08
1411	. .	. . . . . .	140	5.5118	33	1.2992	1	10	0.08
1212	. .	. . . . . .	110	4.3307	22	0.8661	19/32	15	0.06
1312	60	2.3632	130	5.1181	31	1.2205	27/32	12	0.08
1412	. .	. . . . . .	150	5.9055	35	1.3780	17/16	10	0.08

* Radius r indicates maximum fillet radius in housing or on shaft which bearing radius will clear.

TABLE A1-22E.—SINGLE-ROW RADIAL BEARINGS TYPE 1000
(Filling-notch type)
(From New Departure "Handbook")
Radial Load Ratings

The bearing capacities listed on this page are basic radial load ratings in pounds, with rotating inner ring. From these ratings, bearings of the proper size for the service desired can readily be selected by use of data given in Table A1-22D.

N.D. bearing No.	R.p.m. (n)										
	200	300	400	500	600	800	1,000	1,500	2,000	3,000	5,000
1304	1,565	1,365	1,240	1,150	1,085	982	915	800	726	635	535
1404	2,160	1,885	1,720	1,595	1,500	1,370	1,265	1,105	1,005	875	740
1305	1,990	1,735	1,575	1,465	1,380	1,260	1,160	1,015	923	805	680
1405	2,580	2,250	2,050	1,900	1,790	1,625	1,510	1,315	1,195	1,045	883
1206	1,735	1,365	1,420	1,315	1,240	1,125	1,045	917	829	727	811
1306	2,490	2,170	1,970	1,825	1,720	1,570	1,450	1,270	1,150	1,005	886
1406	3,435	2,990	2,720	2,525	2,375	2,145	2,000	1,750	1,590	1,390	1,175
1207	2,530	2,210	2,005	1,865	1,755	1,595	1,480	1,290	1,175	1,025	865
1307	2,855	2,100	2,265	2,100	1,980	1,800	1,670	1,455	1,325	1,155	982
1407	3,980	3,475	3,160	2,935	2,760	2,520	2,330	2,030	1,845	1,610	1,355
1208	2,990	2,610	2,375	2,205	2,075	1,885	1,750	1,530	1,390	1,210	1,020
1308	3,560	3,110	2,830	2,625	2,470	2,245	2,080	1,820	1,650	1,440	1,205
1408	4,550	3,970	3,605	3,350	3,150	2,870	2,660	2,320	2,110	1,840	1,560
1209	3,235	2,825	2,570	2,385	2,245	2,010	1,890	1,650	1,500	1,310	1,100
1309	4,400	3,835	3,480	3,240	3,045	2,770	2,570	2,245	2,040	1,780	1,500
1409	5,535	4,825	4,390	4,075	3,830	3,470	3,230	2,820	2,560	2,240	
1210	3,495	3,055	2,775	2,580	2,430	2,195	2,040	1,785	1,620	1,415	
1310	5,065	4,410	4,020	3,820	3,510	3,190	2,960	2,580	2,345	2,050	
1410	6,190	5,400	4,900	4,550	4,290	3,900	3,615	3,155	2,870	2,505	
1211	4,120	3,650	3,310	3,075	2,895	2,625	2,440	2,135	1,940	1,695	
1311	5,750	5,010	4,560	4,245	3,990	3,605	3,360	2,935	2,665	2,325	
1411	6,870	6,000	5,450	5,060	4,770	4,320	4,015	3,510	3,190	2,785	
1212	4,900	4,300	3,900	3,615	3,400	3,100	2,865	2,615	2,280	1,995	
1312	6,490	5,655	5,145	4,780	4,495	4,100	3,790	3,310	3,010	2,620	
1412	7,575	6,600	6,010	5,585	5,250	4,780	4,430	3,870	3,510	3,065	

TABLE A1-22F.—SINGLE-ROW RADIAL BEARINGS TYPE 1000
(Filling-notch type)
(From New Departure "Handbook")
Principal Dimensions

Provide maximum single-row capacity for radial loads. May be used for combined loads when chosen in accordance with factors *F* (Table A1-22A).

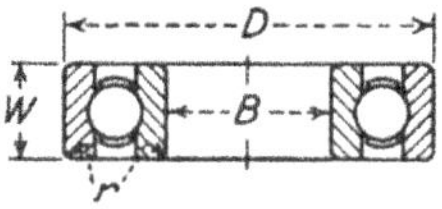

N.D. bearing No.	Bore *B*		Diameter *D*		Width *W*		Balls		Radius r*
	Mm.	In.	Mm.	In.	Mm.	In.	Diam.	No.	
1213	...	......	120	4.7244	23	0.9055	$\frac{21}{32}$	15	0.06
1313	65	2.5591	140	5.5118	33	1.2992	$\frac{29}{32}$	12	0.08
1413	...	......	160	6.2992	37	1.4567	$1\frac{1}{8}$	10	0.08
1214	...	......	125	4.9213	24	0.9449	$\frac{21}{32}$	15	0.06
1314	70	2.7559	150	5.9055	35	1.3780	$\frac{31}{32}$	12	0.08
1414	...	......	180	7.0866	42	1.6535	$1\frac{1}{4}$	10	0.10
1215	...	......	130	5.1181	25	0.9843	$\frac{21}{32}$	16	0.06
1315	75	2.9528	160	6.2992	37	1.4567	1	13	0.08
1415	...	......	190	7.4803	45	1.7717	$1\frac{3}{8}$	10	0.10
1216	...	......	140	5.5118	26	1.0236	$\frac{11}{16}$	17	0.08
1316	80	3.1496	170	6.6929	39	1.5354	$1\frac{1}{16}$	13	0.08
1416	...	......	200	7.8740	48	1.8898	$1\frac{7}{16}$	10	0.10
1217	...	......	150	5.9055	28	1.1024	$\frac{25}{32}$	16	0.08
1317	85	3.3465	180	7.0866	41	1.6142	$1\frac{1}{8}$	13	0.10
1417	...	......	210	8.2677	52	2.0472	$1\frac{1}{2}$	10	0.12
1218	...	......	160	6.2992	30	1.1811	$\frac{27}{32}$	15	0.08
1318	90	3.5433	190	7.4803	43	1.6929	$1\frac{3}{16}$	13	0.10
1418	...	......	225	8.8583	54	2.1260	$1\frac{5}{8}$	10	0.12
1219	...	......	170	6.6929	32	1.2598	$\frac{29}{32}$	15	0.08
1319	95	3.7402	200	7.8740	45	1.7717	$1\frac{1}{4}$	13	0.10
1220	...	......	180	7.0866	34	1.3386	$\frac{31}{32}$	15	0.08
1320	100	3.9370	215	8.4646	47	1.8504	$1\frac{3}{8}$	12	0.10
1221	...	......	190	7.4803	36	1.4173	1	16	0.08
1321	105	4.1339	225	8.8583	49	1.9291	$1\frac{7}{16}$	12	0.10
1222	...	......	200	7.8740	38	1.4961	$1\frac{1}{16}$	16	0.08
1322	110	4.3307	240	9.4488	50	1.9685	$1\frac{1}{2}$	12	0.10

* Radius r indicates maximum fillet radius in housing or on shaft which bearing radius will clear.

TABLE A1-22G.—SINGLE-ROW RADIAL BEARINGS TYPE 1000
(Filling-notch type)
(From New Departure "Handbook")
Radial Load Ratings

The bearing capacities listed on this page are basic radial load ratings in pounds, with rotating inner ring. From these ratings, bearings of the proper size for the service desired can readily be selected by use of data given in Table A1-22F.

N.D. bearing No.	R.p.m. (n)									
	200	300	400	500	600	800	1,000	1,500	2,000	3,000
1213	5,675	4,975	4,500	4,180	3,940	3,585	3,315	2,910	2,640	2,300
1313	7,210	6,300	5,730	5,310	5,000	4,510	4,215	3,680	3,350	2,915
1413	8,295	7,245	6,580	6,110	5,750	5,200	4,850	4,235	3,850	3,360
1214	5,770	5,060	4,575	4,250	4,005	3,645	3,375	2,955	2,680	2,335
1314	8,000	6,980	6,350	5,895	5,550	5,025	4,675	4,080	3,710	3,240
1414	9,720	8,490	7,800	7,160	6,740	6,050	5,685	4,965	4,510	3,940
1215	6,100	5,360	4,850	4,500	4,240	3,860	3,580	3,145	2,845	2,485
1315	9,010	7,860	7,150	6,640	6,250	5,680	5,260	4,600	4,180	3,650
1415	11,010	9,600	8,740	8,100	7,625	6,930	6,430	5,625	5,110	
1216	6,900	6,025	5,490	5,090	4,790	4,360	4,040	3,525	3,200	2,800
1316	9,890	8,635	7,845	7,290	6,875	6,220	5,780	5,050	4,590	
1416	11,880	10,350	9,420	8,750	8,235	7,500	6,945	6,060	5,510	
1217	7,850	6,880	6,220	5,780	5,440	4,940	4,585	4,020	3,645	
1317	10,750	9,395	8,530	7,925	7,465	6,780	6,290	5,495	4,990	
1417	12,700	11,090	10,050	9,350	8,800	8,010	7,420	6,475	5,890	
1218	8,400	7,370	6,680	6,190	5,820	5,300	4,920	4,310	3,900	
1318	11,690	10,200	9,275	8,605	8,110	7,380	6,835	5,970	5,415	
1418	14,200	12,400	11,280	10,460	9,850	8,960	8,300	7,250	6,590	
1219	9,300	8,150	7,400	6,850	6,450	5,880	5,440	4,770	4,320	
1319	12,600	11,000	10,000	9,300	8,745	7,970	7,360	6,440	5,850	
1220	10,225	8,975	8,150	7,540	7,110	6,480	5,990	5,250	4,750	
1320	13,550	11,830	10,750	10,000	9,410	8,550	7,930	6,930	6,295	
1221	11,330	9,900	9,000	8,350	7,860	7,130	6,630	5,785	5,260	
1321	14,510	12,680	11,510	10,700	10,080	9,170	8,495	7,425	6,735	
1222	12,320	10,780	9,790	9,100	8,550	7,750	7,210	6,300	5,715	
1322	15,400	13,600	12,250	11,480	10,700	9,780	9,105	7,870	7,160	

TABLE A1-22H.—SINGLE-ROW RADIAL BEARINGS TYPE 3000

(No filling-notch type)
(From New Departure "Handbook")
Principal Dimensions

For radial or combined loads from either direction where thrust is to be resisted by a single bearing and is not great enough to require use of angular contact type. For capacities under thrust or combined loads, use factors F (Table A1-22A).

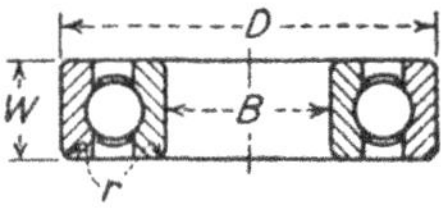

N.D. bearing No.	Bore B		Diameter D		Width W		Balls		Radius r*
	Mm.	In.	Mm.	In.	Mm.	In.	Diam.	No.	
3200	. .	. . . . . .	30	1.1811	9	0.3543	7/32	7	0.025
3300	10	0.3937	35	1.3780	11	0.4331	1/4	7	
3201	. .	. . . . . .	32	1.2598	10	0.3937	0.210	8	0.025
3301	12	0.4724	37	1.4567	12	0.4724	9/32	7	0.04
3202	. .	. . . . . .	35	1.3780	11	0.4331	0.210	9	0.025
3302	15	0.5906	42	1.6536	13	0.5118	5/16	7	0.04
3203	. .	. . . . . .	40	1.5748	12	0.4724	9/32	8	
3303	17	0.6693	47	1.8504	14	0.5512	11/32	7	0.04
3204	. .	. . . . . .	47	1.8504	14	0.5512	5/16	8	
3304	20	0.7874	52	2.0472	15	0.5906	13/32	7	0.04
3205	. .	. . . . . .	52	2.0472	15	0.5906	5/16	9	
3305	25	0.9843	62	2.4409	17	0.6693	13/32	8	0.04
3206	. .	. . . . . .	62	2.4409	16	0.6299	3/8	9	
3306	30	1.1811	72	2.8346	19	0.7480	15/32	8	0.04
3207	. .	. . . . . .	72	2.8346	17	0.6693	7/16	9	0.04
3307	35	1.3780	80	3.1496	21	0.8268	17/32	8	0.06
3208	. .	. . . . . .	80	3.1496	18	0.7087	15/32	9	0.04
3308	40	1.5748	90	3.5433	23	0.9055	19/32	8	0.06
3209	. .	. . . . . .	85	3.3465	19	0.7480	15/32	10	0.04
3309	45	1.7717	100	3.9370	25	0.9843	21/32	8	0.06
3210	. .	. . . . . .	90	3.5433	20	0.7874	15/32	11	0.04
3310	50	1.9685	110	4.3307	27	1.0630	23/32	8	0.08
3211	. .	. . . . . .	100	3.0370	21	0.8268	17/32	11	0.06
3311	55	2.1654	120	4.7244	29	1.1417	25/32	8	0.08

* Radius r indicates maximum fillet radius in housing or on shaft which bearing radius will clear.

TABLE A1-22I.—SINGLE-ROW RADIAL BEARINGS TYPE 3000
(No filling-notch type)
(From New Departure "Handbook")
Radial Load Ratings

The bearing capacities listed on this page are basic radial load ratings in pounds, with rotating inner ring. From these ratings, bearings of the proper size for the service desired can readily be selected by use of data given in Table A1-22H.

N.D. bearing No.	R.p.m.									
	200	300	400	500	600	800	1,000	1,500	2,000	3,000	5,000
3200	419	364	332	307	290	264	244	213	194	169	140
3300	481	419	381	356	333	305	281	245	223	195	163
3201	515	450	410	380	357	319	301	263	239	209	162
3301	603	523	479	441	419	379	352	307	281	244	209
3202	610	533	485	450	424	388	357	312	284	248	200
3302	712	620	564	521	493	448	415	362	330	288	242
3203	788	689	625	581	546	494	460	402	365	319	270
3303	832	729	660	612	578	523	486	425	386	337	284
3204	944	827	749	690	655	595	552	482	438	381	321
3304	1,160	1,010	917	851	802	726	677	593	537	470	398
3205	1,120	976	889	825	775	700	655	571	520	454	383
3305	1,560	1,360	1,235	1,140	1,080	988	910	795	724	631	534
3206	1,475	1,290	1,175	1,085	1,025	930	865	758	683	600	503
3306	1,955	1,700	1,545	1,430	1,350	1,230	1,140	996	902	788	695
3207	2,090	1,825	1,655	1,540	1,450	1,315	1,220	1,065	970	845	714
3307	2,240	1,950	1,775	1,650	1,550	1,410	1,310	1,140	1,040	905	770
3208	2,340	2,040	1,860	1,725	1,620	1,475	1,370	1,200	1,090	947	798
3308	2,790	2,440	2,220	2,060	1,935	1,760	1,630	1,430	1,295	1,130	945
3209	2,580	2,260	2,045	1,900	1,790	1,600	1,510	1,315	1,200	1,045	877
3309	3,260	2,840	2,580	2,400	2,255	2,055	1,905	1,660	1,510	1,320	1,110
3210	2,640	2,455	2,240	2,100	1,980	1,780	1,660	1,450	1,320	1,150	
3310	3,755	3,270	2,980	2,830	2,600	2,360	2,195	1,910	1,735	1,520	
3211	3,390	2,985	2,700	2,500	2,355	2,140	1,980	1,735	1,580	1,380	
3311	4,255	3,715	3,385	3,140	2,960	2,675	2,490	2,175	1,975	1,725	

TABLE A1-22J.—SINGLE-ROW RADIAL BEARINGS TYPE 3000
(No filling-notch type)
(From New Departure "Handbook")
Principal Dimensions

For radial or combined loads from either direction where thrust is to be resisted by a single bearing and is not great enough to require use of angular contact type. For capacities under thrust or combined loads, use factors F (Table A1-22A).

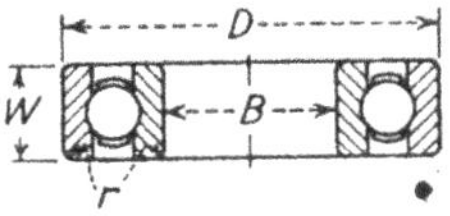

N.D. bearing No.	Bore B		Diameter D		Width W		Balls		Radius r*
	Mm.	In.	Mm.	In.	Mm.	In.	Diam.	No.	
3212	60	2.3622	110	4.3307	22	0.8661	19/32	10	0.06
3312	...	......	130	5.1181	31	1.2205	27/32	8	0.08
3213	...	......	120	4.7244	23	0.9055	21/32	10	0.06
3313	65	2.5591	140	5.5118	33	1.2992	29/32	8	0.08
3214	...	......	125	4.9213	24	0.9449	21/32	11	0.06
3314	70	2.7559	150	5.9055	35	1.3780	31/32	8	0.08
3215	...	......	130	5.1181	25	0.9843	21/32	11	0.06
3315	75	2.9528	160	6.2992	37	1.4567	1	8	0.08
3216	...	......	140	5.5118	26	1.0236	11/16	11	
3316	80	3.1496	170	6.6929	39	1.5354	1 1/16	8	0.08
3217	...	......	150	5.9055	28	1.1024	25/32	11	0.08
3317	85	3.3465	180	7.0866	41	1.6142	1 1/8	8	0.10
3218	...	......	160	6.2992	30	1.1811	27/32	11	0.08
3318	90	3.5433	190	7.4803	43	1.6929	1 3/16	8	0.10
3219	...	......	170	6.6929	32	1.2598	29/32	11	0.08
3319	95	3.7402	200	7.8740	45	1.7717	1 1/4	8	0.10
3220	...	......	180	7.0866	34	1.3386	31/32	11	0.08
3320	100	3.9370	215	8.4646	47	1.8504	1 3/8	8	0.10
3221	...	......	190	7.4803	36	1.4173	1	11	0.08
3321	105	4.1339	225	8.8583	49	1.9291	1 7/16	8	0.10
3222	...	......	200	7.8740	38	1.4961	1 1/16	11	0.08
3322	110	4.3307	240	9.4488	50	1.9685	1 1/2	8	0.10

* Radius r indicates maximum fillet radius in housing or on shaft which bearing radius will clear.

TABLE A1-22K.—SINGLE-ROW RADIAL BEARINGS, TYPE 3000
(No filling-notch type)
(From New Departure "Handbook")
Radial Load Ratings

The bearing capacities listed on this page are basic radial load ratings in pounds, with rotating inner ring. From these ratings, bearings of the proper size for the service desired can readily be selected by use of data given in Table A1-22J.

N.D. No.	R.p.m. (n)									
	200	300	400	500	600	800	1,000	1,500	2,000	3,000
3212	3,750	3,290	2,970	2,750	2,595	2,360	2,180	1,920	1,735	1,520
3312	4,805	4,200	3,815	3,540	3,330	3,040	2,810	2,455	2,230	1,940
3213	4,325	3,800	3,435	3,180	3,000	2,725	2,525	2,215	2,010	1,750
3313	5,350	4,665	4,250	3,940	3,705	3,345	3,130	2,730	2,485	2,160
3214	4,700	4,100	3,720	3,455	3,250	2,960	2,745	2,405	2,180	1,895
3314	5,930	5,170	4,705	4,365	4,110	3,725	3,465	3,025	2,750	2,400
3215	4,770	4,165	3,780	3,510	3,300	3,005	2,790	2,445	2,210	1,920
3315	6,325	5,520	5,020	4,660	4,395	3,980	3,700	3,230	2,935	2,565
3216	5,155	4,500	4,100	3,800	3,575	3,260	3,020	2,635	2,390	2,095
3316	6,945	6,055	5,505	5,110	4,825	4,365	4,055	3,550	3,220	
3217	6,100	5,340	4,845	4,500	4,235	3,845	3,560	3,135	2,835	
3317	7,550	6,600	5,990	5,560	5,250	4,750	4,410	3,860	3,500	
3218	6,820	5,990	5,420	5,010	4,740	4,310	4,000	3,500	3,170	
3318	8,205	7,160	6,510	6,050	5,700	5,185	4,800	4,200	3,810	
3219	7,580	6,620	6,000	5,570	5,230	4,785	4,425	3,880	3,510	
3319	8,845	7,725	7,020	6,530	5,135	5,595	5,160	4,510	4,100	
3220	8,320	7,300	6,620	6,130	5,790	5,260	4,875	4,265	3,865	
3320	10,040	8,760	7,960	7,400	6,970	6,330	5,880	5,140	4,660	
3221	8,780	7,700	6,990	6,480	6,100	5,530	5,150	4,480	4,080	
3321	10,750	9,390	8,535	7,930	7,460	6,790	6,295	5,500	4,985	
3222	9,550	8,390	7,600	7,075	6,630	6,010	5,600	4,890	4,440	
3322	11,410	10,080	9,070	8,500	7,925	7,235	6,745	5,835	5,305	

TABLE A1-23.—ROLLER-BEARING SELECTION

For conditions of extreme load or severe shock loads such as occur in the main bearings of radial engines, roller-type bearings are frequently most applicable. For a given size, roller bearings have a greater load-carrying capacity than ball bearings because they provide "line" contact as against "point" contact. Cylindrical roller bearings have the disadvantage of inability to take appreciable thrust loads. Roller bearings are manufactured in the "metric" and "inch" types; data on the latter are given in Tables A1-22B and A1-22C. Inch-type bearings* as manufactured by the Norma-Hoffmann Bearings Corporation are further classified as "standard," "one-lipped," and "two-lipped" types (see Fig. *A*). Standard and one-lipped types of roller bearings correspond in general to filling-notch types of ball bearings in that they contain the maximum number of rollers. Two-lipped type roller bearings correspond in general to non-filling-notch types of ball bearings, but they cannot take thrust loads. Standard roller bearings have the advantage over one- and particularly two-lipped types in that endwise movement of the shaft does not bind the rollers in the outer race, but this entails provision of means to hold the outer race from slipping out of position. In general, one- and two-lipped types of roller bearings are preferable for radial-engine main bearings because the outer races can be more easily held in place axially.

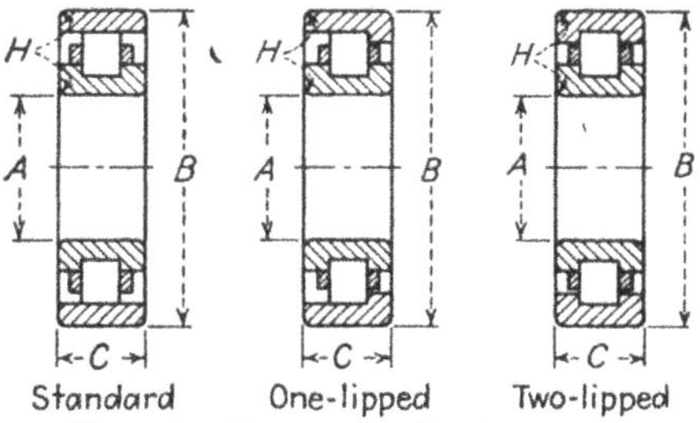

FIG. *A*.—Types of roller bearings.

Radial load ratings for Hoffmann roller bearings are given in Tables A1-23B and A1-23C for the range of bearing sizes most likely to be needed for aircraft-engine main bearings. (Load ratings for standard types are the same as for the one-lipped type. The bearing numbers differ by the deletion of the *L* in the standard type.) These load ratings are for non-shock conditions and are based on a life of 10,000 hr. For aircraft-engine main bearings, the load *C* for entry in the load tables may be determined from

$$C = L \times Z \times K \qquad (1)$$

where C = equivalent radial capacity, lb.
L = calculated radial load on the bearing, lb.
Z = life factor for a desired bearing life other than 10,000 hr. (Fig. *B*).
K = the shock load factor (Table A1-23A).

To determine the proper size of roller bearing, enter C as determined by Eq. (1) in Table A1-23B or A1-23C (as applies) at the proper r.p.m. and read the major bearing dimensions and the bearing number on the left side of the table.

* Norma-Hoffmann roller bearings applicable to use in aircraft engines are also made in standard, one-lipped, and two-lipped metric types, data of Norma-Hoffmann Bearings Corporation.

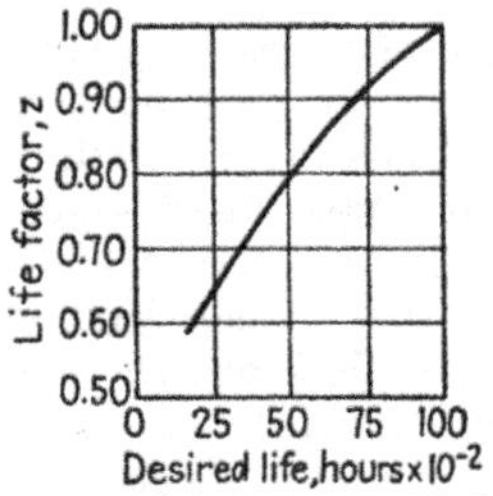

Desired life hours $\times 10^{-2}$.

Fig. B.

Table A1-23A.—Shock-load Factors for Roller Bearings

Type of Service	Shock Factor, K
Uniform and steady load	1.0
Light shock load	1.0
Moderate shock load	1.3
Heavy shock load	1.7
Extreme and indeterminate shock load	2.0

Table A1-23B.—One-lipped Inch-type Hoffmann Precision Roller Bearings

(From engineering data of the Norma-Hoffmann Bearings Corporation)

Bearing No.	Bearing dimensions, in.				Load in pounds at speed in r.p.m.				
	A	*B*	*C*	*H*	500	1,000	1,500	2,000	3,000
RLS-L Type Light Series									
RLS-15-L	2	4	0.8125	0.0937	2,610	2,070	1,810	1,640	1,430
RLS-16-L	2.250	4.500	0.875	0.0937	3,420	2,710	2,370	2,150	1,880
RLS-17-L	2.500	5	0.9375	0.0937	4,050	3,210	2,800	2,550	2,230
RLS-18-L	2.750	5.250	0.9375	0.0937	4,300	3,410	2,980	2,710	2,370
RLS-19-L	3	5.750	1.0625	0.0937	5,290	4,200	3,670	3,330	
RLS-19.5-L	3.250	6	1.0625	0.0937	5,600	4,440	3,880	3,530	
RLS-20-L	3.500	6.500	1.125	0.125	6,750	5,360	4,680	4,250	
RLS-20.5-L	3.750	6.750	1.125	0.125	7,130	5,660	4,940	4,480	
RLS-21-L	4	7.250	1.250	0.125	7,950	6,310	5,510	5,010	
RMS-L Type Medium Series									
RMS-15-L	2	4.50	1.0625	0.0937	5,390	4,280	3,740	3,400	2,970
RMS-16-L	2.250	5	1.250	0.125	6,000	4,760	4,160	3,780	3,300
RMS-17-L	2.500	5.500	1.250	0.125	7,500	5,950	5,200	4,720	4,130
RMS-18-L	2.750	6.250	1.375	0.125	10,300	8,170	7,140	6,490	
RMS-19-L	3	7	1.5625	0.1562	12,620	10,010	8,740	7,940	
RMS-19.5-L	3.250	7.500	1.5625	0.1562	13,520	10,730	9,370	8,510	
RMS-19.75-L	3.375	7.500	1.5625	0.1562	13,520	10,730	9,370	8,510	
RMS-20-L	3.500	8.125	1.750	0.1562	15,920	12,630	11,040	10,030	
RMS-20.5-L	3.750	8.250	1.750	0.1562	15,920	12,630	11,040	10,030	
RMS-21-L	4	8.500	1.750	0.1562	17,060	13,540	11,820	10,740	
RMS-21.5-L	4.500	8.750	1.750	0.1562	17,060	13,540	11,820	10,740	

Note: Bearing dimension symbols correspond to figures on p. 417.

TABLE A1-23C.—TWO-LIPPED INCH-TYPE HOFFMANN PRECISION ROLLER BEARINGS
(From Engineering Data of the Norma-Hoffmann Bearings Corporation)

Bearing No.	Bearing dimensions, in.				Load in pounds at speed in r.p.m.				
	A	*B*	*C*	*H*	500	1,000	1,500	2,000	3,000
				RLS-LL Type Light Series					
RLS-15-LL	2	4	0.8125	0.0937	2,300	1,830	1,600	1,450	1,270
RLS-16-LL	2.250	4.500	0.875	0.0937	3,010	2,390	2,090	1,900	1,660
RLS-17-LL	2.500	5	0.9375	0.0937	3,550	2,820	2,460	2,230	1,950
RLS-18-LL	2.750	5.250	0.9375	0.0937	3,790	3,010	2,630	2,390	2,090
RLS-19-LL	3	5.750	1.0625	0.0937	4,670	3,700	3,230	2,940	
RLS-19.5-LL	3.250	6	1.0625	0.0937	4,980	3,950	3,450	3,130	
RLS-20-LL	3.500	6.500	1.125	0.125	5,810	4,610	4,030	3,660	
RLS-20.5-LL	3.750	6.750	1.125	0.125	6,190	4,910	4,290	3,900	
RLS-21-LL	4	7.250	1.250	0.125	7,070	5,610	4,900	4,460	
				RMS-LL Type Medium Series					
RMS-15-LL	2	4.50	1.0625	0.0937	4,770	3,790	3,310	3,000	2,620
RMS-16-LL	2.250	5	1.250	0.125	5,500	4,370	3,810	3,460	3,030
RMS-17-LL	2.500	5.500	1.250	0.125	6,500	5,160	4,510	4,090	3,580
RMS-18-LL	2.750	6.25	1.375	0.125	8,580	6,810	5,950	5,410	
RMS-19-LL	3	7	1.5625	0.1562	10,810	8,580	7,500	6,810	
RMS-19.5-LL	3.250	7.500	1.5625	0.1562	11,260	8,940	7,810	7,090	
RMS-19.75-LL	3.375	7.500	1.5625	0.1562	11,250	8,940	7,810	7,090	
RMS-20-LL	3.500	8.125	1.750	0.1562	13,660	10,830	9,460	8,600	
RMS-20.5-LL	3.750	8.250	1.750	0.1562	13,660	10,830	9,460	8,600	
RMS-21-LL	4	8.50	1.750	0.1562	14,780	11,730	10,250	9,310	

NOTE: Bearing dimension symbols correspond to figures on page 417.

TABLE A1-23D.—EXTRA-LIGHT INCH-TYPE HOFFMANN PRECISION ROLLER BEARINGS

(From engineering data of the Norma-Hoffmann Bearings Corporation)

RXLS Type Extra Light Series

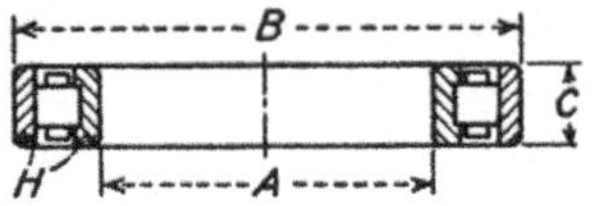

Bearing No.	Bearing dimensions, in.				Loads in pounds at speed in r.p.m.				
	A	B	C	H	500	1,000	1,500	2,000	3,000
RXLS-2.......	2.	3.3125	0.625	0.0625	1,320	1,050	915	830	725
RXLS-2.125...	2.125	3.4375	0.625	0.0625	1,470	1,160	1,020	925	805
RXLS-2.25....	2.25	3.5625	0.625	0.0625	1,470	1,160	1,020	925	805
RXLS-2.375...	2.375	3.75	0.6875	0.0625	1,570	1,250	1,090	990	865
RXLS-2.5.....	2.5	3.875	0.6875	0.0625	1,750	1,390	1,210	1,100	960
RXLS-2.625...	2.625	4.125	0.6875	0.0625	1,750	1,390	1,210	1,100	960
RXLS-2.75....	2.75	4.125	0.6875	0.0625	1,920	1,520	1,330	1,210	1,060
RXLS-2.875...	2.875	4.5	0.75	0.0937	2,250	1,780	1,560	1,420	1,240
RXLS-3.......	3.	4.5	0.75	0.0937	2,250	1,780	1,560	1,420	1,240
RXLS-3.125...	3.125	4.75	0.75	0.0937	2,250	1,780	1,560	1,420	1,240
RXLS-3.25....	3.25	4.75	0.75	0.0937	2,250	1,780	1,560	1,420	1,240
RXLS-3.375...	3.375	5.	0.75	0.0937	2,450	1,950	1,700	1,540	1,350
RXLS-3.5.....	3.5	5.	0.75	0.0937	2,450	1,950	1,700	1,540	1,350
RXLS-3.625...	3.625	5.25	0.75	0.0937	2,660	2,110	1,840	1,670	1,460
RXLS-3.75....	3.75	5.25	0.75	0.0937	2,660	2,110	1,840	1,670	1,460
RXLS-3.875...	3.875	5.625	0.875	0.0937	3,210	2,550	2,230	2,020	1,770
RXLS-4.......	4.	5.625	0.875	0.0937	3,210	2,550	2,230	2,020	1,770

TABLE A1-24.—AVERAGE WEIGHTS AND OVER-ALL DIMENSIONS OF BENDIX-STROMBERG AIRCRAFT CARBURETORS
(From Bendix-Stromberg Carburetor Co.)

Model NA–	Dry weight, lb.	Height flange to flange	Over-all height	Over-all width	Over-all depth	Barrel diameter, in.	Total barrel area, sq. in.	Number of barrels	Discharge nozzles diameter, in.	Total discharge-nozzle area, sq. in.	Maximum vent, in.	Total maximum net vent area, sq. in.
S2	2.55	4 1/4	4 3/4	5 9/16	4 1/4	1 7/16	1.623	1	7/16	0.150	1 1/8	0.844
S3	2.55	4 1/4	4 3/4	5 9/16	4 1/4	1 11/16	2.236	1	7/16	0.150	1 3/8	0.844
R3	4.80	6 1/8	6 5/16	6 7/8	4 1/4	1 11/16	2.236	1	1/2	0.196	1 3/8	1.289
R3A	4.80	6 1/4	6 7/16	6 1/8	4 3/4	1 11/16	2.236	1	1/2	0.196	1 3/8	1.289
DD4	8.00	5 3/8	5 1/2	7 1/8	7 11/16	1 15/16	5.896	2	5/8	0.614	1 1/2	2.921
R4	4.80	6 1/4	6 7/16	6 5/16	4 1/2	1 15/16	2.948	1	1/2	0.196	1 1/2	1.571
R4A	4.80	6 3/8	6 9/16	6 11/16	4 7/8	1 15/16	2.948	1	1/2	0.196	1 9/16	1.721
R4D	5.30	5 3/8	5 1/2	6 1/4	5 3/4	1 15/16	2.948	1	7/16	0.150	1 9/16	1.767
T4B	12.50	7 5/32	7 5/32	9 1/8	7 13/16	1 15/16	8.845	3	21/32	1.015	1 1/2	4.287
U4J	8.75	90°	7 5/8	10 1/8	5 11/16	1 15/16	5.896	2	23/32	0.811	1 1/2	2.723
R5	7.25	6 3/4	7 3/16	7 3/8	6 1/4	2 3/16	3.758	1	27/32	0.559	1 7/8	2.202
R5A	7.50	6 3/4	7 3/16	6 1/4	6 1/4	2 3/16	3.758	1	27/32	0.559	1 7/8	2.202
R5H	6.20	Length 6 1/2	6	7 3/8	6 1/2	2 3/16	3.758	1	7/16	0.494	1 15/16	2.454
S5A	5.50	6 3/16	6 1/2	7 3/16	5 3/4	2 3/16	3.758	1	27/32	0.559	1 7/8	2.202
Y5D	10.25	6 3/32		6 5/16	9 5/16	2 3/16	7.516	2	27/32	1.118	1 7/8	4.404
Y5F	10.25	6 3/32	6 9/16	6 5/8	9 3/8	2 3/16	7.516	2	27/32	1.118	1 7/8	4.404
R6	7.25	6 3/4	7 3/16	7 7/16	6 1/4	2 7/16	4.666	1	27/32	0.559	2 1/16	2.782
R6A	7.25	6 3/4	7 3/16	6 1/4	6 1/4	2 7/16	4.666	1	27/32	0.559	2 1/16	2.782
U6TB	9.50	6 29/32	6 15/16	10 9/16	7 15/16	2 7/16	9.332	2	27/32	1.118	2 1/16	5.564
Y6E	12.70	6	6	7 13/16	11 9/16	2 7/16	9.332	2	7/16	0.301	2 1/16	6.381
Y6F	12.50	7	7	7	9 1/2	2 9/16	10.314	2	1	1.571	2 3/16	6.000
Y6G	12.87	7	7 5/16	7	9 1/2	2 9/16	10.314	2	1	1.571	2 3/16	6.000
Y6O	12.60	7	7	7	9 1/2	2 9/16	10.314	2	1	1.571	2 3/16	6.000
F7	19.80	6 1/2	8 1/2	13 5/8	9 11/16	2 11/16	22.691	4	7/16	3.937	2 1/4	11.967
F7A	18.25	5 5/8	7 1/16	15 7/16	9	2 11/16	22.691	4	7/16	1.079	2 1/4	13.156
F7B	24.50	6 1/2	8 3/4	13 5/8	10 7/16	2 11/16	22.691	4	1/2	4.500	2 1/4	11.404
F7C	27.50	6 1/2	8 1/2	13 5/8	10 5/8	2 11/16	22.691	4	1/2	4.500	2 1/4	11.404

TABLE A1-24.—AVERAGE WEIGHTS AND OVER-ALL DIMENSIONS OF BENDIX-STROMBERG AIRCRAFT CARBURETORS.—(*Continued*)

Model NA–	Dry weight, lb.	Height flange to flange	Over-all height	Over-all width	Over-all depth	Barrel diameter, in.	Total barrel area, sq. in.	Number of barrels	Discharge nozzles diameter, in.	Total discharge-nozzle area, sq. in.	Maximum vent, in.	Total maximum net vent area, sq. in.
F7E	27.60	6 1/2	9	13 5/8	10 5/8	2 11/16	22.691	4	1/2	4.500	2 1/4	11.404
F7F	26.85	6 1/2	9	13 5/8	10 5/8	2 13/16	24.850	4	1/2	0.750	2 3/8	16.970
F7H	28.80	6 1/2	10 5/8	13 5/8	10 5/8	2 11/16	22.691	4	1/2	4.50	2 1/4	11.404
F7J	29.75	6 1/2	10 5/8	13 7/8	10 5/8	2 13/16	24.850	4	1/2	1.656	2 3/8	16.064
F7K	.....	6 1/2	10 5/8	13 7/8	10 5/8	2 13/16	24.850	4	1/2	1.656	2 3/8	16.064
F7L	31.85	6 1/2	11 1/16	14 3/4	10 5/8	2 13/16	24.850	4	1/2	4.750	2 3/8	12.970
F7M	.....	6 1/2	11 1/16	14 3/4	10 5/8	2 11/16	22.691	4	1/2	4.500	2 1/4	11.404
R7	.....	7 1/2	7 13/16	9 1/4	6 5/8	2 11/16	5.673	1	27/32	0.559	2 1/4	3.417
R7A	9.12	7 1/2	7 13/16	9 1/4	6 5/8	2 11/16	5.673	1	27/32	0.559	2 1/4	3.417
R7B	9.68	7 1/2	7 13/16	9 1/4	6 5/8	2 11/16	5.673	1	27/32	0.559	2 1/4	3.417
Y7A	.....	7 11/16	7 11/16	8 5/8	10 1/8	2 11/16	11.346	2	27/32	1.118	2 1/4	6.834
Y7B	.....	7 11/16	7 11/16	8 5/8	10 1/8	2 11/16	11.346	2	27/32	1.118	2 1/4	6.834
U8B	.....	8 3/8	9	12 1/2	8 1/2	2 15/16	13.554	2	1/2	2.500	2 1/2	7.317
U8E	.....	7 13/32	7 13/32	12 5/16	6 5/8	2 15/16	13.554	2	1/2	.....	2 1/2	
U8F	17.4	8 3/8	8 9/16	12 13/16	8 11/16	2 15/16	13.554	2	1/2	2.500	2 1/2	7.317
U8G	.....	8 3/8	9 1/8	13 7/8	8 13/16	2 15/16	13.554	2	1/2	2.500	2 1/2	7.317
U8H	18.45	8 3/8	9 1/8	13 7/8	8 13/16	2 15/16	13.544	2	1/2	2.500	2 1/2	7.317
U8J	17.1	8 21/32	8 21/32	12 5/16	6 3/8	2 15/16	13.554	2	1 1/16	1.773	2 1/2	8.044
Y8A	16.50	7 11/16	7 11/16	10 1/4	9 3/8	2 15/16	13.554	2	1	1.571	2 1/2	8.246
Y8B	15.62	7 11/16	7 11/16	9 7/16	10 1/8	2 15/16	13.544	2	27/32	1.118	2 1/2	8.699
Y8C	15.00	7 11/16	7 11/16	10 1/4	9 3/8	2 15/16	13.544	2	1	1.571	2 1/2	8.246
Y8D	13.70	7 1/8	7 1/8	9	10 9/16	2 15/16	13.554	2	1	1.571	2 1/2	8.246
Y8E	15.00	7 11/16	7 11/16	10 1/4	9 3/8	2 15/16	13.554	2	1	1.571	2 1/2	8.246
Y8F	14.56	7 11/16	7 11/16	10 3/8	9 3/8	2 15/16	13.554	2	1	1.571	2 1/2	8.246
Y8H	.....	7 11/16	7 11/16	10 1/4	9 3/8	2 15/16	13.554	2	1	1.571	2 1/2	8.246

Y8J	17.50	7 11/16	7 11/16	10 1/8	9 3/8	2 15/16	13.554	2	1	1.571	2 1/2	8.246
R9	9.50	7 3/4	8 1/16	9 7/8	6 1/2	3 3/16	7.980	1	1	0.785	2 3/4	5.154
R9A	9.55	7 3/4	8 1/16	9 7/8	6 1/2	3 3/16	7.980	1	1	0.785	2 3/4	5.154
R9B	10.35	7 3/4	8 1/16	9 15/16	6 1/2	3 3/16	7.980	1	1	0.785	2 3/4	5.154
Y9A	17.65	7 11/16	7 11/16	10 1/2	9 5/8	3 3/16	15.960	2	1	1.571	2 3/4	10.308
Y9B	17.75	7 11/16	7 11/16	10 1/2	9 5/8	3 3/16	15.960	2	1	1.571	2 3/4	10.308
Y9C	22.0	9 15/32	9 1/2	11 1/16	10	3 3/16	15.960	2	1	1.571	2 3/4	10.308
Y9D	16.87	7 11/16	7 11/16	10 1/2	9 5/8	3 3/16	15.960	2	1	1.571	2 3/4	10.308
Y9D1	16.87	7 11/16	7 11/16	10 1/2	9 5/8	3 3/16	15.960	2	1	1.571	2 3/4	10.308
Y9E	18.25	7 11/16	7 11/16	10 1/2	9 5/8	3 3/16	15.960	2	1	1.571	2 3/4	10.308
Y9E1	18.60	7 11/16	7 11/16	10 1/2	9 5/8	3 3/16	15.960	2	1	1.571	2 3/4	10.308
Y9F	22.50	8 27/32	8 27/32	11 1/16	9 13/16	3 3/16	15.960	2	1	1.571	2 3/4	10.308
Y9G	22.70	8 27/32	8 27/32	11 1/16	9 13/16	3 3/16	15.960	2	1	1.571	2 3/4	10.308
H10	.....	10	7 9/16	9	10 3/8	3 7/16	9.281	..	1 1/16	2.063	3	5.006
T10A1	37.40	6 15/16	11 1/2	14 1/8	12 1/4	3 7/16	27.843	..	1/2	1.750	3	19.456
T10B1	37.30	8 5/8	11 7/8	18 7/16	11 1/8	3 7/16	27.843	..	1	1.500	3	19.706
U10J	18.00	90°	10 7/8	13 1/2	9 5/16	3 7/16	27.843	..	1	1.571	3	12.566
Y10A	.....	8 9/16	8 13/16	11 3/4	11 1/4	3 7/16	27.843	..	1	1.571	3	12.566
C12	28.3	8 1/4	8 1/4	13 1/8	12 1/4	3 15/16	24.354	..	0.214	0.428	3 1/2	18.814
C12A	28.5	Angle	9 3/8	13 1/8	12 1/2	3 15/16	24.354	..	0.214	0.428	3 1/2	18.814
C12B	27.50	Angle	10 13/16	13 13/16	13 11/16	3 15/16	24.354	..	0.214	1.712	3 1/2	17.530
C12C	.....	Angle	7	13 13/16	12 5/16	3 15/16	24.354	..	0.214	1.712	3 1/2	17.530
C12D1	28.5	Angle	10 13/16	13 13/16	13 11/16	3 15/16	24.354	..	0.214	1.712	3 1/2	17.530
C12E	.....	Angle	7	13 13/16	12 5/16	3 15/16	24.354	..	0.214	1.712	3 1/2	17.530
MODEL P												
PD9A1	Without adapter 29.00 with 33.00	7 11/16	10 1/8	13 3/4	12 3/4	3 3/16	15.960	..	.....	.....	2 3/4	11.879
PD12A1	Without adapter 31.3	6 9/16	10 11/16	14 1/8	12 1/4	3 15/16	24.354	..	.....	.....	3 1/2	19.242
PT13A1	40.0	7 3/4	10 7/8	19 3/16	13 3/4	4 3/16	41.316	..	.....	.....	3 3/4	33.135

Table A1-25.—Carburetor Flanges, Aircraft Types
S.A.E. Standard
(From S.A.E. "Handbook")

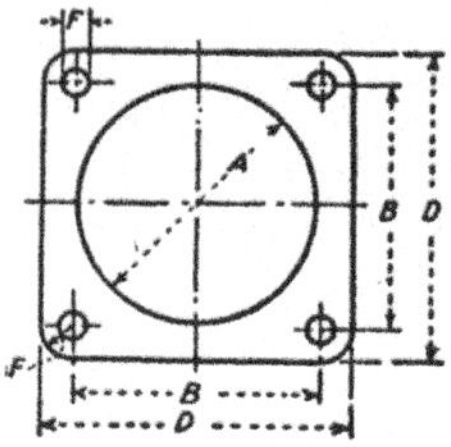

Single barrel (Table *a*)

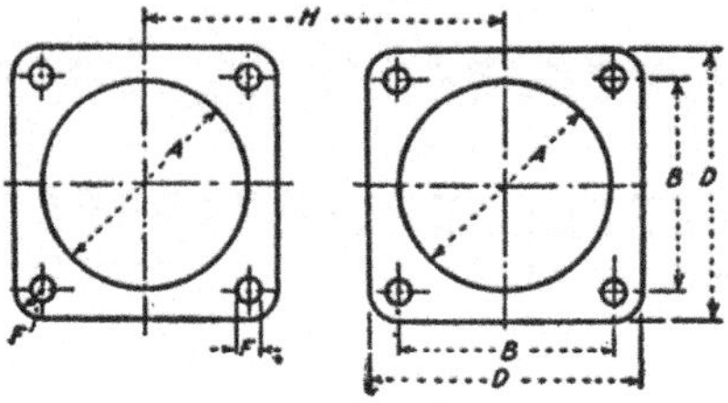

Double barrel (Table *a*)

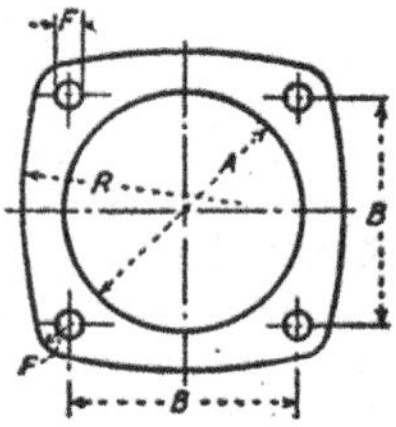

Single barrel (Table *b*)

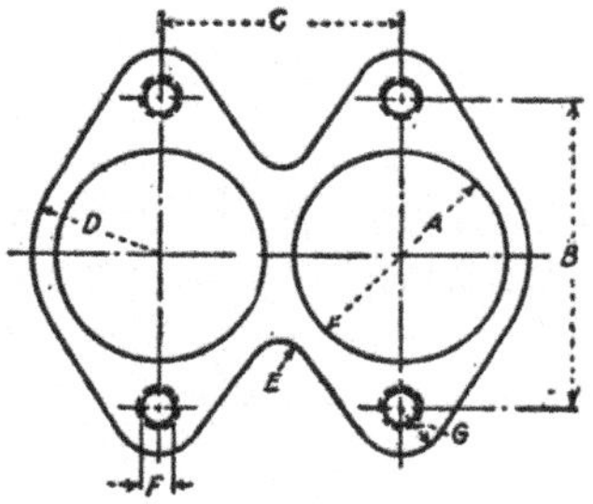

Double barrel (Table *c*)

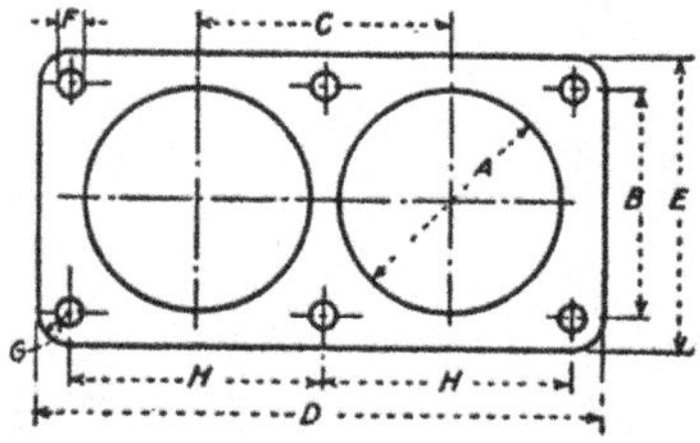

Double barrel (Table *d*)

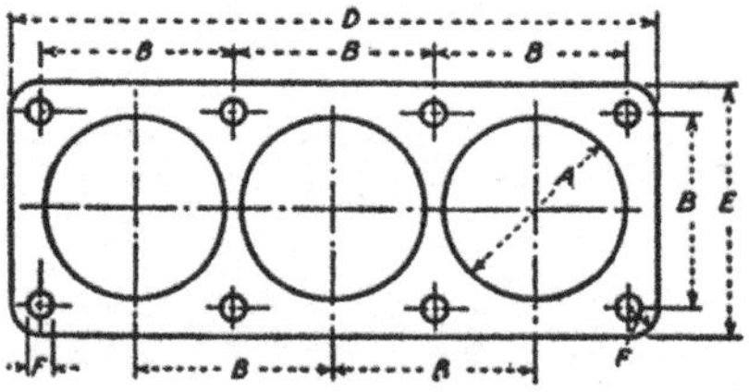

Triple barrel (Table *e*)

Flange Thickness

All types of flange shall be of the following thickness.

Stud Size	Flange Thickness, In.
¼	5/16
5/16	⅜
⅜	13/32

TABLE A1-25.—CARBURETOR FLANGES, AIRCRAFT TYPES.—(*Continued*)

Table *a*. Four- and Eight-bolt Flanges

Nominal size		*A*	*B*	*D*	*F*	*H**
No.	Diam.					
2	1 1/4	1 7/16	1 3/4	2 5/16	9/32	
3	1 1/2	1 11/16	1 15/16	2 1/2	9/32	
4	1 3/4	1 15/16	2 3/16	2 3/4	9/32	4 3/8
5	2	2 3/16	2 11/32	3 1/32	11/32	4 7/8
6	2 1/4	2 7/16	2 1/2	3 3/16	11/32	5 1/8
7	2 1/2	2 11/16	2 3/4	3 7/16	11/32	5 1/2
8	2 3/4	2 15/16	3	3 11/16	11/32	5 5/8
9	3	3 3/16	3 3/16	4	13/32	
10[1]	3 1/4	3 7/16	3 7/16	4 1/4	13/32	6 5/16

* For double-barrel eight-bolt flange type only.

Table *b*. Single-barrel Four-bolt Flange

Nominal size		*A*	*B*	*F*	*R*
No.	Diam.				
10	3 1/4	3 7/16	3 3/8	13/32	2 1/8
11	3 1/2	3 11/16	3 1/2	13/32	2 1/4
12	3 3/4	3 15/16	3 3/4	13/32	2 3/8
13	4	4 3/16	4	13/32	2 1/2

Table *c*. Double-barrel Four-bolt Flange

Nominal size		*A*	*B*	*C*	*D*	*E*	Tap *F*	*G*
No.	Diam.							
4	1 3/4	1 15/16	3 1/8	2 3/16	1 1/4	9/32	7/16–14	13/32
5	2	2 3/16	3 3/8	2 7/16	1 3/8	9/32	7/16–14	13/32
6	2 1/4	2 7/16	3 5/8	2 11/16	1 1/2	9/32	7/16–14	15/32
7	2 1/2	2 11/16	4	2 15/16	1 21/32	9/32	7/16–14	15/32
8	2 3/4	2 15/16	4 1/4	3 3/16	1 25/32	9/32	7/16–14	17/32

Table A1-25.—Carburetor Flanges, Aircraft Types.—(*Continued*)

Table *d*. Double-barrel Six-bolt Flange

Nominal size		*A*	*B*	*C*	*D*	*E*	*F*	*G*	*H*
No.	Diam.								
7	2 1/2	2 11/16	2 13/16	2 15/16	6 5/8	3 11/16	11/32	7/16	2 7/8
8	2 3/4	2 15/16	3	3 3/16	7 1/16	3 7/8	11/32	7/16	3 3/32
9	3	3 3/16	3 3/16	3 7/16	7 9/16	4 1/8	13/32	15/32	3 5/16
10	3 1/4	3 7/16	3 7/16	3 11/16	8 1/16	4 3/8	13/32	15/32	3 9/16
11	3 1/2	3 11/16	3 11/16	3 15/16	8 9/16	4 5/8	13/32	15/32	3 13/16
12	3 3/4	3 15/16	3 15/16	4 3/16	9 1/16	4 7/8	13/32	15/32	4 1/16

Table *e*. Triple-barrel Eight-bolt Flange

Nominal size		*A*	*B*	*D*	*E*	*F*
No.	Diam.					
3	1 1/2	1 11/16	1 15/16	6 3/8	2 1/2	9/32
4	1 3/4	1 15/16	2 3/16	7 1/8	2 3/4	9/32
5	2	2 3/16	2 7/16	8	3 1/8	11/32

(Conforms substantially with AN Standard.)
Aircraft Engine Division report adopted by the Society, January, 1932.

Table A1-26.—Pesco Fuel-pump Characteristics
(From Pump Engineering Service Corporation)

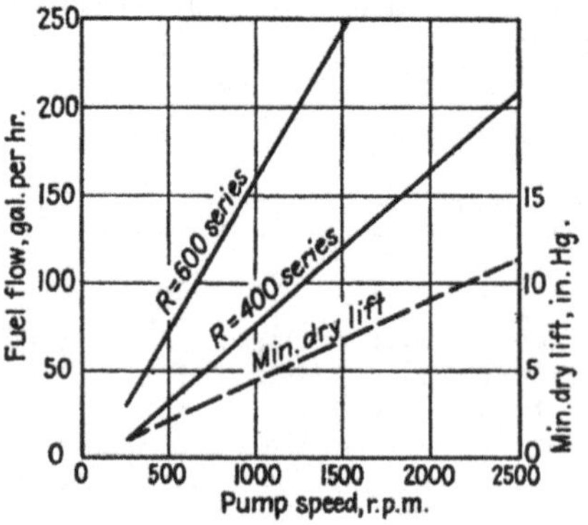

TABLE A1-27.—FUEL-PUMP MOUNTINGS, AIRCRAFT-ENGINE PADS
S.A.E. Standard
(From S.A.E. "Handbook")

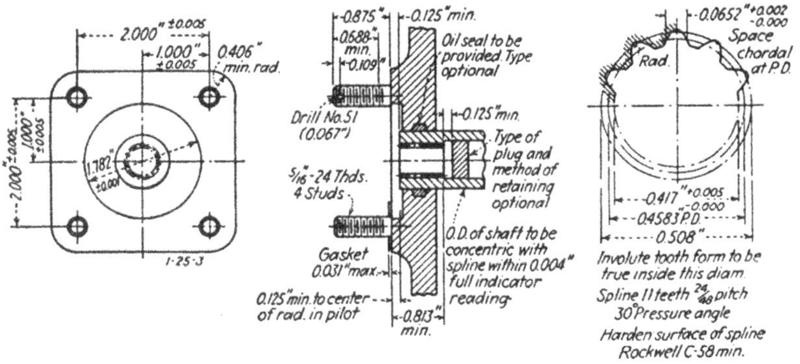

FIG. 1.—Square-type pad. (Conforms substantially with AN Standard, March, 1939.)

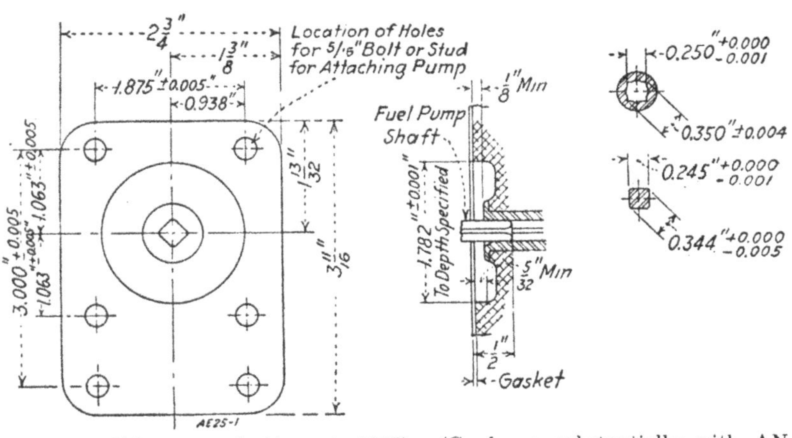

FIG. 2.—Old-type pad (August, 1928). (Conforms substantially with AN Standard, January, 1937.)

Report of Aeronautic Division adopted by the Society, August, 1928. Last revision by Aircraft Engine Division, January, 1940.

TABLE A1-28.—MAGNETO SELECTION

NOTE: The following data should be used for preliminary selections only, and final approval of the selection should be obtained from the manufacturer prior to the starting of construction of the engine.

Number of cylinders	Arrangement	Scintilla magneto type	Ratio magneto to engine speed†	Magneto rotation*	Mounting flange
4	Opposed	SF4R-8	1–1	Clockwise	S.A.E. 2-bolt flanged type
4	In-line	SF4R-8	1–1	Clockwise	S.A.E. 2-bolt flanged type
5	Radial	SF5L-8	1.25–1	Counter-clockwise	S.A.E. 2-bolt flanged type
6	In-line	SF6L-8	1.5–1	Counter-clockwise	S.A.E. 2-bolt flanged type
7	Radial	SF7R-1	0.875–1	1 each	S.A.E. 3-bolt flange single type
9	Radial	SF9L-4	1.125–1	1 each	S.A.E. 3-bolt flange single type
14	2-row radial	SF14L-3	0.875–1	1 each	S.A.E. 3-bolt flange single type

* Viewed from the drive end.

† Magneto drive shafts should be designed to transmit ½ hp.

TABLE A1-29.—S.A.E. MAGNETO MOUNTING-FLANGE DATA
(From S.A.E. "Handbook")

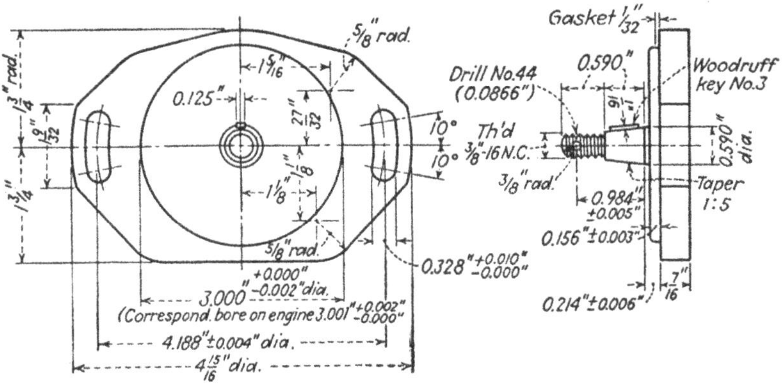

Two-bolt flange, single type (commercial engines to and including six cylinders).

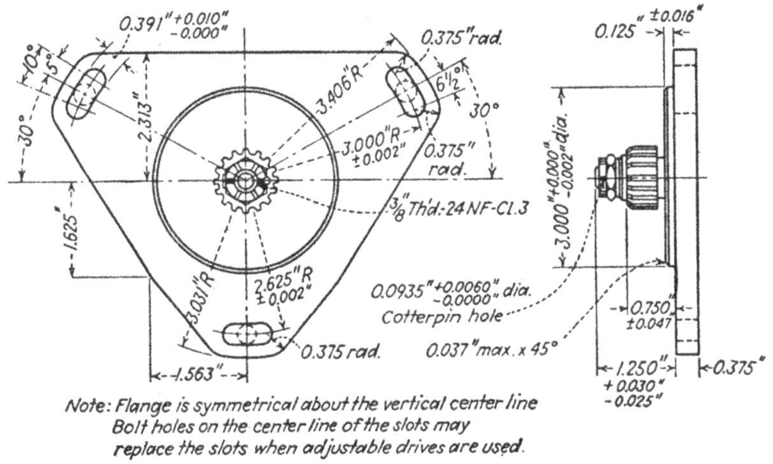

Three-bolt flange, single type (standard flange and drive).

[Conforms substantially with current AN Standard (January, 1941) except bolt slots are wider so 3/8-in. studs can be used on larger size engines, and flange thickness is larger to allow a standard length of engine studs.]

TABLE A1-30.—AIRCRAFT-ENGINE STARTERS
(From Eclipse Aviation Corporation)

NOTE: The selection should be confirmed by the starter manufacturer prior to final approval of the engine design.

Type	Description	Mounting flange	Hand crank available	Weight, lb.	Max. capacity, hp.	Voltage
			Inertia			
Series 6.....	Concentric, hand starter	S.A.E.—6″	Yes	18¼	400	
Series 6.....	Concentric, hand and electric	S.A.E.—6″	Yes	30¼	400	12
Series 6.....	Concentric, hand and electric	S.A.E.—6″	Yes	30¼	400	24
Series 6.....	Concentric, hand with integral booster magneto	S.A.E.—6″	Yes	26½	400	
Series 7.....	Vertical, hand starter	S.A.E.—6″	Yes	31	900	
Series 7.....	Vertical, hand and electric	S.A.E.—6″	Yes	39½	900	12
Series 7.....	Vertical, hand and electric	S.A.E.—6″	Yes	39½	900	24
Series 7A...	Vertical, hand starter	S.A.E.—6″	Yes	34½	1,000	
Series 7A...	Vertical, hand and electric	S.A.E.—6″	Yes	43	1,000	12
Series 7A...	Vertical, hand and electric	S.A.E.—6″	Yes	43	1,000	24
Series 11....	Concentric, hand starter	S.A.E.—6″	Yes	21	900	
Series 11....	Concentric, hand and electric	S.A.E.—6″	Yes	34¾	900	12
Series 11....	Concentric, hand and electric	S.A.E.—6″	Yes	34¾	900	24
Series 11....	Concentric, hand with integral booster magneto	S.A.E.—6″	Yes	28¾	900	
Series 11A..	Concentric, hand starter	S.A.E.—6″	Yes	25	1,000	
Series 11A..	Concentric, hand and electric	S.A.E.—6″	Yes	38¾	1,000	12
Series 11A..	Concentric, hand and electric	S.A.E.—6″	Yes	38¾	1,000	24
Series 16....	Concentric, hand starter	S.A.E.—6″	Yes	21	350	
			Direct-cranking Electric			
Y150.......	Vertical, starter	S.A.E.—5″	No	16¼	150	12
E80........	Concentric, starter	S.A.E.—5″	No	19	250	12
E80........	Concentric, starter	S.A.E.—6″	No	19	250	12
F141.......	Concentric, starter	S.A.E.—6″	No	24½	400	12
E160.......	Concentric, starter	S.A.E.—6″	Yes	32¼	900	12
E160.......	Concentric, starter	S.A.E.—6″	Yes	32¼	1,000	24
			Hand-turning Gear			
4H4........	4:1 ratio H.T.G............	S.A.E.—5″	Yes	8½	115	
BH6........	6:1 ratio H.T.G............	S.A.E.—6″	Yes	12	250	
3HB6......	6:1 ratio H.T.G. with integral booster magneto	S.A.E.—6″	Yes	17	250	
3H8........	8:1 ratio H.T.G............	S.A.E.—6″	Yes	12	300	
3HB8......	8:1 ratio H.T.G. with booster magneto	S.A.E.—6″	Yes	17	300	
3H18.......	18:1 ratio H.T.G.	S.A.E.—6″	Yes	17½	300	
			Combustion			
Type I.....	Concentric-starter...........	S.A.E.—6″	No	25	550	
Type II....	Concentric-starter...........	S.A.E.—6″	No	32	1,250	

TABLE A1-30.—AIRCRAFT-ENGINE STARTERS.—(*Continued*)

Air Injection

Air-injection starter installations normally require the following component units:

1. Main-engine-driven compressor (weight and mounting depends upon make of engine on which installed).
2. Air-storage tank (standard sizes of tanks 6 in. diameter by 20 in. long and 6 in. diameter by 26 in. long).
3. Air-pressure regulating valve.
4. Air-pressure release valve.
5. Air-pressure gage.
6. Primer.
7. Cylinder injector fittings (one furnished for each cylinder).

NOTES TO AIR INJECTION.—Item 1. Compressor will differ depending upon the make of engine on which it is to be installed.

Item 2. Air-storage tank, unless otherwise specified is furnished in the 6- by 20-in. size. Items 3 to 7 inclusive are normally common to all installations. The valves covered by items 3 and 4 are furnished assembled in a fully charged tank (item 2) ready for installation and use.

Weight.—The weight of the air-injector starting equipment installed is approximately 32 lb. for a single-motor application, this weight depending upon the peculiarities of each installation in respect to tubing and fittings required, tank size, etc.

Capacity.—The air-injector starting equipment is recommended for spark-ignition engines of five or more cylinders and rated up to 250 hp.

NOTES TO TABLE.

Mounting Flanges.—See Table A1-31 for dimensional details of the S.A.E. 5 in. and S.A.E. 6 in. diameter engine-starter mounting.

Important.—All cartridge starters incorporate a special 12-tooth engaging jaw requiring a similar engaging member on the engine.

Weights.—*a.* All weights on hand and electric inertia starters include a suitable solenoid relay for the accelerating motor, which is mounted on and connected electrically to the starter.

b. The weights given above on all starters do not include the required hand crank and cranking extension which unless otherwise specified are furnished with all starters providing manual operation.

c. Weights on all electrically operated starters do not include shielding.

d. Weights listed under Combustion starters include all component parts with standard lengths of intake and exhaust tubings with fittings.

Capacity.—The maximum engine horsepower for which any type starter is recommended is approximate and based upon an average installation on a four-cycle spark-ignition engine of conventional design, operating under normal conditions. Factors other than horsepower ratings must be considered in selecting the proper capacity or type starter for a particular engine. It is therefore recommended that unless reliable data are available, and for initial installations, that the manufacturer be consulted so that an analysis of the requirements can be made and definite recommendations submitted.

TABLE A1-30.—AIRCRAFT-ENGINE STARTERS.—(*Continued*)

Voltage.—*a.* All electrically operated starters can be furnished either grounded or ungrounded and shielded or unshielded. Unless otherwise specified, electrically operated starters are furnished as grounded unshielded units. The application of shielding to the various type starters listed above will increase the listed weights as follows:

	Grounded	Ungrounded
Inertia	1/4 lb.	5/16 lb.
Direct-cranking electric	1/16 lb.	1/8 lb.

b. The firing-control switch of the combustion starters and the electric connection at the loading breech incorporates threaded shielding outlets.

c. Integral booster magnetos furnished on series 6 and series 11 inertia starters incorporate threaded shielding at the electrical outlets.

TABLE A1-31.—STARTING-MOTOR MOUNTINGS, AIRCRAFT-ENGINE PADS
S.A.E. Standard
(From S.A.E. "Handbook")

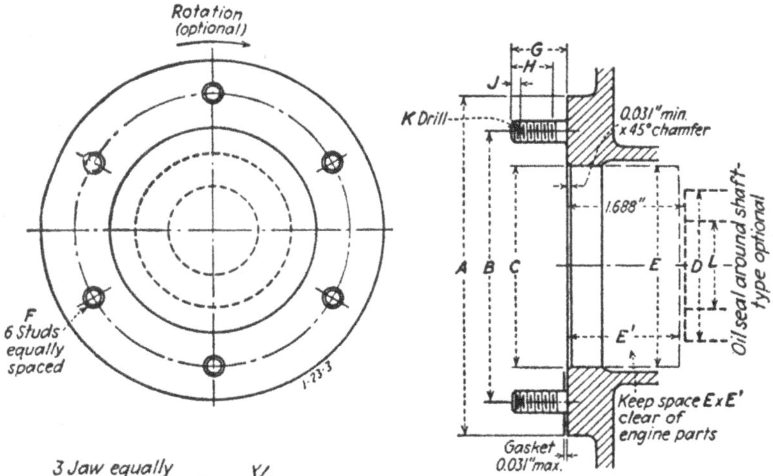

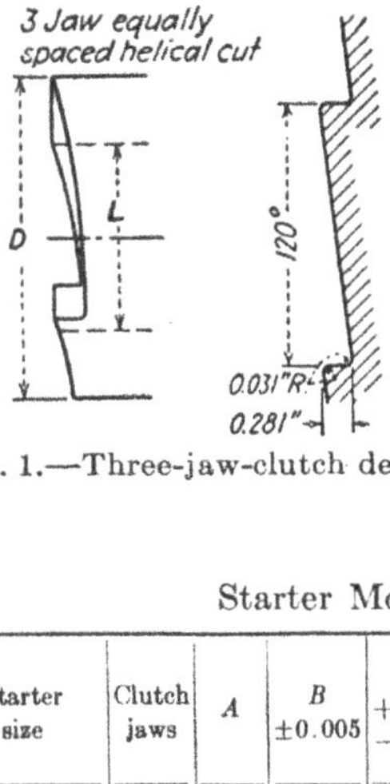

FIG. 1.—Three-jaw-clutch development.

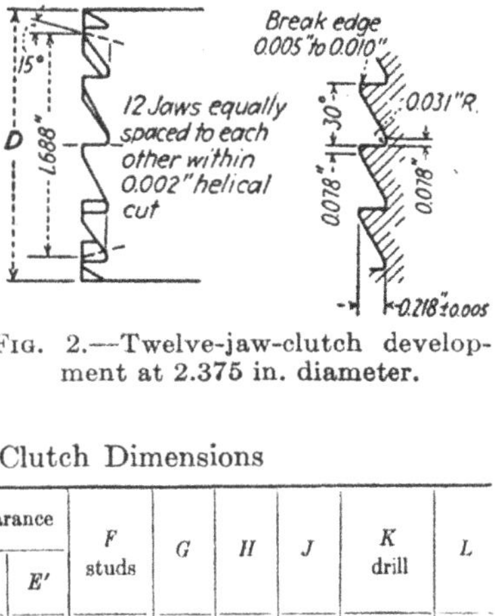

FIG. 2.—Twelve-jaw-clutch development at 2.375 in. diameter.

Starter Mounting and Clutch Dimensions

Starter size	Clutch jaws	A	B ±0.005	C +0.003 −0.000	D	Clearance E	Clearance E'	F studs	G	H	J	K drill	L
Small........	3	5.000	4.000	3.000	2.250	3.000	1.625	5/16–24	0.813	0.625	.....	....	1.000
Medium.....	3 or 12	6.000	5.000	4.125	2.250	4.125	1.500	3/8–24	0.938	0.813	0.125	#36 (0.106)	1.688 max.
Large.......	3 or 12	7.000	5.750	4.625	3.250	4.625	1.500	7/16–20	1.063	0.813	0.125	#36 (0.106)	1.688 max.

Stud threads to be American Standard Fine (NF).

(Conforms substantially with AN Standard, March, 1939.)

Report of Aeronautic Division adopted by the Society, August, 1928. Last revision by Aircraft Engine Division, March, 1939.

TABLE A1-32.—SINGLE-VOLTAGE GENERATORS
(From Eclipse Aviation Corporation)

NOTE: The selection should be confirmed by the generator manufacturer prior to final approval of the engine design.

Ratings

Type	Volts	Amperes	Watts	Weight, lb.	Mounting pad
AL-1 (3d brush)	15	15	225	17¾	4-bolt type
G	15	15	225	15¾	4-bolt type
D	15	25	375	21¼	4-bolt type
E	15	50	750	31¼	4-bolt type
DG-4	30	10	300	21	
E	30	20	600	31¼	4-bolt type

Permanent mount control box (3 unit for 15- or 30-volt operation)	4½ lb.
Permanent mount control box (2 unit for 15-volt operation)	2¼ lb.
Permanent mount control box (2 unit for 3d brush, generator)	2 lb.
Quick detachable control box (3 unit for 15- or 30-volt operation)	5 lb.
Control panel (2 unit for 15-volt operation)	3 lb.
FB-5 filter unit (for 15- or 30-volt operation)	4 lb.

NOTE: The preceding rated outputs are obtained at a generator speed of 2,250 to 4,200 r.p.m. A minimum speed of 2,250 r.p.m. is readily obtainable at cruising speed from the generator drives of all standard types of aircraft engines, as the gearing to the drive shaft is normally at a ratio of 1½ times engine speed. The ratings given as 15 and 30 volts are the values required for operating with 12- and 24-volt battery systems, respectively. The foregoing weights include shielded terminal covers which are standard on all generators, control boxes, and filter unit. The covers are threaded for the attachment of metallic conduit for radio shielding.

AIRCRAFT GENERATOR MOUNTINGS S.A.E. Standard (From S.A.E. "Handbook")

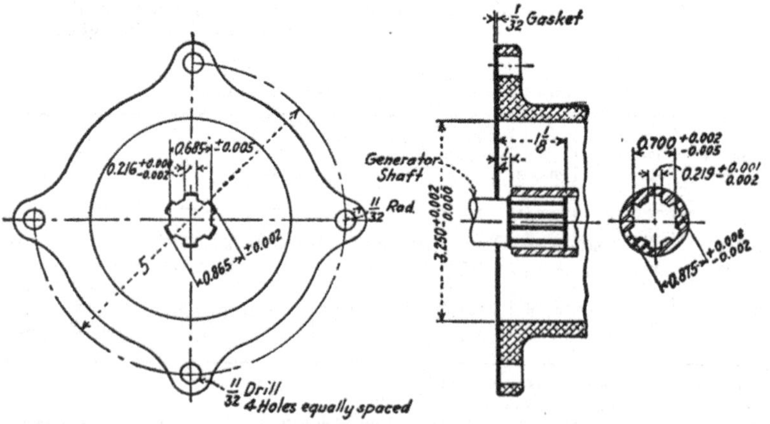

Generator mounting pad, four-bolt type. (Conforms substantially with AN Standard, March, 1939.)

TABLE A1-33.—TACHOMETER DRIVE, AIRCRAFT
S.A.E. Standard
(From S.A.E. "Handbook")
Drive for Mechanical Types

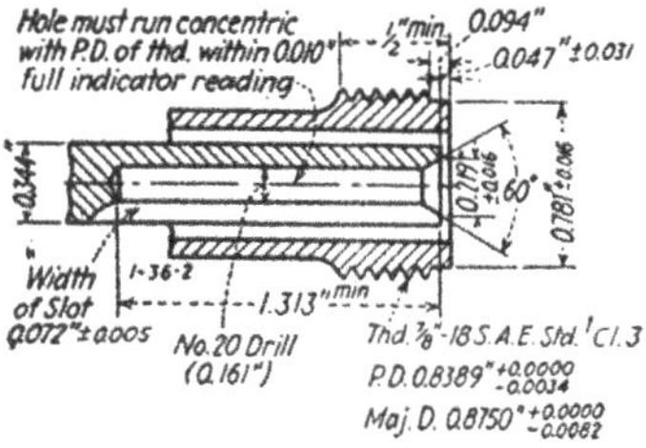

Engine connection.

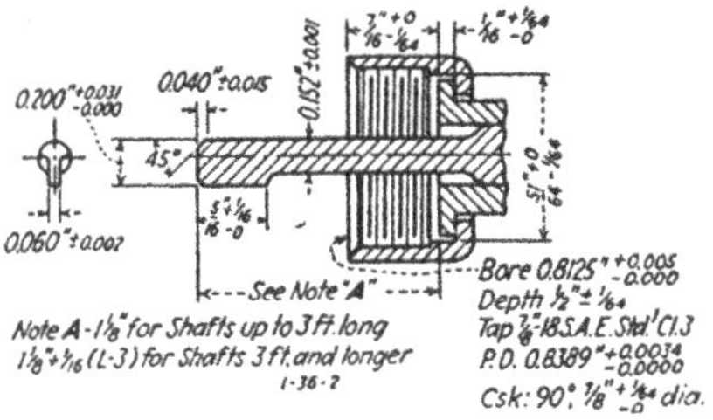

Engine end of drive shaft.

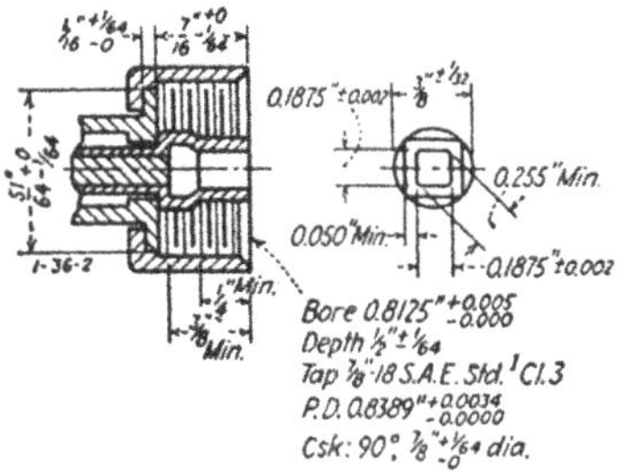

Tachometer end of drive shaft.

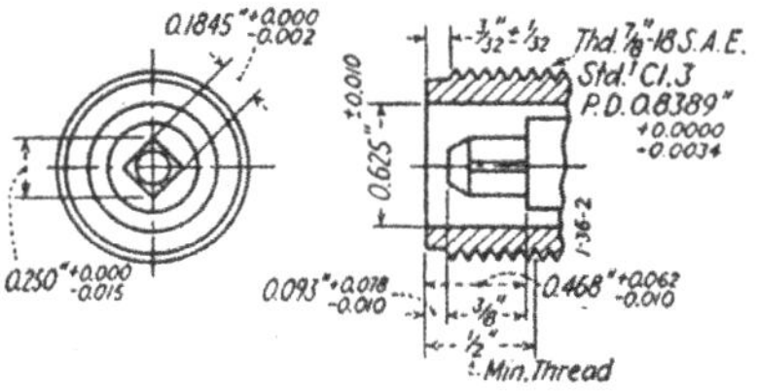

Tachometer connection.

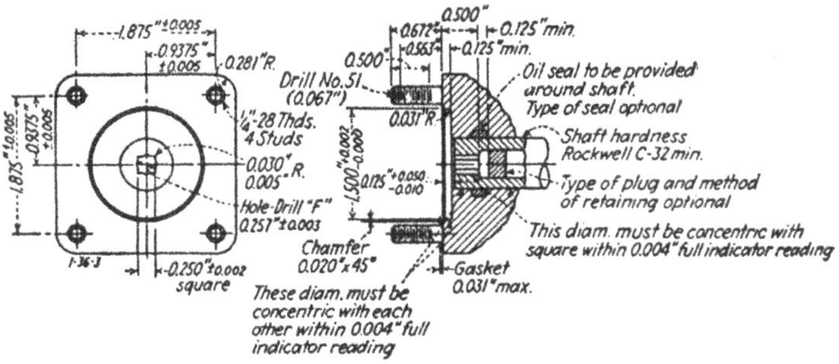

Drive for Electrical Types

NOTE: Tachometer drives to rotate at half engine-crankshaft speed.

Report of the Aeronautic Division adopted by the Society, March, 1918. Last revision by the Aircraft Engine Division, January, 1939.

[1] S.A.E. Special Pitch thread, Class 3.

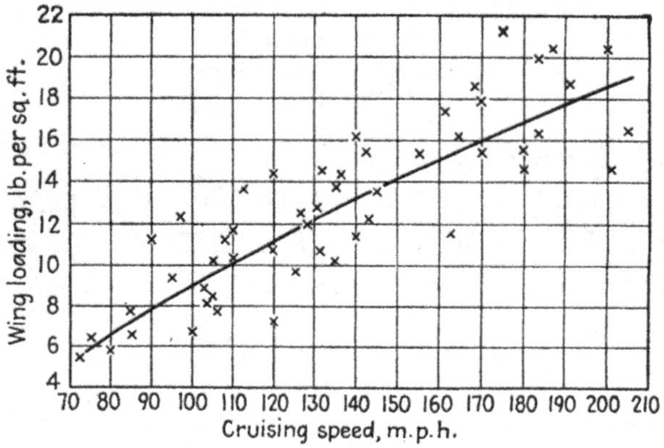

FIG. A1-1.—Wing-loading data for 54 American airplanes.

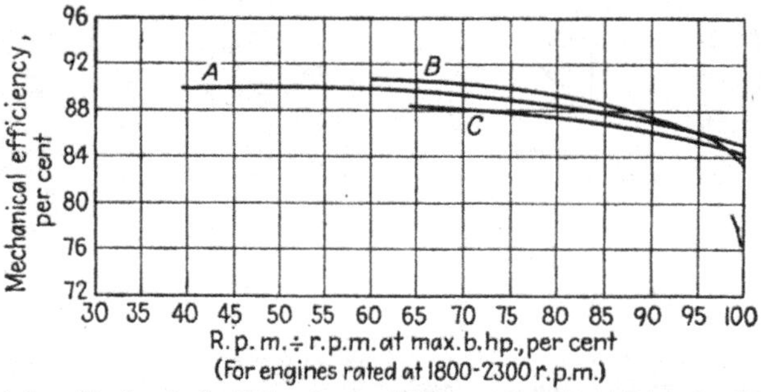

FIG. A1-2.—Mechanical efficiencies at full throttle. [(*a*) *Ricardo, High Speed Internal Combustion Engines*, p. 275; (*b*) *Marks, The Airplane Engine*, p. 24 (*Liberty* 12); (*c*) *Judge, Automobile and Aircraft Engines*, *p*. 410.]

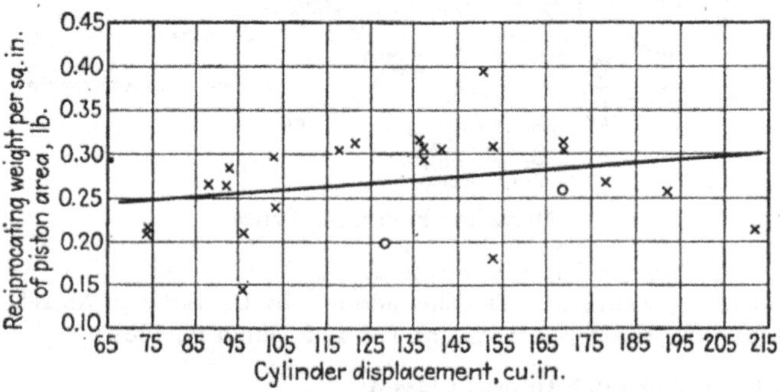

FIG. A1-3.—Reciprocating-weight–cylinder-displacement data for 25 airplane engines.

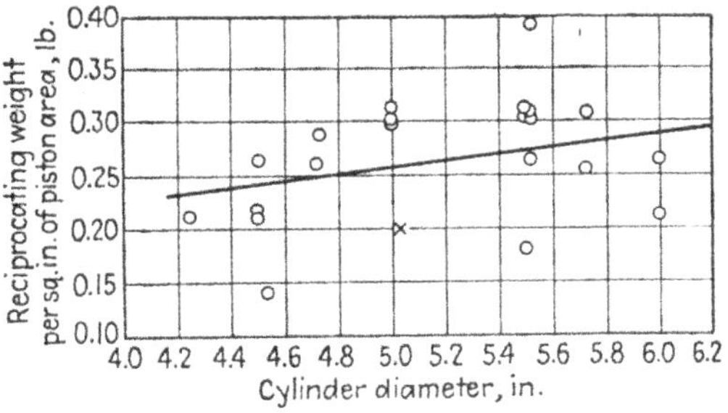

FIG. A1-4.—Reciprocating-weight-cylinder-diameter data for 25 airplane engines.

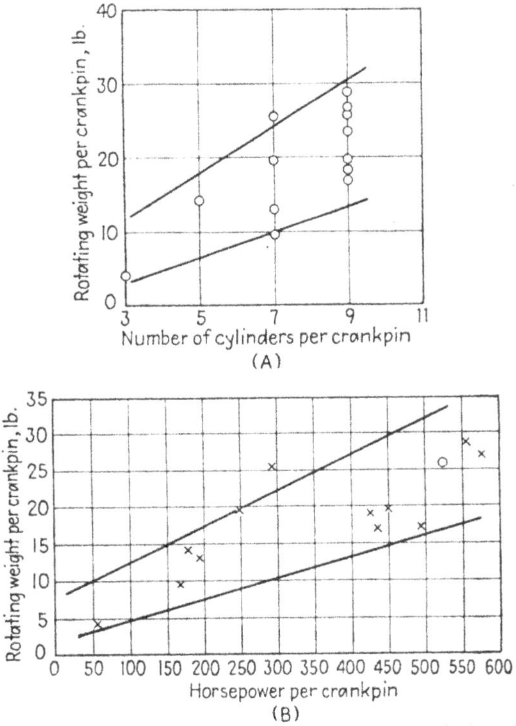

FIG. A1-5.—Rotating weights per crankpin for radial engines.

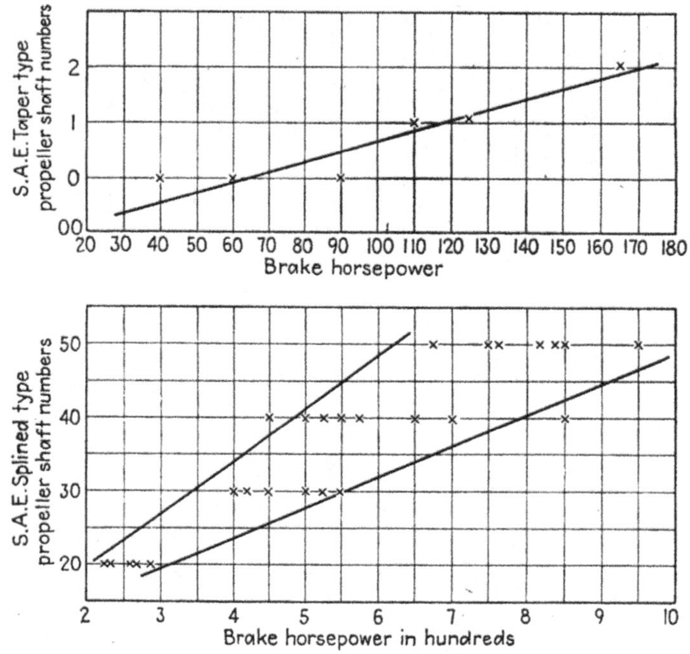

FIG. A1-6.—S.A.E. propeller-shaft sizes vs. rated brake horsepower.

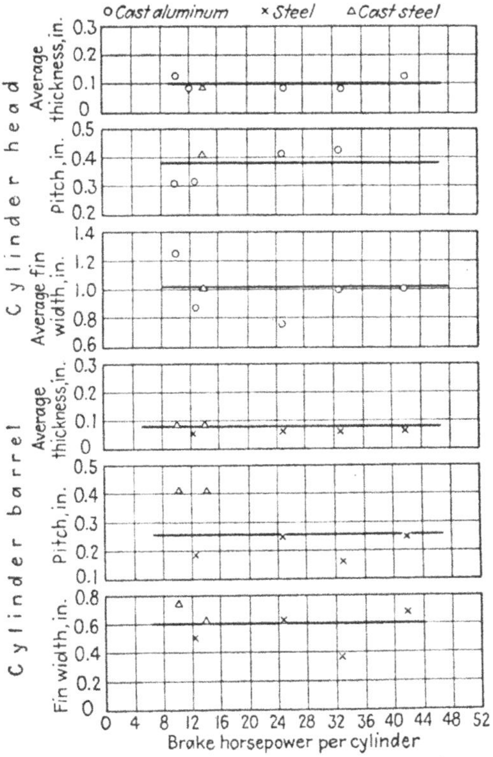

FIG. A1-7.—Cylinder cooling-fin dimensions from several current makes of aircraft engines.

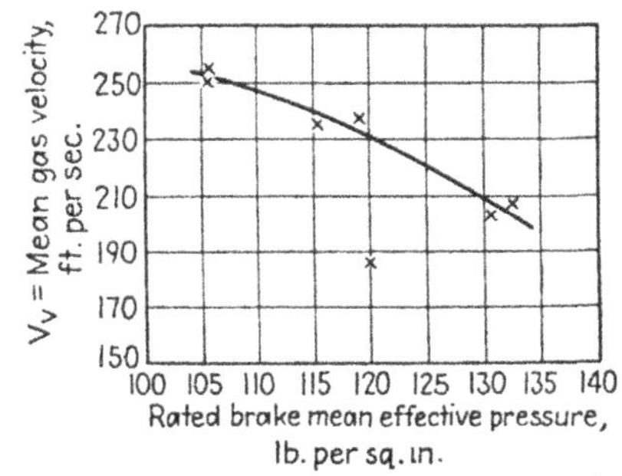

FIG. A1-8.—Effect of mean gas velocity through inlet valve on rated b.m.e.p. for several nonsupercharged aircraft engines.

APPENDIX 2

TABLE A2-1.—CORRELATION OF NUMBERING SYSTEMS FOR ALUMINUM AND MAGNESIUM ALLOYS

(From *Product Engineering*, November 1940, and "Aluminum in Aircraft," 1941)

Alloy trade designation	Federal	Army	Navy	S.A.E.	A.S.T.M.
Aluminum Rolled Sheet (for Baffles, Cowling, Etc.)					
17S	QQ-A-353	Federal	47A3c	26	B78-36T
24S	QQ-A-355	Federal	47A10	24	
Aluminum Sand Castings (for Cylinder Heads, Crankcases, Etc.)					
142	QQ-A-601	AN-QQ-A-379	AN-QQ-A-379	39	B26-37T
355	QQ-A-601	AN-QQ-A-376	AN-QQ-A-376	322	B26-37T
A355	QQ-A-601	Federal	M212a (8-24-34)		
195	QQ-A-601	AN-QQ-A-390	AN-QQ-A-390	38	B26-37T
Aluminum Permanent-mold Castings (for Pistons, Rocker Box Covers, Etc.)					
122	QQ-A-596	....	46A15	34	B108-38T
A132	QQ-A-596	AN-QQ-A-386	AN-QQ-A-386	321	B108-38T
355	QQ-A-596	....	46A15	322	
Aluminum Forgings (for Pistons, Crankcases, Connecting Rods, Impellers, Etc.)					
14S	QQ-A-367a	Federal	46A7b		
18S	QQ-A-367a	Federal	46A7b		
25S	QQ-A-367a	Federal	46A7b	27	
32S	QQ-A-367a	Federal	46A7b		
A51S	QQ-A-367a	Federal	46A7b	280	
Magnesium Sand- and Permanent-mold Castings For (AM 240) Cover Plates, Valve Covers (AM 260) Accessory Drive Housings (AM 265) Crankcases					
AM240 or G	....	....	....	...	B80-36T
AM260	....	....	M-112g (3-15-40)		
AM265 or H	....	57-74-1C	M-112g (3-15-40)	...	B80-36T

NOTE: See references for other alloys.

TABLE A2-2.—NOMINAL COMPOSITION OF ALUMINUM AND MAGNESIUM ALLOYS*
(From "Alcoa Aluminum and Its alloys," and "Mazlo Magnesium Alloys")

Alloy designation	Per cent of alloying elements, aluminum and normal impurities constitute the remainder						
	Copper	Iron	Silicon	Chromium	Magnesium	Nickel	Manganese
Aluminum Sand-casting Alloys							
142	4.0	...	....	....	1.5	2.0	
195	4.0						
355	1.3	...	5.0	....	0.5		
A355	1.4	...	5.0	....	0.5	0.8	0.8
Aluminum Permanent-mold-casting Alloys							
122	10.0	1.2	....	....	0.2		
A132	0.8	0.8	12.0	....	1.0	2.5	
355	1.3	...	5.0	....	0.5		
Aluminum Wrought Alloys							
14S	4.4	...	0.8	....	0.4	...	0.8
17S	4.0	...	....	....	0.5	...	0.5
18S	4.0	...	....	....	0.5	2.0	
24S	4.5	...	....	....	1.5	...	0.6
25S	4.5	...	0.8	....	...	...	0.8
32S	0.9	...	12.5	....	1.0	0.9	
A51S	....	...	1.0	0.25	0.6		

Alloy designation	Per cent of alloying constituents, magnesium constitutes the remainder				
	Aluminum	Manganese	Zinc	Silicon	Total impurities
Magnesium Sand- and Permanent-mold-casting Alloys					
AM240	9.0 -11.0	0.10 min	0.3 max.	0.5 max.	0.3 max.
AM260	8.75- 9.25	0.10 min.	1.8-2.2	0.3 max.	0.3 max.
AM265	5.3 - 6.7	0.15 min.	2.5-3.5	0.5 max.	0.3 max.

* Heat-treatment symbols have been omitted since composition does not vary for different heat-treatment practices.

TABLE A2-3.—PHYSICAL CONSTANTS OF ALUMINUM AND MAGNESIUM ALLOYS
(From "Alcoa Aluminum and Its Alloys," and "Mazlo Magnesium Alloys")

Alloy designation	Density, lb. per cu. in.	Thermal* conductivity at 100°C., c.g.s. units	Average coefficients of thermal expansion per deg. F.		
			68–212°F.	68–392°F.	68–572°F.
Aluminum Sand-casting Alloys					
142	0.099	0.35	0.0000125	0.0000130	0.0000136
195	0.100	0.34	0.0000127	0.0000133	0.0000138
355	0.097	0.33	0.0000122	0.0000127	0.0000133
A355	0.098	0.31	0.0000119	0.0000125	0.0000130
Aluminum Permanent-mold-casting Alloys					
122	0.103	0.31	0.0000122	0.0000127	0.0000130
A132	0.097	....	0.0000105	0.0000111	0.0000116
355	0.097	0.33	0.0000122	0.0000127	0.0000133
Aluminum Wrought Alloys					
14S	0.101	0.37	0.0000122	0.0000130	0.0000138
17S	0.101	0.28	0.0000122	0.0000130	0.0000138
18S	0.101	0.37	0.0000122	0.0000130	0.0000138
24S	0.100	0.28	0.0000122	0.0000130	0.0000138
25S	0.101	0.37	0.0000122	0.0000130	0.0000138
32S	0.097	0.32	0.0000108	0.0000114	0.0000119
A51S	0.097	0.41	0.0000130	0.0000136	0.0000141
Magnesium Sand and Permanent-mold Castings					
AM240 AM260 AM265	0.066 0.066	0.17 0.18 approx.	Mean coefficient of thermal expansion (65 to 750°F.) per deg. F. = 0.000016 for these magnesium alloys		

* For aluminum alloys in the heat-treated condition.

TABLE A2-4.—MECHANICAL PROPERTIES OF ALUMINUM AND MAGNESIUM ALLOYS[1]

(From "Alcoa Aluminum and Its Alloys" and "Mazlo Magnesium Alloys")

Alloy designation	Min. values for specs.		Typical values (not guaranteed)							
	Tension[2]		Tension[2]			Compression[3]			Fatigue[5]	Density
	Ultimate strength, lb. per sq. in.	Elongation, % in 2 in.	Yield strength (set = 0.2%), lb. per sq. in.	Ultimate strength, lb. per sq. in.	Elongation, % in 2 in.	Yield strength (set = 0.2%), lb. per sq. in.	Brinell hardness,[2] 500-kg. load 10-mm. ball	Shearing strength, lb. per sq. in.	Endurance limit, lb. per sq. in.	Lb. per cu. in.
Aluminum Sand-casting Alloys										
142-T61	32,000	[4]	32,000	37,000	0.5	47,000	100	32,000	8,000	0.099
195-T6	32,000	3.0	22,000	36,000	5.0	25,000	80	30,000	6,500	0.100
355-T6	32,000	2.0	25,000	35,000	3.5	29,000	80	30,000	8,500	0.097
A355-T51	25,000	[4]	24,000	28,000	1.5	24,000	65	22,000	8,500	0.099
Aluminum Permanent-mold-casting Alloys										
122-T551	30,000	[4]	35,000	37,000	0.0	40,000	115	27,000	8,500	0.104
A132-T551	31,000	[4]	28,000	36,000	0.5	30,000	105	24,000	......	0.097
355-T6	37,000	1.5	26,000	43,000	4.0	26,000	90	30,000	......	0.097
Magnesium Sand- and Permanent-mold-casting Alloys										
AM240-T4	29,000	6.0	12,000	35,000	9.0	......	52	20,000	11,000	0.066
AM240-T6	29,000	2.0	16,000	35,000	4.0	......	60	21,000	8,000	
AM260-T4	30,000	6.0	14,000	39,000	10.0	......	63	20,000	9,000	
AM265-T4	30,000	6.0	12,000	37,000	9.0	......	51	18,000	11,000	0.066
Aluminum Forging Alloys[6,7]										
14S-T	65,000	10.0	50,000	......	....	......	130	45,000	16,000	0.101
17S-T	55,000	16.0	30,000	......	....	......	100	36,000	15,000	0.101
18S-T	55,000	10.0	35,000	......	....	......	100	......	14,500	0.103
25S-T	55,000	16.0	30,000	......	....	......	100	35,000	15,000	0.101
32S-T	52,000	5.0	40,000	......	....	......	115	38,000	14,000	0.097
A51S-T	44,000	14.0	34,000	......	....	......	90	32,000	10,500	0.097

1 See references for additional data.

2 Tension and hardness values determined from standard ½-in. diameter test specimens.

3 Results of tests on specimens having an L/R ratio of 12.

4 Not specified. The error in determining low elongations is comparable with the value being measured.

5 Endurance limits are based on 500,000,000 reversals.

6 Properties for aluminum forging alloys apply to forgings up to 4 in. in diameter or thickness. Long axis of test specimens taken parallel to direction of grain flow.

7 Yield and hardness data on aluminum forging alloys are minimum specification values.

Modulus of elasticity for aluminum alloys, 10,300,000 lb. per sq. in.

Modulus of elasticity for magnesium alloys, 6,500,000 lb. per sq. in.

TABLE A2-5.—TYPICAL TENSILE PROPERTIES OF ALUMINUM AND MAGNESIUM ALLOYS AT ELEVATED TEMPERATURES
(From "Alcoa Aluminum and Its Alloys" and "Mazlo Magnesium Alloys")

Temp., deg. F.	Sand-casting alloy No. 142-T61 Strength, lb. per sq. in.		Elongation, % in 2 in.	Sand-casting alloy No. 195-T6 Strength, lb. per sq. in.		Elongation, % in 2 in.	Sand-casting alloy No. 355-T6 Strength lb. per sq. in.		Elongation, % in 2 in.	Sand-casting alloy No. A355-T51 Strength, lb. per sq. in.		Elongation, % in 2 in.
	Yield	Tensile		Yield	Tensile		Yield	Tensile		Yield	Tensile	
75	32,000	37,000	0.5	22,000	36,000	5.0	25,000	35,000	3.5	24,000	28,000	1.5
300	28,000	30,000	0.5	13,000	24,000	9.0	25,000	30,000	3.0	20,000	24,000	1.5
400	25,000	27,000	1.0	9,000	15,000	20.0	9,000	13,000	12.0	11,000	16,000	3.5
500	5,000	12,000	9.0	6,000	9,500	25.0	5,000	8,000	22.0	5,000	8,000	18.0
600	3,500	7,500	10.0	3,000	4,000	80.0	3,500	6,000	30.0	4,500	7,000	16.0
	Permanent-mold-casting alloy No. 122-T551			Permanent-mold-casting alloy No. A132-T551			Permanent-mold-casting alloy No. 355-T6			Wrought alloy No. 17S-T		
75	35,000	37,000	0.0	28,000	36,000	0.5	26,000	43,000	4.0	40,000	62,000	20
300	29,000	33,000	0.0	22,000	31,000	1.0	25,000	31,000	3.0	34,000	40,000	16
400	20,000	26,000	1.0	13,500	23,000	2.0	9,000	12,000	20.0	21,000	26,000	25
500	12,500	18,000	3.0	9,500	17,500	2.0	6,000	8,000	25.0	9,500	13,000	35
600	6,000	10,000	10.0	5,000	11,000	8.0	3,000	4,500	50.0	3,500	5,500	90
	Wrought alloy No. 24S-T			Wrought alloy No. 14S-T			Wrought alloy No. 18S-T			Wrought alloy No. 25S-T		
75	45,000	68,000	22	55,000	70,000	14	47,000	63,000	17	35,000	57,000	18
300	35,000	42,000	21	39,000	43,000	14	44,000	49,000	10	28,000	35,000	14
400	23,000	28,000	25	13,000	17,000	28	......	......	....	13,500	19,000	24
500	10,000	14,000	40	8,500	10,500	32	7,000	11,000	32	4,500	6,500	45
600	6,000	7,500	65	4,500	6,000	45	......	......	....	4,000	4,500	50
700	......	......	....	3,500	4,000	55	2,500	4,000	85	3,000	3,500	55
	Wrought alloy No. 32S-T			Wrought alloy No. A51S-T			Sand- and permanent-mold-casting alloy No. AM240-T4			Sand- and permanent-mold-casting alloy No. AM265-T4		
75	46,000	56,000	8	40,000	47,000	20	13,700	32,000	8.2	13,600	36,200	8.5
300	33,000	39,000	9	15,000	19,000	28	......	23,300	9.0	......	22,400	33.2
400	11,000	16,000	34	5,500	7,500	58	......	......	....	......	14,600	35.5
500	6,500	8,500	50	4,500	5,500	59	......	12,500	22.5	......	9,800	28.0
600	3,500	6,000	60	3,500	4,500	60						
700	2,000	3,500	120	3,000	3,500	65						

TABLE A2-6.—S.A.E. STEEL NUMBERING SYSTEM
S.A.E. Standard
(From S.A.E. "Handbook")

Compositions that do not conform to the S.A.E. compositions, or that are not included in the S.A.E. Standard, should not bear the prefix "S.A.E."

(NOTE: For detailed data on chemical composition, grain-size charts, heat-treatments, hardness tests, magnaflux data, physical properties, etc., see Steel Specifications in S.A.E. "Handbook.")

A numeral index system is used to identify the compositions of the S.A.E. steels, which makes it possible to use numerals on shop drawings and blueprints that are partly descriptive of the composition of material covered by such numbers. The first digit indicates the type to which the steel belongs; thus "1-" indicates a carbon steel; "2-" a nickel steel and "3-" a nickel chromium steel. In the case of the simple alloy steels, the second digit generally indicates the approximate percentage of the predominant alloying element. Usually the last two or three digits indicate the average carbon content in "points" or hundredths of 1 per cent. Thus "2340" indicates a nickel steel of approximately 3 per cent nickel (3.25 to 3.75) and 0.40 per cent carbon (0.35 to 0.45).

In some instances, in order to avoid confusion it has been found necessary to depart from this system of identifying the approximate alloy composition of a steel by varying the second and third digits of the number. An instance of such departure is the steel numbers selected for several of the corrosion- and heat-resisting alloys.

The basic numerals for the various types of S.A.E. steel are as follows:

Type of Steel	Numerals (and Digits)
Carbon steels	1xxx
Plain carbon	10xx
Free cutting, (screw stock)	11xx
Free cutting, manganese	X13xx*
Manganese	13xx
Nickel steels	2xxx
3.50 per cent nickel	23xx
5.00 per cent nickel	25xx
Nickel chromium steels	3xxx
1.25 per cent nickel, 0.60 per cent chromium	31xx
1.75 per cent nickel, 1.00 per cent chromium	32xx
3.50 per cent nickel, 1.50 per cent chromium	33xx
3.00 per cent nickel, 0.80 per cent chromium	34xx
Corrosion and heat-resisting steels	30xxx

* The prefix "X" is used in numerous instances to denote variations in the range of elements.

Report on Iron and Steel Division adopted January, 1912. Last revision January, 1941.

TABLE A2-6.—S.A.E. STEEL NUMBERING SYSTEM.—(*Continued*)

Molybdenum steels	4xxx
Chromium	41xx
Chromium-nickel	43xx
Nickel	46xx and 48xx
Chromium steels	5xxx
Low-chromium	51xx
Medium-chromium	52xxx
Corrosion- and heat-resisting	51xxx
Chromium-vanadium steels	6xxx
Silicon-manganese steels	9xxx

TABLE A2-7.—MAIN AND CONNECTING-ROD BEARINGS
S.A.E. Standard
(From S.A.E. "Handbook")

During the past few years, new materials for high-duty bearings have been developed which it is believed are now sufficiently stabilized for standardization. Accordingly, the previous specifications have been revised, new ones added, and all grouped under the heading main and connecting-rod bearings.

The choice of the material to use for main and connecting-rod bearings depends upon a number of factors. The resistance to fatigue of bearing materials depends to a great extent upon the design of the bearing, the strength and rigidity of the supporting structure, the thickness of the backing metal (steel or bronze), the thickness of the bearing material, and the physical properties of the bond between the bearing material and the backing. The resistance to corrosion depends upon the chemical composition and characteristics of both lubricant and bearing alloy and upon temperature and other operating conditions.

The S.A.E. tin- and lead-base babbitts have nonscoring and nonwearing properties and are resistant to corrosion from organic acidity of the type normal to lubricating oil, but they are low in resistance to fatigue. Copper-lead bearings are inferior to tin- and lead-base babbitts in nonscoring but they are greatly superior in resistance to fatigue. Cadmium bearings may approach the tin- or lead-base babbitts in nonscoring and the copper-lead bearings in resistance to fatigue, depending upon design and operating conditions. However, the S.A.E. copper-lead and the S.A.E. cadmium bearings may corrode if operated at sufficiently high temperatures using lubricants containing animal or vegetable oil additions or using mineral oils which develop acidic compounds on oxidation. Numerous satisfactory applications of these two alloys are made in cases where such acidity is not present and does not develop in service.

Many other alloys are sometimes used for main and connecting-rod bearings, depending upon design and operating conditions. Reference should be made to the S.A.E. brass, bronze and copper-alloy specifications for data in regard to the composition and properties of these other bearing materials.

TABLE A2-7.—MAIN AND CONNECTING-ROD BEARINGS.—(*Continued*)

The analysis of finished bearings shall be made from a sample taken between the bearing surface and a point midway between the bearing surface and the bonding material.

Babbitt Metals

Composition in percentage	S.A.E. 10		S.A.E. 110		S.A.E. 11		S.A.E. 13		S.A.E. 14	
	Bearing	Ingot	Bearing	Ingot	Bearing	Ingot	Bearing	Ingot	Bearing	Ingot
Tin, min	90.0	90.75	87.75	89.0	86.0	87.25	4.5–5.5	4.75–5.25	9.25–10.75	9.75–10.25
Antimony	4.0–5.0	4.25–4.75	7.0–8.5	7.25–8.25	6.0–7.5	6.5–7.0	9.75–10.75	9.75–10.25	14.0–16.0	14.75–15.25
Lead, max	0.35*	0.35	0.35*	0.35	0.35	0.35	86.0	85.5	76.0	75.25
Copper, max	4.0–5.0	4.25–4.75	2.25–3.75	2.5–3.5	5.0–6.5	5.5–6.0	0.50	0.50	0.50	0.50
Iron, max	0.08	0.08	0.08	0.08	0.08	0.08				
Arsenic, max	0.10	0.10	0.10	0.10	0.10	0.10	0.60	0.60	0.60	0.60
Bismuth, max	0.08	0.08	0.08	0.08	0.08	0.08				
Zinc and aluminum	None	None	None	None	None	None	None	None	None	None

* A maximum of 0.60 per cent lead is permissible in finished-steel or bronze-backed bearings provided a lead-tin solder has been used in bonding the bearing material to the backing metal.

Typical compositions of a rolled-bronze split bushing and a composite cast alloy on a steel back are as follows:

Composition in percentage	S.A.E. 791 rolled or wrought	S.A.E. 792 cast lining
Copper	86.0 –88.0	78.0–82.0
Tin	3.50– 4.50	9.0–11.0
Zinc	3.00– 5.00	
Lead	3.50– 4.50	9.0–11.0
Iron, max	0.10	0.30
Phosphorus, max	....	0.01
Other impurities, max	0.20	0.30

S.A.E. 791 is a rolled split-bushing alloy and is similar in many of its properties to the cast alloy S.A.E. 40.

TABLE A2-7.—MAIN AND CONNECTING-ROD BEARINGS.—(*Continued*)

S.A.E. 792 is the cast-bronze liner for a rolled split bushing made with a steel backing, usually S.A.E. 1010 or S.A.E. 1015. This bronze is similar in many of its properties to the cast alloy S.A.E. 64.

GRAPHITE BRONZE AND SINTERED ALLOY BEARINGS

General Information

Graphite bronze bearings may, in some installations, replace cast or wrought bearings. One type is a split bushing with indentations rolled into its inner surface which, in the case of graphite bearings, are filled under pressure with graphite paste. Four alloys are in general use as follows:

Composition in percentage	Alloys			
	A	B	C	D
Copper	89.0 –91.0	Remainder	65.5–68.5	Remainder
Tin	0.25– 0.75	3.5 –4.5	....	4.0 –6.0
Lead	....	3.5 –4.5	3.5– 4.5	0.75–1.25
Zinc	Remainder	3.25–4.75	Remainder	0.35 max.
Iron, max	0.10	0.10	0.10	0.05
Other impurities, max.	0.20	0.20	0.20	

The temper of the wrought strip metal used for these bushings is usually 1 to 3 Brown and Sharpe gage numbers hard, depending on the hardness specified by the purchaser.

The minimum Rockwell hardness of the finished bushing is 95 for alloy A, 90 for alloy B, and 80 for alloy C on the B scale, using a 1/8-in. diameter ball and 100-kg. load.

The split graphite bronze bushings are used where the motion is oscillatory and not rotary, such as in rocker arms, steering knuckles and arms, and piston pins, and for starting motors, distributor shafts, and places where rotary speed is not high and unit load is low. They are also used on the chassis frame, in brake systems, etc.

Another type of graphite bronze bearing is made by molding into the desired shape mixtures of graphite with powdered metals or metallic compounds and heat-treating to effect alloying of the metallic ingredients. The bushings are then sized to definite finished dimensions, which results in a porous structure capable of absorbing appreciable quantities of the lubricant and having graphite uniformly dispersed throughout the mass. These bushings can be manufactured in various porosities, densities, and load-carrying capacities, depending upon the use to which they are to be put. The porosity can be varied from practically zero to as high as 30 per cent by volume. These bushings are usually finished to the final dimensions and require no finish machining, but they must be properly supported.

TABLE A2-7.—MAIN AND CONNECTING-ROD BEARINGS.—(*Continued*)

Other bearing materials, with or without graphite and having the properties of cast or wrought bearing alloys, may be made by the powder and sintering processes.

Report on Non-ferrous Metals Division adopted June, 1911. Last revision January, 1940.

TABLE A2-8.—BRASS AND BRONZE CASTINGS SUITABLE FOR BUSHINGS, ETC.

(From S.A.E. "Handbook")

Spec. No. 63—Leaded Gun Metal Castings

Composition	%
Copper	86.00–89.00
Tin	9.00–11.00
Phosphorus, max	0.25
Zinc and other impurities, max	0.50
Lead	1.00– 2.50

General Information. The following minimum properties should be attained from standard tension test specimens poured from this alloy in sand molds and tested without machining, provided that proper foundry practices are used.

Tensile strength, lb. per sq. in	30,000
Yield point, lb. per sq. in	12,000
Elongation in 2 in. or proportionate gage length, per cent	10

Combining strength with fair machining qualities, this general utility bronze is especially good for bushings subject to heavy loads and severe working conditions.

Spec. No. 64—Phosphor-bronze Castings

Composition	%
Copper	78.50–81.50
Tin	9.00–11.00
Lead	9.00–11.00
Phosphorus	0.05– 0.25
Zinc, max	0.75
Other impurities, max	0.25

General information. The following minimum properties should be attained from standard tension test specimens poured from this alloy in sand molds and tested without machining, provided that proper foundry practices are used.

Tensile strength, lb. per sq. in	25,000
Yield point, lb. per sq. in	12,000
Elongation in 2 in. or proportionate gage length, per cent	8

TABLE A2-8.—BRASS AND BRONZE CASTINGS SUITABLE FOR BUSHINGS, ETC.—(*Continued*)

This metal is an excellent composition for use where antifraction qualities are desired standing up exceedingly well under heavy loads and severe usage.

This specification practically conforms in composition with ASTM B66-38.

Spec. No. 66—Bronze Backing for Lined Bearings

Composition	%
Copper	83.00–86.00
Tin	4.50– 6.00
Lead	8.00–10.00
Zinc, max	2.00
Other impurities, max	0.25

General Information. The following minimum properties should be attained from standard tension test specimens poured from this alloy in sand molds and tested without machining, provided that proper foundry practices are used.

Tensile strength, lb. per sq. in	25,000
Yield point, lb. per sq. in	12,000
Elongation in 2 in. or proportionate gage length, per cent	8

This composition is recommended as an inexpensive but suitable alloy for bronze-backed bearings.

Spec. No. 660—Bronze Bearing Castings

Composition	%
Copper	81.00–85.00
Tin	6.50– 7.50
Lead	6.00– 8.00
Zinc	2.00– 4.00
Iron, max	0.20
Antimony, max	0.20
Aluminum	None
Other impurities, max	0.50

General Information. This alloy is one of the widely used compositions for bronze bearings. In the automotive industry, it is used extensively in applications such as spring bushings, torque-tube bushings, steering-knuckle bushings, piston-pin bushings, and washers.

The following minimum properties should be attained from standard tension test specimens poured in sand molds and tested without machining, provided that proper foundry practices are used.

TABLE A2-8.—BRASS AND BRONZE CASTINGS SUITABLE FOR BUSHINGS, ETC.—(*Continued*)

Tensile strength, lb. per sq. in.	30,000
Yield point, lb. per sq. in.	14,000
Elongation in 2 in. or proportionate gage length, per cent	18

This alloy is essentially the same in chemical and physical properties as alloy No. 3B of ASTM B30-40T.

NOTE: For other brass, bronze, and copper alloys, see S.A.E. "Handbook."

TABLE A2-9.—FERROUS METALS USED IN ENGINE CONSTRUCTION
(From *S.A.E. Jour.*, Vol. 40, No. 4, April, 1937)

Chemical composition									S.A.E. No. or trade designation
Carbon	Manganese	Silicon	Nickel	Chromium	Tungsten	Molybdenum	Vanadium	Other elements	
0.05	0.45	....	.....	.....	....	....	...	....	1010
0.15	0.45	....	.....	.....	....	....	...	....	1015
0.20	0.45	....	.....	.....	....	....	...	....	1020
0.35	0.65	....	.....	.....	....	....	...	....	1035
0.50	0.65	....	.....	.....	....	....	...	....	1050
0.90	0.35	....	.....	.....	....	....	...	....	1090
0.95	0.35	....	.....	.....	....	....	...	....	1095
0.20	0.75	....	.....	.....	....	....	...	S-0.11	1120
0.30	0.65	....	3.50	.....	....	....	...	....	2330
0.40	0.65	....	3.50	.....	....	....	...	....	2340
0.15	0.45	....	1.25	0.60	....	....	...	....	3115
0.40	0.75	....	1.25	0.80	....	....	...	....	X-3140
0.40	0.45	....	1.75	1.10	....	....	...	....	3240
0.50	0.45	....	1.75	1.10	....	....	...	....	3250
0.15	0.45	....	5.00	.....	....	....	...	....	2515
0.12	0.45	....	3.50	1.50	....	....	...	....	3312
0.30	0.50	....	.....	0.95	....	0.2	...	....	X-4130
0.40	0.65	....	1.75	0.70	....	0.35	...	....	4340
1.0	0.30	....	.....	1.35	....	....	...	....	52100
0.35	0.75	....	.....	0.95	....	....	0.2	....	6135
0.50	0.75	....	.....	0.95	....	....	0.2	....	6150
0.90	0.35	....	.....	0.90	....	....	0.2	....	6190
0.43	0.55	....	.....	1.6	....	0.4	...	Al. 1.2	Nitralloy
0.35	0.45	....	.....	1.2	....	0.2	...	Al. 1.2	Nitralloy
0.07*	0.45	....	7.0*	17*	....	....	...	Ti-0.2	30905
0.60	0.3	....	.....	3.5	13	....	...	....	71360
0.65	0.3	....	.....	3.5	16	....	...	....	71665
0.45	0.40	3.0	.....	8.0	....	1.5	...	....	Cr-Si-W
0.30	0.30	2.50	8.0	12.5	....	...	...	....	Cr-Ni-Si
0.45	0.50	1.25	14.0	14.0	2.5	....	...	....	Cr-Ni-W-Si
1.20	0.50	1.2	.....	18	....	0.6	...	....	Cr-Si-Mo
1.2	0.5	0.5	.....	13	....	0.9	...	Co-1.3	Co-Cr
1.0	0.5	1.5	0.6	13.5	3.5	0.6	...	....	13.5 Cr

* Tensile strength at 1200°F. All other tensile properties are typical specification values.

TABLE A2-9.—FERROUS METALS USED IN ENGINE CONSTRUCTION.—(*Continued*)

Mechanical properties				Application
Tensile strength, lb. per sq. in.	Yield strength, lb. per sq. in.	Elongation 2 in. %	Reduction in area, %	
38,000	20,000	35	..	Ferrules, clips, lock wire
45,000	25,000	22	..	Camshaft, washers, ball ends
55,000	36,000	22	..	Nuts, screws, counterweights, flanges
80,000	50,000	20	..	Shafts, sleeves, nuts, rivets
100,000	75,000	16	..	Cylinder barrels, keys
225,000	350,000	..	..	Springs
....	....	..	..	Shims, wearing parts, valve mechanism
55,000	....	..	..	Screws, nuts, dowels for minor attachments
125,000	100,000	15	50	Bolts, studs, nuts, shafts
135,000	110,000	15	50	Connecting rods, gears
130,000	90,000	16	40	Gears, piston pins
130,000	100,000	17	50	Bolts, studs, shafts
135,000	110,000	15	50	Crankshaft, drive shafts
225,000	200,000	10	40	Gears, pins
170,000	145,000	14	45	Gears, cams, crankshaft
160,000	135,000	15	50	Gears, drive shafts, cams
95,000	75,000	12	..	Washers, shims, spacers, tubes
160,000	140,000	15	50	Crankshaft, connecting rods
....	....	..	..	Balls, bearings, knuckle pins
150,000	120,000	15	50	Gears, shafts, propeller hubs
220,000	200,000	10	..	Piston pins, gears, drive shafts, tappets, springs, ball ends, dowels, tappets, bolts, studs
135,000	100,000	18	55	Gears, cylinders
120,000	80,000	15	45	Piston pins, shafts, pump liners, bushings
100,000	35,000	40	..	Exhaust manifolds, supercharger casing
55,000*	....	..	..	Valve, inlet
60,000*	....	..	..	Valve, inlet tips
50,000*	....	..	..	Valves, inlet
60,000*	....	..	..	Valves, inlet
80,000*	....	..	..	Valves, inlet and exhaust, supercharger buckets
50,000*	....	..	..	Valves, inlet and exhaust
60,000*	....	..	..	Valves, exhaust
50,000*	....	..	..	Valves, exhaust

* Tensile strength at 1200°F. All other tensile properties are typical specification values.

TABLE A2-10.—NONFERROUS METALS USED IN ENGINE CONSTRUCTION
(From *S.A.E. Jour.*, Vol. 40, No. 4, April, 1937)

	Chemical composition											S.A.E. No. or trade designation
	Aluminum	Copper	Zinc	Tin	Lead	Silicon	Iron	Nickel	Magnesium	Manganese	Chromium	
Aluminum base	94	4	..	....	....	....	....	....	0.5	0.5	....	26
	93	4.2	..	....	....	....	....	....	1.5	0.6	....	24
Bar	97	....	..	....	....	....	....	....	2.5	......	0.25	NF-1
Sheet	91	4.0	..	....	....	....	....	2.0	0.6	......	....	NF-2
Tube	96.5	....	..	....	....	1.0	....	....	0.7	......	0.25	28
Forgings	85	1.0	..	....	....	12.0	....	1.0	1.0	......	....	NF-3
	92	8	..	....	....	....	....	....	......	......	....	30
Sand castings	88	10.0	..	....	....	....	....	....	0.2	......	....	34
	93	4	..	....	....	....	....	2.0	1.5	......	....	39
Weight lb. per	92	4.5	..	....	....	1.2	....	....	......	0.3	....	38
cu. in.	84	1.0	..	....	....	14.0	....	2.0	1.0	......	....	321
0.092–0.105	92	1.2	..	....	....	5.0	....	....	0.5	......	....	322
	94	....	..	....	....	5.0	....	....	......	......	....	35
	93	....	..	....	....	1.5	....	....	3.8	......	....	320
	88	....	..	....	....	....	....	....	10.0	......	....	324
Die castings	87	4.0	..	....	....	5.0	....	....	......	......	....	307
	87	....	..	....	....	12	....	....	......	......	....	305
Magnesium base	6.5	0.5	..	....	....	....	....	....	......	0.2	....	NF-4
Forgings	10	....	..	....	....	....	....	....	88	0.1	....	NF-5
Castings	8	....	..	....	....	....	....	....	90	0.2	....	NF-6
Weight, lb. per												
cu. in.												
0.064–0.066	6	....	3	....	....	....	....	....	......	0.2	....	NF-7
Copper base	....	99.5	..	....	....	....	....	....	......	......	....	71–75
		60	37	....	2.0	....	....	....	......	......	....	72–88
Wire	....	66	34	....	....	....	....	....	......	......	....	70
Sheet	....	94	..	5.0	....	....	....	....	......	P0.4	....	77 & 81
Tube	....	98	..	....	....	....	....	....	......	Be2.3	....	NF-8
	9	88	..	....	....	....	3	....	......	......	....	701
Weight, lb. per												
cu. in.	10	81	..	....	....	....	2.5	5	......	1	....	NF-9
0.27–0.35	....	58	42	....	....	....	....	....	......	1	....	NF-10
		85	5	5	5	....	....	....	......	......	....	40
		88	2	10	....	....	....	....	......	......	....	62
		88	2	10	1.5	....	....	....	......	......	....	63
	....	84	..	5	9	....	....	....	......	......	....	66
Castings	10	89	..	....	....	....	1	....	......	......	....	68
	11	79	..	5	....	....	....	5	......	......	....	NF-11
		80	..	12	5	....	....	....	......	......	....	NF-12
		70	..	5	25	....	....	....	......	......	....	NF-13
		83	..	10	3	....	....	4	......	......	....	NF-14
Tin base	....	4.5	..	90	....	....	....	....	......	......	Sb7	10
				50	50	....	....	....	......	......	....	......
Silver base	....	30	25	....	....	....	....	....	......	Ag45	....	......
Lead base	....	....	..	90	....	....	....	....	......	Ag6	....	......
Nickel base	....	....	..	....	....	....	....	97	......	3	....	......
	....	....	..	....	....	....	10	70	......	......	15	......
Cobalt base	....	....	..	....	....	1	....	....	W-14	Co-52	27	......
	....	....	..	....	....	2	....	....	W-4	Co-65	27	......

* Brinell unless preceded by letter for appropriate Rockwell scale.

TABLE A2-10.—NONFERROUS METALS USED IN ENGINE CONSTRUCTION.—(*Continued*)

Mechanical properties					Application
Tensile strength, lb. per sq. in.	Yield strength, lb. per sq. in.	Elongation 2 in., %	Hardness*	Fatigue limit, lb. per sq. in. 5×10^8 cycles	
55,000	32,000	16	90	13,000	Washers, rivets, deflectors
62,000	40,000	15	100	14,000	Washers, rivets, deflectors small forgings
29,000	14,000	20	60	14,000	Tubes
55,000	35,000	8	95	....	Pistons
43,000	34,000	12	90	11,000	Large forgings
55,000	40,000	5	100	14,000	Pistons
18,000	....	....	......	....	Cases, coverplates, oil sump
30,000	18,000	....	100	....	Cylinder heads
32,000	18,000	1	95	8,000	Cylinder heads, pistons, bearings
32,000	18,000	3	80	9,000	Crankcase, gear housing
32,000	18,000	....	100	....	Pistons
32,000	20,000	2	90	7,500	Miscellaneous castings
16,000	7,000	2	40	5,000	Oil pans, gear-case covers
18,000	10,000	....	......	....	Castings requiring high corrosion resistance
40,000	22,000	11	85	7,500	High stressed castings
32,000	....	2	......	....	Accessory housings, covers
29,000	....	3	......	....	Accessory housings, covers
42,000	24,000	5	......	15,000	Nose section, forging
30,000	18,000	1	65	9,000	Nose section, gear-case covers, housings, rear-end crankcase
29,000	10,000	6	48	7,500	Crankcase
32,000	18,000	4	65	9,000	Miscellaneous castings
34,000	10,000	50	......	11,000	Tube, shims
55,000	25,000	30	60	....	Spacers, dowels
....	....	....	......	....	Bolts, shims
55–100,000	....	15.1	......	20,000	Bushings
160,000	130,000	5	300	30,000	Bushings, springs
80,000	40,000	15	160	25,000	Valve guides, bushings
....	....	.....	200	....	Valve seats
85,000	40,000	20	......	....	Counterweights
26,000	12,000	15	B-80	....	Fuel- and oil-line connections
30,000	15,000	14	B-85	....	Bushings
30,000	12,000	10	......	....	Bushings
25,000	12,000	8	......	....	Bushings
80,000	50,000	4	170	....	Valve seats, propeller cones
85,000	....	3.5	230	....	Valve seats, propeller cones
....	....	....	......	....	Oil-seal rings
....	....	....	45	....	Bushings
....	....	....	......	....	Exhaust valve guides
....	....	....	......	....	Bearings
440†	....	....	......	360°F.‡	Solder
				1250°F.‡	Solder
1550†	....	....	......	580°F.‡	Solder
					Spark-plug electrodes
80,000	30,000	25	......	....	Exhaust equipment
			600	....	Stellite No. 1 valve tips
			450	....	Stellite No. 6 valve seat facing

† Shearing strength at 350°F.—Base metal, copper. ‡ Melting starts, deg. F.

Table A2-11.—Principal Engine Parts and Representative Specifications*
(From *S.A.E. Jour.*, Vol. 40, No. 4, April, 1937)

Name of part	Spec. No.	Hardness	Spec. No.	Hardness	Spec. No.	Hardness	Spec. No.	Hardness
Cylinder barrels	1050	225	4140	300	Nitralloy	900		
Cylinder heads	39	65	34	65				
Piston	321	100	NF-2	100	NF-3	115	39	110
Piston rings	Cast iron	B-100						
Valve, intake	Cr-Si-W	C-42	Cr-Ni-Si	C-40	Cr-Si-Mo	C-50	71360	C-45
Valve, exhaust	Cr-Si-Mo	C-50	Cr-Ni-W-Si	C-15	Co-Cr	C-55	13.5 Cr	C-50
Valve, guide	68	175	701	160	Co-Cr	C-50	62	B-85
Valve, spring	1095	C-44	6150	C-44				
Valve, spring washer	3135	C-35	6150	C-40				
Valve, spring retainer	3250	C-50	6150	C-40				
Valve seat	68	225	NF-9	200	Cr-Ni-W-Si	C-15	701	160
Valve lock wire	1085							
Rocker arm	3140	C-30	6150	C-30	2330	C-30		
Rocker-arm hub bolt	3140	C-32	6150	C-25	3312	C-60		
Rocker-arm bearing	Ball		Roller					
Rocker-arm cup	10115	C-62	3250	C-42	52100	C-60	6180	C-55
Push rod, tube	26		X-4130	C-35	1025			
Push rod, ball end	1015	C-62	3250	C-46	1095	C-50		
Push rod, roller	3115	C-60	3215	C-60				
Push rod, roller pin	3140	C-40	3250	C-50	6150	C-52		
Cam	3250	C-55	2515	C-60				
Camshaft	1015	C-60						
Cam bearing	64	B-75	62	B-85				
Cam drive shaft and gear	3312	C-60	3140	C-38	2515	C-55		
Crankcase, main	322	80	38	80	28	115		
Crankcase, front; rear; blower section	322	80	38	80	NF-5 and 6	45		
Impeller	26	115	27	100				
Impeller shaft	2515	C-60	3312	C-60	Nitralloy	900		
Impeller-shaft bearing	Ball							
Crankshaft	X-3140	260	3240	280	4340	320	2515	C-60
Crankshaft counterweights	1035	160	43	140	640			
Crankshaft extension	3140	C-40	3250	C-45	2515	C-55		
Propeller-hub nut	6135	250	3312	C-60	2330	C-30		
Propeller-hub cone	68	210	65	160				
Rod, connecting	3140	270	4340	350	2340	340		
Pin, piston	6150	C-50	3312	C-60	3250	C-47	Nitralloy	900
Pin, knuckle	3312	C-60	3120	C-60	52100	C-60		
Pin, bushings	62	150	63	130				
Bolts, connecting rod, crankcase	3250	C-35	6150	C-35	3140	C-30		
Studs, cylinder	6150	C-26	3140	C-30				
Nuts	6150	B-95	3140	C-25	2330	C-20		
Gears, reduction; camshaft drive; accessory drive	3250	C-45	3312	C-60	2515	C-60	Nitralloy	900
Housings, accessory drive	30	70	322	50–80	38	50–70	NF-5 and 6	45
Housings, accessory drive covers	30	70	322	50–80	35	45	38	50–70
Sump, oil; scoops, air	30	70	35	45	NF-5	45		

* By S.A.E. number except as noted.

Hardness values are taken from service parts and are Brinell (10-mm. ball—3000-kg.-load for steel and aluminum-bronze—500 kg. for other nonferrous alloys) unless preceded by C or B for Rockwell—Cone and 1/16 ball, respectively.

TABLE A2-12.—COPPER ALLOYS SUITABLE FOR MISCELLANEOUS ENGINE PARTS
(From S.A.E. "Handbook")

Spec. No. 62—Hard Bronze Castings

Composition	%
Copper	86.00–89.00
Tin	9.00–11.00
Lead, max	0.20
Iron, max	0.06
Zinc	1.00– 3.00

General Information. The following minimum properties should be attained from standard tension test specimens poured from this alloy in sand molds and tested without machining, provided that proper foundry practices are used.

Tensile strength, lb. per sq. in	30,000
Yield point, lb. per sq. in	15,000
Elongation in 2 in. or proportionate gage length, per cent	14

This alloy is suitable wherever a strong general utility bronze is required. It may be used for severe working conditions where heavy pressures obtain, as in gears and bearings.

This specification conforms in composition with A.S.T.M. B60-36.

Spec. No. 68—Cast Aluminum Bronze

Composition	% Grade A	% Grade B
Copper	87.00–89.00	89.50–90.50
Aluminum	7.00– 9.00	9.50–10.50
Iron	2.50– 4.00	Not over 1.00
Tin, max	0.5	0.2
Total other impurities	1.0	0.5

General Information. Standard test bars cast in sand and tested without machining should give the following minimum mechanical properties:

	Grade A	Grade B
Tensile strength, lb. per sq. in. (as cast)	65,000	65,000
Tensile strength, lb. per sq. in. (as heat-treated, quenched and drawn)	......	80,000
Yield point (as cast)	25,000	25,000
Yield point (as heat-treated)	......	50,000
Elongation in 2 in., per cent (as cast)	20	15
Elongation in 2 in., per cent (as heat-treated)	......	4

This is a corrosion-resistant alloy of great strength with a hardness equal to that of manganese bronze and under certain conditions has good bearing qualities. It is used for worm wheels, gears, and similar parts.

Table A2-12.—Copper Alloys Suitable for Miscellaneous Engine Parts.—(*Continued*)

This specification conforms with the composition and physical properties of A.S.T.M. B59-39.

Report on Non-ferrous Metals Division adopted June, 1911. Last revision, January, 1940.

Spec. No. 701—Wrought Aluminum Bronze
(Annealed, Hot Rolled, or Forged)

Note: Prior to January, 1933, this specification was published as S.A.E. Specification 69.

Composition	%
Copper	88.00–95.00
Aluminum	4.50–10.00
Iron, max	4.00
Other additions including nickel, tin, and manganese, max	2.00
Other impurities including zinc and lead, max	0.25

Mechanical requirements

Tensile Test Data

Diameter or thickness, in.		Width, in.	Ultimate strength, lb. per sq. in. min.	Yield point, lb. per sq. in. min.	Elongation in 2 in., per cent min.
Rods and bars:					
Over	To and inc.				
0	0.50	....	80,000	40,000	15
0.50	1.0	....	75,000	37,500	15
1.0	....	....	72,000	35,000	20
Shapes, all sizes		....	75,000	30,000	15
Plates, sheets, and strips:					
0	0.5	Less than 30	60,000	24,000	25
0	0.50	Over 30	55,000	22,000	25
0.50	....	All	50,000	20,000	30

General Information. This is a corrosion-resistant alloy of great strength with a hardness equal to manganese bronze. It has good bearing and antifrictional properties. Wrought aluminum bronze is used for diaphragms, gears, forgings, for its color, strength, resistance to corrosion and wear and for its properties at elevated temperature. Valve seats and bushings are hot forged for use on internal-combustion engines. The 10 per cent alloy can be heat-treated in a manner similar to steel. Its physical properties are improved to some extent by heating and quenching.

Note: For other data on this alloy, see S.A.E. "Handbook."

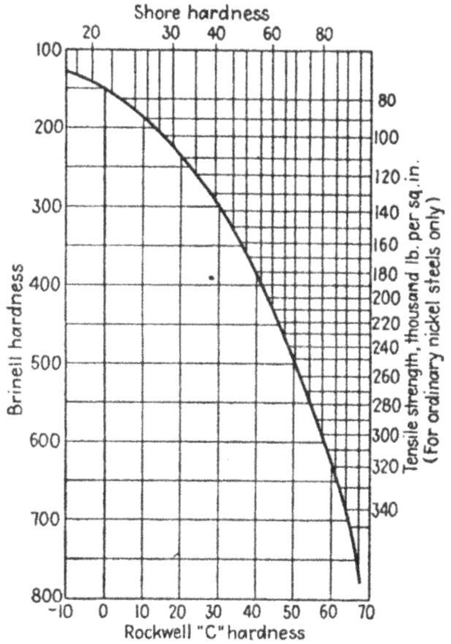

FIG. A2-1.—Approximate relation between Brinell, Rockwell, and Shore hardnesses and the strength of structural alloy steel. (*From data of the International Nickel Company.*)

APPENDIX 3

Table A3-1.—Properties of Various Beam Cross Sections
(From Marks, "Mechanical Engineers' Handbook")

Section	Moment of inertia (= I)	Section modulus (= I/C)	Radius of gyration ($k = \sqrt{I/A}$)
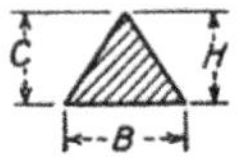	$\frac{BH^3}{12}$	$\frac{BH^2}{12}$	$\frac{H}{\sqrt{6}} = 0.408H$
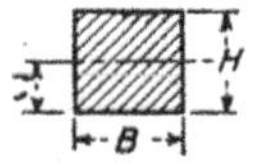	$\frac{BH^3}{12}$	$\frac{BH^2}{6}$	$\frac{H}{\sqrt{12}} = 0.289H$
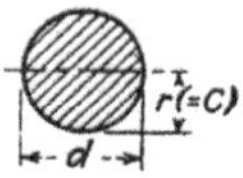	$\frac{\pi d^4}{64}$ $\left(= \frac{\pi r^4}{4}\right)$	$\frac{\pi d^3}{32}$ $\left(= \frac{\pi r^3}{4}\right)$	$\frac{d}{4}$ $\left(= \frac{r}{2}\right)$
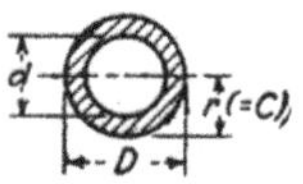	$\frac{\pi}{64}(D^4 - d^4)$	$\frac{\pi}{32}\left(\frac{D^4 - d^4}{D}\right)$	$\frac{\sqrt{D^2 + d^2}}{4}$
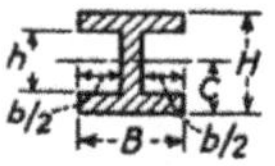	$\frac{BH^3 - bh^3}{12}$	$\frac{BH^3 - bh^3}{6H}$	$\sqrt{\frac{BH^3 - bh^3}{12(BH - bh)}}$
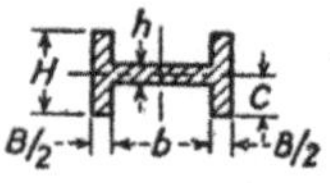	$\frac{BH^3 + bh^3}{12}$	$\frac{BH^3 + bh^3}{6H}$	$\sqrt{\frac{BH^3 + bh^3}{12(BH + bh)}}$
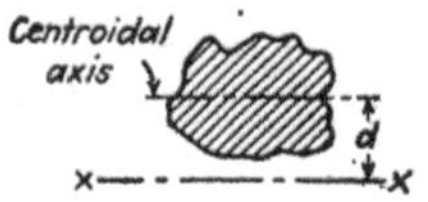	$I_x = \bar{I} + Ad^2$ $k_x^2 = \bar{k}^2 + d^2$ $\bar{I}$ = moment of inertia with respect to the centroidal axis. A = area. $\bar{k}$ = radius of gyration with respect to the centroidal axis.		

TABLE A3-2.—DEFINITION OF SPUR-GEAR TOOTH PARTS
(From data of Foote Brothers Gear and Machine Company)

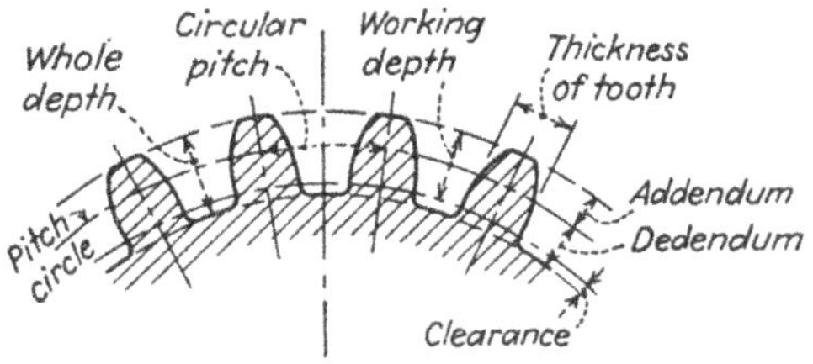

The **addendum circle** is the circle that limits the tops of the teeth.

The **pitch circle** is the trace of the pitch cylinders on which the tooth curves are formed.

The **dedendum, or root circle,** is the circle that limits the bottom of the tooth.

The **clearance** is the amount by which the tops of the teeth of one wheel clear the bottoms of the spaces of the other, as they pass the line of centers.

The **working-depth circle** is a circle of radius equal to that of the dedendum circle plus the clearance.

The **face** is that part of the tooth lying between the pitch and addendum circles.

The **flank** is that part of the tooth lying between the pitch and dedendum circles.

The **thickness of the tooth** is its width measured on the pitch circle.

The **width of space** is the space between the teeth measured on the pitch circle.

The **backlash** is the difference between the thickness of a tooth and the space into which it meshes, measured on the pitch circles.

The **pitch diameter** is the diameter of the pitch circle. It is the diameter that is used in making all calculations for the size of gears.

The **circular pitch** is the distance in inches between similar points of adjacent teeth, measured along the pitch circle. It is the thickness of the tooth plus the width of a space, measured on the pitch circle.

The **diametral pitch** is the number of teeth on the gear per inch of diameter of the pitch circle. Circular pitch times diametral pitch $= \pi$.

The **diametral pitch, or module,** is the distance in inches obtained by dividing the diameter of the pitch circle by the number of teeth in the gear.

The **chordal pitch** is the length of a straight line drawn from the pitch points of two adjacent teeth. Chordal pitch = diameter of pitch circle times sine (180 deg. + number of teeth). Chordal pitch is used with chain or sprocket wheels, to conform to the pitch of the chain.

The **angles of action** are the angles through which the gears turn while a pair of teeth are in action. If the diameters are not alike, these angles will be inversely as the radii of the pitch circles, since the arcs which subtend them are equal.

The **angle of approach** is the angle through which a gear turns from the beginning of contact of a pair of teeth until contact reaches the pitch point.

The **angle of recess** is the angle through which a gear turns while the contact point of a pair of teeth moves from the pitch point to the point where the teeth pass out of contact. Angle of approach + angle of recess = angle of action.

TABLE A3-3.—STANDARD 14½- AND 20-DEG. INVOLUTE GEARS[1,2]

As the result of work by the American Standards Association, gear-tooth profiles and dimensions have been standardized. Gears having a 14½-deg. pressure angle are made either in the composite (part involute and part cycloidal) type or in the 14½-deg. full-depth involute type. The composite type is stronger but more difficult to machine accurately. The two types are *not* interchangeable.[1] Gears of the 20-deg. full-depth involute type have teeth that are thicker at the base and therefore they are stronger than the 14½-deg. pressure angle gears. However, the force tending to separate mating gears is greater with the 20-deg. pressure angle than with the 14½-deg. pressure angle and this may be a disadvantage in some instances. Still greater gear tooth strength may be had by using the 20-deg. involute stub tooth (see Table A3-4).

The American Standards Association recommends the following proportions for 14½-deg. composite, 14½-deg. full-depth involute and 20-deg. full-depth involute gears.[1]

Dimensions for 14½-deg. Composite, 14½-deg. Full-depth, 20-deg. Full-depth Gears

Part	In terms of diametral pitch, in.	In terms of circular pitch, in.
Addendum	$1.0/P_d$	$0.3183 \times P_c$
Minimum dedendum	$1.157/P_d$	$0.3683 \times P_c$
Working depth	$2.0/P_d$	$0.6366 \times P_c$
Minimum total depth	$2.157/P_d$	$0.6866 \times P_c$
Minimum clearance	$0.157/P_d$	$0.0500 \times P_c$
Pitch diameter	N/P_d	$0.3183 \times M \times P_c$
Outside diameter	$\dfrac{N+2}{P_d}$	$0.3183 \times (N + 2) \times P_c$

N = number of teeth.
P_d = diametral pitch.
P_c = circular pitch.
$P_d \times P_c = \pi$.

[1] Norman, Ault, and Zarobsky, "Fundamentals of Machine Design."

[2] Data of Foote Brothers Gear and Machine Company.

Table A3-4.—Stub-tooth Gears[1,2]

The advantages of the stub-tooth gear as compared with the standard involute 14½-deg.-pressure-angle spur gears are as follows:[2]

1. Equal arc of rolling action.
2. Reduction of sliding action.
3. More even wearing contact.
4. Longer life.
5. Greater strength.
6. Possibility of decreasing pitch.
7. Reduction of noise.
8. Less distortion in hardening.

Strength is the factor that is generally uppermost in the designer's mind when he is laying out a gear-tooth form of gearing that will be subjected to shocks or intermittent loads. It is the fact of increased strength that has been largely instrumental in advancing the popularity of the stub-tooth gear. This type of tooth has been successfully applied in the manufacture of automobile and airplane gear drives.[2]

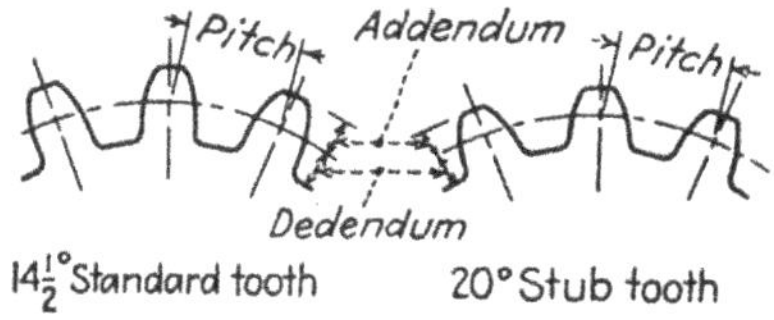

A shortening and widening of the tooth at the base results from increasing the pressure angle from 14½ to 20 deg. As a general rule, stub-tooth gears with less than 25 teeth are about 25 per cent stronger than standard 20-deg. full-depth involute teeth and 40 per cent stronger than 14½-deg. involute teeth. For larger numbers of teeth, the gain in strength is somewhat less.[2]

The American Standards Association recommends the following proportions for 20-deg. stub-tooth gears:[1]

Dimensions for 20-deg. Stub Gears

Part	In terms of diametral pitch, in.	In terms of circular pitch, in.
Addendum	$0.8/P_d$	$0.2546 \times P_c$
Minimum dedendum	$1.0/P_d$	$0.3183 \times P_c$
Working depth	$1.6/P_d$	$0.5092 \times P_c$
Minimum total depth	$1.8/P_d$	$0.5729 \times P_d$
Minimum clearance	$0.2/P_d$	$0.0637 \times P_c$
Pitch diameter	N/P_d	$0.3183 \times N \times P_c$
Outside diameter	$\frac{N + 1.6}{P_d}$	D + (2 addendums)

N = number of teeth.
P_d = diametral pitch.
D = pitch diameter.
P_c = circular pitch.
$P_d \times P_c = \pi$.

[1] Norman, Ault, and Zarobsky, "Fundamentals of Machine Design."

[2] Data of Foote Brothers Gear and Machine Company.

TABLE A3-5.—INTERNAL GEARS[1,2]

Internal gear dimensions may be found by the same formulas as for external gearing except for modifications made necessary by the fact that the center distance in internal gearing is equal to the *difference* between the two pitch radii instead of their *sum*. In addition, the term inside diameter takes the place of outside diameter in external gearing. Inside diameter is the diameter of the hole in the gear blank before the teeth are cut.

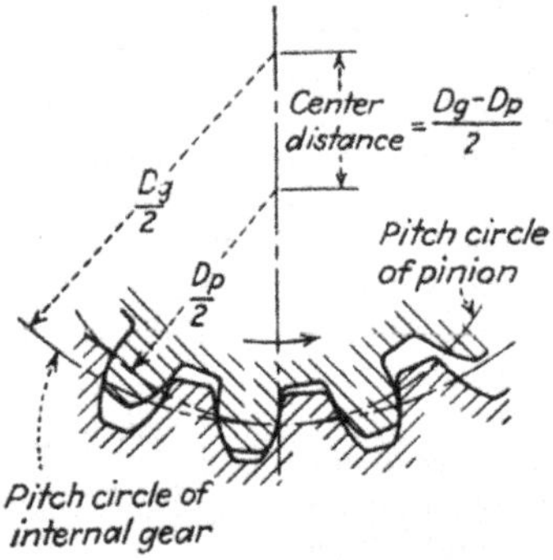

Internal gear details.

Internal gears have several desirable characteristics:[1]

1. Much stronger teeth due to greater base width.
2. Less sliding action, hence less wear.
3. Higher efficiency.
4. More teeth in contact.
5. Smoother operation.
6. More compact design.

For internal gears of the same material as their pinions, it is unnecessary to calculate the strength because, (*a*) the torque arm is greater, hence the torque force is less than on the pinion, and (*b*) the teeth are much stronger than the pinion teeth because of greater base width.

[1] Norman, Ault, and Zarobsky, "Fundamentals of Machine Design."
[2] Data of Foote Brothers Gear and Machine Company.

TABLE A3-6.—SIMPLE EPICYCLIC GEARING[1,2,3]

The principle of operation of the simple epicyclic gear train is illustrated in the accompanying diagram where S is the sun pinion, P is the planet pinion, G is the internal gear, and A is the arm carrying P.

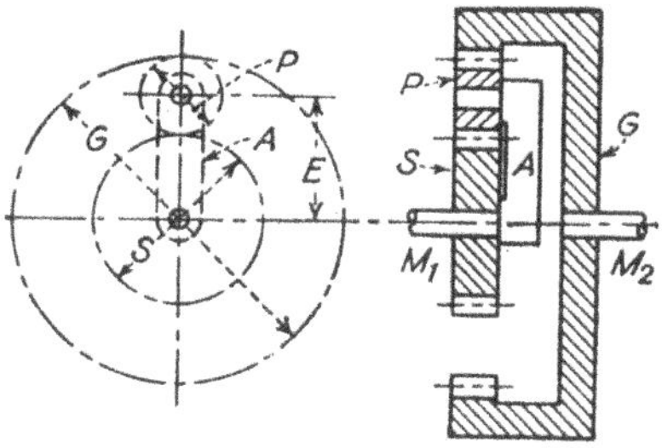

The following formulas give the speeds, pitch-line velocities, and loads for the six most commonly used cases. In these formulas:

N_G = number of teeth in internal gear G.
N_P = number of teeth in planet pinion P.
N_S = number of teeth in sun pinion S.
E = center distance of gears S and P.
V = pitch-line velocity of transmitted load, f.p.m.
V_G = pitch-line velocity of tooth engagement on G, f.p.m.
V_S = pitch-line velocity of tooth engagement on S, f.p.m.
V_A = velocity of center of planet pinion P, f.p.m.
n = r.p.m. of driving member.
W = transmitted load, lb.
W_G = tooth load on G, lb.
W_S = tooth load on S, lb.
W_A = load at center of pinion, lb.
Q = driving torque, in. lb.
hp. = number of horsepower transmitted.

[1] Data of Foote Brothers Gear and Machine Company.
[2] Schwamb, Merill, and James, "Elements of Mechanism."
[3] Buckingham, "Manual of Gear Design."

TABLE A3-6.—INTERNAL GEARS.—(*Continued*)

Case No.	1	2	3	4	5	6
Stationary member	A	A	G	S	G	S
Driving member	S	G	A	A	S	G
Driven member	G	S	S	G	A	A
Reduction ratio	$\frac{N_S}{N_G}$	$\frac{N_G}{N_S}$	$\frac{N_G+N_S}{N_S}$	$\frac{N_G+N_S}{N_G}$	$\frac{N_S}{N_G+N_S}$	$\frac{N_G}{N_G+N_S}$
R.p.m. of A	0	0	n	n	$n\left(\frac{N_S}{N_G+N_S}\right)$	$n\left(\frac{N_G}{N_G+N_S}\right)$
R.p.m. of S	n	$n\left(\frac{N_G}{N_S}\right)$	$n\left(\frac{N_G+N_S}{N_G}\right)$	0	n	0
R.p.m. of P	$n\left(\frac{N_S}{N_P}\right)$	$n\left(\frac{N_G}{N_S}\right)$	$n\left(\frac{N_G}{N_P}\right)$	$n\left(\frac{N_S}{N_P}\right)$	$n\left(\frac{N_G}{N_P}\right)\left(\frac{N_S}{N_G+N_S}\right)$	$n\left(\frac{N_S}{N_P}\right)\left(\frac{N_G}{N_G+N_S}\right)$
R.p.m. of G	$n\left(\frac{N_S}{N_G}\right)$	n	0	$n\left(\frac{N_G+N_S}{N_G}\right)$	0	n
V	$0.5236nE\left(\frac{N_S}{N_S+N_P}\right)$	$0.5236nE\left(\frac{N_G}{N_G-N_P}\right)$	$0.5236nE$	$0.5236nE$	$0.5236nE\left(\frac{N_S}{N_S+N_P}\right)$	$0.5236nE\left(\frac{N_G}{N_G-N_P}\right)$
V_S	$0.5236nE\left(\frac{N_S}{N_S+N_P}\right)$	$0.5236nE\left(\frac{N_G}{N_S+N_P}\right)$	$0.5236nE\left(\frac{N_G}{N_S+N_P}\right)$	$0.5236nE\left(\frac{N_S}{N_S+N_P}\right)$	$0.5236nE\left(\frac{N_S}{N_S+N_P}\right)\left(\frac{N_G}{N_G+N_S}\right)$	$0.5236nE\left(\frac{N_S}{N_S+N_P}\right)\left(\frac{N_G}{N_G+N_S}\right)$
V_A	0	0	$0.5236nE$	$0.5236nE$	$0.5236nE\left(\frac{N_S}{N_G+N_S}\right)$	$0.5236nE\left(\frac{N_G}{N_G+N_S}\right)$
V_G	$0.5236nE\left(\frac{N_S}{N_G-N_P}\right)$	$0.5236nE\left(\frac{N_G}{N_G-N_P}\right)$	$0.5236nE\left(\frac{N_G}{N_G-N_P}\right)$	$0.5236nE\left(\frac{N_S}{N_G-N_P}\right)$	$0.5236nE\left(\frac{N_G}{N_G-N_P}\right)\left(\frac{N_S}{N_G+N_S}\right)$	$0.5236nE\left(\frac{N_G}{N_G-N_P}\right)\left(\frac{N_S}{N_G+N_S}\right)$
$W=\frac{33,000\text{ hp.}}{V}$ or →	$\frac{Q}{E}\left(\frac{N_S+N_P}{N_S}\right)$	$\frac{Q}{E}\left(\frac{N_G-N_P}{N_G}\right)$	$\frac{Q}{E}$	$\frac{Q}{E}$	$\frac{Q}{E}\left(\frac{N_S+N_P}{N_S}\right)$	$\frac{Q}{E}\left(\frac{N_G-N_P}{N_G}\right)$
W_S	W	$W\left(\frac{N_S+N_P}{N_G-N_P}\right)$	$W\left(\frac{N_S+N_P}{N_S+N_G}\right)$	$W\left(\frac{N_S+N_P}{N_S+N_G}\right)$	W	$W\left(\frac{N_S+N_P}{N_G-N_P}\right)$
W_A	$W\left(\frac{N_S+N_G}{N_S+N_P}\right)$	$W\left(\frac{N_S+N_G}{N_G-N_P}\right)$	W	W	$W\left(\frac{N_G+N_S}{N_S+N_P}\right)$	$W\left(\frac{N_G+N_S}{N_G-N_P}\right)$
W_G	$W\left(\frac{N_G-N_P}{N_S+N_P}\right)$	W	$W\left(\frac{N_G-N_P}{N_S+N_G}\right)$	$W\left(\frac{N_G-N_P}{N_S+N_G}\right)$	$W\left(\frac{N_G-N_P}{N_S+N_P}\right)$	W

TABLE A3-7.—COMPOUND EPICYCLIC GEARING[1,2,3]

The principle of operation of the compound epicyclic gear train is illustrated in the accompanying diagram where S is the sun pinion, G is the internal gear, A is the arm carrying P_S and P_G, P_G is the planet gear meshing with G, P_S is the planet gear meshing with S P_G and P_S are both keyed to shaft M_3.

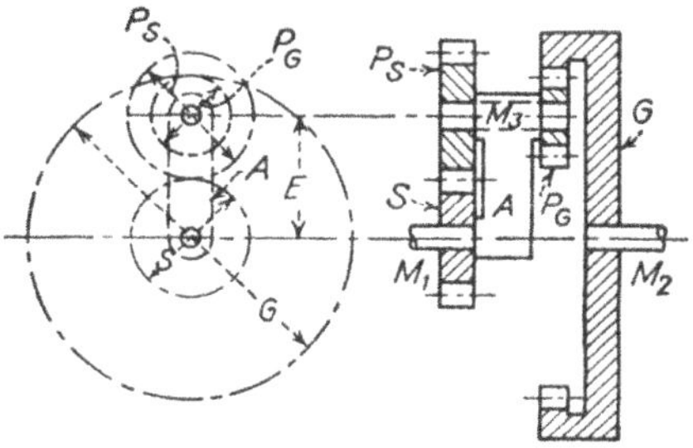

The following formulas give the speeds, pitch-line velocities, and loads for the six most commonly used cases. In these formulas:

N_G = number of teeth in internal sun gear.
N_{PS} = number of teeth in planet pinion meshing with S.
N_{PG} = number of teeth in planet pinion meshing with G.
N_S = number of teeth in spur sun gear.
E = center distance of gears, in.
V = velocity of the transmitted load, f.p.m.
V_S = pitch-line velocity of tooth engagement on S, f.p.m.
V_G = pitch-line velocity of tooth engagement on G, f.p.m.
V_A = velocity of center of planet pinions, f.p.m.
Q = driving torque, in.-lb.
$hp.$ = number of horsepower transmitted.
n = r.p.m. of driving member.
W = transmitted load, lb.
W_G = tooth load on G, lb.
W_S = tooth load on S, lb.
W_A = load at center of pinions, lb.

Potential power (ft.-lb. per min.) $= W_G V_G = W_{PS} V_{PS}$.
Transmitted power (ft.-lb. per min.) $= WV$.

[1] Data of Foote Brothers Gear and Machine Company.
[2] Schwamb, Merrill, and James, "Elements of Mechanism."
[3] Buckingham, "Manual of Gear Design."

TABLE A3-7.—COMPOUND EPICYCLIC GEARING.—(*Continued*)

Case No.	1	2	3	4	5	6
Fixed member	A	A	G	S	G	S
Driving member	S	G	A	A	S	G
Driven member	G	S	S	G	A	A
Reduction ratio	$\frac{N_S N_{PG}}{N_G N_{PS}}$	$\frac{N_G N_{PS}}{N_S N_{PG}}$	$\frac{N_S N_{PG}+N_G N_{PS}}{N_S N_{PG}}$	$\frac{N_S N_{PG}+N_G N_{PS}}{N_G N_{PS}}$	$\frac{N_S N_{PG}}{N_S N_{PG}+N_G N_{PS}}$	$\frac{N_G N_{PS}}{N_S N_{PG}+N_G N_{PS}}$
R.p.m. of A	0	0	n	n	$n\left(\frac{N_S N_{PG}}{N_S N_{PG}+N_G N_{PS}}\right)$	$n\left(\frac{N_G N_{PS}}{N_S N_{PG}+N_G N_{PS}}\right)$
R.p.m. of S	n	$n\left(\frac{N_G N_{PS}}{N_S N_{PG}}\right)$	$n\left(\frac{N_S N_{PG}+N_G N_{PS}}{N_S N_{PG}}\right)$	0	n	0
R.p.m. of P_G and P_S	$n\left(\frac{N_S}{N_{PS}}\right)$	$n\left(\frac{N_G}{N_{PG}}\right)$	$n\left(\frac{N_G}{N_{PG}}\right)$	$n\left(\frac{N_S}{N_{PS}}\right)$	$n\left(\frac{N_G N_S}{N_S N_{PG}+N_G N_{PS}}\right)$	$n\left(\frac{N_S N_G}{N_S N_{PG}+N_G N_{PS}}\right)$
R.p.m. of G	$n\left(\frac{N_S N_{PG}}{N_G N_{PS}}\right)$	n	0	$n\left(\frac{N_S N_{PG}+N_G N_{PS}}{N_G N_{PS}}\right)$	0	n
V	$0.5236nE\left(\frac{N_S}{N_S+N_{PS}}\right)$	$0.5236nE\left(\frac{N_G}{N_G-N_{PG}}\right)$	$0.5236nE$	$0.5236nE$	$0.5236nE\left(\frac{N_S}{N_S+N_{PS}}\right)$	$0.5236nE\left(\frac{N_G}{N_G-N_{PG}}\right)$
V_S	$0.5236nE\left(\frac{N_S}{N_S+N_{PS}}\right)$	$0.5236nE\left[\frac{N_G N_{PS}}{N_{PG}(N_S+N_{PS})}\right]$	$0.5236nE\left[\frac{N_G N_{PS}}{N_{PG}(N_S+N_{PS})}\right]$	$0.5236nE\left(\frac{N_S}{N_S+N_{PS}}\right)$	$0.5236nE\left(\frac{N_G N_{PS}}{N_S N_{PG}+N_G N_{PS}}\right)\left(\frac{N_S}{N_S+N_{PS}}\right)$	$0.5236nE\left(\frac{N_G N_{PS}}{N_S N_{PG}+N_G N_{PS}}\right)\left(\frac{N_S}{N_S+N_{PS}}\right)$
V_A	0	0	$0.5236nE$	$0.5236nE$	$0.5236nE\left(\frac{N_S N_{PG}}{N_S N_{PG}+N_G N_{PS}}\right)$	$0.5236nE\left(\frac{N_G N_{PS}}{N_S N_{PG}+N_G N_{PS}}\right)$
V_G	$0.5236nE\left[\frac{N_S N_{PG}}{N_{PS}(N_G-N_{PG})}\right]$	$0.5236nE\left(\frac{N_G}{N_G-N_{PG}}\right)$	$0.5236nE\left(\frac{N_G}{N_G-N_{PG}}\right)$	$0.5236nE\left[\frac{N_S N_{PG}}{N_{PS}(N_G-N_{PG})}\right]$	$0.5236nE\left(\frac{N_S N_{PG}}{N_S N_{PG}+N_G N_{PS}}\right)\left(\frac{N_G}{N_G-N_{PG}}\right)$	$0.5236nE\left(\frac{N_S N_{PG}}{N_S N_{PG}+N_G N_{PS}}\right)\left(\frac{N_G}{N_G-N_{PG}}\right)$
$W = \frac{33{,}000 \text{ hp.}}{V}$ or →	$\frac{Q}{E}\left(\frac{N_S+N_{PS}}{N_S}\right)$	$\frac{Q}{E}\left(\frac{N_G-N_{PG}}{N_G}\right)$	$\frac{Q}{E}$	$\frac{Q}{E}$	$\frac{Q}{E}\left(\frac{N_S+N_{PS}}{N_S}\right)$	$\frac{Q}{E}\left(\frac{N_G-N_{PG}}{N_G}\right)$
W_S	W	$W\left[\frac{N_{PG}(N_S+N_{PS})}{N_{PS}(N_G-N_{PG})}\right]$	$W\left[\frac{N_{PG}(N_S+N_{PS})}{N_S N_{PG}+N_G N_{PS}}\right]$	$W\left[\frac{N_S(N_S+N_{PS})}{N_S N_{PG}+N_G N_{PS}}\right]$	W	$W\left[\frac{N_{PG}(N_S+N_{PS})}{N_{PS}(N_G-N_{PG})}\right]$
W_A	$W\left[\frac{N_S N_{PG}+N_G N_{PS}}{N_{PG}(N_S+N_{PS})}\right]$	$W\left[\frac{N_S N_{PG}+N_G N_{PS}}{N_{PS}(N_G-N_{PG})}\right]$	W	W	$W\left[\frac{N_S N_{PG}+N_G N_{PS}}{N_{PG}(N_S+N_{PS})}\right]$	$W\left[\frac{N_S N_{PG}+N_G N_{PS})}{N_{PS}(N_G-N_{PG})}\right]$
W_G	$W\left[\frac{N_{PS}(N_G-N_{PG})}{N_{PG}(N_S+N_{PS})}\right]$	W	$W\left[\frac{N_{PS}(N_G-N_{PG})}{N_S N_{PG}+N_G N_{PS}}\right]$	$W\left[\frac{N_{PS}(N_G-N_{PG})}{N_S N_{PG}+N_G N_{PS}}\right]$	$W\left[\frac{N_{PS}(N_G-N_{PG})}{N_{PG}(N_S+N_{PS})}\right]$	W

TABLE A3-8.—ENGINEERING INFORMATION COVERING THE DESIGN AND APPLICATION OF BEVEL AND MITER GEARING

(From data of Foote Brothers Gear and Machine Company)

The elements to be considered in the design and selection of bevel gearing, calculation of tooth strengths and horsepower ratings are very similar to those already covered in Tables A3-2 and A3-3 on spur gearing, with the exception that we must consider here pitch angles, face angles, cutting angles, etc.

The following table and diagrams define the various elements that are generally used in the solution of bevel and miter-gear problems, where shafts run at 90-deg. angles.

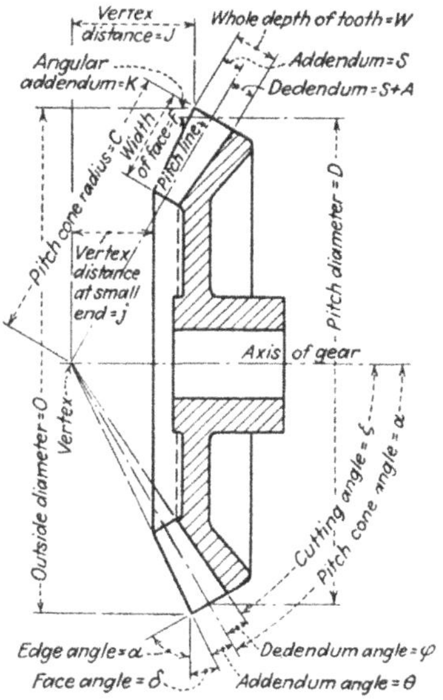

N = number of teeth.
P = diametral pitch.
P' = circular pitch.
π = 3.1416.
α = pitch cone angle and edge angle.
γ = center angle.
D = pitch diameter.
S = addendum.
$S + A$ = dedendum (A = clearance).
W = whole depth of tooth space.
T = thickness of tooth at pitch line.
C = pitch cone radius.
F = width of face.
s = addendum at small end of tooth.
t = thickness of tooth at pitch line at small end.
θ = addendum angle.
ϕ = dedendum angle.
δ = face angle.
ζ = cutting angle.
K = angular addendum.
O = outside diameter (edge diameter for internal gears).
J = vertex distance.
j = vertex distance at small end.
N' = number of teeth for which to select cutter, also called "number of teeth in equivalent spur gear."

TABLE A3-8.—ENGINEERING INFORMATION COVERING THE DESIGN AND APPLICATION OF BEVEL AND MITER GEARING.—(*Continued*)

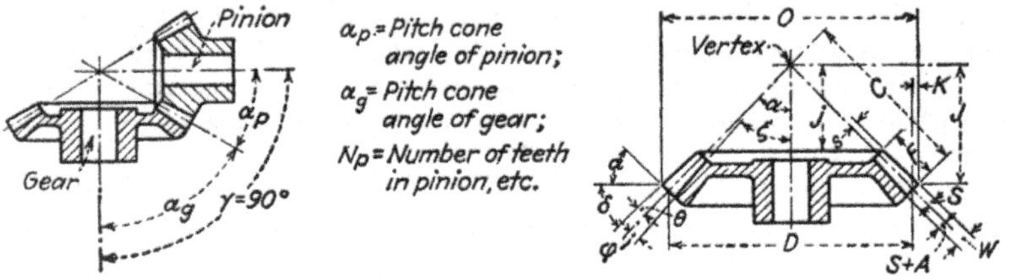

Use rules and formulas 1 to 21 in order given.

No.	To find	Rule	Formula
1	Pitch-cone angle (or edge angle) of pinion	Divide the number of teeth in the pinion by the number of teeth in the gear to get the tangent	$\tan \alpha_p = \frac{N_p}{N_g}$
2	Pitch-cone angle (or edge angle) of gear	Divide the number of teeth in the gear by the number of teeth in the pinion to get the tangent	$\tan \alpha_g = \frac{N_g}{N_p}$
3	Proof of calculations for pitch-cone angles	The sum of the pitch-cone angles of the pinion and gear equals 90 deg.	$\alpha_p + \alpha_g = 90°$
4	Pitch diameter	Divide the number of teeth by the diametral pitch; or multiply the number of teeth by the circular pitch and divide by 3.1416	$D = \frac{N}{P} = \frac{NP'}{\pi}$
5*	Addendum	Divide 1.0 by the diametral pitch; or multiply the circular pitch by 0.318	$S = \frac{1.0}{P} = 0.318P'$
6	Dedendum	Divide 1.157 by the diametral pitch; or multiply the circular pitch by 0.368	$S + A = \frac{1.157}{P} = 0.368P'$
7	Whole depth of tooth space	Divide 2.157 by the diametral pitch; or multiply the circular pitch by 0.687	$W = \frac{2.157}{P} = 0.687P'$

* Nos. 5 to 13 are the same for both gear and pinion.

TABLE A3-8.—ENGINEERING INFORMATION COVERING THE DESIGN AND APPLICATION OF BEVEL AND MITER GEARING.—(*Continued*)

No.	To find	Rule	Formula
8	Thickness of tooth at pitch line	Divide 1.571 by the diametral pitch; or divide the circular pitch by 2	$T = \frac{1.571}{P} = \frac{P'}{2}$
9	Pitch-cone radius	Divide the pitch diameter by twice the sine of the pitch-cone angle	$C = \frac{D}{2 \times \sin \alpha}$
10	Addendum of small end of tooth	Subtract the width of face from the pitch-cone radius, divide the remainder by the pitch-cone radius, and multiply by the addendum	$s = S \times \frac{C - F}{C}$
11	Thickness of tooth at pitch line at small end	Subtract the width of face from the pitch-cone radius, divide the remainder by the pitch-cone radius, and multiply by the thickness of the tooth at the pitch line	$t = T \times \frac{C - F}{C}$
12	Addendum angle	Divide the addendum by the pitch-cone radius to get the tangent	$\tan \theta = \frac{S}{C}$
13	Dedendum angle	Divide the dedendum by the pitch-cone radius to get the tangent	$\tan \phi = \frac{S + A}{C}$
14	Face angle	Subtract the sum of the pitch-cone and addendum angles from 90 deg.	$\delta = 90° - (\alpha + \theta)$
15	Cutting angle	Subtract the dedendum angle from the pitch-cone angle	$\zeta = \alpha - \phi$
16	Angular addendum	Multiply the addendum by the cosine of the pitch-cone angle	$K = S \times \cos \alpha$
17	Outside diameter	Add twice the angular addendum to the pitch diameter	$O = D + 2K$
18	Vertex distance	Multiply one-half the outside diameter by the tangent of the face angle	$J = \frac{O}{2} \times \tan \delta$

TABLE A3-8.—ENGINEERING INFORMATION COVERING THE DESIGN AND APPLICATION OF BEVEL AND MITER GEARING.—(*Continued*)

No.	To find	Rule	Formula
19	Vertex distance at small end of tooth	Subtract the width of face from the pitch-cone radius, divide the remainder by the pitch-cone radius, and multiply by the vertex distance	$j = J \times \frac{C - F}{C}$
20	Number of teeth for which to select cutter	Divide the number of teeth by the cosine of the pitch-cone angle	$N' = \frac{N}{\cos \alpha}$
21	Proof of calculations by rules 9, 12, 14, 16 and 17	The outside diameter equals twice the pitch-cone radius multiplied by the cosine of the face angle and divided by the cosine of the addendum angle	$O = \frac{2C \times \cos \delta}{\cos \theta}$

Recommended Practice for Bevel Gearing

The American Gear Manufacturers Association has adopted as recommended practice the following rules:

The maximum length of face of bevel gears should not be over one-third of the cone distance for gears up to 3-in. pitch diameter and not over one-fourth of the cone distance for gears from 3- to 20-in. pitch diameter, assuming that the pitch in every case will be in proper proportion to the size of the gears. A safe rule is to make the face 1½ to 2½ times the circular pitch.

The minimum length of bearing along the face is to be at least one-half the length of the face when the gears are held in correct alignment.

Bevel gears with generated involute teeth of standard addendum having a pressure angle of 14½ deg. may be used according to the following rule:

Ratio	No. of Teeth
1:1	14 and over
1½:1	18 and over
2:1	19 and over
3:1 and over	21 and over

This rule is given applying mainly to gears up to 20-in. pitch diameter and to average industrial machine design as distinguished from automobiles.

Recommended Practice for Backing Dimensions of Bevel and Miter Gears

The following formulas are recommended for the calculation of backing dimensions of bevel and miter gears.

Table A3-8.—Engineering Information Covering the Design and Application of Bevel and Miter Gearing.—(*Continued*)

Bevel Gears and Pinions

$$\text{Backing in inches of pinion} = \frac{\text{pitch diameter of gear} \times 0.250}{(\text{ratio of gear to pinion}) + 1}$$

Backing in inches of gear = (pitch diameter of gear × 0.250) − backing of pinion

Miter Gears

Backing in inches of gear = pitch diameter × 0.125

This does not allow for set screws in pinions below 4 in. pitch diameter.

Strength of Bevel and Miter Gears

The method for the calculation of strength of bevel gearing is based on the Lewis formula for the strength of gear teeth and is practically the same as is used for spur gears.

The accompanying tables of rules and formulas for the strength of bevel gears with the tables "Working Stresses Used in the Lewis Formula for the Strength of Gear Teeth" in tables on Spur Gearing give all necessary data for calculating the strength of bevel gears.

The following tables and formulas are based on the use of the diametral pitch of the gears, and if the circular pitch is given it should be transformed into diametral pitch by dividing 3.1416 by the circular pitch.

Rules and Formulas for the Strength of Bevel Gears

D = pitch diameter of gear, in.
R = r.p.m.
V = velocity, ft. per min. at pitch diameter.
S_s = allowable static unit stress for material.
S = allowable unit stress for material at given velocity.
F = width of face.
N' = number of teeth in equivalent spur gear (see diagram in table, page 475).
Y = outline factor (see table, page 475).
P = diametral pitch (if circular pitch is given, divide 3.1416 by circular pitch to obtain diametral pitch).
C = pitch-cone radius.
W = maximum safe tangential load in pounds at pitch diameter.
hp. = maximum safe horsepower.

TABLE A3-8.—ENGINEERING INFORMATION COVERING THE DESIGN AND APPLICATION OF BEVEL AND MITER GEARING.—(*Continued*)
Use rules and formulas 1 to 4 in the order given.

No.	To find	Rule	Formula
1	Velocity in feet per minute at the pitch diameter	Multiply the product of the diameter in inches and the number of revolutions per minute by 0.262	$V = 0.262DR$
2	Allowable unit stress at given velocity	Multiply the allowable static stress by 600, and divide the result by the velocity in feet per minute plus 600	$S = S_s \times \frac{600}{600 + V}$
3	Maximum safe tangential load at pitch diameter	Multiply together the allowable stress for the given velocity, the width of face, the tooth outline factor, and the difference between the pitch-cone radius and the width of face; divide the result by the product of the diametral pitch and the pitch-cone radius	$W = \frac{SFY(C - F)}{PC}$
4	Maximum safe horsepower	Multiply the safe load at the pitch line by the velocity in feet per minute, and divide the result by 33,000	$\text{hp.} = \frac{WV}{33{,}000}$

TABLE A3-8.—ENGINEERING INFORMATION COVERING THE DESIGN AND APPLICATION OF BEVEL AND MITER GEARING.—(*Continued*)

Factors for Calculating Strength of Bevel Gears

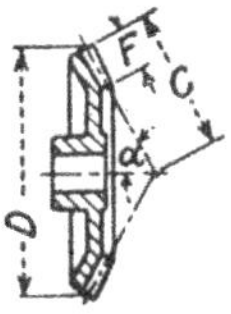

$$N' = \frac{\text{number of teeth}}{\cos \alpha}$$

Table of outline factors (Y) for 14½ and 20 deg. involute

N'	Outline factor = Y: 14½ deg. involute (std.)	Outline factor = Y: 20 deg. involute	N'	Outline factor = Y: 14½ deg. involute (std.)	Outline factor = Y: 20 deg. involute
12	0.210	0.245	27	0.314	0.349
13	0.220	0.261	30	0.320	0.358
14	0.226	0.276	34	0.327	0.371
15	0.236	0.289	38	0.336	0.383
16	0.242	0.295	43	0.346	0.396
17	0.251	0.302	50	0.352	0.408
18	0.261	0.308	60	0.358	0.421
19	0.273	0.314	75	0.364	0.434
20	0.283	0.320	100	0.371	0.446
21	0.289	0.327	150	0.377	0.459
23	0.295	0.333	300	0.383	0.471
25	0.305	0.339	Rack	0.390	0.484

End Thrust on Bevel Gears

In designing bearings to be used with bevel and miter gears, it is important to know the end thrust exerted by the bevel gears and pinions so that proper end-thrust bearings may be provided.

The method of calculation of end thrusts is as follows:

A = pressure angle of the gear teeth.
P = tooth pressure at middle of tooth face.
F = separating force = $P \times \tan A$.
B = pitch angle of pinion.
T = thrust on pinion = $P \times \tan A \times \sin B$.
T_1 = thrust on gear = $P \times \tan A \times \cos B$.

The following table gives the factors by which the tooth pressure is multiplied to find the thrust, giving practically the same values found by solving the formulas for T and T_1, given above.

TABLE A3-8.—ENGINEERING INFORMATION COVERING THE DESIGN AND APPLICATION OF BEVEL AND MITER GEARING.—(*Continued*)

Gear ratio	Pressure angle *A*							
	14½ deg.		15 deg.		20 deg.		22 deg.	
	Gear	Pinion	Gear	Pinion	Gear	Pinion	Gear	Pinion
1 :1	0.183	0.183	0.189	0.189	0.257	0.257	0.286	0.286
1½:1	0.215	0.143	0.223	0.148	0.303	0.202	0.336	0.224
2 :1	0.232	0.116	0.239	0.120	0.325	0.163	0.361	0.181
2½:1	0.240	0.096	0.249	0.100	0.338	0.135	0.375	0.150
3 :1	0.246	0.082	0.254	0.085	0.345	0.115	0.383	0.128
3½:1	0.249	0.071	0.258	0.074	0.350	0.100	0.389	0.111
3¾:1	0.250	0.067	0.259	0.069	0.352	0.094	0.390	0.104
4 :1	0.251	0.062	0.260	0.065	0.353	0.088	0.392	0.097
4½:1	0.253	0.056	0.262	0.058	0.355	0.079	0.394	0.087
5 :1	0.254	0.051	0.263	0.053	0.357	0.072	0.396	0.080
5½:1	0.255	0.046	0.264	0.048	0.358	0.065	0.398	0.072

TABLE A3-9.—BEARING LOADS DUE TO STRAIGHT SPUR GEARS
(From New Departure "Handbook")
Bearing loads due to tangential and separating forces at tooth contact.

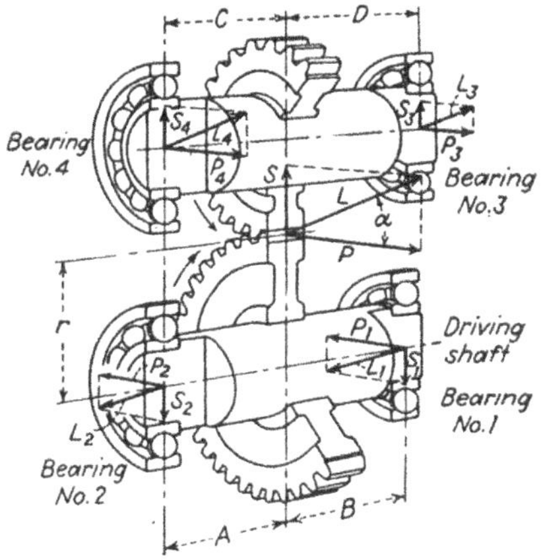

$$\text{Torque input} = Q\ \frac{63{,}025 \times \text{hp. transmitted}}{\text{r.p.m. of driving gear}},\ \text{lb. inches.}$$

$$\text{Tangential force} = P = \frac{Q}{r} = \frac{\text{torque input, lb.-in.}}{\text{pitch radius of gear, in.}};\ \text{lb.}$$

Separating force $= S = P \tan \alpha$, lb. (α = tooth pressure angle)
L = total load on both shafts due to gears

Due to	On bearing 1	On bearing 2	On bearing 3	On bearing 4
P	$P_1 = P\ \frac{A}{A+B}$	$P_2 = P\ \frac{B}{A+B}$	$P_3 = P\ \frac{C}{C+D}$	$P_4 = P\ \frac{D}{C+D}$
S	$S_1 = S\ \frac{A}{A+B}$	$S_2 = S\ \frac{B}{A+B}$	$S_3 = S\ \frac{C}{C+D}$	$S_4 = S\ \frac{D}{C+D}$
L	$L_1 = \sqrt{P_1^2 + S_1^2}$	$L_2 = \sqrt{P_2^2 + S_2^2}$	$L_3 = \sqrt{P_3^2 + S_3^2}$	$L_4 = \sqrt{P_4^2 + S_4^2}$

TABLE A3-10.—BEARING LOADS DUE TO STRAIGHT SPUR GEARS IN TRAIN

(From New Departure "Handbook")

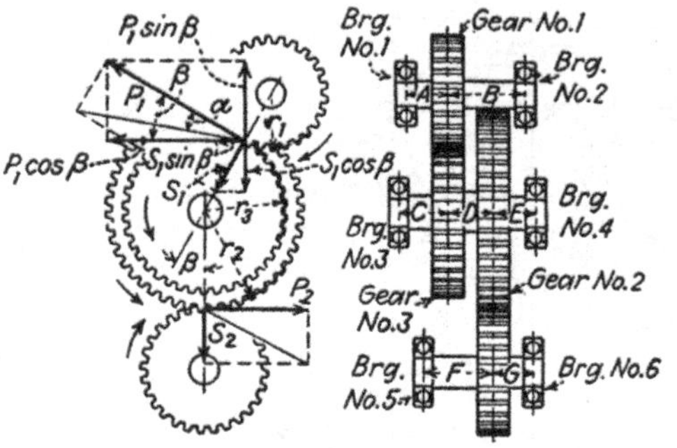

Bearings supporting a shaft having gears in mesh with gears on other shafts (as bearings 3 and 4) carry loads due to both pairs. When the shafts do not lie in the same plane, the forces P_1 and S_1 must be divided so the components of each lie in the same plane as those due to the other gears.

Torque input $= Q = \dfrac{63{,}025 \times \text{hp. transmitted}}{\text{r.p.m. of driving gear}}$, lb.-in.

Tangential force (drive gear 1) $= P_1 = \dfrac{Q}{r_1} = \dfrac{\text{torque input, lb.-in.}}{\text{pitch radius of gear, in.}}$, lb.

Separating force (drive gear 1) $= S_1 = P_1 \tan \alpha$, lb. (α = tooth pressure angle)

Tangential force (drive gear 2) $= P_2 = P_1 \times \dfrac{r_3}{r_2}$ (r_2 and r_3 = pitch radii of gears 2 and 3)

Separating force (drive gear 2) $= S_2 = P_2 \tan \alpha$

Bearing loads

Due to	On bearing 1	On bearing 2	On bearing 3	On bearing 4
P_1	$P_1 = P_1 \frac{B}{A+B}$	$P_2 = P_1 \frac{A}{A+B}$	$P_{3V}^* = P_1 \sin\beta \frac{D+E}{C+D+E}$ $P_{3H}^* = P_1 \cos\beta \frac{D+E}{C+D+E}$	$P_{4V} = P_1 \sin\beta \frac{C}{C+D+E}$ $P_{4H} = P_1 \cos\beta \frac{C}{C+D+E}$
S_1	$S_1 = S_1 \frac{B}{A+B}$	$S_2 = S_1 \frac{A}{A+B}$	$S_{3V} = S_1 \cos\beta \frac{D+E}{C+D+E}$ $S_{3H} = S_1 \sin\beta \frac{D+E}{C+D+E}$	$S_{4V} = S_1 \cos\beta \frac{C}{C+D+E}$ $S_{4H} = S_1 \sin\beta \frac{C}{C+D+E}$
	On bearing 5	On bearing 6		
P_2	$P_5 = P_2 \frac{G}{F+G}$	$P_6 = P_2 \frac{F}{F+G}$	$P_{3,2} = P_2 \frac{E}{C+D+E}$	$P_{4,2} = P_2 \frac{C+D}{C+D+E}$
S_2	$S_5 = S_2 \frac{G}{F+G}$	$S_6 = S_2 \frac{F}{F+G}$	$S_{3,2} = \frac{E}{C+D+E}$	$S_{4,2} = S_2 \frac{C+D}{C+D+E}$

* V = vertical, H = horizontal.

Total load on bearing 1 $= \sqrt{P_1^2 + S_1^2}$, on bearing 2 $= \sqrt{P_2^2 + S_2^2}$

On bearing 3 $= \sqrt{(P_{3V} + S_{3.2} - S_{3V})^2 + (P_{3H} + S_{3H} + P_{3.2})^2}$

On bearing 4 $= \sqrt{(P_{4V} + S_{4.2} - S_{4V})^2 + (P_{4H} + S_{4H} + P_{4.2})^2}$

On bearing 5 $= \sqrt{P_5^2 + S_5^2}$, on bearing 6 $= \sqrt{P_6^2 + S_6^2}$

TABLE A3-11.—BEARING LOADS DUE TO PLANETARY GEARING
(From New Departure "Handbook")

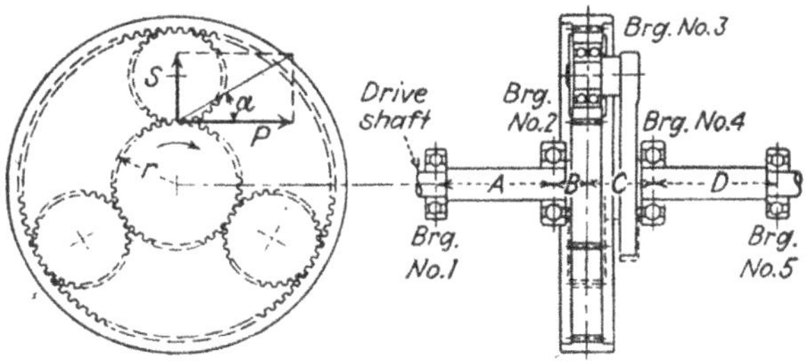

When the system employs two or more planet gears (usually three, as shown by dotted lines in diagram), the tangential and separating forces, due to the input torque, counterbalance each other in so far as they can produce appreciable loads on bearings 1, 2, 4, and 5. However, bearing 3 will be loaded because of the torque transmitted.

$$Q = \frac{\text{hp.} \times 63{,}025}{N} = \text{torque input, lb.-in.,}$$

where hp. = horsepower transmitted
and N = r.p.m. of driving gear.

$$P = \frac{Q}{r} = \text{tangential force, in pounds of driving sun gear,}$$

where r = pitch radius of gear, in.

$$S = P \tan \alpha = \text{separating force in pounds of the sun gear,}$$

where α = tooth pressure angle.

For three planet gears, the load on each planet pin bearing, position 3, due to driving torque and reaction, will (with equally distributed torque) be

$$\frac{2P}{3}$$

When torque is transmitted through a single planet gear, the following loads are produced:

Bearing Loads

Due to	On bearing 1	On bearing 2	On bearing 3	On bearing 4	On bearing 5
P	$P\frac{B}{A} = P_1$	$P\frac{A+B}{A} = P_2$	$2P$	$2P\frac{C+D}{D}$	$2P\frac{C}{D}$
S	$S\frac{B}{A} = S_1$	$S\frac{A+B}{A} = S_2$			
Total load	$\sqrt{P_1^2 + S_1^2}$	$\sqrt{P_2^2 + S_2^2}$	$2P$	$2P\frac{C+D}{D}$	$2P\frac{C}{D}$

TABLE A3-12.—BEARING LOADS DUE TO PLAIN BEVEL GEARING
(From New Departure "Handbook")

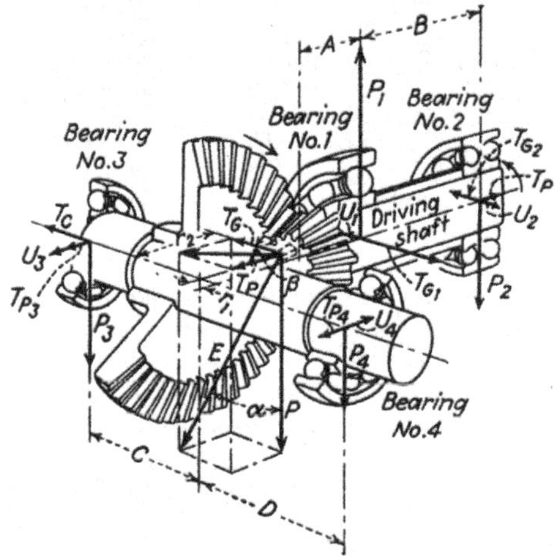

In the process of determining bearing loads with this type of gearing, the force E normal to the driving tooth contact is resolved into three forces. The first, P, is directed vertically, the second, T_G, horizontally, both in a plane at right angles to the pinion shaft, and the third, T_P, parallel to the pinion axis.

$$Q = \frac{\text{hp.} \times 63{,}025}{N} = \text{torque input, lb.-in.}$$

where hp. = horsepower transmitted.
N = r.p.m. of pinion.

$$P = \frac{Q}{r_1} = \text{tangential force}$$

where r_1 = mean pinion pitch radius in inches = ½ (pinion pitch diameter − tooth face × sin β), angle β being defined below.

$$r_2 = \text{mean gear pitch radius} = r_1 \times \frac{\text{number of teeth in gear}}{\text{number of teeth in pinion}}$$

$$T_G = P \tan \alpha \cos \beta = \text{gear thrust}$$

where α = tooth pressure angle.

$$\beta = \tfrac{1}{2} \text{ pinion pitch cone angle.}$$

$$= \tan^{-1} \frac{\text{number of teeth in pinion}}{\text{number of teeth in gear}}$$

$$T_P = P \tan \alpha \sin \beta = \text{pinion thrust.}$$

Note derivation from diagram.

TABLE A3-12.—BEARING LOADS DUE TO PLAIN BEVEL GEARING.—(*Continued*)

Bearing Loads

Due to	On bearing 1	On bearing 2
P..............	$P\frac{A+B}{B} = P_1$	$P\frac{A}{B} = P_2$
T_G.............	$T_G\frac{A+B}{B} = T_{G1}$	$T_G\frac{A}{B} = T_{G2}$
T_P.............	$T_P\frac{r_1}{B} = U_1$	$T_P\frac{r_1}{B} = U_2 = U_1$
Total radial load...........	$\sqrt{P_1^2 + (T_{G1} - U_1)^2}$	$\sqrt{P_2^2 + (T_{G2} - U_2)^2} = R_2$
Thrust load.....		T_P
Total load......	$\sqrt{P_1^2 + (T_{G1} - U_1)^2}$	$\sqrt{R_2^2 + T_P^2}$

Due to	On bearing	On bearing 4
P..............	$P\frac{D}{C+D} = P_3$	$P\frac{C}{C+D} = P_4$
T_G.............	$T_G\frac{r_2}{C+D} = U_3$	$T_G\frac{r_2}{C+D} = U_4 = U_3$
T_P.............	$T_P\frac{D}{C+D} = T_{P3}$	$T_P\frac{C}{C+D} = T_{P4}$
Total radial load...........	$\sqrt{P_3^2 + (T_{P3} + U_3)^2} = R_3$	$\sqrt{P_4^2 + (U_4 - T_{P4})^2}$
Thrust load.....	T_G	
Total load......	$\sqrt{R_3^2 + T_G^2}$	$\sqrt{P_4^2 + (U_4 - T_{P4})^2}$

INDEX

The Aviation Collection by Sportsman's Vintage Press

www.SportsmansVintagePress.com

Aircraft Construction Handbook	by Thomas A. Dickinson
Aircraft Sheet Metal Work	by C. A. LeMaster
The Aircraft Apprentice	by Leslie MacGregor
Aircraft Woodwork	by Col. R. H. Drake
Aircraft Welding	by Col. R. H. Drake
Aircraft Sheet Metal	by Col. R. H. Drake
Aircraft Engines	by Col. R. H. Drake
Aircraft Electrical and Hydraulic Systems, and Aircraft Instruments	by Col. R. H. Drake
Aircraft Engine Maintenance and Service	by Col. R. H. Drake
Aircraft Maintenance and Service	by Col. R. H. Drake
Wings of Tomorrow: The Story of the Autogiro	by Juan De La Cierva & Don Rose
Aeroplane Engines in Theory and Practice	by John B. Rathburn
Aircraft Radio and Electrical Equipment	by H. K. Morgan
Aircraft Engine Design	by Joseph Liston

Made in United States
Orlando, FL
05 November 2026

90231081R00275